AI, Nanobiotechnology, and the Future of Precision Antibiotics

Reyed M. Reyed
https://orcid.org/0000-0001-5027-3478
Genetic Engineering and Biotechnology Research Institute (GEBRI), Egypt & City of Scientific Research and Technological Applications (SRTA-City), Alexandria, Egypt

IGI Global
Scientific Publishing
Publishing Tomorrow's Research Today

Vice President of Editorial	Melissa Wagner
Director of Acquisitions	Mikaela Felty
Director of Book Development	Jocelynn Hessler
Production Manager	Mike Brehm
Cover Design	Jose Rosado

Published in the United States of America by
IGI Global Scientific Publishing
701 East Chocolate Avenue
Hershey, PA, 17033, USA
Tel: 717-533-8845 | Fax: 717-533-7115
Website: https://www.igi-global.com E-mail: cust@igi-global.com

Library of Congress Cataloging-in-Publication Data

Names: Reyed, Reyed M., 1970- editor
Title: AI, nanobiotechnology, and the future of precision antibiotics / edited by Reyed M. Reyed.
Description: Hershey, PA : IGI Global Scientific Publishing, [2026] | Includes bibliographical references and index. | Summary: "This book charts a clear, multidisciplinary path toward next-generation antimicrobial therapies by seamlessly uniting cutting-edge computational design with advanced nano-enabled delivery platforms"-- Provided by publisher.
Identifiers: LCCN 2026002164 (print) | LCCN 2026002165 (ebook) | ISBN 9798337352688 hardcover | ISBN 9798337352695 paperback | ISBN 9798337352701 ebook
Subjects: LCSH: Antibiotics--Biotechnology | Nanobiotechnology | Drug resistance in microorganisms | Artificial intelligence--Medical applications
Classification: LCC RM267 .A46 2026 (print) | LCC RM267 (ebook)
LC record available at https://lccn.loc.gov/2026002164
LC ebook record available at https://lccn.loc.gov/2026002165

British Cataloguing in Publication Data
A Cataloguing in Publication record for this book is available from the British Library.

To the cherished memory of my beloved father and mother, may Allah's mercy and blessings be upon them. Their enduring guidance and quiet strength continue to illuminate my path.

To my devoted wife and children, whose steadfast support, patience, and encouragement remain a constant source of inspiration and meaning.

This work stands as a humble tribute to the pillars of my life — the foundation of my strength, purpose, and perseverance.

Editorial Advisory Board

Table of Contents

Detailed Table of Contents

This chapter presents the Reyed Precision Convergence Framework for Antimicrobial Resistance (RPCF-AMR) as a systemic redefinition of AMR, beyond microbial-centric views. Integrating nanobiotechnology, AI-driven modeling (NexDi), and multi-scale architectures (SCF–DNPS), it addresses dynamics across host, microbial, environmental, and policy domains. Molecular dynamics, machine learning, and nanocarrier design converge to anticipate intracellular drug availability, bypass classical resistance, and optimize precision. Translational pathways, including programmable nanomaterials and stimuli-responsive carriers, operationalize One Health stewardship while reducing selective pressures in human, animal, and ecological reservoirs. RPCF-AMR reconciles historical surveillance gaps with multi-modal interventions, establishing a deployable reference standard. Through modeling and continuous feedback, it ensures predictive reliability, clinical relevance, and global scalability, defining a trajectory toward effective AMR control by 2040.

via stimuli-responsive nanocarriers for site-specific delivery and ARG silencing. Global collaboration and ethical data-sharing are emphasized to ensure accessibility, concluding that these advancements can reduce AMR burdens and sustain antibiotic efficacy through predictive, personalized stewardship.

An emerging convergence of artificial intelligence and nanobiotechnology is transforming precision antibiotic discovery and targeted antimicrobial delivery. This chapter per the authors analyzes how machine learning models and bioinformatics enable data-driven drug discovery through pathogen genomics and molecular targeting strategies. The discussion further highlights the use of nanocarriers for controlled drug release and smart drug delivery, supporting improved therapeutic optimization while minimizing off-target effects and toxicity. Emphasis is placed on resistance mitigation through predictive modeling and adaptive delivery mechanisms. By integrating computational intelligence with systems biology and translational medicine principles, the chapter presents an interdisciplinary framework demonstrating how AI-enabled nanobiotechnological platforms can enhance the accuracy, effectiveness, and sustainability of next-generation antimicrobial therapies.

Antibiotics have been essential in treating bacterial infections and assisting with life-saving medical procedures; however, the development of antibiotic-resistant bacteria as a result of overuse of antibiotics presents serious problems. Artificial intelligence (AI) is revolutionizing precision medicine and drug discovery through quick compound screening, biological interaction modeling, and drug repurposing. Nanocarriers serve as advanced delivery platforms that address limitations of conventional therapies such as poor bioavailability, systemic toxicity, and inadequate penetration into complex infection sites like biofilms. By incorporating patient data, AI improves diagnostic capabilities and allows for customized treatment plans. The fusion of AI and nanobiotechnology promise enormous scope for tailoring antibiotic therapy to individuals and microbial types. To use these potentials to their full potential in the improvement of global health, scientists, physicians and policymakers must work and cooperate together in the future.

Antimicrobial resistance (AMR) poses an escalating threat to world health by diminishing the efficacy of common antibiotics. Antimicrobial timing, use, and placement must be carefully planned to combat AMR. These are crucial to advanced infection management. This chapter discusses clever nanocarriers that distribute

Preface

INTRODUCTION

Antimicrobial resistance (AMR) is often described as a problem of diminishing drug efficacy. While this characterization remains valid, it appears increasingly insufficient as a comprehensive explanation. A growing body of work suggests that AMR may be more appropriately understood as a system-level condition, shaped by the interaction of microbial evolution, host biological variability, environmental pressures, and the structural limitations of current therapeutic and policy frameworks. In this sense, resistance does not emerge in isolation; rather, it develops within a broader network of biological and institutional processes that reinforce its persistence over time.

This perspective invites a reconsideration of how AMR has traditionally been approached. For decades, responses have been guided largely by a microbe-centered logic, focusing on the relationship between pathogen and drug. Although this approach has generated important advances, it has also tended to constrain the scope of intervention. Increasingly, attention is being drawn to additional dimensions, including microbiome dynamics, host–pathogen interactions, environmental reservoirs, and the practical complexities of antimicrobial stewardship across diverse contexts. Taken together, these factors suggest that effective responses may require strategies that are not only molecularly precise, but also context-aware and system-oriented.

The present volume was developed in response to this expanding understanding. It brings together contributions that explore the convergence of artificial intelligence (AI) and nanobiotechnology as a potential pathway toward more adaptive and targeted antimicrobial strategies. Rather than presenting these domains as isolated technological solutions, the book considers how their interaction may support a more integrated approach to antimicrobial discovery, delivery, and application.

At the conceptual center of this volume is the Reyed Precision Convergence Framework for Antimicrobial Resistance (RPCF-AMR), introduced as a guiding reference point. The framework is not intended as a prescriptive model, but as a structured lens through which relationships between predictive modeling, molecular design, delivery systems, and systems-level biology may be examined. Its purpose is to provide coherence across diverse approaches while preserving the flexibility required for ongoing scientific development. In this respect, it reflects a gradual shift away from broadly applied antimicrobial strategies toward more selective, coordinated, and programmable forms of intervention.

From a technological perspective, advances in AI are considered in terms of their capacity to integrate complex datasets and identify patterns in resistance development that may not be readily accessible through conventional analytical methods. These tools may support earlier recognition of resistance trajectories, more informed prioritization of therapeutic candidates, and iterative refinement of intervention strategies. At the same time, such approaches remain dependent on data quality, model transparency, and continued validation.

Parallel developments in nanobiotechnology are examined with a focus on targeted and stimuli-responsive delivery systems. These platforms may enable more localized therapeutic action, potentially improving drug availability at sites of infection while limiting unintended systemic exposure. Their relevance becomes particularly evident in complex biological environments—such as biofilms or intracellular infections—where conventional delivery mechanisms often encounter limitations. When considered alongside predictive modeling, these advances suggest a shift toward interventions that are more closely aligned with the biological context in which they operate.

The chapters in this volume are organized to reflect a progression from conceptual reframing to technological exploration and, ultimately, to considerations of clinical and global implementation. Early contributions revisit AMR as a multidimensional challenge and examine how convergence-oriented thinking may extend beyond conventional antibiotic paradigms. Subsequent chapters engage with biological complexity and computational modeling, followed by detailed discussions of nanobiotechnological platforms and targeted delivery systems. The later sections consider how these developments may be translated into diverse healthcare settings, with attention to scalability, accessibility, and ethical considerations. Below is a chapter by chapter summary that you may use to accompany your reading.

CHAPTER OVERVIEWS

Chapter 1 introduces the Reyed Precision Convergence Framework for Antimicrobial Resistance (RPCF-AMR), a comprehensive systems-based model that redefines AMR beyond a purely microbial phenomenon. By integrating AI-driven predictive modeling, nanobiotechnology, and multi-scale biological architectures, this chapter establishes a foundational framework for precision therapeutics aligned with the One Health approach. It emphasizes predictive reliability, translational applicability, and global scalability toward AMR control by 2040.

Chapter 2 expands the discussion beyond traditional antibiotics, presenting the synergistic potential of AI and nanotechnology in developing sustainable antimicrobial solutions. It explores how AI accelerates drug discovery and how nanotechnology enhances delivery and efficacy, while also addressing sustainability through lifecycle assessment, ethical data governance, and alignment with global development goals.

Chapter 3 focuses on the integration of microbiome science and genomics in the development of precision antibiotics. It highlights the role of the gut microbiome as a reservoir of resistance genes and advocates for targeted therapies that preserve microbial balance. The chapter demonstrates how AI, multi-omics, and nanotechnology converge to enable personalized, microbiome-sparing interventions.

Chapter 4 examines AI-enabled nanobiotechnological platforms for antibiotic discovery and targeted delivery. It details how machine learning and bioinformatics facilitate molecular targeting and drug design, while nanocarriers enable controlled release and reduced toxicity. The chapter underscores the role of interdisciplinary integration in improving therapeutic outcomes.

Chapter 5 explores the integration of AI and nanobiotechnology in precision antibiotic delivery. It discusses how AI enhances diagnostic accuracy and treatment personalization, while nanocarriers address limitations such as poor bioavailability and systemic toxicity. The chapter highlights the importance of collaboration among scientists, clinicians, and policymakers.

Chapter 6 presents AI-guided stimuli-responsive nanocarriers that deliver antibiotics in response to infection-specific signals such as pH changes and enzymatic activity. This approach minimizes unnecessary drug exposure, preserves beneficial microbiota, and enhances treatment precision, demonstrating a shift toward intelligent and adaptive antimicrobial therapies.

Chapter 7 redefines AMR as not only a biological challenge but also a limitation in drug delivery systems. It explores AI-driven release mechanisms and smart nanocarriers designed to overcome barriers such as biofilms and intracellular pathogens, addressing critical shortcomings in traditional antibiotic therapies.

Chapter 8 provides a comprehensive overview of nanocarrier technologies for targeted drug delivery. It discusses various platforms, including liposomes, polymeric nanoparticles, dendrimers, and biomimetic systems, and their roles in improving pharmacokinetics and therapeutic specificity, while also aligning with sustainable development principles.

Chapter 9 highlights precision-driven strategies for treating multidrug-resistant pathogens through the integration of AI, genomics, and nanobiotechnology. It addresses real-world implementation challenges, particularly in resource-limited settings, and emphasizes the importance of multidisciplinary collaboration and global accessibility.

The perspective outlined here is further developed in the following Executive Summary, where the conceptual foundations of the volume are examined through a more structured analytical lens.

Looking ahead, it seems likely that addressing AMR will require approaches that move beyond incremental refinement of existing tools. The convergence explored in this volume suggests a gradual transition toward strategies that are more predictive, more targeted, and more attentive to context. While the trajectory remains uncertain, such approaches may contribute to a more integrated and sustainable response to antimicrobial resistance.

Reyed M. Reyed
Genetic Engineering and Biotechnology Research Institute (GEBRI), Egypt &
City of Scientific Research and Technological Applications (SRTA-City), Egypt

Acknowledgement

Scholarly work of this scope does not emerge in isolation. It is shaped, refined, and, at times, challenged through careful and critical engagement. The reviewers of this volume have played a decisive role in that process. Their evaluations did more than assess submissions—they clarified arguments, strengthened methodological rigor, and helped align individual contributions with the broader intellectual direction of the book.

I am sincerely grateful for the time, expertise, and professional judgment they have invested. Their contributions have been instrumental in advancing this work from a collection of chapters into a coherent and credible academic volume.

I also extend my sincere appreciation to IGI Global Scientific Publishing for providing a distinguished platform that supports rigorous scholarship, fosters innovation, and enables the global dissemination of impactful academic work. Their longstanding commitment to excellence in scholarly publishing has created a supportive and dynamic environment in which ideas can be developed, refined, and shared with the wider research community.

I would like to express my particular gratitude to Lindsay Wertman, President of IGI Global Scientific Publishing, and Melissa Wagner, Vice President of Publishing Ethics & Research Integrity. Their professional guidance, openness, and collaborative approach were instrumental throughout the publication process. Their support not only facilitated the progression of the manuscript but also contributed meaningfully to the successful finalization of the book, including its overall presentation and cover design.

I am also deeply grateful to Jordan Hildebrand, Editorial Intern at IGI Global Scientific Publishing, whose consistent follow-up, attention to detail, and clear communication played a vital role in ensuring a smooth and efficient production process. His dedication helped navigate practical challenges and maintained steady progress toward timely completion.

More broadly, I wish to acknowledge the entire IGI Global team for their technical expertise, responsiveness, and commitment to quality. Their efforts have been essential in managing the complexities of contemporary academic publishing and in ensuring that this volume reaches an international audience engaged in advancing AI-driven and precision-based biomedical research.

It is through such thoughtful and professional collaboration that academic projects are able to reach their full potential. For this, I remain sincerely appreciative.

Reyed M. Reyed
Genetic Engineering and Biotechnology Research Institute (GEBRI), Egypt & City of Scientific Research and Technological Applications (SRTA-City), Alexandria, Egypt

Introduction

THE CONVERGENCE PARADIGM: RE-ENGINEERING THE ANTIMICROBIAL LANDSCAPE

The global trajectory of healthcare is currently navigating one of its most precarious periods, defined by the persistent and adaptive threat of antimicrobial resistance (AMR). While the scientific community has long recognized AMR as a primary challenge to modern medicine, the traditional responses—largely centered on the intermittent discovery of individual antibiotic classes—appear increasingly insufficient. The evidence suggests that we are not merely facing a biological hurdle, but rather the limitations of a fragmented scientific paradigm. This volume, *AI, Nanobiotechnology, and the Future of Precision Antibiotics*, originates from the necessity to move beyond these traditional silos toward a more integrated, convergent approach to infection control.

The Crisis of Fragmentation

To understand the necessity of this work, one must first acknowledge the nature of the current impasse. For decades, the development of antimicrobial agents and the engineering of delivery systems have often proceeded along independent tracks. Similarly, the computational revolution, despite its profound impact on data science, has only recently begun to be deeply embedded into the biological fabric of drug discovery. This fragmentation has resulted in therapies that, while molecularly potent, often fail to account for the complex, context-dependent realities of the human body and the microbial ecosystems it hosts.

The introduction of the Reyed Precision Convergence Framework for Antimicrobial Resistance (RPCF-AMR) within these pages is a direct response to this lack of coordination. By positioning antimicrobial efficacy as an "emergent property" of a coordinated system—rather than an intrinsic quality of a drug—this book seeks to reframe the debate. It suggests that the future of medicine lies in the seamless alignment of three critical domains: predictive intelligence, precision delivery, and biological context.

The Triad of Innovation: AI, Nano, and the Polybiome

The conceptual architecture of this volume is built upon a triad of converging technologies and perspectives:

First, Artificial Intelligence (AI) is presented not merely as an auxiliary tool for automation, but as a foundational enabler of "predictive intelligence". In the context of AMR, AI allows us to navigate the immense complexity of microbial genomics and protein-ligand interactions with unprecedented speed. However, the value of AI lies in its ability to move from "big data" to "actionable insight," providing a navigational map for the design of the next generation of antibiotics.

Second, Nanobiotechnology serves as the precision-engineering platform that brings these predictive insights into the physical realm. The challenge is not solely about finding a compound that kills bacteria; it is about ensuring that the compound reaches the site of infection at the right concentration, at the right time, while sparing the host's healthy tissues. Through the lens of nanobiotechnology, we examine how "smart" delivery architectures can respond to local physiological cues, effectively bypassing resistance mechanisms that have rendered legacy antibiotics obsolete.

Third, the concept of the Polybiome introduces a necessary ecological dimension. We are moving away from the reductive "one-pathogen-one-drug" model toward a more sophisticated understanding of microbial communities. This volume emphasizes that any intervention must be "context-aware"—mindful of its impact on the host's commensal flora and the broader environmental gene pool. By integrating a One Health perspective, we acknowledge that the health of humans, animals, and ecosystems is inextricably linked.

Navigating the Volume: System Architecture

To visually consolidate this complex integration, the volume includes a central conceptual figure, Figure 1: AI, Nanobiotechnology, and the Future of Precision Antibiotics Across the Book. This figure illustrates the book's logic as a structured progression from problem formulation to system integration and translational application. It maps the convergence of scientific and technological layers, providing a navigational guide to the thematic clusters discussed throughout the chapters.

The structure of this book is designed to guide the reader through a logical progression from conceptual reframing to global implementation:

> Cluster I (Chapter 1): Establishes the conceptual and systemic frameworks, providing the theoretical grounding for the convergence model and the RPCF-AMR.
> Cluster II (Chapter 2): Situates these approaches within socio-environmental and sustainability contexts, ensuring that innovations are policy-aligned and equitable.
> Cluster III (Chapters 3–8): Delves into the technical nuances of AI-driven modeling, multi-omics integration, and advanced nanocarrier architectures, offering a deep dive into the engineering of precision.
> Cluster IV (Chapter 9): Expands the scope to translational challenges, regulatory landscapes, and global stewardship, ensuring that the innovations are grounded in the realities of clinical practice.

A Generative Reference for the Future

It is important to emphasize that this introduction does not propose a definitive end to the AMR crisis. Rather, it offers a "structured pathway" for inquiry. The integration of AI and nanomedicine is a frontier characterized by both immense potential and significant uncertainty. As such, this book adopts a tone of "disciplined optimism"—celebrating the breakthroughs while remaining rigorously critical of the hurdles that remain, from scalability and biosafety to the ethics of autonomous intelligence in healthcare.

This work is intended for a diverse global audience: for the researcher seeking methodological rigor, the clinician looking for the next frontier of precision medicine, and the policymaker striving for sustainable health security. As you navigate these chapters, we invite you to look past the boundaries of your individual discipline and consider how the convergence of data, engineering, and biology might finally turn the tide in our long-standing encounter with the microbial world.

Figure 1. Illustrates a convergence-based conceptual framework integrating artificial intelligence and nanobiotechnology in the development of precision antibiotics. The model positions antimicrobial resistance (AMR) as a systemic challenge addressed through a multi-layered architecture spanning reframing (Chapter 1), biological contextualization (Chapter 2), technological integration (Chapters 3–8), and global translational application (Chapter 9). Rather than a linear progression, the framework emphasizes interconnectedness between predictive intelligence, precision delivery systems, and biological context within a One Health paradigm.

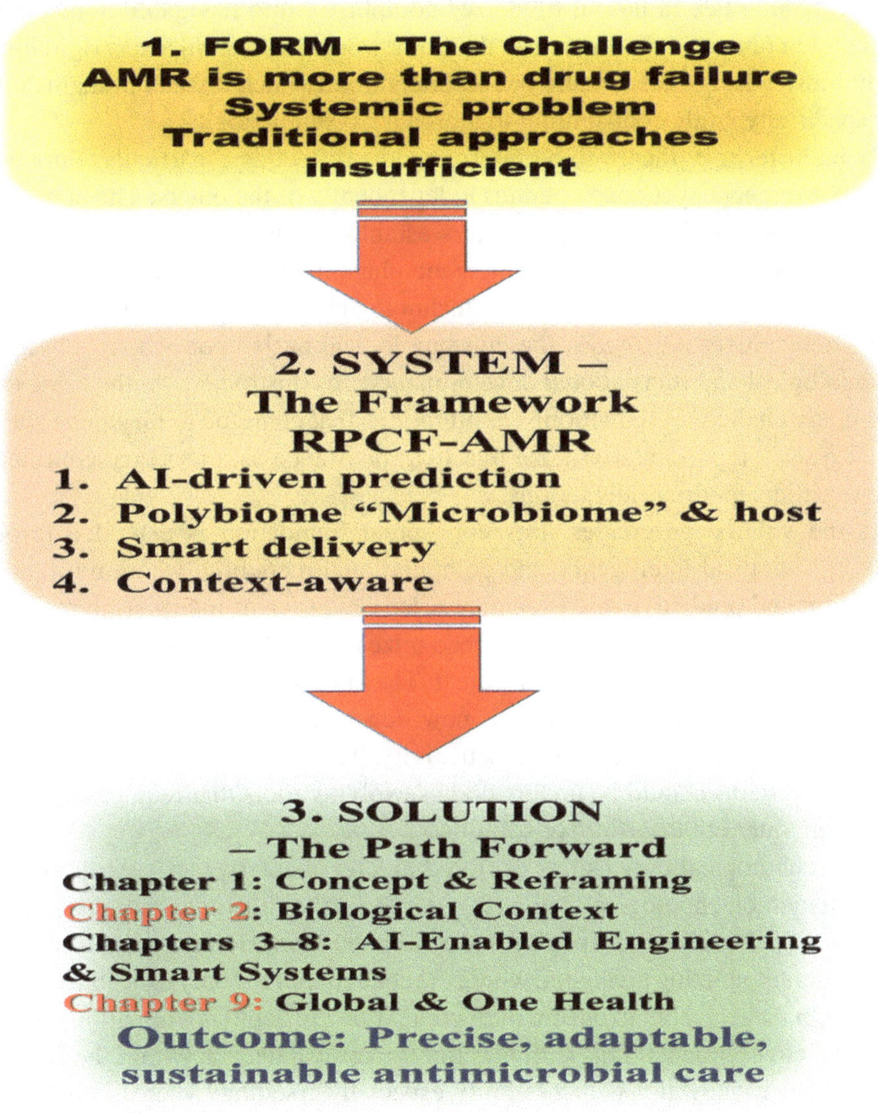

Source: *Reyed, R. M. (Ed.). (2026). AI, Nanobiotechnology, and the Future of Precision Antibiotics. IGI Global Scientific Publishing. https://doi.org/10.4018/979-8-3373-5268-8*

FROM CONCEPTUAL CONVERGENCE TO STRUCTURED NAVIGATION: INTERPRETING THE FRAMEWORK ACROSS THE VOLUME

Building on the conceptual structure illustrated above, the chapters that follow may be approached as a deliberate and layered exploration of convergence within antimicrobial resistance (AMR) research, rather than as isolated thematic contributions. While the graphical representation offers a unified visual synthesis, its analytical value lies in how it organizes complexity into navigable domains that remain interconnected. In this sense, the framework does not impose a rigid hierarchy, but instead suggests a relational architecture in which distinct domains retain their specificity while contributing to a broader systemic logic.

Each cluster may therefore be interpreted as reflecting a particular dimension of this convergence, yet none operates independently of the others. The conceptual framing introduced in the early chapters establishes the conditions under which AMR can be understood as an emergent phenomenon shaped by interactions across biological, environmental, and technological systems. This framing is not merely descriptive; it provides the epistemological basis upon which subsequent methodological and translational developments are constructed. At the same time, it implicitly challenges reductionist assumptions that continue to influence antimicrobial research, particularly those that treat resistance as a property confined to discrete pathogen–drug interactions.

As the volume progresses, this conceptual foundation is extended through layers of biological intelligence and computational modeling, where multi-omics integration and predictive analytics may offer new ways of interpreting resistance dynamics. However, it is important to recognize that such approaches do not eliminate uncertainty; rather, they reconfigure it. The integration of artificial intelligence introduces both analytical power and new dependencies, including data quality, model generalizability, and context sensitivity. These considerations suggest that predictive systems should be understood as evolving tools that require continuous calibration, rather than definitive solutions.

The technological dimensions explored in later chapters, particularly those related to nanobiotechnological delivery systems, further illustrate how convergence operates at the interface between theory and application. Here, the emphasis shifts toward operationalizing precision, where the challenge is not only to design effective antimicrobial agents, but to ensure their context-aware deployment within complex biological environments. This shift highlights a critical tension within the field: the need to balance innovation with feasibility, especially in settings where infrastructural and regulatory constraints may limit implementation.

Finally, the translational layer extends the framework beyond the laboratory, situating antimicrobial strategies within global health systems and a One Health perspective. This progression underscores that the effectiveness of any intervention is contingent upon its alignment with ecological, clinical, and policy contexts. In this regard, the framework may be read not as a fixed model, but as a guiding structure that evolves through iterative feedback between discovery, application, and real-world constraints.

Taken together, the organization of this volume reflects an attempt to move from conceptual convergence toward structured navigation. Rather than simplifying complexity, it seeks to make it interpretable and actionable, while acknowledging that the pathways toward precision antimicrobial strategies remain contingent, context-dependent, and open to refinement.

CHAPTER SYNOPSIS & TECHNICAL HIGHLIGHTS: A QUICK REFERENCE FOR READERS

This section provides a concise, chapter-by-chapter overview of the contributions included in *AI, Nanobiotechnology, and the Future of Precision Antibiotics*. Each entry highlights the core focus, key methodologies, and primary outcomes of the chapters, offering a practical reference for readers seeking a rapid understanding of the volume's scope.

While this synopsis emphasizes technical and conceptual clarity, it is designed to complement, rather than replace, the Preface and Executive Summary, which provide a more integrated, narrative-driven perspective on the Reyed Precision Convergence Framework for Antimicrobial Resistance (RPCF-AMR). Researchers, practitioners, and policy-makers may use this section to quickly identify the technological approaches employed and how each chapter contributes to the overall convergence strategy.

FROM SYSTEMS TO SOLUTIONS: A METHODOLOGICAL ROADMAP FOR PRECISION ANTIMICROBIALS

The contributions within this volume are structured to illustrate a progressive transition from conceptual reframing to technological innovation and systemic application. Collectively, these chapters delineate a convergent trajectory, integrating predictive modeling, nanobiotechnology, and One Health principles as organized under the Reyed Precision Convergence Framework (RPCF-AMR).

Cluster 1: Conceptual Foundations and Ecological Framing

Chapter 1 – Nanobiotechnology and AI Convergence for Antimicrobial Resistance and Predictive Precision Therapeutics

Author: Reyed M. Reyed

This opening contribution establishes the intellectual baseline by framing AMR as a systems-level challenge defined by the interdependence of microbial evolution, host variability, and ecological pressures. Framework-level constructs—notably NexDi and SCF–DNPS—are introduced to support predictive methodologies. The analysis situates these innovations within a One Health-oriented perspective, providing the analytic grounding for the subsequent volume.

Chapter 2 – Beyond Antibiotics: AI–Nanotech Synergy for a Sustainable Antimicrobial Future

Authors: Arun Agrawal, Y. Rami Reddy, K. Hussain, Srinivasan Nagaraj, B. Venkateswarlu, Rekha Gangula, Deepak Gupta, Deshdeepak Shrivastava

The discourse expands here into the intersection of artificial intelligence and nanotechnology as a platform for sustainable innovation. By aligning genomics and microbiome analysis with predictive modeling, the focus shifts toward macroecological sustainability, emphasizing equitable access and the necessity of interdisciplinary synergy within global health infrastructures.

Chapter 3 – Harnessing Microbiome and Genomics for Precision Antibiotics: AI and Nanobiotechnology Innovations

Author: Tarun Patel

Biological complexity is examined through the lens of the host–microbiome interface, identifying the gut as a primary reservoir of resistance-associated genes (ARGs). The discussion evaluates CRISPR-based strategies and responsive nanocarriers as molecular routes toward microbiome-sensitive therapy. Such an approach suggests that effective antimicrobial design appears increasingly dependent on preserving ecological stability alongside pathogen control.

Cluster 2: Biological and Predictive Intelligence

Chapter 4 – AI-Enabled Nanobiotechnological Platforms for Precision Antibiotic Discovery and Targeted Antimicrobial Delivery

Author: Ruba M.

In evaluating the computational-physical interface, this chapter explores AI-enabled discovery platforms for identifying novel antimicrobial candidates. Machine learning and bioinformatics are presented as auxiliary tools to enhance therapeutic selectivity. The emphasis remains on how computational intelligence facilitates the initial identification and screening of high-potential molecular architectures.

Chapter 5 – Integrating Artificial Intelligence and Nanobiotechnology for Precision Antibiotic Delivery and Targeted Antimicrobial Therapy

Authors: Dhanya S, Dhanushiya R, Veena Gayathri K

Transitioning from discovery to therapeutic optimization, this work addresses AI-guided drug repurposing and refinement. The convergence of computational selection with advanced nanocarriers appears likely to refine pharmacokinetic profiles and tissue penetration. The narrative establishes a methodology for customizing treatment protocols to account for real-world biological variability.

Cluster 3: Technological Delivery and Advanced Intervention

Chapter 6 – AI-Guided Stimuli-Responsive Nanocarriers for Precision Antimicrobial Therapy

Authors: Mona Khalifa, Ahmed Mohammed, Halah Hafiz, Mariam A. Ahmed, Mohamed Al Refaei, Eman Mohammed, Hagar Abdelgwad

Engineering solutions are examined here to address the constraints of conventional drug administration. Through the development of stimuli-responsive systems, the authors illustrate how AI-assisted optimization enables site-specific release at infection loci. These mechanisms are designed to improve precision while potentially minimizing host exposure and mitigating post-treatment dysbiosis.

Chapter 7 – Smart Nanocarriers and AI-Driven Release Systems for Targeted Antibiotic Delivery

Authors Rabia Yasa Kostas, Pinar Bozbeyoglu, Elif Altuner, Muzaffer Elmas

AMR is addressed here through the lens of delivery architecture and formulation efficiency. The analysis demonstrates how AI-guided release systems may resolve persistent challenges related to drug stability and intracellular transport. The discussion provides a technical overview of how engineered platforms can be dynamically aligned with the physiological requirements of targeted intervention.

Chapter 8 – Nanocarrier Technologies for Targeted Delivery

Authors: Jyotishman Tamuli, Nirbanya Devi, Jharna Tamuli, Khemnath Patir, Sarbani Giri

A comprehensive review of representative platforms—including liposomes, polymers, and biomimetic systems—focuses on site-specific delivery. The authors evaluate pharmacokinetic optimization and the translational considerations necessary for clinical relevance, establishing the technological foundation required for next-generation antimicrobial intervention.

Cluster 4: Translational Realities and Global Integration

Chapter 9 – AI, Nanobiotechnology, and the Future of Precision Antibiotics: Tailored Therapies for MDR Pathogens

Authors: Dini Agustina and Bagus Hermansyah

The final chapter situates these innovations within the context of multidrug-resistant (MDR) pathogens and global health frameworks. It addresses implementation constraints in resource-limited settings, suggesting scalable strategies for deployment and stewardship. This synthesis reinforces the principle that antimicrobial innovation is inherently systemic, necessitating a bridge between scientific discovery and societal application.

Cumulative Synthesis

The collected chapters articulate a structured, convergence-driven approach to antimicrobial innovation:

Conceptual Foundation: Outlines the systemic nature of AMR and establishes a multi-scalar analytical perspective.

Predictive Layer: Offers computational tools to anticipate resistance patterns and prioritize therapeutic candidates.
Technological Layer: Deploys engineered, context-sensitive intervention platforms to translate predictive insights into physical reality.
Translational Layer: Evaluates the operational, regulatory, and ecological conditions that shape real-world implementation.

This arrangement achieves a refined methodological coherence, suggesting a linkage between high-level conceptual reframing and actionable, system-oriented solutions for the future of precision medicine.

THE REYED PRECISION CONVERGENCE FRAMEWORK ON ANTIMICROBIAL RESISTANCE (RPCF-AMR): A META-METHODOLOGICAL ARCHITECTURE

1. Epistemic Positioning and Conceptual Rationale

The Reyed Precision Convergence Framework on Antimicrobial Resistance (RPCF-AMR) is introduced as a robust methodological scaffold designed to address the fragmented nature of contemporary antimicrobial research. In the current landscape, a persistent tension exists between reductionist mechanistic treatments—which provide granular molecular detail but often lack systemic context—and systems-based or AI-driven interpretations, which offer breadth but remain contingent on data completeness and algorithmic transparency.

RPCF-AMR occupies an intermediate epistemic space, functioning as a convergence-oriented structure. Rather than attempting to eliminate the inherent stochasticity of microbial evolution, the framework organizes such complexity into analytically navigable layers. By establishing "convergence" as a rigorous methodological principle rather than a descriptive metaphor, the framework facilitates the harmonization of heterogeneous evidence—from molecular pathways to socio-ecological drivers—without reducing them to a single explanatory hierarchy.

2. The Architecture of Convergence: A Layered Taxonomy

The framework is organized into five interdependent analytical domains. These layers are not static tiers but constitute a continuously adaptive feedback structure, where each domain informs and refines the others:

The Basal Biological Layer: Focuses on the fundamental substrate of AMR—microbial genomic evolution, pathogen-host dynamics, and environmental selection pressures.

The Systems Biology Integration Layer: Acts as the primary interface for multi-omics data, transforming raw biological inputs into structured network-level representations of resistance.

The Computational Intelligence Layer: Deploys Artificial Intelligence (AI) and Machine Learning (ML) as inference-support systems. Within this architecture, AI serves pattern recognition and predictive hypothesis generation, with its accuracy strictly dependent on data quality and contextual interpretability.

The Nanotechnology & Precision Therapeutics Layer: Represents the domain of intervention. It encompasses stimuli-responsive nanocarriers and targeted delivery platforms designed to function within biologically complex environments, subject to physiological and regulatory constraints.

The Integrative Translational & One Health Layer: This layer situates clinical and technological interventions within the broader context of environmental systems and policy frameworks, highlighting that functional antimicrobial efforts are inextricably linked to socio-structural and ecological determinants.

3. Dynamics of the Framework: Iterative Feedback and Non-Linearity

As illustrated in Figure 2, the RPCF-AMR architecture operates through an iterative feedback architecture. Unlike traditional linear pipelines, this framework posits that biological understanding and technological capability co-evolve through constant interaction.

The core interface facilitates a "context-sensitive design" where computational predictions are continuously refined by biological feedback and nanobiotechnological performance. This creates a dynamically interconnected circuit of prediction, intervention, and ecological response, positioning AMR as an emergent system-level phenomenon that requires a multi-scalar and adaptive interpretive approach.

4. Protocol for Scholarly Application and Citation

To maintain methodological consistency and ensure standardized adoption in interdisciplinary research, authors are encouraged to adhere to the following protocols:

Conceptual Status: RPCF-AMR should be referenced as a methodological framework rather than a parameterized empirical model.

Analytical Specificity: Researchers should specify which analytical layer(s) are being engaged (e.g., *Computational* vs. *Translational*) to enhance clarity in interdisciplinary synthesis.

Standard Citation Forms:

"As conceptualized within the RPCF-AMR framework (Reyed, 2026)..."

"Within the methodological structure of the Reyed Precision Convergence Framework (RPCF-AMR)..."

Editorial Note

The framework presented in Figure 2 constitutes a central conceptual framework guiding the thematic integration of the volume *"AI, Nanobiotechnology, and the Future of Precision Antibiotics."* It serves as a structured reference for interpreting the convergence of computational, biological, and technological domains in the global effort to mitigate antimicrobial resistance.

RPCF-AMR: Conceptual Definition

The Reyed Precision Convergence Framework on Antimicrobial Resistance (RPCF-AMR) may be understood as a conceptual and methodological architecture developed to support the interpretation and integration of antimicrobial resistance (AMR) across biological, computational, technological, and One Health contexts. Rather than representing a deterministic or parameter-driven model, it appears more appropriately framed as a convergence-oriented scaffold through which diverse forms of evidence can be aligned without being constrained to a single explanatory hierarchy.

Positioned between reductionist and systems-oriented traditions, RPCF-AMR occupies what might be described as an intermediate epistemic space. While mechanistic approaches have provided important insights into molecular resistance processes, and systems or AI-based methods offer broader integrative potential, each carries its own limitations. In this context, the framework does not seek to resolve these tensions directly; instead, it organizes them into analytically tractable layers that may facilitate comparison, interpretation, and synthesis.

Structurally, the framework brings together several interacting domains, including biological processes, systems-level integration, computational intelligence, nanobiotechnological intervention, and translational health systems. These domains are not presented as strictly hierarchical. Rather, they appear to be linked through iterative and potentially bidirectional relationships, suggesting an ongoing interplay between biological complexity and technological development.

As a citable construct, RPCF-AMR is intended to contribute to methodological clarity, particularly in interdisciplinary settings. It is best applied as a guiding structure for interpretation and synthesis, while primary empirical and clinical evidence remains central to validation and application.

Figure 2. A Convergence-Based Conceptual Framework for Precision Antimicrobial Innovation (RPCF-AMR

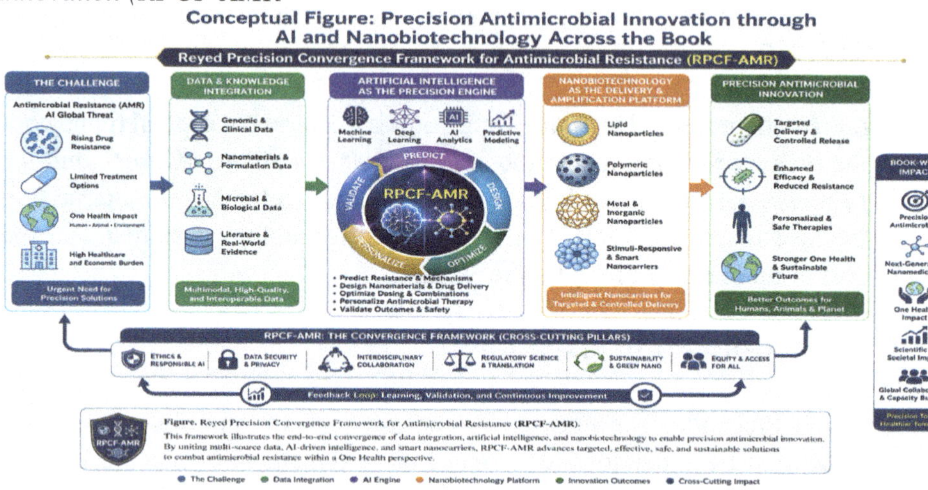

This figure delineates the RPCF-AMR as a multi-layered conceptual architecture that enables integration of computational intelligence, nanobiotechnology, and biological systems dynamics. Within this structure, Antimicrobial Resistance (AMR) is conceptualized as an emergent phenomenon resulting from the continuous interplay between microbial evolution, host variability, and environmental selection pressures.

Biological processes constitute the foundational layer, supported by systems-level and multi-omics data. This is augmented by a computational intelligence layer—utilizing machine learning for pattern recognition—and a nanobiotechnology intervention layer focused on adaptive therapeutic delivery. The uppermost layer represents the Integrative Translational and One Health domain, which aligns clinical, environmental, and policy determinants.

The framework adopts a non-linear, feedback-driven architecture where prediction, intervention, and ecological response are dynamically interconnected. It serves as an interpretive scaffold for harmonizing heterogeneous evidence across diverse scientific domains.

Reyed M. Reyed

Genetic Engineering and Biotechnology Research Institute (GEBRI), Egypt

& City of Scientific Research and Technological Applications (SRTA-City), Alexandria, Egypt

Chapter 1

Nanobiotechnology and AI Convergence for Antimicrobial Resistance and Predictive Precision Therapeutics:

Precision Therapeutics and One Health

Reyed M. Reyed
http://orcid.org/0000-0001-5027-3478
Institute of Genetic Engineering and Biotechnology (GEBRI), City of Scientific Research and Technological Applications (SRTA-City), Alexandria, Egypt

ABSTRACT

This chapter presents the Reyed Precision Convergence Framework for Antimicrobial Resistance (RPCF-AMR) as a systemic redefinition of AMR, beyond microbial-centric views. Integrating nanobiotechnology, AI-driven modeling (NexDi), and multi-scale architectures (SCF–DNPS), it addresses dynamics across host, microbial, environmental, and policy domains. Molecular dynamics, machine learning, and nanocarrier design converge to anticipate intracellular drug availability, bypass classical resistance, and optimize precision. Translational pathways, including programmable nanomaterials and stimuli-responsive carriers, operationalize One Health stewardship while reducing selective pressures in human, animal, and ecological reservoirs. RPCF-AMR reconciles historical surveillance gaps with multi-modal

DOI: 10.4018/979-8-3373-5268-8.ch001

interventions, establishing a deployable reference standard. Through modeling and continuous feedback, it ensures predictive reliability, clinical relevance, and global scalability, defining a trajectory toward effective AMR control by 2040.

1. INTRODUCTION: THE EVOLUTIONARY SHIFT TOWARD SUSTAINABLE NEXT NANOBIOTECHNOLOGY

Figure 1. Graphical Abstract

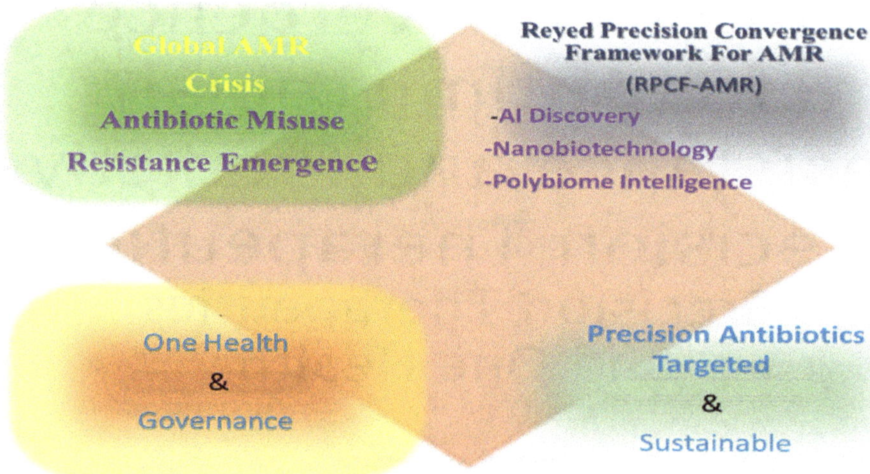

The framework presents a convergence-oriented pathway that connects the systemic drivers of antimicrobial resistance with AI-assisted discovery, microbiome- and polybiome-informed precision delivery, and targeted nanobiotechnology interventions. These elements appear to operate within a broader governance context that aligns with One Health principles and sustainability considerations. Collectively, the model suggests a gradual transition from fragmented antimicrobial practices toward more adaptive, targeted, and context-sensitive strategies for precision antibiotics. (Franklin et al., 2024).

Antimicrobial resistance (AMR) has evolved into the defining existential challenge for contemporary global health systems, extending far beyond a localized clinical burden to encompass profound economic instability and systemic biological vulnerability (Hernando-Amado et al.,2020; Elbaiomy et al., 2025). Despite decades of pharmaceutical innovation, the persistent and widening divergence between therapeutic advancements and microbial adaptation signals a deeper structural mismatch in our current epistemic frameworks (Agu et al., 2023). As we navigate the dawn of the next nanobiotechnology era in 2026, the urgency to deploy nontraditional antimicrobials has reached a critical juncture, necessitating a total departure from the stagnant, monotarget antibiotic pipelines of the past century.

According to Pathak et al. (2026), nanoparticles have effectively transcended their historical role as passive delivery vehicles to become the primary frontier in the global fight against antimicrobial challenges. By offering a multitarget, programmable approach, these nanobiotechnology architectures bypass existing resistance markers through diverse physicochemical mechanisms—such as mechanical membrane disruption, oxidative stress induction, and metal ion homeostasis interference—that operate outside the selective pressure thresholds of traditional small-molecule chemistry (Pathak et al., 2026).

This strategic transition toward high-fidelity intervention is further catalyzed by the emergence of nanoantimicrobial peptides (nano AMPs). As synthesized by Saleem et al. (2026), the strategic integration of these peptides with next-generation nanotechnology specifically targets the complex, recalcitrant outer membranes of multidrug-resistant gram-negative bacteria, effectively addressing a historical void in the antimicrobial armamentarium (Mubeen et al., 2021). This synergy ensures enhanced molecular stability and controlled release, shifting the paradigm from broad-spectrum biological destruction to targeted molecular disruption, a core functional requirement for precision therapeutics in the postantibiotic stage (Saleem et al., 2026).

However, the efficacy of these nanoarchitectures is not only a product of their physical properties but also of their sustainable and intelligent design. In this context, Abbas et al. (2025) provided a profound advance in research by integrating bacterial colicin-mediated silver nanoparticles. This approach utilizes sustainable nanotechnology to bridge the gap between natural antimicrobial proteins and metallic nanoparticles (Alprol et al., 2026; Do et al., 2024; Hetta et al., 2023). By employing rigorous in silico and dynamic simulations, this methodology allows for a predictive understanding of how novel nanocomplexes interact within DNPS (Distributed Networked Pathogen Systems) (Sarkar et al., 2023; Newman et al., 2011; Shin et al., 2022).

This shift from descriptive observation to dynamic, predictive modeling, as operationalized by the Reyed Precision Convergence Framework (RPCF-AMR), is essential for ensuring that nanointerventions can outpace microbial evolutionary trajectories (Di Salvatore et al., 2025; Reyed, 2026g; Abbas et al., 2025; Pathak et al., 2026). The implication is clear: the future of AMR governance lies in a multiscale synthesis where sustainable nanocomplexes are simulated and optimized within coherent computational frameworks (Hosny et al., 2026). This integrative approach ensures that the next generation of antibiotic armamentariums is not only biologically potent but also epidemiologically resilient and context specific, aligning with the 2040 vision of anticipatory biogovernance (Al-Asmi et al., 2024).

1.1. Sovereign computational frameworks and marine armamentariums

To operationalize this technological leap, the future of AMR governance must pivot toward a sovereign computational framework (SCF) capable of simulating these complex interactions across diverse ecological reservoirs (Navarro-Muñoz et al., 2020). As established by Afroze et al. (2026), the exploration of a marine-derived prospective antibiotic armamentarium leaving biometabolites that exhibit antiquorum sensing and antibiofilm properties provides a nonbactericidal route to neutralize resistance (Dzoyem et al., 2025). By focusing on the disruption of microbial communication (quorum quenching) rather than immediate lethality, this nanobiotchnology strategy mitigates the evolutionary arms race that has historically compromised pharmaceutical efficacy (Lu, et al., 2026). Within the RPCF-AMR architecture, these biometabolites are integrated as dynamic inputs into the NexDi AI system, where predictive analytics link marine-derived molecular precision to real-time clinical intervention (Afroze et al., 2026; Duncan et al., 2015; Reyed, 2026a).

1.2. Digital Nanoplastic Science (DNPS) and Sustainable Design

The efficacy of these nanoarchitectures is not only a product of their physical properties but also of their sustainable and intelligent design. Abbas et al. (2025) provided a profound improvement in the use of bacterial colicin-mediated silver nanoparticles. This approach utilizes sustainable nanotechnology to bridge the gap between natural antimicrobial proteins and metallic nanoparticles (Törős et al., 2025). By employing rigorous in silico and dynamic studies, this methodology allows for a predictive understanding of how novel nanocomplexes interact within the digital nanoplastic science (DNPS) layer (Hosny et al., 2026).

This transition from descriptive observation to dynamic, mechanistic modeling allows for the precise mapping of system-level perturbations at the nanoscale, ensuring that nanointerventions can outpace microbial evolutionary trajectories through the perturbation theory machine learning (PTML) engine (Abbas et al., 2025; Le Piane et al., 2024).

1.3. Conclusion: Epistemic Integration and Bioinorganic Interfacing

This epistemic integration represents the core rationale for the Reyed Precision Convergence Framework (RPCF-AMR). By moving beyond the socioeconomic paradoxes and fragmented methodologies of the past century, the framework unites polybiome intelligence, next nanobiotechnology, and AI-driven predictive modeling into a single, system-aware architecture. The implication is a total reconfiguration of the bioinorganic interface, where sustainable nanobiocomplexes are simulated, optimized, and deployed with unprecedented accuracy (Xia et al., 2023). This integrative approach ensures that the next generation of global health defenses is not only biologically potent but also epidemiologically resilient and context specific, anchoring national healthcare sovereignty in the era of biointelligent medicine (Pathak et al., 2026; Saleem et al., 2026; Abbas et al., 2025; Afroze et al., 2026).

1.4. Chapter Objectives and Narrative Architecture: Multiscale Synthesis

This chapter is engineered as a progressive, high-resolution analytical trajectory that transitions from a rigorous epistemic critique to a robust operational blueprint. Its primary objective is to dismantle the prevailing linear-reductionist view of resistance and replace it with a next nanobiotechnology paradigm that is both technologically advanced and system aware (Gatherer et al., 2010). The objectives are structured across four strategic pillars:

- Pillar I: Ontological Redefinition of AMR as a System-Level Phenomenon

AMR is recharacterized not as a localized microbiological failure but as an emergent property of the Polybiome, a complex, multiscalar interaction between biological, environmental, and socioeconomic domains (Salehi et al., 2025; Kang et al., 2024; Reyed, 2023). By shifting the focus from pathogens in isolation to pathogens in the system, we bridge the critical gap in current surveillance models that ignore metabolic and transcriptomic rewiring at the bioinorganic interface (Sultana et al., 2022).

- Pillar II: Epistemic Deconstruction of Disciplinary Fragmentation

A critical evaluation of the innovation deficit paradox highlights how siloed methodologies constrain predictive capacity, causing over 90% of lead compounds to fail in clinical pipelines (Denton et al.,2021). The limitations of classical GWASs and phenotypic assays illustrate the necessity for cross-domain integration between pharmacology and environmental genomics.

- Pillar III: Formal introduction of the RPCF-AMR convergence architecture.

The RPCF-AMR framework orchestrates the sovereign computational framework (SCF), the digital nanoplastic science (DNPS) mechanistic layer, and the NexDi AI system (Cheung et al., 2025). In silico dynamics and PTML are leveraged to map nanocarrier–biological interactions with atomic precision, transforming raw data into actionable, biointelligent interventions (He et al.,2024).

- Pillar IV: Strategic Alignment with One Health and Sovereign Governance

NexDi outputs are aligned with One Health principles to support national healthcare sovereignty (Li, X., et al., 2024). This includes a sovereign armamentarium of marine-derived biometabolites and programmable nanocomplexes, ensuring epidemiological resilience and context-specific interventions for long-term global health security until 2040 (Fernandes, et al. 2025; Wu, et al., 2025).

2. NARRATIVE STRUCTURE: RECURSIVE EVOLUTION AND DIALECTIC REFINEMENT

The chapter adopts a recursive and dialectic structure, revisiting and stress-testing earlier assumptions with emerging 2026 evidence (Pathak et al., 2026; Saleem et al., 2026; Abbas et al., 2025). This ensures that conceptual development and methodological refinement evolve in tandem, creating a self-correcting analytical loop:

- Phase 1: Diagnostic Critique – Exposes socioeconomic and biological vulnerabilities in current global health systems (Wimalawansa, 2025).
- Phase 2: Mechanistic Synthesis – Explaining the molecular mechanics of the next nanobiotechnology, illustrating how DNPS-targeted disruption bypasses resistance markers (Patra et al., 2018).
- Phase 3: Predictive Operationalization– Translates theoretical insights into real-time predictive models via the NexDi AI layer (Berros et al.,2025).

By bridging high-level theory and practical applicability, RPCF-AMR is positioned not only as a conceptual model but also as a National Operational Model for bioinorganic interface reconfiguration, anticipatory action, and sovereign resilience.

2.1. The Polybiome as a Regulatory Interface: The Rise of Fermentobiometrics

Within this systems-level perspective, the polybiome emerges as a central regulatory interface, requiring a shift from descriptive taxonomic catalogs to functional, predictive intelligence (Yang et al., 2025). Moving beyond the static "microbiome" paradigm, the polybiome reflects a dynamic, multikingdom network encompassing bacteria, viruses, fungi, and archaea, characterized by sophisticated, nonlinear cross-talk with host physiology (Wegener 2015; Bass 2019; Koide 2023; Reyed & Prabhakar, 2025). The structural and methodological trajectory of this paradigm is synthesized in Figure 2, illustrating the shift from theoretical foundations to the operational maturity required for sovereign health governance.

Figure 2. Evolutionary trajectory of the polybiome framework (2023–2040).

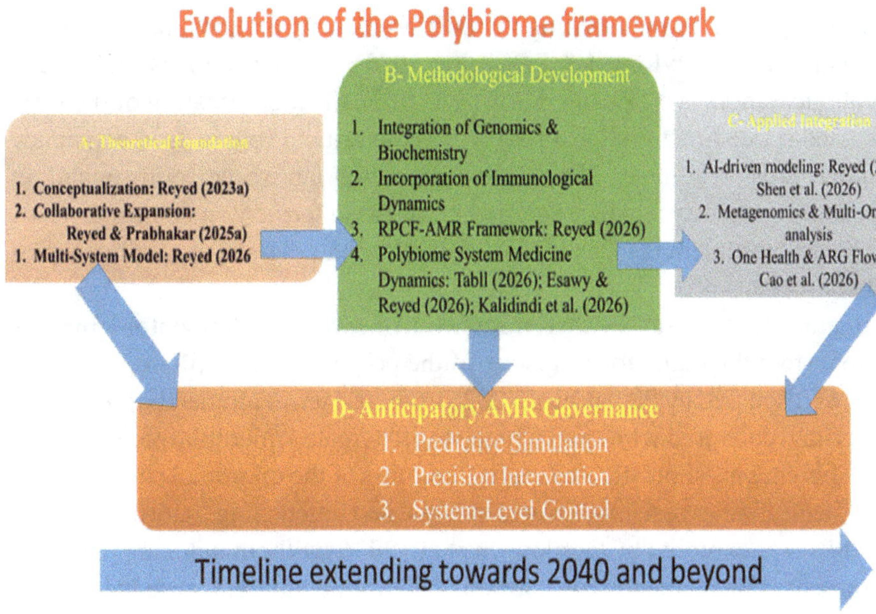

As demonstrated in Figure 2, the polybiome acts as a biointelligent buffer and a primary checkpoint for modulating infection susceptibility, the antimicrobial response, and immunological priming (Nithyanand et al., 2025). Disruptions in this multikingdom diversity—frequently driven by antibiotic overpressurization and xenobiotic accumulation—destabilize this interface, creating "ecological vacuums" and metabolic niches that favor the expansion of resistant lineages (Zou et al., 2026; Hrncir, 2022). As evidenced by recent investigations into metabolic signaling, this process is not merely compositional; it is fundamentally functional. Research by Tabll (2026) suggested that shifts in metabolic pathways and cellular signaling networks are the primary drivers of resistance persistence, transforming the host environment into a pro-evolutionary landscape for pathogens.

The pivotal innovation in this domain is the formal introduction of fermentometabolites—a novel, high-affinity class of biometabolites derived from specific microbial fermentation pathways that function as "biosignal modulators" at the host–microbe interface (Rusu et al.,2023). As pioneered by Reyed & Prabhakar (2025), the emerging field of fermentobiometrics represents a revolutionary strategy that leverages these specific molecules not only for direct antimicrobial defense but also for the systemic blockade of complex, convergent pathological pathways, including those involved in cancer progression and chronic multidrug resistance.

This "fermentometabolite–cell" interaction functions as a homeostatic stabilizer. Microbial byproducts, such as short-chain fatty acids (SCFAs), serve as critical metabolic checkpoints for maintaining gut barrier integrity and host immune equilibrium (Silva et al., 2020; Reyed & Prabhakar 2025). Furthermore, as explored in diverse biological models by Zhang, **C.**, et al. (2025), the identification of the fecal microbiota and related metabolites underscores the critical role of feed efficiency and metabolic cross-feeding in shaping the microbial landscape (Nithyanand, et al., 2025). In a state of dysbiosis, the absence of these fermentometabolites leads to metabolic collapse, leaving the host vulnerable to the collective intelligence of microbial consortia.

To counteract this, the RPCF-AMR framework integrates the fermentobiometrics approach to restore the regulatory capacity of the polybiome. By utilizing nanobiotechnology, as discussed by Pathak et al. (2026), specific fermentometabolites can be encapsulated and delivered with spatial precision to targeted microbial niches (Wu et al., 2024). This enables sophisticated reprogramming of the microbial environment, effectively neutralizing resistance mechanisms and quenching quorum-sensing (QS) networks—as explored by Gerilovych et al. (2026) without inducing the lethal selective pressure that characterizes traditional antibiotic failure. Consequently, the integration of marine-derived antibiotic armamentariums—targeting anti-quorum-sensing and antibiofilm fermentometabolites (Afroze et al., 2026)—offers a sustainable, nonbactericidal path to resistance neutralization, anchoring the future of next nanobiotechnology in the restoration of biological system dynamics (Pathak et al., 2026; Afroze et al., 2026).

2.2 Conclusion: From Ecological Breakdown to Sustainable Nano-Sovereignty

The convergence of environmental contamination and microbial adaptation has rendered traditional, reactive healthcare models obsolete. The persistent presence of antibiotic residues in wastewater and soil creates a global incubator for resistance genes, which, as highlighted by Wu et al. (2025) and Gerilovych et al. (2026), re-enter human populations through integrated food and water systems. This cyclical environmental exchange creates a feedback loop that cannot be broken by incremental pharmaceutical innovation alone (Diaz et al., 2016; Maitreya et al., 2022).

The RPCF-AMR framework provides the only viable path forward by linking this environmental complexity to the sovereign computational framework (SCF). By integrating the holistic genomic understanding provided by Pham et al. (2024) with the mechanistic modeling of the digital nanoplastic science (DNPS) layer, the framework enables the predictive simulation of how subsequent nanobiotechnology interventions interact with the polybiome. This ensures that the deployment of fermentometabolites and nanocomplexes validated through the dynamic studies of Abbas et al. (2025) is not only biologically effective but also epidemiologically resilient. In this light, the shift toward Fermentobiometrics and sustainable nanobomplexes is not merely a scientific advancement; it is a strategic necessity for securing national healthcare sovereignty and ensuring a resilient global health architecture toward 2040.

2.3. Selection pressure, adaptation, and evolutionary acceleration

Selective pressure remains the foundational driver of resistance, yet its manifestations within the polybiome are neither uniform nor strictly linear. Subtherapeutic dosing, incomplete clinical courses, and pervasive environmental antimicrobial gradients create "selection mosaics" that accelerate adaptive responses within microbial populations (Zhu, 2020). These pressures do not act in isolation; rather, they intersect with microbial community architectures, host immune status, and environmental flux to reshape evolutionary trajectories. This intersection often triggers nonlinear dynamics, where marginal changes in antimicrobial exposure produce disproportionately large, systemic shifts in resistance patterns.

A profound challenge to traditional passive adaptation models is the recent evidence of anticipatory behavior in microbial populations. As demonstrated by Zhao et al. (2024), microbial systems can proactively adjust gene expression in response to environmental cues before direct exposure to lethal antimicrobial agents. This adaptive foresight suggests that microbial networks function as active, biointelligent entities rather than passive recipients of genetic change (Papo & Buldú, 2026). Such findings complicate conventional intervention strategies, as simple reductions in antibiotic use—while necessary—may fail to reverse deeply entrenched, anticipatory resistance patterns. Consequently, disrupting these evolutionary pathways requires a shift toward interventions that can redirect microbial behavior at the signaling level, a core objective of the Fermentobiometrics approach (Zhao et al., 2024; Silva et al., 2020; Reyed & Prabhakar, 2025; Chatterjee et al.,2014).

2.4. Epistemic implications for modeling and system-aware intervention

The biological and ecological complexities outlined above necessitate a fundamental reconfiguration of how AMR is modeled and addressed. Linear, reductionist models that prioritize isolated causality are fundamentally ill-equipped to capture the multiscalar interactions and emergent behaviors inherent in the polybiome. This epistemic gap is the primary cause of limited predictive accuracy and the subsequent failure of many traditional interventions to achieve sustained impact (Pattapulavar et al., 2026).

This systemic failure reinforces the clinical and operational rationale for convergence frameworks such as the Reyed Precision Convergence Framework (RPCF-AMR). By integrating the biological depth of the polybiome with the computational precision of the sovereign computational framework (SCF), such architectures move beyond retrospective, descriptive analysis toward predictive and adaptive intervention. This transition, however, is contingent upon overcoming significant constraints: data heterogeneity, the complexity of model validation, and the necessity for computational scalability.

Despite these challenges, the shift toward system-aware modeling represents a mandatory progression for 2026 and beyond. It aligns with the overarching objective of this chapter: to transition from fragmented, siloed knowledge to a state of epistemic integration. By leveraging the NexDi AI system to synthesize high-fidelity inputs—including the GWAS and machine learning insights established by Pham et al. (2024)—the framework enables the precise mapping of nanocarrier–biological interactions within the DNPS layer. This integrative structure ensures that the next generation of antimicrobial strategies is not only technologically superior but also epidemiologically resilient and anchored in sovereign healthcare governance (Zhao et al., 2024; Pham et al., 2024; Reyed, 2026c; Dai, et al., 2025; Yoon & Yoon, 2018).

3. CONCEPTUAL TAXONOMY OF BIOLOGICAL SYSTEMS: FROM STATIC RESERVOIRS TO DYNAMIC INTERFACING

3.1. Beyond Phenotypes: The Emergence of a Systems-Genetic Taxonomy

The strategic re-evaluation of antimicrobial resistance (AMR) necessitates an immediate departure from the descriptive classification of resistant phenotypes toward a structured, ecological–genetic taxonomy grounded in nonlinear system dynamics (Charani et al., 2022; Nusrat et al., 2025). This transition recognizes that resistance is not an isolated microbial trait but rather an emergent property of interacting biological layers that span microbial consortia, mobile genetic elements, and host environments. However, historically foundational, organism-centered models increasingly constrain our evolutionary interpretation by obscuring the distributed nature of resistance propagation (Mustafa et al., 2022). Consequently, a multikingdom, interaction-driven taxonomy is required to enable the precise mapping of these resistance networks (Zhang **Z.**, et al., 2025). This epistemic shift forms the critical infrastructure for biointelligent and predictive intervention strategies operationalized within the RPCF-AMR framework.

3.2. The Microbiome as an Ecological Conduit: Evidence from Genomic Surveillance

From this systems perspective, the microbiome has been redefined as an ecological conduit—a dynamic interface that mediates the storage, exchange, and amplification of antimicrobial resistance genes (ARGs) across clinical and environmental boundaries. Recent metagenomic analyses of urban wastewater have shown that resistome composition correlates more strongly with socioeconomic and structural variables than with direct antibiotic exposure (Arnold et al., 2025).

This empirical reality confirms that resistance is embedded within interconnected networks, which traditional bacteria-centric models fail to capture. By treating the microbiome as a multisystem network that integrates bacterial, viral, and fungal interactions, the NexDi AI system receives the high-fidelity data required for predictive modeling. This systemic intelligence is critical for identifying resistance "hotspots" before they manifest as clinical failures (Zhu et al., 2026; Reyed, 2023; Liu et al., 2022; Helliwell et al., 2025).

3.3. Resistome and Mobilome: The Evolutionary Architecture of Genetic Flux

To progress from ecological observation to mechanistic control, the Resistome and Mobilome form the analytical core of the digital nanoplastic science (DNPS) layer. The resistome encompasses the full repertoire of ARGs, including latent and nonclinical reservoirs, shifting focus from observed resistance to evolutionary potential (Zhang T. et al., 2026; Bilal et al., 2025).

The mobilome—comprising plasmids, transposons, and bacteriophages—serves as the vehicle for horizontal gene transfer (HGT), enabling rapid dissemination of traits across species boundaries (Elgin, 2025). Evidence of ARG-bearing plasmids in environmental matrices confirms that circulation at a scale requires a trans-kingdom approach (He et al., 2026). Viral communities, particularly bacteriophages, act as vectors for ARG transduction, expanding the ecological scope of the distributed resistance pangenome (Kumavath et al., 2025; Fan et al., 2025).

Recognizing these elements as part of a continuous genetic flux allows RPCF-AMR to simulate and intercept pathogen evolutionary trajectories in real time.

3.4. The Structural Transition: Polybiome Paradigm and the Rise of Fermentobiometrics

This conceptual progression culminates in a structural transition to the polybiome paradigm (Reyed, 2026b). Unlike fragmented microbiome approaches, the polybiome defines the biological system as a multiscalar interaction network, where function emerges from connectivity. This paradigm is operationalized through Fermentobiometrics—the strategic use of fermentometabolites as biosignal modulators (Wu et al., 2024).

These fermentometabolites serve as the biochemical currency through which microbial networks achieve the anticipatory behavior documented by Zhao et al. (2024). By integrating these signaling molecules into the DNPS layer, surveillance shifts from passive observation to active bioinorganic interfacing. The NexDi AI system leverages this functional flux to predict how specific nanocomplexes can be programmed to neutralize quorum-sensing (QS) networks, thereby bypassing microbial adaptive foresight (Gerilovych et al., 2026; Pathak et al., 2026).

This integration ensures that environmental genomic data are not merely recorded but are actively processed as a computational input for sovereign governance. It secures the resilience of national healthcare systems against the evolving AMR landscape (Elgin, 2025).

Table 1. Evolutionary trajectory of the polybiome paradigm (2023–2026)

Phase	Paradigm Layer	Core Conceptual Innovation	Key Scientific Drivers & Methodology	Epistemic Outcome for AMR
I. Foundations (2023)	Taxonomic Microbiome	Establishing the link between host-microbiome interactions and clinical phenotypes (Li, **J.**, et al., 2024; Rusu et al., 2023; Reyed, 2023)	Metagenomic profiling and descriptive baselines of microbial populations	Identification of the microbiome as a static reservoir of resistance genes
II. Expansion (2024–2025)	Functional Fermentobiometrics	Introduction of Fermentometabolites as biosignal modulators (Rusu et al., 2023; Reyed & Prabhakar, 2025 ; Thakur et al., 2025)	Integration of metabolomics and pathway blocking strategies in cancer and AMR	Shift from who is there to what they are doing (Functional flux)
III. Convergence (Late 2025)	DNPS Layer	Mechanistic modeling of the bioinorganic interface (Abbas et al., 2025; Pathak et al., 2026)	In silico dynamic studies and PTML (Perturbation Theory Machine Learning)	Programmable intervention at the molecular level to bypass microbial adaptation
IV. Sovereignty (2026)	RPCF-AMR/ NexDi AI	Predictive Governance and anticipatory resistance control (Elgin, 2025; Salehi et al., 2025; Pattapulavar et al., 2026; Miller et al., 2024)	AI-driven synthesis of multiscale data (GWAS, socioeconomic, Polybiome flux (van Pinxteren & Jespers, 2025).	Transition to a National Sovereign Framework for long-term epidemiological resilience

3.5. Polybiome Systems Medicine (PSM): Operational Realization

In the context of the transition from the microbiome to the polybiome, polybiome system (PSM) medicine represents the operationalization of this biological governance model (Salehi et al., 2025; Reyed, 2026e). PSM translates multikingdom interactions into measurable, reproducible, and intervention-ready systems. The analytical focus shifts from amplifying complexity to structuring it into functional modules that support prediction, simulation, and targeted intervention (Pattapulavar et al., 2026).

Unlike conventional microbiome approaches, which remain descriptive and retrospective (Lloyd-Price et al., 2016), PSM enables forward-looking modeling. This includes system perturbation, resilience testing, and adaptive intervention design. By integrating multiomics data with environmental and ecological context, PSM provides computational blueprints for the NexDi AI system to execute precise metabolic governance.

Table 2. Operational Dimensions of Polybiome Systems Medicine (PSM)

Dimension	Operational Definition	Systemic Innovation	Critical Analytical Position	References
Biodiversity Translation	Conversion of ecological diversity into stratified, validated modules	Links genomic, metabolomic, and ecological data into deployable units	Conventional probiotics lack network-level resolution; PSM enables functional reproducibility	Salehi et al., 2025
Heritage-Informed Intelligence	Contextual microbial selection using heritage-linked functional indices	Integrates environmental, cultural, and biological variables	Generic strain use is reductionist; PSM introduces contextual and sovereign precision	Agboola et al., 2025; Zielińska et al., 2025 ; Reyed, 2026f; Yang et al.,2025)
Multi-Omics Integration	Unified analytical system (Genomics, Proteomics, Metabolomics, Transcriptomics)	Prioritizes interaction and functional flux over isolated biomarkers	Single-omics approaches are insufficient for modeling AMR system dynamics	Abdelsattar et al., 2021; Reyed, 2026e; Pattapulavar et al., 2026 ; Wörheide et al., 2021
Digital Twin Biocultures	Computational modeling of host–microbiome–environment systems	Enables in silico simulation, perturbation, and resilience testing	Traditional models are retrospective; PSM supports predictive intervention	Satheeskumar, 2025; Alharthi et al., 2025; Zhang, J. et al., 2026
Intervention-First Causality	Iterative simulation combined with controlled biological perturbation	Accelerates translation from theoretical experiment to sovereign application	Observational models lag system dynamics; PSM aligns causality with biological function	Bilal et al., 2025; Sultana et al., 2022
Precision Nutrition 2040	AI-driven adaptive nutritional systems targeting microbiome and ARG modulation	Integrates diet and resistance control as a metabolic governance tool	Generic nutrition lacks system impact; PSM reduces selective pressure on the resistome	Greathouse et al., 2025; Reyed, 2026f ; Quan et al., 2025; Mills et al.,2019; Zielińska et al., 2025
Ethical & Regulatory Governance	Embedded frameworks for bioethics, data sovereignty, and compliance	Integrates governance structurally to ensure reproducibility and safety	Governance is often neglected; PSM makes it a structural prerequisite for deployment	Muñoyerro-Muñiz et al., 2025; McNair, 2022

3.6. Epistemic Integration: Securing the Resistome–Mobilome Interface

The limitations of legacy microbiome models are most evident in the context of antimicrobial resistance (AMR), where resistance emerges from multiscale interactions that extend far beyond host boundaries (Hernando-Amado et al.,2020;

Sabih Ur Rehman et al., 2025). The polybiome paradigm, developed through the collaborative work of Reyed & Prabhakar (2025), addresses these constraints by conceptualizing biological organization as an interconnected network. In this model, function emerges from interdependent ecological and genomic interactions, which is consistent with the principles of network biology (Barabási et al., 2011; Hasin et al., 2017).

By integrating genomic, biochemical, and immunological dimensions, the polybiome provides a traceable pathway from theoretical formulation to operational implementation (Tabll, 2026; Yang et al., 2025). Coupled with the predictive capabilities of the NexDi AI system, it enables computational replication and optimization of biological networks. This framework supports the highly precise prediction of antimicrobial resistance gene (ARG) flow across human, animal, and environmental domains, adhering to One Health principles while projecting system stability toward 2040 (Altevogt et al., 2025; Shen et al., 2026; Cao et al., 2026). By establishing this conceptually robust and empirically contingent pathway, RPCF-AMR provides the foundational architecture for anticipatory, system-level AMR governance (Chen et al., 2026).

3.6.1. **Strategic Consequences of the Multikingdom Approach: Operationalizing the Polybiome**

Transitioning to a multikingdom, a polybiome-informed framework carries operational consequences across surveillance, modeling, policy, and innovation. By building on the biological taxonomy outlined previously, AMR has been reconceptualized as a distributed, interaction-driven system that requires coordinated strategies across ecological, clinical, and environmental domains.

3.6.1.1. Epistemic Shift in Surveillance: Beyond Taxonomic Myopia

Traditional AMR surveillance is constrained by taxonomic myopia, focusing primarily on clinically isolated bacterial strains while neglecting latent genetic reservoirs in environmental and cross-kingdom systems. This limitation undermines early detection and underestimates the scale of resistance propagation.

Polybiome-informed surveillance integrates multiscalar data, encompassing wastewater resistomes, environmental viromes, and soil mycobiomes (Fan et al., 2025). Such integration enhances anticipatory capacity, allowing resistance signals to be detected before clinical manifestation (Arnold et al., 2025; Hasin et al., 2017). Recent studies highlight that global governance frameworks risk underrepresenting atmospheric and ecological ARG transmission if these multiomics datasets are excluded (Chen et al., 2026). By repositioning surveillance as a predictive tool, RPCF-AMR transforms environmental "noise" into actionable intelligence.

3.6.1.2. Evolutionary Dynamics and the Mobilome: The Velocity of Adaptation

From this expanded perspective, evolutionary dynamics and horizontal gene transfer (HGT) have become central to understanding resistance. ARGs propagate actively through mobile genetic elements (MGEs), which facilitate rapid adaptation across species and ecological boundaries (Sum et al., 2025). These elements operate as high-velocity networks that coevolve with resistomes under selective pressure (Elgin, 2025; Kumavath et al., 2025).

Plasmids, as molecular cornerstones of AMR, drive shared resistomes that extend across urban water systems and clinical environments. Monitoring these communities of plasmids necessitates Networked Diagnostics (NexDi) approaches (Castañeda-Barba et al., 2024; He et al., 2026; Smyth et al., 2025). By mapping the velocity of genetic flux, the framework identifies critical nodes where interventions can most effectively disrupt transmission chains.

3.6.1.3. Computational innovation and targeted intervention

Advanced computational strategies—including molecular docking, nanoparticle interaction modeling, and multiomics integration—enable the identification of mechanisms that disrupt gene transfer and microbial communication (Abdelsattar et al., 2021). The advent of SynOmics and explainable AI (XAI) has facilitated systematic mapping of the gut–brain–microbiome axis and its response to environmental stressors (Alif & Zhang, 2025; Cao et al., 2026; Zhang, J et al., 2026).

Applied to sepsis management and precision medicine, these tools decrypt distributed networked pathogen systems (DNPSs), advancing toward network medicine frameworks that move beyond linear models (Shen et al., 2026; Barabási et al., 2011). These innovations ensure that interventions are both biologically potent and mathematically optimized for the host environment.

3.6.1.4. Policy and the One Health Paradigm: A Sovereign Blueprint

By uncovering resistance reservoirs beyond clinical settings—including food production and natural ecosystems—the polybiome approach aligns directly with One Health principles (Al-Khalaifah et al., 2025). Advanced nanomedicine strategies, such as stimuli-responsive nanocarriers, allow targeted microbiome restoration and oral biofilm control without collateral damage caused by broad-spectrum antibiotics (Farah et al., 2026; AlQurashi et al., 2025; Chen et al., 2026; Mi, 2020).

Mathematical modeling of transmission dynamics ensures that policy interventions address the networked nature of AMR rather than isolated pathways (Changruenngam et al., 2022; Sabih Ur Rehman et al., 2025). Integrating XAI within the One Health framework ensures that interventions are epidemiologically transparent, targeting plasmid-driven resistomes that span urban and environmental water systems (Smyth et al., 2025; He et al., 2026). This alignment is critical for the NexDi blueprint, securing national health policies under principles of Genomic Sovereignty and predictive governance.

3.6.2. Conclusion: From Conceptual Taxonomy to Actionable Intelligence

The RPCF-AMR framework is grounded in a redefined taxonomy of biological systems, integrating microbiome and polybiome perspectives into a unified analytical structure. This integration resolves key limitations in conventional AMR models by bridging descriptive analysis with predictive and operational capabilities.

Expanding the analytical scope to include polybiome dynamics enables several critical advancements:

1. High-Fidelity Simulation: Feasible through host-specific modeling, including PBPK-based antibiotic optimization (Almajidi et al., 2025).
2. Molecular Precision: Systematic exploration of drug–target interactions via computational techniques such as molecular dynamics simulations (Aulifa et al., 2024).
3. Predictive Surveillance: AI can be leveraged to detect nonobvious resistance pathways across interconnected biological and environmental domains (Arnold et al., 2025).
4. Integrated Innovation: Development of dual-target inhibitors capable of disrupting both biofilm formation and genetic replication (Agboola et al., 2025). Collectively, these developments mark a transition from descriptive frameworks to precision-oriented, convergence-based models grounded in biointelligent foresight. Nevertheless, challenges related to data integration, validation, and cross-sector implementation remain and must be addressed to ensure scalability and real-world impact. AMR is not merely a pharmacological problem; it is a cross-domain, multiscale phenomenon. Subsequent sections examine the mechanistic potential of nanobiotechnology, the epistemic role of computational intelligence systems, and the cross-sectoral implications of deploying integrated frameworks at scale.

4. COMPUTATIONAL AND AI SOLUTIONS: EPISTEMIC AMPLIFICATION AND SYSTEM INTEGRATION

The mechanistic constraints of nanotechnology underscore the necessity of a shift toward computational intelligence and artificial intelligence (AI), which function not only as tools for drug discovery but also as epistemic amplifiers of system understanding (Ezike et al., 2023). Machine learning techniques can identify novel antimicrobial candidates and optimize molecular interactions with unprecedented efficiency (Durrant & Amaro, 2015). However, predictive reliability remains structurally bounded by model assumptions and dataset biases (Gheorghita et al., 2025). This limitation reinforces the need for a sovereign intelligence paradigm, exemplified by the NexDi blueprint. AI does not replace expert reasoning; rather, it operates within a cooperative framework, where algorithmic outputs are iteratively refined through deep domain expertise (Satheeskumar, 2025; Jena et al., 2024; Golriz Khatami et al., 2021). This hybrid architecture mitigates overfitting and false predictive confidence. Beyond molecular modeling, AI frameworks provide a unifying analytical layer capable of integrating epidemiological, environmental, and socioeconomic datasets. This integrative capacity transforms AI from a standalone analytical instrument into a system-level coordinator, enabling anticipatory, cross-domain strategies that outpace microbial evolutionary flux (Forli et al., 2016).

4.1. Cross-Sector Dynamics: From Localized Failure to Systemic Propagation

Building on this integrated computational perspective, AMR must be understood as a system-level propagation phenomenon rather than a series of localized clinical failures. Pathogens such as *Escherichia coli* and *Klebsiella pneumoniae* exemplify this transition, where resistance profiles compromise routine interventions and necessitate reliance on second-line therapies with significant economic and biological costs (Guégan et al., 2018; Sultana et al., 2022; Reyed, 2020). This clinical burden is amplified by nonclinical pathways. Agricultural and animal health systems act as persistent sources of antimicrobial exposure, facilitating the dissemination of resistance into the soil, water, and food chains and thereby establishing dynamic environmental reservoirs (Sharma et al., 2024; Elgin, 2025; Zhitlov et al., 2025). Wastewater treatment systems further illustrate a systemic paradox: when designed to reduce the microbial load, they may inadvertently catalyze horizontal gene transfer (HGT), increasing the diversity of antibiotic resistance genes (ARGs) even after advanced processing (Bilal et al., 2025; Altammar, 2023). Interventions targeting isolated components—either clinical or environmental—cannot fully account for the networked nature of resistance propagation.

4.2. Path-to-Sovereign Coordination: Overcoming Operational Fragmentation

The systemic propagation of resistance is structurally reinforced by operational fragmentation. Despite advances in nanotechnology and AI, AMR management remains compartmentalized, with the clinical, agricultural, and environmental sectors frequently operating in isolation. Limited data exchange and fragmented decision-making architectures exacerbate this disconnect (Esawy & Reyed, 2026).

While one-health models conceptually acknowledge interdependence, they often fail at the implementation stage owing to the absence of integrated, real-time surveillance. Figure 3 illustrates these structural constraints, highlighting the reactive nature of current containment strategies.

Figure 3. Fragmented systems and disciplinary silos in AMR

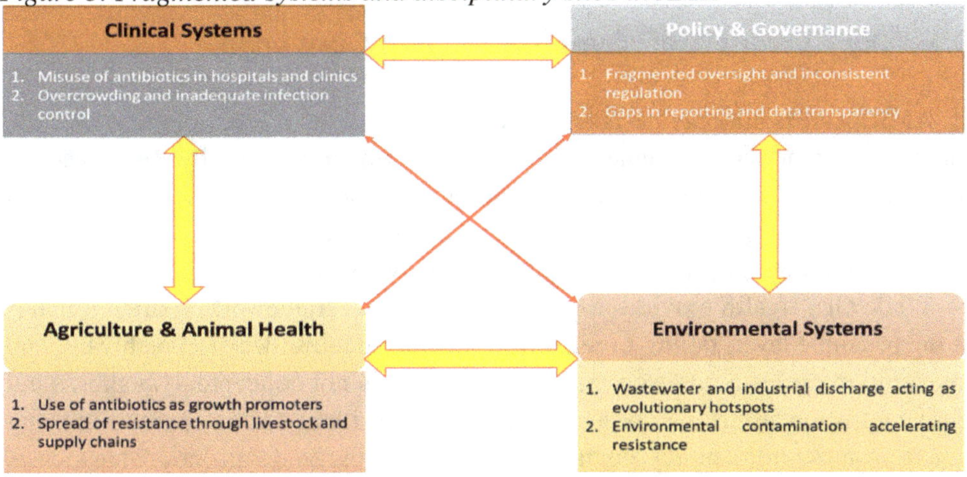

This schematic depicts three parallel domains—clinical, agricultural, and environmental—with minimal cross-domain integration. Arrows indicate limited interaction pathways, failing to capture the complex, multidirectional flow of resistance across the polybiome. Highlighted nodes denote resistome reservoirs exploiting systemic discontinuities.

The analytical insight from Figure 2 confirms that technological interventions remain siloed. Strategies pioneered through next nanobiotechnology and fermentobiometrics reinforce fragmentation rather than resolving systemic vulnerabilities. Conventional surveillance lacks the epistemic resolution to map multiscale interactions, obscuring the distributed nature of resistance propagation. Consequently, resistomes circulate freely between hospitals and environmental reservoirs, creating exploitable gaps for microbial anticipatory behavior (Zhao et al., 2024).

Recent studies leveraging GWAS and resistome reconstruction (Pham et al., 2024) underscore that addressing AMR under these conditions requires a fundamental structural transformation—moving from reactive containment toward predictive coordination as offered by the RPCF-AMR architecture. By aligning molecular precision with sovereign computational intelligence and cross-sector governance (Pattapulavar et al., 2026; Tabll, 2026), AMR can be repositioned from an escalating crisis to a manageable, system-level challenge, ensuring long-term resilience toward 2040 and beyond.

5. STRUCTURAL CONVERGENCE: RPCF-AMR AS AN EXECUTABLE ARCHITECTURE

5.1. Why Current AMR Systems Do Not Converge

AMR persists not only because of technological insufficiency but also because the systems that should connect interventions remain structurally disconnected. Sections 2.1–2.2.5 establish the recognition of clinical surveillance, agricultural governance, environmental monitoring, and computational analysis. However, this recognition rarely translates into operational convergence.

While One Health provides a coherent intellectual framework, it often remains at the principle level, failing to reliably align data streams, decision pathways, and intervention cycles across sectors. The limitation is not conceptual weakness but implementation failure. Consequently, clinical surveillance often operates in isolation from agricultural policy, environmental monitoring, and translational response. Resistomes can circulate across hospitals, food systems, and ecological reservoirs without coordinated containment.

From this perspective, AMR is perpetuated not only by technical gaps but also by the inability of interdependent systems to function as a single adaptive order. Advanced tools—AI, bioinformatics, and nanotechnology—are incomplete when deployed in isolation and are capable of generating information without achieving convergence. In some cases, they exacerbate fragmentation by expanding surveillance without synchronizing responses, providing microbial populations with additional adaptive opportunities. Policy gaps, surveillance discontinuities, and operational delays thereby translate into evolutionary advantages for pathogens (McCubbin et al., 2022).

The implication is clear: isolated innovation is insufficient. What is required is an executable architecture that integrates monitoring, interpretation, and intervention across clinical, agricultural, and environmental domains—realizing the structural convergence necessary for long-term AMR governance.

5.2. RPCF-AMR as a convergence architecture

RPCF-AMR is explicitly designed to address structural discontinuities by transforming convergence from an abstract goal into an operationally executable process. Rather than treating surveillance, prediction, and response as independent tasks, the framework organizes them into a layered architecture, where each stage informs and conditions the next. The limitation of many contemporary models is their tendency to describe integration conceptually without operationalizing it. RPCF-AMR, in contrast, functions as a mechanism for structured coordination, not merely a slogan.

The first layer, the multisource sensing layer, integrates clinical, agricultural, and environmental resistome data through interoperable, standardized pipelines. Its role extends beyond mere data accumulation; it ensures that heterogeneous information becomes comparable within a unified analytical environment. The limitation of data collection alone is that it does not produce interpretive unity. Hence, sensing must be designed for cross-domain usability from the outset.

The second layer, the AI-guided inference layer, transforms heterogeneous signals into synchronized risk maps and predictive intervention triggers. Here, the raw data are converted into operational intelligence. Retrospective surveillance waits for resistance to become clinically visible before action; RPCF-AMR enables early inference, identifying emerging risks before they consolidate into measurable spread.

The third layer, the feedback-oriented mechanistic layer, connects the prediction to the action. Therapeutics, nanobiotechnology, and ecological interventions are treated as elements of a continuous feedback cycle rather than as isolated solutions. A delayed response allows microbial adaptation to exploit latency; RPCF-AMR ensures that the framework is both diagnostic and responsive, reducing the time between recognition and intervention.

This substantive operationalization distinguishes RPCF-AMR from conventional One Health models. While One Health remains primarily a coordination paradigm, RPCF-AMR involves collaboration within the architecture itself, enabling detection, interpretation, and intervention to evolve simultaneously and in real time. Principle-based frameworks rely on external compliance; RPCF-AMR treats fragmentation as a system parameter that can be actively managed.

5.3. Critical, Evidence-Based Application of RPCF-AMR

The practical value of RPCF-AMR emerges when its principles are tested against real-world AMR dynamics. The framework is not a detached theoretical construct; it is designed to respond to the conditions under which resistance emerges, spreads, and persists. Many integrative models remain internally coherent but empirically thin; RPCF-AMR is explicitly intended as an applied architecture, not merely a conceptual proposal.

5.4. Core Principles

1. Systems Thinking: AMR arises from nonlinear interactions among antibiotic use, environmental contamination, social behavior, and technical infrastructure (Pham & Wozniak, 2025). Single-factor interventions cannot control a multicausal problem. Clinical progress, while necessary, is insufficient without simultaneous ecological and policy interventions.
2. Multiscalar integration: RPCF-AMR combines molecular, host–, clinical, agricultural, and sociopolitical data within a single interpretive structure (Redman White et al., 2026). Scale-bound analysis misinterprets a distributed phenomenon as if it were confined to one level. Resistance must be interpreted across genetic, organizational, institutional, and environmental registers simultaneously.
3. Predictive, application-oriented analytics: Machine learning, dynamic modeling, and real-time data assimilation support proactive interventions rather than delayed reactions (Sakagianni et al., 2023). Descriptive systems explain only post hoc; anticipatory analytics shift AMR governance from reaction to prevention.

5.5. Case Study: Hospital–Environment Interface

The hospital–environment interface exemplifies RPCF-AMR in practice. Carboreenem-resistant *Klebsiella pneumoniae* (CRKP) was detected in a tertiary hospital and its surrounding ecosystem. Clinical and environmental systems interact continuously; compartmentalized analysis obscures cross-boundary transmission pathways. Hence, hospital resistance cannot be fully understood without reference to wider resistome ecology.

Recent modeling underscores this point. Changruenngam et al. (2022) developed transmission dynamics models for CRKP and demonstrated that resistance trajectories are highly sensitive to small changes in transmission structure and intervention intensity. Liu et al. (2025) applied machine learning-based risk evaluation in neonatal units and reported that CRKP dissemination is shaped by the interaction of patient flow, clinical exposure, and the environmental context. Individually, these studies fail to capture the full ecology of spread; together, they show that resistance is better understood as a coupled system rather than as isolated infection events.

Implementation requires three convergent data streams: genomic surveillance (e.g., detection of blaKPC), real-time antibiotic use monitoring, and wastewater-based environmental surveillance (Hendriksen et al., 2019). A single stream leaves blind spots; convergence across streams improves both detection and intervention. Within this context, a 10–15% reduction in effective transmission, when embedded in coordinated clinical and environmental interventions, can plausibly shift prevalence trajectories downward over 6–12 months. However, outcomes remain model-dependent and context-sensitive; the directionality of effect is supported by transmission modeling and machine-learning evaluation (Changruenngam et al., 2022; Liu et al., 2025).

RPCF-AMR thus moves AMR management from reactive containment toward proactive, system-conscious intervention.

5.6 Framework Architecture: The Triad of Functional Integration

RPCF-AMR operationalizes three interacting structural levels, bridging the theoretical depth with the implementation capacity:

1. Systemic Drivers' Level: This level addresses structural constraints—siloed practices and surveillance gaps—that allow resistance to persist despite technical capacity. Conventional models treat these drivers as background noise; RPCF-AMR identifies them as primary intervention targets, recognizing that resistance persists due to environmental and policy conditions.
2. Mechanistic Layer: This layer focuses on the dynamic functional flux of the polybiome. By integrating continuous feedback and resistome interactions, this layer accounts for adaptive evolutionary trajectories. RPCF-AMR is a model of ongoing sociotechnical and biological coupling, rather than fixed, linear causation.
3. Convergence and Action Layer: where analysis becomes executable. AI-driven predictive analytics and One Health governance converge to transform policy into synchronized operational response. The structural maturity of this integration is illustrated in Figure 4.

Figure 4. Convergence-driven approach to AMR: RPCF-AMR integration

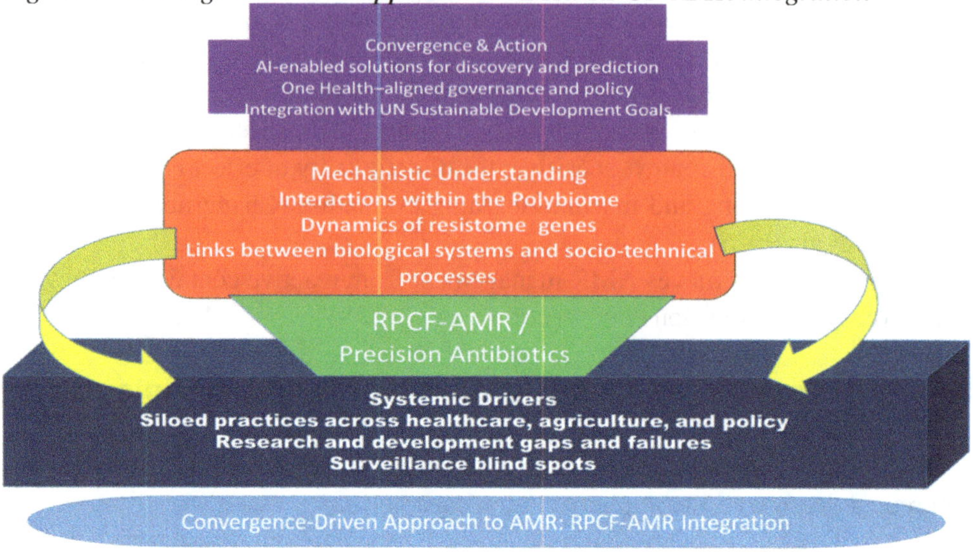

Figure 4 illustrates the vertical, three-layer architecture of the Reyed Precision Convergence Framework (RPCF-AMR) (Reyed, 2026 g). The blue base layer (systemic drivers) represents structural barriers spanning the clinical and environmental domains. The red middle layer (Mechanistic Understanding) captures polybiome–resistome dynamics, providing the operational signals necessary for designing interventions. At the apex, the purple layer (Convergence and Action) integrates AI-driven analytics with international policy alignment, including the UN Sustainable Development Goals, ensuring that interventions are embedded within a real-time, closed-loop operational system.

Central to this architecture is the green node (RPCF-AMR/Precision Antibiotics), which functions as the executable convergence engine. The yellow recursive arrows indicate critical systemic innovation: feedback from the mechanistic and intervention layers is continuously routed upstream to adaptively reshape systemic drivers. This recursive structure enables the framework to function as a living system that evolves in response to emergent microbial threats and ecological shifts. This structure aligns with emerging network-based models of disease propagation and systems medicine approaches (e.g., Barabási et al., 2011), reinforcing its conceptual and operational plausibility.

5.7. Analytical limitations and epidemic design

No framework of this complexity should be treated as deterministic. Predictive outputs within RPCF-AMR remain inherently probabilistic, reflecting the stochastic and context-dependent nature of biological evolution (Redman-White et al., 2026; Holmes et al., 2016). Accordingly, RPCF-AMR functions as a decision-support architecture, providing an epistemic scaffold that amplifies human judgment rather than substituting it for it.

RPCF-AMR does not replace existing stewardship models; rather, it strengthens them by providing a higher-order structure for cross-domain integration (Pennisi et al., 2025; McEwen & Collignon, 2018). The implication is a transition from replacement thinking to integrative coherence, whereby current clinical, environmental, and policy systems are unified within a sovereign computational framework to increase epidemiological resilience.

5.8. Conclusion Synthesis: From Fragmentation to Sovereign Resilience

The central insight of this analysis is consequential: AMR persists not because of a shortage of technological tools but because of architectural fragmentation. Existing solutions, while technically sound, remain misaligned across disconnected sectors. RPCF-AMR transcends conceptual modeling; it is an executable convergence framework that links sensing, interpretation, and response across the human–animal–environment triad.

The future of AMR management does not hinge solely on isolated technological advances—such as a novel antibiotic or a standalone AI platform—but rather on the capacity of these tools to operate within the coherent, adaptive, and interoperable architecture established by the NexDi blueprint (Cao et al., 2026). By treating fragmentation as the core challenge, RPCF-AMR establishes a pathway capable of outpacing microbial evolution, safeguarding national and global health sovereignty toward 2040 and beyond.

5.9 Framing What Comes Next: Toward Convergence

AMR is an inherently interdisciplinary phenomenon shaped by continuous interactions across clinical, agricultural, and environmental systems. Its persistence is a natural consequence of a dominant but fragmented response architecture—reactive, siloed, and institutionally separated. Individual improvements may reduce local burdens but fail to disrupt the systemic machinery that regenerates resistance.

Partial gains, therefore, cannot address the structural conditions that permit AMR to circulate and reconstitute itself across interconnected sociobiological networks.

The primary obstacle to containment is not a deficit of knowledge or tools but rather the persistence of fragmentation as an organizing principle. Enhanced surveillance, stewardship, and monitoring remain structurally limited when deployed in isolation; they treat symptoms rather than root causes. Their effectiveness depends entirely on embedding these measures within a predictive architecture capable of integrating observation, interpretation, and intervention across domains.

A decisive shift is needed from passive monitoring to adaptive governance and from isolated control to coordinated system administration. RPCF-AMR formalizes this transition, converting dispersed monitoring, policy logic, and intervention pathways into a continuous, adaptive operational pipeline. Conventional models are constrained by the separation of data collection, decision-making, and execution; RPCF-AMR overcomes these limitations by making systemic complexity actionable, linking functions into a convergent operational sequence (Dini, 2026). AMR mitigation thus evolves from partial correction to sustained, cross-sectoral anticipation. Failure to converge guarantees that long-term control remains elusive; success depends not on isolated excellence but on system-wide coherence.

6. NANOBIOTECHNOLOGY, AI, AND ONE HEALTH CONVERGENCE

AMR represents a distributed, feedback-mediated system that integrates molecular adaptation, clinical pressures, and environmental circulation. Localized interventions may reduce burden but cannot alter the systemic machinery that perpetuates resistance. Effective control requires a convergent architecture, not parallel, uncoordinated technical fixes. As recognized by the World Health Organization (WHO), AMR is a One Health challenge, traversing healthcare, animals, food systems, and environmental reservoirs, making fragmented intervention structurally insufficient.

6.1. Strategic Convergence of Nanobiotechnology, AI, and One Health

Operational convergence is achieved through high-fidelity integration of nanobiotechnology and AI. Advanced nanocarrier platforms—such as PLGA, PEGylated systems, liposomes, and micelles—enable site-specific therapeutic delivery and the disruption of resistant architectures (Altammar, 2023; Ezike et al., 2023; Torchilin, 2006). However, molecular precision alone cannot resolve systemic scale mismatches. Human physiological variability and inconsistent clinical translation—often referred to as the "translational valley of death"—limit efficacy (Trucillo, 2021; Zeng et al., 2011; Farah et al., 2026). Nanotechnology is mechanistically necessary but insufficient; it requires embedding within higher-order frameworks to ensure that selection, delivery, and postintervention monitoring converge.

Predictive modeling and AI extend mechanistic capacity by simulating molecular interactions, pharmacokinetics (PBPK), and systemic responses with high granularity (D'Hondt et al., 2025; Satheeskumar, 2025; Almajidi et al., 2025). Computational inference, however, remains structurally bounded by model assumptions, overfitting, and scale incompatibility (Barnard, 2015; Brooks & Thompson, 2017; Gheorghita et al., 2025). AI achieves operational impact only when structurally aligned with biological and ecological realities, specifically polybiome and mobilome dynamics.

RPCF-AMR provides this structural embedding, aligning molecular precision, predictive analytics, and One Health governance into a real-time, anticipatory intervention strategy (Reyed, 2026 g). Through fermentobiometrics and digital nanoplastic science (DNPS), the framework identifies metabolic vulnerabilities within the distributed resistance pangenome before clinical manifestation (Tabll, 2026). This convergence transforms One Health from a conceptual ideal into a sovereign computational reality, ensuring that the intervention velocity matches microbial evolutionary dynamics across all reservoirs (Cao et al., 2026; Shen et al., 2026; Pattapulavar et al., 2026).

6.2. Convergence of Bacterial Resilience and Precision Engineering

Bacterial resistance is increasingly recognized as a dynamic, adaptive process rather than a static defensive state. Gram-positive organisms, particularly *Staphylococcus aureus*, exhibit notable structural plasticity, adjustable membrane charge, adjustable cell wall architecture, and envelope-associated functions in response to sustained selective pressures. Interpreting these adaptive events in isolation, however, risks obscuring their broader role within the host's internal ecological network.

Evidence from Zhang, **Z.**, et al. (2025) indicates that specific fecal microbiota and their metabolites are closely linked to host metabolic efficiency and systemic health outcomes. These findings suggest that resistance mechanisms operate within an integrated metabolic landscape rather than independently. Consequently, effective surveillance strategies may require a shift from simple taxonomic cataloging toward approaches that incorporate metabolite signaling and functional flux, providing a more holistic view of microbial adaptation (Zhang Z., et al. 2025).

These adaptive architectures often rely on precision therapeutic agent delivery systems (TADSs). These systems are designed to circumvent efflux mechanisms, increase intracellular drug accumulation, and optimize target engagement beyond conventional clinical thresholds (Sultana et al., 2022). The functionality of TADS depends on a complex interplay of molecular forces, including hydrophobic interactions, hydrogen bonding, and π–π stacking, all operating within constrained thermodynamic landscapes (Hwang et al., 2020; Makadia & Siegel, 2011; Zhuang et al., 2019).

A critical challenge in nanoscale optimization is the so-called "scale-propagation" failure, where molecular-level efficacy does not consistently translate into systemic impact. This limitation underscores the importance of integrating nanocarrier design with systemic architecture to influence resistance trajectories meaningfully. Moreover, well-established resistance mechanisms—such as target alteration and enzymatic inactivation—reinforce the notion that AMR is fundamentally multiscalar and internally systemic, requiring convergent solutions that span molecular, cellular, and ecological scales (Huovinen et al., 1995; Floss & Yu, 2017; Redgrave et al., 2014; Alcala et al., 2020; Morar & Wright, 2010).

The RPCF-AMR framework provides a practical pathway for such convergence, aligning molecular precision with cross-sectoral governance to address resistance as a distributed phenomenon across the polybiome spectrum (Reyed, 2026b). By integrating a mechanistic understanding with systemic coordination, this approach enables more anticipatory and coherent management of AMR across clinical, agricultural, and environmental domains.

7. COMPUTATIONAL PRINCIPLES OF PRECISION NANOCARRIER DESIGN

Next-generation nanocarriers increasingly rely on AI-enabled designs and high-fidelity computational modeling. Given that AMR involves simultaneous microbial adaptation, ecological feedback, and clinical pressures, traditional trial–and–error approaches are generally insufficient. Computational strategies thus play a translational role, aligning with the WHO's emphasis on innovation and multisectoral governance.

Modern nanocarrier design incorporates structure-based drug design (SBDD), molecular docking, and AI/ML-guided anticipatory optimization to increase therapeutic specificity and delivery efficiency (Forli et al., 2016; Gheorghita et al., 2025). Complementary strategies, including quality by design (QbD) and design of experiments (DoE), are used to control critical parameters such as ligand density, pH responsiveness, and payload release kinetics (Wang et al., 2025).

However, these methodologies may fail to achieve systemic impact if applied in isolation. The implication is that design and optimization must form a coordinated, computationally integrated pipeline, where scalability, process control, and biological relevance are considered from the outset. This approach enables nanocarrier development to move beyond proof-of-concept methods to predict and translatable outcomes, potentially enhancing both clinical efficacy and ecological compatibility.

Taken together, these principles suggest that computational precision and multiscalar convergence are not optional enhancements but are essential requirements for next-generation nanocarrier strategies targeting complex, distributed AMR systems.

7.1. Framework Architecture: Hierarchical Intelligence

The RPCF-AMR architecture is structured hierarchically but maintains functional distinction across its three principal layers, ensuring translational continuity from mechanistic modeling to clinical application.

1. Sovereign computational framework (SCF): Serving as the foundation layer, the SCF supports multiscale simulation and data harmonization. While computational infrastructure alone does not generate insight, it establishes the essential backbone enabling mechanistic and clinical layers to function coherently (Le Piane et al., 2024).
2. Digital Nanoplastic Science (DNPS) Paradigm: This mechanistic layer models nanocarrier-mediated perturbations of proteostasis and biofilm architecture (Reyed, 2026c). DNPS bridges theoretical nanoscale behavior with practical translational design, providing a predictive link between molecular intervention and systemic effects (Zhang C., et al., 2025; Zhang et al., 2022; Farah et al., 2026).
3. NexDi AI System: In the clinical intelligence layer, NexDi integrates genomic, phenotypic, and resistance-expression profiles. By interfacing with context-aware clinical decision support, NexDi enables patient-specific therapeutic planning that can adapt dynamically to observed resistome profiles (Al Mazrouei et al., 2026; Scaglione et al., 2026; Elalouf et al., 2025; Reyed, 2026e).

Together, these layers ensure interoperability and nonredundancy, producing a coherent precision nanomedicine ecosystem capable of linking computational prediction to actionable clinical insight.

7.2. The Data Ecosystem: Enabling Precision through Information

Precision nanomedicine relies on a multitiered data infrastructure. Public repositories such as PubChem (~100 M compounds) and ChEMBL (~15 M curated activity pairs) provide the chemical and bioactivity space necessary for high-throughput virtual screening (Zdrazil, 2025). However, raw data are not inherently interoperable; curation, standardization, and harmonization are critical prerequisites for predictive validity.

Phenotypic integration extends across chemical descriptors, electronic health records, and biochemical assays, facilitating the anticipation of drug–drug interactions and off-target effects (Tatonetti et al., 2012). Because phenotype-rich datasets are heterogeneous in quality, predictions must be context-calibrated within the RPCF-AMR ecosystem. This approach ensures that nanocarrier deployment is interpreted across molecular, clinical, and ecological scales, aligning technological optimization with the broader system-level dynamics of AMR.

7.3. Strategic One: Health Implications and Systematic Validation

The convergence of computationally optimized nanobiotechnology with AI-guided modeling may enhance bacterial clearance while reducing health and environmental risks. Nevertheless, interventions cannot be considered ecologically neutral. Effective strategies must address human, animal, and environmental reservoirs concurrently to mitigate the potential "rebound effect" of resistance (Gupta et al., 2026).

Within this context, One Health logic operationalizes precision nanomedicine into a scalable AMR mitigation strategy, which is consistent with UN Sustainable Development Goals 3 (Good Health) and 9 (Industry, Innovation, and Infrastructure) (Malik et al., 2026). The RPCF-AMR serves as the operative framework, integrating molecular precision, computational inference, and cross-sector governance into a coordinated, anticipatory intervention pipeline.

7.4. Strategy of Validation and Benchmarking: The Hybrid Rigor

The predictive reliability of molecular docking and molecular dynamics (MD) simulations underpins this methodological framework, providing estimates of binding affinity and conformational stability, which are essential for assessing nanocarrier–drug compatibility (Sulimov et al., 2019; Ferreira et al., 2015). To overcome the inherent limitations of purely in silico models, the framework adopts a hybrid validation strategy comprising three complementary tiers:

1. In silico benchmarking: Curated bioactivity datasets such as ChEMBL are leveraged to evaluate model performance against established chemical–biological relationships (Zdrazil, 2025).

2. Retrospective Clinical Correlation: Predictions are compared with electronic health record (EHR)-derived outcome patterns to provide real-world validation of phenotypic forecasts (Tatonetti et al., 2012).
3. Prospective Experimental Design: Using computational predictions to guide wet-laboratory validation and microbiome-level testing. While large-scale microbiome experiments remain complex, this approach aligns with best practices by reducing cost, duration, and translational uncertainty (Tavares Luiz et al., 2021).

This hybrid pipeline ensures that computational outputs are contextually grounded, maintaining relevance across clinical and industrial applications.

7.5. Biological Complexity and the Computational Imperative

AMR represents a higher-order adaptive phenotype, extending beyond simple genetic mutations. Pathogens can reorganize metabolic and regulatory networks under therapeutic pressure; for example, *Escherichia coli* may adjust metabolic flux through secA gene mutations in response to antibiotic-induced oxidative stress (Lopatkin et al., 2021). Such reprogramming confers phenotypic persistence, allowing survival despite nongenetic interventions (Liu et al., 2020; Brauner et al., 2016).

Additionally, biofilms provide physical and chemical protection (Guerra et al., 2022), whereas niche-specific behaviors, such as intracellular survival, further increase resilience (Bhattacharyya et al., 2020). Conventional reductionist models are often inadequate for capturing these multiscale adaptations.

The DNPS paradigm addresses this by computationally modeling nanocarrier-mediated perturbation of bacterial proteostasis, whereas NexDi AI dynamically classifies resistance phenotypes across multiomic datasets. This ensures that docking simulations and MD analyses account for microbial heterogeneity and host-specific factors, positioning computational modeling as a proactive guide for translational design rather than a methodological endpoint.

7.6. Integrative Closure: Transitioning to Translational Readiness

By integrating multiscale modeling, data-driven intelligence, and biologically grounded simulation, RPCF-AMR establishes a robust computational foundation for high-precision nanocarrier design. SCF, DNPS, and NexDi AI operate as interoperable layers, translating predictive insights into experimental priorities.

Addressing AMR at the systems level—considering molecular, cellular and ecological complexities—reframes nanocarriers as adaptive therapeutic interfaces capable of destabilizing resistance trajectories across multiple scales. This convergence of computation, mechanistic modeling, and clinical intelligence positions precision nanomedicine for effective translational application within integrated, health-informed AMR strategies.

Figure 5. Integrated MD–AI–Nanobiotechnology Pipeline for Translational Drug Development

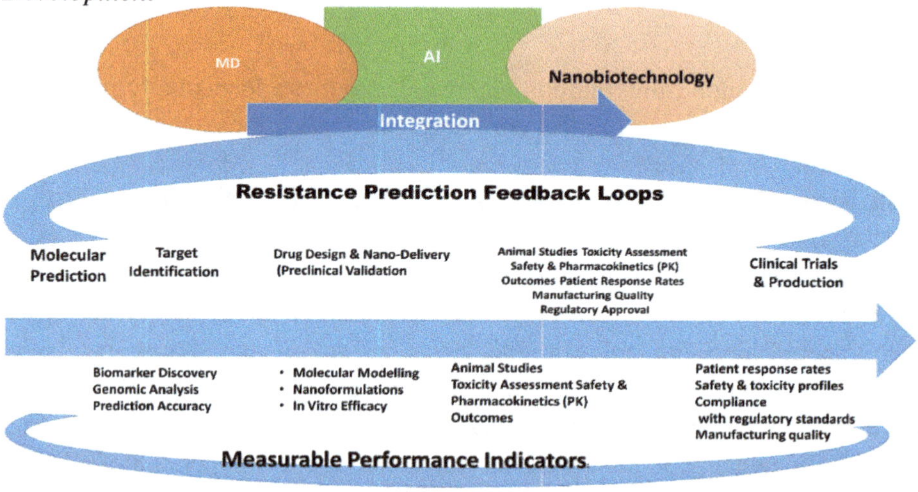

Figure 5 illustrates a unified translational pipeline that operationalizes the RPCF-AMR framework, extending from molecular prediction to industrial validation. The process begins with AI-driven prediction, encompassing biomarker discovery and genomic analysis, and progresses through the drug design and nanodelivery phase, which integrates molecular modeling with in vitro validation (Wang et al., 2024). The pipeline culminates in clinical translation and production, including toxicity evaluation, pharmacokinetics, and regulatory compliance. Crucially, resistance prediction feedback loops connect downstream clinical data back to upstream computational models, enabling iterative refinement, adaptive optimization, and anticipatory control over resistance dynamics.

8. PRECISION NANOCARRIER DELIVERY: BIOPHYSICAL AND ALGORITHM FRAMEWORKS

Building on the architectural foundations of Sections 2 through 7, this framework emphasizes the biophysical and algorithmic characterization of nanocarriers. By combining quantitative modeling with a systems-level design, RPCF-AMR ensures a coherent translational chain. Molecular dynamics (MD), AI-enhanced biointelligence, and nanobiotechnology converge not as isolated disciplines but as a unified response to the global AMR challenge. This approach reframes AMR as a tractable system within a global One Health perspective, enabling computationally guided interventions that support long-term health sovereignty (Reyed, 2019; Shen et al., 2026; Pattapulavar et al., 2026).

8.1. Molecular dynamics (MD) trajectory analysis: the quantitative triad

Molecular dynamics simulations provide high-resolution, time-resolved insights for bridging nanocarrier design and biological performance. To ensure therapeutic stability en route to the target, the sovereign computational framework (SCF) employs a stability–interaction–permeability triad, comprising the following:

- Global structural integrity: Metrics such as the root mean square deviation (RMSD) and radius of gyration (Rg) quantify overall structural stability and compactness, ensuring that nanocarrier geometry is maintained under thermal and mechanical stress (Maiorov & Crippen, 1994; Bagewadi et al., 2023).
- Residue-level plasticity: Root mean square fluctuation (RMSF) is used to assess molecular flexibility, whereas the solvent accessible surface area (SASA) reflects potential membrane interactions and penetration efficiency (Salo-Ahen et al., 2021; Sraphet & Javadi, 2022).
- Chemical Anchoring and Resilience: Hydrogen bond analysis and the radial distribution function (RDF) estimate the chemical anchoring strength and the nanocarrier's resilience to enzymatic or ionic perturbations (Goharshadi, 2025; Rath et al., 2025).

While atomistic MD delivers unparalleled detail, it is limited in capturing long-range immune interactions or large-scale conformational changes over extended timescales. RPCF-AMR addresses this issue by integrating high-resolution MD with multiscale in silico bioprocess modeling, approximating emergent system-level behavior and ensuring that molecular stability translates into therapeutic fidelity across the Polybiome Systems Medicine landscape.

8.2. ML-Driven Precision: From Universal to Patient-Centered Phenotypes

The evolution of clinical practice necessitates moving from universal antibiotic application to precision interventions guided by machine learning (ML). NexDi AI functions as an intelligence coordinator, integrating multiomic inputs—genomic, transcriptomic, and metabolomic—to dynamically adjust physics-based simulations (Sakagianni et al., 2023). This approach identifies strain-specific mutations and resistance-expression profiles that are often obscured in aggregated datasets (Reyed, 2026a).

Translational reliability depends on dataset completeness. NexDi AI mitigates overfitting by repeated cross-validation across diverse clinical populations, producing granular outputs such as predicted minimum inhibitory concentration (MIC) shifts and resistance score changes per strain. These predictions guide targeted in vitro and in vivo validation, transitioning workflows from reactive trial-and-error to anticipatory, data-driven precision.

8.3. Addressing the Structural Vacuum: Computational Imperative

Classical antibiotic development pipelines, particularly for gram-negative pathogens, face a high attrition rate (>90%) due to a challenging toxicity–potency balance (Shinu et al., 2022), with development costs exceeding US$1.5 billion per compound (Boyd et al., 2021). A critical limitation is the inability to model complex interactions across human, animal, and environmental reservoirs.

RPCF-AMR addresses this gap by implementing nature-inspired, multitarget strategies. As Afroze et al. (2026) highlighted, marine-derived metabolites with antiquorum-sensing and antibiofilm properties offer promising avenues for neutralizing resistance without imposing lethal selective pressures. Computational modeling predicts survival fractions of target bacteria and microbiome composition shifts (%), operationalizing the framework as a governance tool for national healthcare strategies (Afroze et al., 2026).

8.4. Rational Design of Biologically Assisted Intelligent Antibiotics

Adaptive resistance requires the full integration of computational design, synthetic biology, and nanobiotechnology. The rational design protocol within RPCF-AMR is guided by three core principles:

1. NexDi Profiling: Targeting specific resistance signatures identified through dynamic phenotypic modeling.
2. Synthetic Biology Integration: Creating novel druggable targets and biological sensors for real-time resistance detection (Reyed et al., 2026a).
3. Environmental Compatibility: Incorporating self-degradable nanocarrier features to reduce selective pressure and ecological accumulation (Chakraborty et al., 2022).

This intelligent antibiotic paradigm remains dependent on empirical validation across controlled microbial communities. Outcomes—including pathogen survival reduction (%) and microbiome taxon abundance shifts—link computational design directly to clinical and ecological effectiveness, supporting the SDG 3 and SDG 9 objectives (Giacobbe et al., 2026).

8.5. Integrative Synthesis: The Computational–Nanobiotechnological Umbrella

The convergence of MD simulations, AI-driven predictive architectures, and biointelligent design principles establishes a precision antibiotic delivery system that surpasses traditional pharmacology (Sakagianni et al., 2023). Within this umbrella architecture:

- SCF provides a structural backbone and data harmonization.
- DNPS models nanostructure–biological interface dynamics, elucidating mechanistic pathways of resistance.
- NexDi AI enables patient-adaptive prioritization, linking submolecular interactions with macroscale environmental and organismal dynamics (Giacobbe et al., 2026).

A key analytical challenge is harmonizing multiscale data ranging from femtosecond MD trajectories to longitudinal clinical outcomes. Data latency and incomplete ecological sampling can generate conflicting signals. Continuous recalibration against longitudinal datasets ensures actionable precision. Output metrics include predicted pathogen reduction (%), selective pressure mitigation scores, and microbiome stability indices, collectively enabling operational evaluation of One Health outcomes (Kang et al., 2024).

8.6. Translational Convergence and Industrial Scale-Up

The progression from laboratory-scale innovation to industrial production necessitates precise computational coordination of molecular planning coupled with predictive benchmarking. Within this context, the RPCF-AMR framework operates as the central convergence platform, integrating biological, computational, and industrial domains. Nevertheless, the lab-to-market trajectory is inherently limited by scale-dependent physicochemical variability. As production volumes increase, factors such as shear stress, heat dissipation, and batch-to-batch inconsistencies can compromise the reproducibility and functional performance of nanocarriers (Agüero-Chapin et al., 2026).

To address these challenges, iterative calibration and mechanistic validation across all scales are essential to ensure that laboratory insights reliably translate into high-volume manufacturing. Key quantifiable metrics—such as the particle size distribution (nm), encapsulation efficiency (%), and release kinetics (μg/mL)—are continuously monitored. The use of digital twins to simulate these parameters enables a reproducible industrial pipeline, where pharmacokinetic and pharmacodynamic (PK/PD) profiles are optimized well before the production of the initial physical batch (Maharjan et al., 2025).

9. MECHANISTIC INTERVENTIONS AND PREDICTIVE NANOMEDICINE: A MULTISCALE SYNTHESIS

9.1. Bridging the Lab-to-Market Gap: The Computational–Translational Engine

The integration of molecular dynamics (MD) simulations and high-resolution molecular docking constitutes the primary mechanistic engine for investigating nanocarrier–pathogen interactions (Pawar & Rohane, 2021). These simulations provide high-fidelity inputs for an integrative data layer that merges process analytical technology (PAT) measurements with heterogeneous experimental datasets.

Navigating this high-dimensional, nonlinear landscape requires predictive architectures that go beyond conventional descriptive statistics (Waykar et al., 2025). Traditional models such as random forests and SVMs capture basic patterns but often fail to accommodate the dynamic variability inherent in multisystem environments (Sarker, 2021; Jia et al., 2024). In contrast, perturbation theory machine learning (PTML) introduces a deeper analytical perspective by treating experimental variations—such as concentration gradients, microbial diversity, and environmental stressors—as perturbation variables, thereby providing a unified mathematical framework linking chemical structure with biological response (Cabrera-Andrade et al., 2021).

Within the polybiome paradigm, PTML enables decryption of complex multi-target interactions in the digital nanoplastic science (DNPS) domain, which models the "plasticity" and "programmability" of nanoscale interventions within biological systems. This approach ensures that predictive modeling remains robust against the stochastic nature of biological evolution, thereby facilitating the transition from reactive therapy to anticipatory precision interventions, in alignment with the Nex-Di blueprint for predictive health care governance. Platforms such as GastroPlus further enable PK/PD extrapolation across species, supporting translational fidelity (Almajidi et al., 2025).

9.2. Programmable Metal Ion Release: The Bioinorganic Interface

The controlled and programmable release of metal ions from inorganic nanoparticles represents a mechanobiological strategy designed to systematically disrupt microbial homeostasis. Divalent and monovalent cations, notably Zn^{2+} and Ag^{+}, bind selectively to thiolated proteomic residues, triggering cascades of enzymatic failure and irreversible structural destabilization (Jung et al., 2008 ; Srivastava & Kowshik, 2013).

This ion-mediated effect is synergistic rather than additive: the corelease of multiple ionic species saturates molecular binding sites, accelerates adsorption kinetics, and bypasses conventional bacterial defense mechanisms. As highlighted by Salmani-Zarchi et al. (2024), this shift from classical antibiotic mechanisms to nanoparticle-based modalities constitutes a critical translational pivot. The surface chemistry and physical properties of nanoparticles dictate the trajectory of antimicrobial efficacy. In addition, metal oxide NPs—particularly ZnO and FeO—act as catalysts for localized ROS generation, facilitating rapid membrane permeabilization and intracellular penetration and thereby reinforcing multimodal antimicrobial action (Jia et al., 2008; Lemire et al., 2013).

Advances in nanoparticle architecture further enhance therapeutic sophistication, allowing NPs to act as active agents that orchestrate context-specific responses rather than as passive carriers (Nechaeva et al., 2026). Within the RPCF-AMR framework, this convergence of ionic and oxidative stress constitutes a system-level destabilization of distributed networked pathogen systems (DNPSs), which target networked pathogen clusters rather than isolated cells, thus maintaining efficacy against evolutionary adaptation (Salmani-Zarchi et al., 2024; Nechaeva et al., 2026).

9.3. Critical Limitation & Dialectic Analysis: The Stochastic Bottleneck

Despite the progress in nanoinorganic interfaces, a major translational challenge persists: ion dissolution kinetics are inherently nonlinear and stochastic in complex biological fluids. This unpredictability can create a "kinetic vacuum," where silver or zinc ions remain subtherapeutic, potentially promoting adaptive resistance, or conversely accumulate locally, causing cytotoxicity (Pati et al., 2016).

As emphasized by Nechaeva et al. (2026), translational reliability thus requires coupling mechanistic dissolution assays with dynamic physiological models. Only by accounting for these nonlinear kinetics can context-specific efficacy be achieved within the polybiome landscape. This limitation has motivated the development of the sovereign computational framework (SCF), which simulates dissolution trajectories in real time, ensuring that therapeutic windows remain optimized and robust.

9.4. Precision Nanobiotechnology: Mechanistic Interference of Microbial Systems

Precision nanobiotechnology leverages ion-mediated effects for multilevel microbial targeting, dismantling the structural defenses of resistant lineages. Nanocarriers function as active interfering agents, addressing membrane remodeling, efflux pump activity, and the dense architectures of biofilm matrices (Torchilin, 2006).

Careful tuning of physicochemical properties—zeta potential optimization and pH responsiveness—facilitates selective accumulation and triggered destabilization within the acidic microenvironments of mature biofilms. Mechanistic efficacy was validated through high-resolution crystal violet staining and confocal laser scanning microscopy (CLSM). However, the structural heterogeneity and adaptive memory of microbial polybiomes necessitate integration with predictive population dynamics to ensure that interventions remain as adaptive as the pathogens they target.

9.5. Transcriptomic Rewiring and Genomic Destabilization: The Nanobiological Frontier

In addition to causing immediate structural damage, nanomaterials impose genotoxic stress and systemic transcriptomic rewiring, reshaping microbial metabolism and proteostasis. These effects depend on the physicochemical state of the nanoparticle at the nanobio interface. For example, polymer-encapsulated AgNPs selectively suppress the TCA cycle and amino acid metabolism genes in *E. coli*, whereas unencapsulated variants induce broader genomic destabilization (Monteiro et al., 2012; Ashmore et al., 2018).

Additional disruptions include interference with riboflavin metabolism and ribosomal-tRNA interactions, resulting in terminal ATP depletion, which is consistent with DNPS projections of system-wide energetic collapse (Leung et al., 2014; Cui et al., 2012). Translation into specialized clinical contexts, notably endodontics, represents a paradigm shift, with high surface-to-volume ratios enabling deep penetration into dentinal tubules, overcoming the spatial limitations of traditional irrigants (Afkhami et al., 2024).

Nonetheless, transcriptomic signatures do not always align with phenotypic survival because of compensatory metabolic pathways and the persister cell phenomenon. Accordingly, efficacy must be evaluated through the lens of polybiome resilience, ensuring that interventions exceed the microbial capacity for homeostatic recovery (Afkhami et al., 2024; Reyed, 2026a).

9.6. Antimicrobial Nanomaterial Classification: Mechanistic Taxonomy

In addition to simple compositional catalogs, nanomaterials are classified according to their dominant mechanistic signatures:

- Inorganic (AgNPs, AuNPs, FeONPs): ROS-mediated cytotoxicity and electrostatic interference (Wadan et al., 2026).
- Organic (liposomes, PLGA/chitosan): targeted intracellular delivery and sustained release.
- Hybrid & nanozymes: Enzyme-mimetic activity or hyperthermia-enabled versatility (Okkeh et al., 2021; Yu et al., 2020; Cao et al., 2019).

Traditional taxonomies overemphasize static composition, neglecting dynamic functional behavior in biological fluids. Within RPCF-AMR, we propose functional descriptors aligned with metabolic vulnerabilities—for example, bioenergetic disruptors versus membrane permeabilizers—as a rational basis for clinical selection.

9.7. Predictive nanomedicine: Modeling translational bottlenecks

Computational modeling underpins rational design, yet translation faces temporal and spatial bottlenecks. MD simulations are often limited to nanosecond timescales, constraining the observation of slow conformational dynamics essential for controlled drug release (Voter, 1997; Barnard, 2015). Nonlinear solvation dynamics further complicate drug-loading predictions (Spasic, 2018).

Regularization techniques such as LASSO and ridge regression reduce data noise (Xiaolin et al., 2023) but cannot substitute for closed-loop validation pipelines. RPCF-AMR addresses this via iterative benchmarking against PK/PD and functional assays, linking multiscale stochastic outputs to produce precision nanocarriers grounded in predictive, verifiable science that is resilient to the evolutionary dynamics of the global resistome (Jamkhande et al., 2017).

9.8 Clinical implications and translational relevance

The translation of precision nanomedicine into clinical practice is increasingly guided by machine learning (ML)-assisted optimization, particularly for pharmacokinetic parameters such as LogP and the topological polar surface area (TPSA). These computationally informed adjustments may increase the targeting efficiency of stimuli-responsive carriers, promoting environment-specific payload release (Arnold et al., 2025; Bilal et al., 2025; Majumder & Minko, 2021). Integrating in vitro–in vivo correlation protocols with IoT-enabled monitoring can provide a feedback loop for real-time assessment of therapeutic performance (Lu et al., 2011; Jena et al., 2024).

Nonetheless, clinical implementation is limited by substantial interpatient variability, which complicates the standardization of release kinetics. This inherent heterogeneity suggests that uniform dosing strategies may be insufficient. Instead, adaptive, patient-specific interventions—such as those conceptualized in the NexDi profile—may be required to achieve consistent and reproducible outcomes across diverse physiological contexts.

9.9 Experimental Design (DoE): Constraints on Predictive Precision

The predictive reliability of any nanomedicine platform fundamentally depends on the robustness of its experimental design. Conventional "one-factor-at-a-time" approaches are generally inadequate, as they fail to capture nonlinear and synergistic interactions between nanoparticles, microbial targets, and their bioenvironmental contexts (Tavares Luiz et al., 2021). This limitation is reflected in the relatively low reported success rates (10–15%) of ML-guided antibiotic development (Konagurthu et al., 2026; Xie et al., 2025).

Within the RPCF-AMR framework, these observations underscore the need for more complex, multiparameter experimental designs. Although resource intensive, such designs are likely essential for refining algorithmic predictions and improving the translational reliability of preclinical findings, which in turn is critical for successful industrial scale-up.

9.10 Critical Review: Multi-Scale Integration and One Health Alignment

A multiscale integration approach may be necessary to align molecular design with broader One Health principles. By evaluating the impact of nanomaterials across human, animal, and environmental reservoirs, this strategy has the potential to reduce selective pressures that contribute to antimicrobial resistance (Malik et al., 2026).

A major impediment is the fragmentation of data standards, which hinders interoperability between clinical and ecological datasets. The adoption of unified data frameworks—such as those proposed in the sovereign computational framework (SCF)—appears to be a key step toward ensuring that precision nanomedicine remains both clinically effective and environmentally friendly.

9.11 Integrative Conceptualization: RPCF-AMR Sovereign Umbrella

The RPCF-AMR Umbrella represents an integrative model that seeks to unify mechanistic precision, ecological stewardship, and translational continuity. By explicitly incorporating biological and ecological heterogeneity, it maintains a coherent logic from theoretical conception to practical implementation (Esawy & Reyed, 2026; Malik et al., 2026).

While empirical validation in diverse global health contexts remains limited, the framework provides a conceptual basis for controlled, predictable, and ecologically integrated nanomedicine. This has encouraged a shift from reactive treatment approaches toward anticipatory, biointelligent strategies aimed at preserving human and environmental health.

Table 3. Comparative Critical Analysis of Mechanisms, Limitations, and Targeted Applications in Predictive Nanomedicine

Subsection	Argument	Critical Limitation	Strategic Implication
9.1.1 Ion Release	Controlled Zn^{2+} and Ag^{+} release disrupts thiolated proteins and promotes ROS-mediated damage (Reyed & Prabhakar, 2026d; Jung et al., 2008).	Nonlinear dissolution in biological fluids may reduce efficacy or cause localized cytotoxicity (Lemire et al., 2013).	Dynamic physiological models coupled with mechanistic assays are needed to ensure context-specific outcomes.
9.2 Nanobiotech	Engineered nanocarriers bypass efflux pumps and penetrate biofilms through size/charge/pH triggers (Torchilin, 2006).	Microbial polybiomes may adapt structurally or functionally, limiting uniform efficacy (Jung et al., 2008).	Integration with predictive ecological modeling can enhance robustness across diverse microbial contexts.
9.3 Genotoxicity	Encapsulated AgNPs and MgO/AuNPs can destabilize DNA and suppress metabolic/energy genes (Monteiro et al., 2012).	Transcriptomic signatures alone may not predict survival due to compensatory metabolic pathways (Leung et al., 2014).	Multiomic validation, including proteomics and metabolomics, is recommended to confirm irreversible antimicrobial effects.
9.4 Classification	Nanoparticles are categorized as inorganic, organic, or hybrid based on mechanistic "signatures" (Thakur et al., 2025).	Current taxonomies emphasize static composition rather than dynamic functional behavior in biological fluids (Yu, 2020).	A shift toward functional descriptors aligned with metabolic vulnerabilities supports rational selection.
9.5 Predictive MD	Computational modeling (MD, LASSO/Ridge) guides NP design and drug-loading predictions (Voter, 1997; Xiaolin, 2023).	Simulations are limited by nanosecond timescales and solvation dynamics complexity (Spasic, 2018).	Iterative benchmarking against PK/PD and functional assays bridges in silico predictions with in vivo behavior.
9.6 Clinical Relevance	ML-guided carriers optimize PK/PD and enable IoT-enabled real-time monitoring (Arnold et al., 2025; Bilal, 2025).	High interpatient variability complicates standardization of release kinetics (Majumder & Minko, 2021).	Personalized, adaptive strategies are essential to achieve reproducible clinical outcomes (Jena et al., 2024).
9.7 DoE Strategy	Multiparameter DoE captures synergistic interactions among NPs and environments (Tavares Luiz et al., 2021).	One-factor-at-a-time approaches miss critical nonlinear interactions and are resource-intensive (Xie, 2025).	Systematic, ML-informed DoE is crucial for improving preclinical reliability.
9.8 One Health	Unified frameworks link molecular design to ecological monitoring, mitigating global AMR (Malik, 2026).	Fragmented data standards hinder interoperability across clinical and environmental datasets (Reyed, 2023).	Standardized integrated data frameworks are necessary for ecologically responsible nanomedicine.
9.9 RPCF-AMR	Integrates mechanistic precision with stewardship to create a predictable nanomedicine ecosystem (Esawy & Reyed, 2026).	Empirical validation in heterogeneous global health contexts remains limited.	Supports proactive metabolic governance, shifting from reactive to anticipatory interventions.

Analytical Note: This table synthesizes mechanistic strengths and inherent limitations in predictive nanomedicine, highlighting a pathway toward rational, structurally convergent strategies that integrate computational modeling, clinical precision, and ecological stewardship.

10 SYSTEM-LEVEL ML INTEGRATION: POLYBIOME–RESISTOME MAPPING

Within the RPCF-AMR framework, machine learning has been extended to multidomain polybiome analysis, enabling the prediction of resistome dynamics across human, animal, and environmental reservoirs. By integrating high-dimensional datasets—including microbial metagenomics, metabolomics, and environmental chemical exposures—ML approaches may quantify cross-domain resistance transmission and identify potential high-risk ecological niches (Sodagari, et al., 2025).

Nonetheless, analytical challenges persist. Data heterogeneity and incomplete temporal sampling introduce uncertainty, complicating longitudinal resistome projections. This aligns with broader observations in ML-AMR applications, where variations in genomic and phenotypic data quality often constrain clinical translation (Ali et al., 2023). Incorporating formal uncertainty quantification and probabilistic inference may help transform raw surveillance data into risk-weighted predictions while retaining operational utility.

RPCF-AMR also considers polybiome connectivity by mapping interactions among the gut, soil, wastewater, and hospital microbial communities. Correlating resistome signatures with environmental vectors may allow early anticipation of emergent resistance patterns, which is consistent with wastewater-based epidemiology approaches under the One Health paradigm. Conventional network models often assume static interactions, which may overlook rapid horizontal gene transfer and stochastic shifts in mobile genetic elements. Temporal graph modeling and adaptive learning are therefore incorporated to account for dynamic polybiome evolution and enable proactive interventions rather than purely retrospective analysis.

Finally, ML-informed polybiome modeling can guide the rational design of multitarget antimicrobial interventions. Simulations of nanoparticle-mediated ion release and metabolic interference reveal potential synergistic mechanisms that may be difficult to isolate experimentally. While computational demands remain substantial, scalable, modular cloud-based ML pipelines offer a practical route to iteratively refine predictions and translate in silico insights into field-relevant strategies.

10.1 Predictive Nanocarrier Optimization: ML-Driven Feature Engineering

Within the RPCF-AMR framework, machine learning employs nanocarrier design as a multiobjective optimization challenge. Physicochemical, structural, and functional attributes—such as particle size, zeta potential, LogP, and TPSA—are represented as feature vectors to predict uptake and release kinetics, aiming to maximize therapeutic efficacy while minimizing host toxicity (Kehrein et al., 2024; Agboola et al., 2025).

High-dimensional feature spaces carry an inherent risk of overfitting, particularly when training datasets are sparse or biased. To mitigate this, RPCF-AMR couples ML-based feature engineering with systematic experimental design (DoE), imposing mechanistic constraints that improve model fidelity across diverse microbial and host contexts.

The framework also simulates the dynamic interplay between nanocarriers and adaptive microbial defenses. Variables such as efflux pump activity and biofilm matrix density are treated as context-dependent boundaries influencing penetration and release kinetics (Gao et al., 2021). Recognizing that deterministic algorithms may fail to capture stochastic phenotypic fluctuations typical of chronic infections, RPCF-AMR incorporates Bayesian probabilistic simulations, producing context-sensitive strategies that may preemptively adjust carrier properties in response to microbial evasion.

Structural predictions derived from tools such as AlphaFold reveal potential binding sites and the likelihood of membrane traversal. While these predictions are often highly accurate for static molecular conformations, they may not fully account for local microenvironmental effects such as pH-induced conformational shifts. Integrating these structural insights with high-throughput experimental validation creates a closed-loop optimization cycle, progressively refining nanocarrier designs toward maximal clinical relevance.

10.2 Hospital–environment interface: Predictive surveillance and adaptive stewardship

At the interface between hospital and environmental ecosystems, RPCF-AMR leverages ML to enable real-time surveillance by integrating genomic, pharmacological, and wastewater resistome datasets. Hospital effluents and wastewater treatment plants function as interconnected reservoirs where shifts in resistome composition may reflect broader AMR dynamics (Wu et al. 2025). Historically, fragmented data standards and reporting delays have limited the proactive use of such information. Within RPCF-AMR, ML facilitates early warning indicators, allowing interventions that anticipate resistance emergence before it becomes entrenched in clinical populations (Gheorghita et al., 2025).

The framework quantifies complex relationships among antimicrobial usage, resistance development, and environmental dissemination. This approach aligns with a One Health perspective, capturing extrinsic influences such as wastewater-mediated spillover. Although predictive capacity remains constrained by partial observability and stochastic factors—including climatic variability—probabilistic inference and adaptive feedback loops help maintain optimized antimicrobial deployment, balancing patient outcomes with ecological stewardship.

By integrating individual pharmacokinetic data with aggregate resistome metrics, ML continuously monitors deviations from expected therapeutic performance, enabling dynamic adjustments in dosing or nanocarrier application. While patient heterogeneity may limit generalization, this closed-loop system transforms reactive stewardship into predictive governance, supporting interventions that are context sensitive, resilient, and aligned with global AMR control priorities.

11 STRATEGIC CLOSURE: EPISTEMIC IRREVERSIBILITY AND THE ONE HEALTH SOVEREIGN IMPERATIVE

The Reyed Precision Convergence Framework (RPCF-AMR) represents epistemic recalibration, which refers to AMR from a series of isolated clinical problems to a systems-level phenomenon (Mitchell et al., 2021; Sambaza & Naicker, 2023). By 2040, traditional reactive pharmacology is expected to be supplanted by predictive architectures that treat AMR as an emergent state within the One Health continuum, establishing a benchmark for anticipatory biogovernance anchored in the interconnectedness of human, animal, and environmental health (Kasse et al., 2025; Abdelbaky et al., 2025; Reyed, 2026).

11.1 One Health Centrality: A Multidomain Defense Shield

At the core of RPCF-AMR lies the One Health Sovereign Imperative. AMR is conceived not as a discrete clinical event but as a bioecological contagion, which is circulated through anthropogenic, veterinary, and environmental reservoirs. Within this model, a clinical cure alone is insufficient if the environmental resistome continues to act as a reservoir for horizontal gene transfer (HGT) (Mitchell et al., 2021; Sambaza & Naicker, 2023).

Using the NexDi AI engine, RPCF-AMR maps the kinetics of resistome spillover across hospital effluents, intensive livestock operations, and terrestrial aquatic ecosystems. This multidomain mapping enables interventions such as fermentobiomics-based modulation not only at the point of care but also at the point of origin, limiting the dissemination of resistance determinants before they re-enter human populations (Polianciuc et al., 2020; Musicha et al., 2024).

11.2 Seven Pillars of the 2040 Predictive Landscape

RPCF-AMR operationalizes predictive intelligence through seven interconnected dimensions:

1. Biodiversity Translation: Beyond species cataloging, microbial functional diversity is translated into actionable therapeutic targets across one health domain (Jacob et al., 2026; Topol, 2019; Elalouf et al., 2025; Reyed, 2026f).
2. Heritage-Informed Intelligence: Leveraging the evolutionary memory of polybiomes to anticipate shifts in resistance under novel selective pressures (Rolff et al., 2024).
3. Multiomics Integration: Real-time alignment of genomic, transcriptomic, and metabolomic data streams via NexDi AI to capture the full metabolic landscape of microbial communities (Eicher et l., 2020).
4. Digital Twin Biocultures: High-resolution AI-driven modeling of fermentative ecosystems to simulate "digital twin" environments, allowing preclinical evaluation of nanocarrier efficacy across human and environmental niches (Wu & Koelzer 2024).
5. Intervention-First Causality: Prioritizing proactive interventions over descriptive correlations ensures that computational predictions translate into measurable, actionable outcomes (Salehi et al., 2025).
6. Precision Nutrition 2040: Incorporating dietary–microbiome interactions to modulate the resistome, advancing nutrition as a medicine paradigm for humans and livestock (Vivarelli et al.,2026).
7. Ethical & Regulatory Governance: Establishing a ssovereign shield through adaptive policies to ensure that AI-driven antimicrobial deployment remains safe, transparent, and ecologically responsible (Mennella et al., 2024).

11.3 Irreversible Advancement: Precision Nanomedicine 2040

RPCF-AMR defines a reference architecture for patient-specific nanomedicine, integrating molecular descriptors (LogP, TPSA) to circumvent classical resistance pathways. Context-responsive nanocarriers facilitate precise intracellular delivery, overcoming multidrug resistance through adaptive, environmentally sensitive design principles. Anchoring these strategies in the DNPS paradigm ensures harmonization with host biology while minimizing environmental impact (Baig et al., 2024; Mills et al., 2019; Silva & Queirós, 2025).

11.4 Forward Trajectory: Global Stewardship and Policy Alignment

RPCF-AMR envisions governance through four intertwined transitions:

1. Systemic One Health Integration: Dissolving historical silos between clinical and environmental monitoring to treat AMR as a trans-domain ecological threat (Altevogt et al., 2025).
2. Anticipatory Predictive Intelligence: Utilizing WBE and AI to detect resistance emergence at its source provides a temporal advantage over conventional reactive measures (Villanueva-Miranda et al., 2025).
3. Surgical precision modulation: Targeted nanoparticle-mediated interventions induce localized metabolic collapse in pathogenic lineages while preserving commensal polybiomes (Pochet et al., 2026)
4. Adaptive Sovereign Stewardship: Continuous feedback from NexDi AI dynamically calibrates clinical protocols and environmental policies, transforming static oversight into proactive governance (Lee et al., 2026).

11.5 Policy Statement and Strategic Closure

AMR is fundamentally an ecological and evolutionary phenomenon. RPCF-AMR operationalizes Fermentobiomics and NexDi AI to transition global health from reactive treatment to anticipatory biogovernance (Wimalawansa, 2025). By centering on One Health, the framework positions humanity not as a passive recipient of microbial evolution but as an active governor, achieving sustained translational, clinical, and ecological resilience through 2040 and beyond (Perini et al., 2024; Ginnan et al., 2025).

Table 4. Summary of Strategic Pillars and One Health Priorities

Pillar	One Health Strategic Contribution	Future Priority (2030–2040)
Fermentobiomics	Decodes metabolic crosstalk in gut-soil–water digital twins	Real-time simulation of cross-domain microbial axes
Biodiversity	Translates environmental diversity into drug targets	Mapping the dark matter of the global resistome
One Health AI	Predicts resistance spillover across reservoirs	Integration of WBE-ML (Wastewater-Based Epidemiology)
Governance	Ensures ethical and sovereign ecological control	International harmonization of AI-AMR regulations

Concluding Remarks:

RPCF-AMR signals the end of the reactive era of antimicrobial treatment. By anchoring interventions in Fermentobiomics, NexDi AI predictive intelligence, and the One Health paradigm, it offers a future in which antimicrobial efficacy is maintained through informed governance rather than solely chemical intervention—creating a resilient, postantibiotic sovereign shield (Kasse et al., 2025).

REFERENCES

Abbas, I. M., Mahdi, S. A., & Jabir, M. S. (2025). Insights into the antibacterial and anticancer activity of novel silver nanoparticles mediated bacterial colicin: Sustainable nanotechnology, in silico and dynamic study. *Plasmonics*, *20*, 10985–10999. DOI: 10.1007/s11468-025-03153-7

Abdelsattar, A. S., Dawoud, A., & Helal, M. A. (2021). Interaction of nanoparticles with biological macromolecules: A review of molecular docking studies. *Nanotoxicology*, *15*(1), 66–95. DOI: 10.1080/17435390.2020.1842537

Afkhami, F., Chen, Y., Walsh, L. J., Peters, O. A., & Xu, C. (2024). Application of nanomaterials in endodontics. *BME Frontiers, 5*, 0043. https://doi.org/DOI: 10.34133/bmef.0043

Afroze, C. A., Ahmed, M. N., Jahan, R., & Rahmatullah, M. (2026). Marine-derived prospective antibiotic armamentarium: From anti-quorum sensing to anti-biofilm and anti-biofouling biometabolites. *Heliyon*, *12*(2), e44524. DOI: 10.1016/j.heliyon.2026.e44524

Agboola, O. E., Agboola, O., Ayinla, Z. A., Agboola, S. S., Olaiya, O. E., Oyinloye, O. M., Fajana, O. M., Idowu, O. T., Omotuyi, O. I., Ilesanmi, O. S., & Oyinloye, B. E. (2025). Integrated virtual screening for anti-caries compounds from neem: Dual-target inhibition of biofilm formation and bacterial DNA replication. *Biomedicines*, *13*(9), 2202. DOI: 10.3390/biomedicines13092202

Agu, P. C., Afiukwa, C. A., Orji, O. U., Ezeh, E. M., Ofoke, I. H., Ogbu, C. O., Ugwuja, E. I., & Aja, P. M. (2023). Molecular docking as a tool for the discovery of molecular targets of nutraceuticals in diseases management. *Scientific Reports*, *13*(1), 13398. DOI: 10.1038/s41598-023-40160-2

Agüero-Chapin, G., Antunes, A., & Marrero-Ponce, Y. (2026). A 2026 update on computational approaches to the discovery and design of antimicrobial peptides. *Antibiotics (Basel, Switzerland)*, *15*(2), 203. DOI: 10.3390/antibiotics15020203

Al-Asmi, A., Al-Jabri, F. S., Al-Amrani, S. A., Al Farsi, Y., Al Sabahi, F., Al-Futaisi, A., & Al-Adawi, S. (2024). Medical tourism and neurological diseases: Omani patients' experience seeking treatment abroad. *Oman Medical Journal*, *39*(3), e628. DOI: 10.5001/omj.2024.69

Al-Khalaifah, H., Rahman, M. H., Al-Surrayai, T., Al-Dhumair, A., & Al-Hasan, M. (2025). A One-Health perspective of antimicrobial resistance (AMR): Human, animals and environmental health. *Life*, *15*(10), 1598. DOI: 10.3390/life15101598

Al Mazrouei, N., Ahmed Elnour, A., Badi, S., Abdalla, N. M., Mohamed, M. G., Ali, A. A., Sheikh, A., & Ashmawi, S. (2026). The impact of artificial intelligence on the prescribing, selection, resistance, and stewardship of antimicrobials: A scoping review. *BMC Infectious Diseases*, *26*, 222. DOI: 10.1186/s12879-025-12320-4

Alharthi, S. (2025). AI-powered in silico twins: Redefining precision medicine through simulation, personalization, and predictive healthcare. *Saudi Pharmaceutical Journal : SPJ : The Official Publication of the Saudi Pharmaceutical Society*, *34*(1), 1. DOI: 10.1007/s44446-025-00055-x

Ali, T., Ahmed, S., & Aslam, M. (2023). Artificial intelligence for antimicrobial resistance prediction: Challenges and opportunities toward practical implementation. *Antibiotics (Basel, Switzerland)*, *12*(3), 523. DOI: 10.3390/antibiotics12030523

Alif, M. N., & Zhang, W. (2025). SynOmics: Integrating multiomics data through feature interaction networks. *Briefings in Bioinformatics*, *26*(6), bbaf595. Advance online publication. DOI: 10.1093/bib/bbaf595

Almajidi, Y. Q., Faisal, R. R., Yaseen, A., & Al-hakeem, M. A. L. (2025). GastroPlus® as simulating/conducting software for modeling physiologically based pharmacokinetics & physiologically based biopharmaceutics. *Maaen Journal for Medical Sciences*, *4*(3), 1078. DOI: 10.55810/2789-9136.1078

Alprol, A. E., Hamad, T. M., Sharaf, H. E. R., El-Sayed, H. S., Barakat, K. M., Abouelwafa, A., & Khairy, H. M. (2026). Toxicological impacts of biogenic zinc oxide nanoparticles on blue parrotfish using multibiomarker assessment. *Scientific Reports*, *16*(1), 6546. DOI: 10.1038/s41598-026-36870-y

AlQurashi, D. M., AlQurashi, T. F., Alam, R. I., Shaikh, S., & Tarkistani, M. A. M. (2025). Advanced nanoparticles in combating antibiotic resistance: Current innovations and future directions. *Journal of Nanotheranostics*, *6*(2), 9. DOI: 10.3390/jnt6020009

Altammar, K. A. (2023). A review on nanoparticles: Characteristics, synthesis, applications, and challenges. *Frontiers in Microbiology*, *14*, 1155622. DOI: 10.3389/fmicb.2023.1155622

Altevogt, B. M., Taylor, P., Akwar, H. T., Graham, D. W., Ogilvie, L. A., Duffy, E., & Essack, S. Y. (2025). A One Health framework for global and local stewardship across the antimicrobial lifecycle. *Communications Medicine*, *5*(1), 414. DOI: 10.1038/s43856-025-01090-4

Aminov, R. (2021). Acquisition and Spread of Antimicrobial Resistance: A *tet*(X) Case Study. *International Journal of Molecular Sciences*, *22*(8), 3905. DOI: 10.3390/ijms22083905

An, A. Y., Choi, K. G., Baghela, A. S., & Hancock, R. E. W. (2021). An Overview of Biological and Computational Methods for Designing Mechanism-Informed Anti-biofilm Agents. *Frontiers in Microbiology*, *12*, 640787. DOI: 10.3389/fmicb.2021.640787

Arnold, A., McLellan, S., & Stokes, J. M. (2025). How AI can help us beat AMR. *npj antimicrobials and resistance, 3*(1), 18. https://doi.org/DOI: 10.1038/s44259-025-00085-4

Ashmore, D., Chaudhari, A., Barlow, B., Barlow, B., Harper, T., Vig, K., Miller, M., Singh, S., Nelson, E., & Pillai, S. (2018). Evaluation of *E. coli* inhibition by plain and polymer-coated silver nanoparticles. *Revista do Instituto de Medicina Tropical de São Paulo*, *60*, e18. DOI: 10.1590/s1678-9946201860018

Aulifa, D. L., Al Shofwan, A. A., Megantara, S., Fakih, T. M., & Budiman, A. (2024). Elucidation of Molecular Interactions Between Drug-Polymer in Amorphous Solid Dispersion by a Computational Approach Using Molecular Dynamics Simulations. *Advances and Applications in Bioinformatics and Chemistry : AABC*, *17*, 1–19. DOI: 10.2147/AABC.S441628

Bagewadi, Z. K., Yunus Khan, T. M., Gangadharappa, B., Kamalapurkar, A., Mohamed Shamsudeen, S., & Yaraguppi, D. A. (2023). Molecular dynamics and simulation analysis against superoxide dismutase (SOD) target of *Micrococcus luteus* with secondary metabolites from *Bacillus licheniformis* recognized by genome mining approach. *Saudi Journal of Biological Sciences*, *30*(9), 103753. DOI: 10.1016/j.sjbs.2023.103753

Baig, M. S., Ahmad, A., Pathan, R. R., & Mishra, R. K. (2024). Precision Nanomedicine with Bio-Inspired Nanosystems: Recent Trends and Challenges in Mesenchymal Stem Cells Membrane-Coated Bioengineered Nanocarriers in Targeted Nanotherapeutics. *Journal of Xenobiotics*, *14*(3), 827–872. DOI: 10.3390/jox14030047

Barabási, A. L., Gulbahce, N., & Loscalzo, J. (2011). Network medicine: A network-based approach to human disease. *Nature Reviews. Genetics*, *12*(1), 56–68. DOI: 10.1038/nrg2918

Barnard, A. S. (2015). Challenges in modeling nanoparticles for drug delivery. *Journal of Physics. Condensed Matter : An Institute of Physics Journal*, *28*(2), 023002. DOI: 10.1088/0953-8984/28/2/023002

Bass, D., Stentiford, G. D., Wang, H. C., Koskella, B., & Tyler, C. R. (2019). The Pathobiome in Animal and Plant Diseases. *Trends in Ecology & Evolution, 34*(11), 996–1008. DOI: 10.1016/j.tree.2019.07.012

Berros, N., Filaly, Y., El Mendili, F., & El Bouzekri El Idrissi, Y. (2025). Scalable predictive modeling for hospitalization prioritization: A hybrid batch–streaming approach. *Big Data and Cognitive Computing, 9*(11), 271. DOI: 10.3390/bdcc9110271

Bhattacharyya, S., Walker, D. M., & Harshey, R. M. (2020). Dead cells release a 'necrosignal' that activates antibiotic survival pathways in bacterial swarms. *Nature Communications, 11*(1), 4157. DOI: 10.1038/s41467-020-17709-0

Bilal, H., Khan, M. N., Khan, S., Shafiq, M., Fang, W., Khan, R. U., Rahman, M. U., Li, X., Lv, Q. L., & Xu, B. (2025). The role of artificial intelligence and machine learning in predicting and combating antimicrobial resistance. *Computational and Structural Biotechnology Journal, 27*, 423–439. DOI: 10.1016/j.csbj.2025.01.006

Boyd, N. K., Teng, C., & Frei, C. R. (2021). Brief Overview of Approaches and Challenges in New Antibiotic Development: A Focus On Drug Repurposing. *Frontiers in Cellular and Infection Microbiology, 11*, 684515. DOI: 10.3389/fcimb.2021.684515

Brauner, A., Fridman, O., Gefen, O., & Balaban, N. Q. (2016). Distinguishing between resistance, tolerance and persistence to antibiotic treatment. *Nature Reviews. Microbiology, 14*(5), 320–330. DOI: 10.1038/nrmicro.2016.34

Brooks, C., & Thompson, C. (2017). Predictive modeling in teaching and learning. In *Handbook of Learning Analytics* (pp. 61–68). Society for Learning Analytics Research (SoLAR). https://doi.org/DOI: 10.18608/hla17.005

Cabrera-Andrade, A., López-Cortés, A., Munteanu, C. R., Pazos, A., Pérez-Castillo, Y., Tejera, E., Arrasate, S., & González-Díaz, H. (2020). Perturbation-Theory Machine Learning (PTML) multilabel model of the ChEMBL dataset of preclinical assays for antisarcoma compounds. *ACS Omega, 5*(42), 27211–27220. DOI: 10.1021/acsomega.0c03356

Cao, F., Zhang, L., Wang, H., You, Y., Wang, Y., Gao, N., Ren, J., & Qu, X. (2019). Defect-Rich Adhesive Nanozymes as Efficient Antibiotics for Enhanced Bacterial Inhibition. *Angewandte Chemie International Edition, 58*(45), 16236–16242. DOI: 10.1002/anie.201908289

Cao, Y., Lancaster, E., Lee, J., & Wu, J. (2026). Integrating explainable AI and One Health: A new frontier in combating infectious diseases. *EBioMedicine, 126*, 106207. DOI: 10.1016/j.ebiom.2026.106207

Castañeda-Barba, S., Top, E. M., & Stalder, T. (2024). Plasmids, a molecular cornerstone of antimicrobial resistance in the One Health era. *Nature Reviews. Microbiology*, *22*(1), 18–32. DOI: 10.1038/s41579-023-00926-x

Chakraborty, N., Jha, D., Roy, I., Kumar, P., Gaurav, S. S., Marimuthu, K., Ng, O. T., Lakshminarayanan, R., Verma, N. K., & Gautam, H. K. (2022). Nanobiotics against antimicrobial resistance: Harnessing the power of nanoscale materials and technologies. *Journal of Nanobiotechnology*, *20*(1), 375. DOI: 10.1186/s12951-022-01573-9

Changruenngam, S., Modchang, C., & Bicout, D. J. (2022). Modeling of the transmission dynamics of carbapenem-resistant *Klebsiella pneumoniae* in hospitals and design of control strategies. *Scientific Reports*, *12*(1), 3805. DOI: 10.1038/s41598-022-07728-w

Charani, E., McKee, M., Balasegaram, M., Mendelson, M., Singh, S., & Holmes, A. H. (2022). Global burden of antimicrobial resistance: Essential pieces of a global puzzle. *Lancet*, *399*(10344), 2346–2347. DOI: 10.1016/S0140-6736(22)00935-7

Chatterjee, A. K., Chakraborty, R., & Basu, T. (2014). Mechanism of antibacterial activity of copper nanoparticles. *Nanotechnology*, *25*(13), 135101. DOI: 10.1088/0957-4484/25/13/135101

Chen, D., Li, D., Su, L., Wang, D. Y., Ren, Y., van der Mei, H. C., Shi, L., Chen, T., & Busscher, H. J. (2026). Stimuli-responsive, antimicrobial-loaded nanocarriers for oral biofilm control and microbiome restoration. *International Journal of Oral Science*, *18*(1), 17. DOI: 10.1038/s41368-025-00422-3

Cheung, H. Y. F., Tantiwong, C., Kale, D., Gibbins, J. M., Watson, S. P., Heemskerk, J. W. M., Sickmann, A., Ahrends, R., & Dunster, J. L. (2025). A computational framework for the investigation of phosphoinositide regulation. *PLoS Computational Biology*, *21*(9), e1013477. DOI: 10.1371/journal.pcbi.1013477

Cui, Y., Zhao, Y., Tian, Y., Zhang, W., Lü, X., & Jiang, X. (2012). The molecular mechanism of action of bactericidal gold nanoparticles on *Escherichia coli*. *Biomaterials*, *33*(7), 2327–2333. DOI: 10.1016/j.biomaterials.2011.11.057

D'Hondt, S., Oramas, J., & De Winter, H. (2025). A beginner's approach to deep learning applied to VS and MD techniques. *Journal of Cheminformatics*, *17*, 47. DOI: 10.1186/s13321-025-00985-7

Dai, J., Ouyang, Y., Gupte, R., Liu, X. J. A., Li, Y., Yang, F., Chen, S., Provin, T., Van Schaik, E., Samuel, J. E., Jayaraman, A., Zhou, A., de Figueiredo, P., Zhou, J., & Han, A. (2025). Microfluidic droplets with amended culture media cultivate a greater diversity of soil microorganisms. *Applied and Environmental Microbiology*, *91*(3), e0179424. DOI: 10.1128/aem.01794-24

Denton, N., Molloy, M., Charleston, S., Lipset, C., Hirsch, J., Mulberg, A. E., Howard, P., & Marsh, E. D. (2021). Data silos are undermining drug development and failing rare disease patients. *Orphanet Journal of Rare Diseases*, *16*(1), 161. DOI: 10.1186/s13023-021-01806-4

Di Salvatore, V., Cernuto, F., Russo, G., & Pappalardo, F. (2025). A computational framework for optimizing mRNA vaccine delivery via AI-guided nanoparticle design and in silico gene expression profiling. *Frontiers in Immunology*, *16*, 1628583. DOI: 10.3389/fimmu.2025.1628583

Diaz, R., Ramalheira, E., Afreixo, V., & Gago, B. (2016). Methicillin-resistant *Staphylococcus aureus* carrying the new mecC gene—a meta-analysis. *Diagnostic Microbiology and Infectious Disease*, *84*(2), 135–140. DOI: 10.1016/j.diagmicrobio.2015.10.014

Dini, I. (2026). Mechanisms of Bacterial Resistance and Innovative Strategies to Overcome Antimicrobial Resistance. *Antibiotics (Basel, Switzerland)*, *15*(3), 319. DOI: 10.3390/antibiotics15030319

Do, T., Vaculciakova, S., Kluska, K., Peris-Díaz, M. D., Priborsky, J., Guran, R., Krężel, A., Adam, V., & Zitka, O. (2024). Antioxidant-related enzymes and peptides as biomarkers of metallic nanoparticles (eco)toxicity in the aquatic environment. *Chemosphere*, *364*, 142988. DOI: 10.1016/j.chemosphere.2024.142988

Duncan, K. R., Crüsemann, M., Lechner, A., Sarkar, A., Li, J., Ziemert, N., Wang, M., Bandeira, N., Moore, B. S., Dorrestein, P. C., & Jensen, P. R. (2015). Molecular networking and pattern-based genome mining improves discovery of biosynthesis-related gene clusters and their products from *Salinispora* species. *Chemistry & Biology*, *22*(4), 460–471. DOI: 10.1016/j.chembiol.2015.03.010

Durrant, J. D., & Amaro, R. E. (2015). Machine-learning techniques applied to antibacterial drug discovery. *Chemical Biology & Drug Design*, *85*(1), 14–21. DOI: 10.1111/cbdd.12423

Dzoyem, J. P., Tsemeugne, J., Pone Kamdem, B., Foyou Meupiap, R., Kuate, B. A., Mkounga, P., Fekam Boyom, F., & McGaw, L. J. (2025). Antibacterial, antibiofilm and anti-quorum sensing activities of 1,2,3,5-tetrazine derivatives linked to a benzothiazole moiety. *PLoS One*, *20*(6), e0318135. DOI: 10.1371/journal.pone.0318135

Eicher, T., Kinnebrew, G., Patt, A., Spencer, K., Ying, K., Ma, Q., Machiraju, R., & Mathé, A. E. A. (2020). Metabolomics and multiomics integration: A survey of computational methods and resources. *Metabolites, 10*(5), 202. DOI: 10.3390/metabo10050202

Elalouf, A., Elalouf, H., & Rosenfeld, A. (2025). Artificial intelligence in drug resistance management. *3 Biotech, 15*(4), 126. https://doi.org/DOI: 10.1007/s13205-025-04282-w

Elbaiomy, R. G., El-Sappah, A. H., Guo, R., Luo, X., Deng, S., Du, M., Jian, X., Bakeer, M., Li, Z., & Zhang, Z. (2025). Antibiotic resistance: A genetic and physiological perspective. *MedComm, 6*(11), e70447. DOI: 10.1002/mco2.70447

Elgin, C. (2025). Anticipatory macroeconomic governance: Exploring future-oriented strategies for economic resilience and sustainability. *European Journal of Futures Research, 13*(4). Advance online publication. DOI: 10.1186/s40309-025-00248-6

Esawy, M. A., & Reyed, R. M. (2026). Biointelligent reframing of colorectal cancer: From somatic mutation to systemic coherence. In Reyed, R., Prabhakar, P., & Haghi, A. (Eds.), *Advancing Personalized Medicine with AI-Driven Microbiome and Nutrigenomics: Innovations and Future Perspectives* (pp. 173–218). IGI Global., DOI: 10.4018/979-8-3373-3121-8.ch003

Ezike, T. C., Okpala, U. S., Onoja, U. L., Nwike, C. P., Ezeako, E. C., Okpara, O. J., Okoroafor, C. C., Eze, S. C., Kalu, O. L., Odoh, E. C., Nwadike, U. G., Ogbodo, J. O., Umeh, B. U., Ossai, E. C., & Nwanguma, B. C. (2023). Advances in drug delivery systems, challenges and future directions. *Heliyon, 9*(6), e17488. DOI: 10.1016/j.heliyon.2023.e17488

Fan, L., Chen, C., Zhang, H., Zeng, Y., Li, T., Gao, R., Li, J., Ren, Y., Wu, Z., Bi, F., Chu, Y., Li, H., Hu, J., Xu, J., & Xu, Y. (2025). Atmospheric detection, prevalence, transmission, health and ecological consequences of antibiotic resistance genes and resistant bacteria: A comprehensive review. *Emerging Contaminants, 11*(3), 100514. DOI: 10.1016/j.emcon.2025.100514

Farah, H., Kadhim-Abosaoda, M., Mohaisen-Mousa, H., Renuka Jyothi, S., Priyadarshini-Nayak, P., Bethanney Janney, J., Singh, G., Singh-Chauhan, A., & Kumar-Mishra, M. (2026). Nanomedicine Strategies Against Biofilm-Associated Infections: Advances, Challenges, and Translational Barriers. *MicrobiologyOpen, 15*(1), e70210. DOI: 10.1002/mbo3.70210

Fernandes, F. O. L., Crima, S., & Ferrinho, P. (2025). Is it feasible to develop a one health surveillance system in a fragile state? The case of Guinea-Bissau. *One health (Amsterdam, Netherlands), 21*, 101269. DOI: 10.1016/j.onehlt.2025.101269

Ferreira, L. G., Dos Santos, R. N., Oliva, G., & Andricopulo, A. D. (2015). Molecular docking and structure-based drug design strategies. *Molecules (Basel, Switzerland), 20*(7), 13384–13421. DOI: 10.3390/molecules200713384

Floss, H. G., & Yu, T. W. (2005). Rifamycin-mode of action, resistance, and biosynthesis. *Chemical Reviews, 105*(2), 621–632. DOI: 10.1021/cr030112j

Forli, S., Huey, R., Pique, M. E., Sanner, M. F., Goodsell, D. S., & Olson, A. J. (2016). Computational protein–ligand docking and virtual drug screening with the AutoDock suite. *Nature Protocols, 11*(5), 905–919. DOI: 10.1038/nprot.2016.051

Franklin, A. M., Weller, D. L., Durso, L. M., Bagley, M., Davis, B. C., Frye, J. G., Grim, C. J., Ibekwe, A. M., Jahne, M. A., Keely, S. P., Kraft, A. L., McConn, B. R., Mitchell, R. M., Ottesen, R. R., Sharma, M., Strain, E. A., Tadesse, D. A., Tate, H., Wells, J. E., & Garland, J. L. (2024). A One Health approach for monitoring antimicrobial resistance: Developing a national freshwater pilot effort. *Frontiers in Water, 6*, 1359109. DOI: 10.3389/frwa.2024.1359109

Gao, F., Shao, T., Yu, Y., Xiong, Y., & Yang, L. (2021). Surface-bound reactive oxygen species generating nanozymes for selective antibacterial action. *Nature Communications, 12*(1), 745. DOI: 10.1038/s41467-021-20965-3

Gatherer, D. (2010). So what do we truly mean when we say that systems biology is holistic? *BMC Systems Biology, 4*, 22. DOI: 10.1186/1752-0509-4-22

Gerilovych, A., Shevchenko, N., Pishchanskyi, O., Aliekseieva, H., Rosada, M., Gerilovych, I., & Okaievych, O. (2026). Operationalizing the One Health approach in a conflict-affected setting: A scientometric review of policy foundations, systemic gaps, and future pathways in Ukraine. *Veterinary World, 19*(1), 389–408. DOI: 10.14202/vetworld.2026.389-408

Gheorghita, F. I., Bocanet, V. I., & Iantovics, L. B. (2025). Machine learning-based drug–drug interaction prediction: A critical review of models, limitations, and data challenges. *Frontiers in Pharmacology, 16*, 1632775. DOI: 10.3389/fphar.2025.1632775

Giacobbe, D. R., Grossi, A. A., Bassetti, M., & de la Fuente-Nunez, C. (2026). The future of antibiotics and artificial intelligence: Some thoughts from discovery to bedside. *Infectious Diseases and Therapy, 15*(2), 461–475. DOI: 10.1007/s40121-025-01288-y

Ginnan, N., Crandall, S. G., Imchen, M., Dini-Andreote, F., Miyashiro, T. I., Singh, V., Ganda, E., & Bordenstein, S. R. (2025). Ecologically expanding the One Health framework to unify the microbiome sciences. *mBio*, *16*(6), e03147–e24. DOI: 10.1128/mbio.03147-24

Goharshadi, E. K. (2025). A review on the radial distribution function: Insights into molecular structure, intermolecular interactions, and thermodynamic properties. *Journal of Molecular Liquids*, *433*, 127900. DOI: 10.1016/j.molliq.2025.127900

Golriz Khatami, S., Mubeen, S., Bharadhwaj, V. S., Kodamullil, A. T., Hofmann-Apitius, M., & Domingo-Fernández, D. (2021). Using predictive machine learning models for drug response simulation by calibrating patient-specific pathway signatures. *NPJ Systems Biology and Applications*, *7*(1), 40. DOI: 10.1038/s41540-021-00199-1

Greathouse, K. L., & Choudhury, A. (2025). Precision nutrition and the gut microbiome: Harnessing AI to revolutionize cancer prevention and therapy. *Cell Host & Microbe*, *33*(6), 766–776. DOI: 10.1016/j.chom.2025.05.011

Guégan, J.-F., Suzán, G., Kati-Coulibaly, S., Bonpamgue, D. N., & Moatti, J.-P. (2018). Sustainable Development Goal #3, "health and well-being," and the need for more integrative thinking. *Veterinaria México OA*, *5*(2), 443. Advance online publication. DOI: 10.21753/vmoa.5.2.443

Guerra, M. E. S., Destro, G., Vieira, B., Lima, A. S., Ferraz, L. F. C., Hakansson, A. P., Darrieux, M., & Converso, T. R. (2022). *Klebsiella pneumoniae* biofilms and their role in disease pathogenesis. *Frontiers in Cellular and Infection Microbiology*, *12*, 877995. DOI: 10.3389/fcimb.2022.877995

Gupta, V. P., Reyed, M. R., & Haghi, A. K. (Eds.). (2026). *The blue economy and environmental sustainability: Advancing global governance, innovation, and finance for a resilient future*. Springer Cham., DOI: 10.1007/978-3-032-13638-1

Hasin, Y., Seldin, M., & Lusis, A. (2017). Multiomics approaches to disease. *Genome Biology*, *18*(1), 83. DOI: 10.1186/s13059-017-1215-1

He, S., Nader, K., Abarrategi, J. S., Bediaga, H., Nocedo-Mena, D., Ascencio, E., Casanola-Martin, G. M., Castellanos-Rubio, I., Insausti, M., Rasulev, B., Arrasate, S., & González-Díaz, H. (2024). NANO.PTML model for read-across prediction of nanosystems in neurosciences: Computational model and experimental case of study. *Journal of Nanobiotechnology*, *22*(1), 435. DOI: 10.1186/s12951-024-02660-9

He, W., Yu, Z., Wu, Z., Olesen, A. K., Madsen, J. S., Dechesne, A., Smets, B. F., Nesme, J., & Sørensen, S. J. (2026). Beyond borders: Plasmids drive a shared antibiotic resistome in European urban water systems. *Microbiome*, *14*(1), 39. DOI: 10.1186/s40168-025-02253-0

Helliwell, R., Ewin, I., Williams, A. D., Levine, D. T., Singer, A. C., Raman, S., Morris, C., & Stekel, D. J. (2025). Rethinking the words hotspot reservoir and pristine in the environmental dimensions of antimicrobial resistance. *npj antimicrobials and resistance, 3*(1), 11. https://doi.org/DOI: 10.1038/s44259-025-00080-9

Hendriksen, R. S., Munk, P., Njage, P., van Bunnik, B., McNally, L., Lukjancenko, O., Röder, T., Nieuwenhuijse, D., Pedersen, S. K., Kjeldgaard, J., Kaas, R. S., Clausen, P. T. L. C., Vogt, J. K., Leekitcharoenphon, P., van de Schans, M. G. M., Zuidema, T., de Roda Husman, A. M., Rasmussen, S., Petersen, B., & Aarestrup, F. M. (2019). Global monitoring of antimicrobial resistance based on metagenomics analyses of urban sewage. *Nature Communications*, *10*(1), 1124. DOI: 10.1038/s41467-019-08853-3

Hernando-Amado, S., Coque, T. M., Baquero, F., & Martínez, J. L. (2020). Antibiotic Resistance: Moving From Individual Health Norms to Social Norms in One Health and Global Health. *Frontiers in Microbiology*, *11*, 1914. DOI: 10.3389/fmicb.2020.01914

Hetta, H. F., Ramadan, Y. N., Al-Harbi, A. I., Ahmed, A., Battah, E., Abd, B., Ellah, N. H., Zanetti, S., & Donadu, M. G. (2023). Nanotechnology as a promising approach to combat multidrug resistant bacteria: A comprehensive review and future perspectives. *Biomedicines*, *11*(2), 413. DOI: 10.3390/biomedicines11020413

Hosny, S., Alakib, M., Bendary, S. H., Abu-Dief, A. M., & Abbass, L. M. (2026). Eco-friendly synthesis of novel nanocomplexes: Integrated in vitro evaluation, advanced molecular dynamics simulations, network pharmacology, and in silico ADME/toxicity assessment. *Inorganic Chemistry Communications*, *188*(Part 1), 116479. DOI: 10.1016/j.inoche.2026.116479

Hrncir, T. (2022). Gut microbiota dysbiosis: Triggers, consequences, diagnostic and therapeutic options. *Microorganisms*, *10*(3), 578. DOI: 10.3390/microorganisms10030578

Huovinen, P., Sundström, L., Swedberg, G., & Sköld, O. (1995). Trimethoprim and sulfonamide resistance. *Antimicrobial Agents and Chemotherapy*, *39*(2), 279–289. DOI: 10.1128/AAC.39.2.279

Jacob, M. C. M., de Carvalho, A. M., Batista, Â. G., Santos Júnior, A. F., Carioca, A. A. F., de Morais, C. M. M., Cazarin, C. B. B., Tregidgo, D., de Oliveira, D. V. B., Marchioni, D. M. L., Giuntini, E. B., de Menezes-Neto, E. J., Pereira, F. O., Moura, G. F., El Bizri, H. R., Bezerra, I. W. L., Aquino, J. S., Silva, J. V. M., Steluti, J., & Albuquerque, U. P. (2026). Prioritizing neglected food species in nutritional studies using expert-knowledge and explainable AI. *Scientific Reports*. Advance online publication. DOI: 10.1038/s41598-026-39484-6

Jamkhande, P. G., Ghante, M. H., & Ajgunde, B. R. (2017). Software based approaches for drug designing and development: A systematic review on commonly used software and its applications. *Bulletin of Faculty of Pharmacy, Cairo University*, *55*(2), 203–210. DOI: 10.1016/j.bfopcu.2017.10.001

Jena, G. K., Patra, C. N., Jammula, S., Rana, R., & Chand, S. (2024). Artificial intelligence and machine learning implemented drug delivery systems: A paradigm shift in the pharmaceutical industry. *Journal of Bio-X Research, 7*, 0016. https://doi.org/DOI: 10.34133/jbioxresearch.0016

Jia, H., Hou, W., Wei, L., Xu, B., & Liu, X. (2008). The structures and antibacterial properties of nano-SiO2 supported silver/zinc-silver materials. *Dental Materials : Official Publication of the Academy of Dental Materials*, *24*(2), 244–249. DOI: 10.1016/j.dental.2007.04.015

Jia, Y., Naceur, H., Saadlaoui, Y., Dubar, L., & Bergheau, J. M. (2024). A comprehensive comparison of modeling strategies and simulation techniques applied in powder-based metallic additive manufacturing processes. *Journal of Manufacturing Processes*, *110*, 1–29. DOI: 10.1016/j.jmapro.2023.12.048

Johnson, N. C., & Marín, C. (2025). Functional team selection as a framework for local adaptation in plants and their belowground microbiomes. *The ISME Journal*, *19*(1), wraf137. Advance online publication. DOI: 10.1093/ismejo/wraf137

Jung, W. K., Koo, H. C., Kim, K. W., Shin, S., Kim, S. H., & Park, Y. H. (2008). Antibacterial activity and mechanism of action of the silver ion in *Staphylococcus aureus* and *Escherichia coli. Applied and Environmental Microbiology*, *74*(7), 2171–2178. DOI: 10.1128/AEM.02001-07

Kang, S., Antoniewicz, M. R., & Hay, N. (2024). Metabolic and transcriptomic reprogramming during contact inhibition-induced quiescence is mediated by YAP-dependent and YAP-independent mechanisms. *Nature Communications*, *15*(1), 6777. DOI: 10.1038/s41467-024-51117-y

Kasse, G. E., Cosh, S. M., Humphries, J., & Islam, M. S. (2025). Leveraging artificial intelligence for One Health: Opportunities and challenges in tackling antimicrobial resistance - scoping review. *One Health Outlook, 7*(1), 51. DOI: 10.1186/s42522-025-00170-8

Kehrein, J., Bunker, A., & Luxenhofer, R. (2024). POxload: Machine learning estimates drug loadings of polymeric micelles. *Molecular Pharmaceutics, 21*(7), 3356–3374. DOI: 10.1021/acs.molpharmaceut.4c00086

Koide, R. T. (2023). On holobionts, holospecies, and holoniches: The role of microbial symbioses in ecology and evolution. *Microbial Ecology, 85*(4), 1143–1149. DOI: 10.1007/s00248-022-02005-9

Konagurthu, S., Ranathunga, D. T. S., Buchanan, S., Mehta, N. M., & Reynolds, T. (2026). The predictive edge: Modeling and simulation in drug product development. *Advanced Drug Delivery Reviews, 230*, 115784. DOI: 10.1016/j.addr.2026.115784

Kumavath, R., Gupta, P., Tatta, E. R., Mohan, M. S., Salim, S. A., & Busi, S. (2025). Unraveling the role of mobile genetic elements in antibiotic resistance transmission and defense strategies in bacteria. *Frontiers in Systems Biology, 5*, 1557413. DOI: 10.3389/fsysb.2025.1557413

Le Piane, F., Vozza, M., Baldoni, M., & Mercuri, F. (2024). Integrating high-performance computing, machine learning, data management workflows, and infrastructures for multiscale simulations and nanomaterials technologies. *Beilstein Journal of Nanotechnology, 15*, 1498–1521. DOI: 10.3762/bjnano.15.119

Lee, J. H., Choi, B., Jeong, K., Suh, S. W., Rhee, H., Kim, J. H., & Son, D. S. (2026). Diverging regulatory DNA in adaptive medical AI: US agility and EU accountability in lifecycle governance. *Frontiers in Medicine, 13*, 1758708. DOI: 10.3389/fmed.2026.1758708

Lemire, J. A., Harrison, J. J., & Turner, R. J. (2013). Antimicrobial activity of metals: Mechanisms, molecular targets and applications. *Nature Reviews. Microbiology, 11*(6), 371–384. DOI: 10.1038/nrmicro3028

Leung, Y. H., Ng, A. M., Xu, X., Shen, Z., Gethings, L. A., Wong, M. T., Chan, C. M., Guo, M. Y., Ng, Y. H., Djurišić, A. B., Lee, P. K., Chan, W. K., Yu, L. H., Phillips, D. L., Ma, A. P., & Leung, F. C. (2014). Mechanisms of antibacterial activity of MgO: Non-ROS mediated toxicity of MgO nanoparticles toward *Escherichia coli*. *Small, 10*(6), 1171–1183. DOI: 10.1002/smll.201302434

Li, J., Wang, Y., Tang, M., Zhang, C., Fei, Y., Li, M., Li, M., Gui, S., & Guo, J. (2024). New insights into nanotherapeutics for periodontitis: A triple concerto of antimicrobial activity, immunomodulation and periodontium regeneration. *Journal of Nanobiotechnology*, *22*(1), 19. DOI: 10.1186/s12951-023-02261-y

Li, X., Zhang, Y., Zhang, Q., Liu, J., Zhu, Z., Feng, X., Han, L., & Zhang, X. (2024). Strategy and mechanism of One Health governance: Case study of China. *Science in One Health*, *4*, 100098. DOI: 10.1016/j.soh.2024.100098

Liu, J., Gefen, O., Ronin, I., Bar-Meir, M., & Balaban, N. Q. (2020). Effect of tolerance on the evolution of antibiotic resistance under drug combinations. *Science*, *367*(6474), 200–204. DOI: 10.1126/science.aay3041

Liu, N. N., Jiao, N., Tan, J. C., Wang, Z., Wu, D., Wang, A. J., Chen, J., Tao, L., Zhou, C., Fang, W., Cheong, I. H., Pan, W., Liao, W., Kozlakidis, Z., Heeschen, C., Moore, G. G., Zhu, L., Chen, X., Zhang, G., & Wang, H. (2022). Multikingdom microbiota analyses identify bacterial-fungal interactions and biomarkers of colorectal cancer across cohorts. *Nature Microbiology*, *7*(2), 238–250. DOI: 10.1038/s41564-021-01030-7

Liu, X., Jiang, M., Zhang, J., Li, H., Liu, Y., Zhang, J., Chen, X., Bu, J., Xie, S., Zhang, M., Dong, N., Cao, Q., & Zhou, Z. (2025). Machine learning-based evaluation of risk factors for carbapenem-resistant *Klebsiella pneumoniae* dissemination in neonatal units. *mSystems*, *10*(10), e00909–e00925. DOI: 10.1128/msystems.00909-25

Lloyd-Price, J., Abu-Ali, G., & Huttenhower, C. (2016). The healthy human microbiome. *Genome Medicine*, *8*, 51. DOI: 10.1186/s13073-016-0307-y

Lopatkin, A. J., Bening, S. C., Manson, A. L., Stokes, J. M., Kohanski, M. A., Badran, A. H., Earl, A. M., Cheney, N. J., Yang, J. H., & Collins, J. J. (2021). Clinically relevant mutations in core metabolic genes confer antibiotic resistance. *Science*, *371*(6531), eaba0862. Advance online publication. DOI: 10.1126/science.aba0862

Lu, Y., Kim, S., & Park, K. (2011). In vitro-in vivo correlation: Perspectives on model development. *International Journal of Pharmaceutics*, *418*(1), 142–148. DOI: 10.1016/j.ijpharm.2011.01.010

Lu, Y., Zhang, B., Wang, Z., He, Y., Ge, H., Ma, H., & Cui, P. (2026). From host-derived pressures to the environmental anti-antimicrobial peptides resistome: Mechanisms, reservoirs and implications for therapeutic peptide design. *Marine Drugs*, *24*(2), 76. DOI: 10.3390/md24020076

Maharjan, R., Kim, N. A., Kim, K. H., & Jeong, S. H. (2025). Transformative roles of digital twins from drug discovery to continuous manufacturing: Pharmaceutical and biopharmaceutical perspectives. *International Journal of Pharmaceutics: X*, *10*, 100409. DOI: 10.1016/j.ijpx.2025.100409

Maiorov, V. N., & Crippen, G. M. (1994). Significance of root-mean-square deviation in comparing three-dimensional structures of globular proteins. *Journal of Molecular Biology*, *235*(2), 625–634. DOI: 10.1006/jmbi.1994.1017

Maitreya, A., Pal, S., Qureshi, A., Reyed, M. R., & Purohit, H. J. (2022). Nitric oxide–secreting probiotics as sustainable biocleaners for reverse osmosis membrane systems. *Environmental Science and Pollution Research International*, *29*(4), 4911–4929. DOI: 10.1007/s11356-021-17289-6

Majumder, J., & Minko, T. (2021). Multifunctional and stimuli-responsive nanocarriers for targeted therapeutic delivery. *Expert Opinion on Drug Delivery*, *18*(2), 205–227. DOI: 10.1080/17425247.2021.1828339

Makadia, H. K., & Siegel, S. J. (2011). Poly Lactic-co-Glycolic Acid (PLGA) as biodegradable controlled drug delivery carrier. *Polymers*, *3*(3), 1377–1397. DOI: 10.3390/polym3031377

Malik, A. K., Kushwah, S. V., Aslam, M. S., Reyed, M. R., & Haghi, A. K. (Eds.). (2026). *Aligning ESG goals with the UN's sustainable development goals planning: Strategies for equitable financing and world sustainability*. Springer Cham., DOI: 10.1007/978-3-032-01715-4

McCubbin, K. D., Barkema, H. W., Babujee, A., Forseille, J., Naum, K., Buote, P., Dalton, D., Checkley, S. L., Lehman, K., Morris, T., Smilski, K., Wilkins, W. L., Anholt, R. M., Larose, S., Saxinger, L. M., Blue, D., & Otto, S. J. G. (2022). One health and antimicrobial stewardship: Where to go from here? *The Canadian Veterinary Journal = La Revue Vétérinaire Canadienne*, *63*(2), 198–200.

McNair, L. (2022). Ethical and regulatory oversight of clinical research: The role of the Institutional Review Board. *Experimental Biology and Medicine (Maywood, N.J.)*, *247*(7), 561–566. DOI: 10.1177/15353702221078216

Mennella, C., Maniscalco, U., De Pietro, G., & Esposito, M. (2024). Ethical and regulatory challenges of AI technologies in healthcare: A narrative review. *Heliyon*, *10*(4), e26297. DOI: 10.1016/j.heliyon.2024.e26297

Mi, P. (2020). Stimuli-responsive nanocarriers for drug delivery, tumor imaging, therapy and theranostics. *Theranostics*, *10*(10), 4557–4588. DOI: 10.7150/thno.38069

Miller, C., Portlock, T., Nyaga, D. M., & O'Sullivan, J. M. (2024). A review of model evaluation metrics for machine learning in genetics and genomics. *Frontiers in Bioinformatics*, *4*, 1457619. Advance online publication. DOI: 10.3389/fbinf.2024.1457619

Mills, S., Stanton, C., Lane, J. A., Smith, G. J., & Ross, R. P. (2019). Precision nutrition and the microbiome, part I: Current state of the science. *Nutrients*, *11*(4), 923. DOI: 10.3390/nu11040923

Mitchell, M. J., Billingsley, M. M., Haley, R. M., Wechsler, M. E., Peppas, N. A., & Langer, R. (2021). Engineering precision nanoparticles for drug delivery. *Nature Reviews. Drug Discovery*, *20*(2), 101–124. DOI: 10.1038/s41573-020-0090-8

Monteiro, D. R., Gorup, L. F., Takamiya, A. S., de Camargo, E. R., Filho, A. C., & Barbosa, D. B. (2012). Silver distribution and release from an antimicrobial denture base resin containing silver colloidal nanoparticles. *Journal of Prosthodontics*, *21*(1), 7–15. DOI: 10.1111/j.1532-849X.2011.00772.x

Morar, M., & Wright, G. D. (2010). The genomic enzymology of antibiotic resistance. *Annual Review of Genetics*, *44*, 25–51. DOI: 10.1146/annurev-genet-102209-163517

Mubeen, B., Ansar, A. N., Rasool, R., Ullah, I., Imam, S. S., Alshehri, S., Ghoneim, M. M., Alzarea, S. I., Nadeem, M. S., & Kazmi, I. (2021). Nanotechnology as a novel approach in combating microbes providing an alternative to antibiotics. *Antibiotics (Basel, Switzerland)*, *10*(12), 1473. DOI: 10.3390/antibiotics10121473

Muñoyerro-Muñiz, D., Villegas, R., de la Oliva, V., Esteban-Medina, A., Fernández Del Valle, P., Sánchez, A., Susin, M. B., Gutierrez-Alvarez, I., Reboredo, M., Alejos, L., Loucera, C., & Dopazo, J. (2025). Ethical and secure evidence generation from regionwide clinical data through a collaborative environment for advancing predictive care. *Frontiers in Public Health*, *13*, 1630351. DOI: 10.3389/fpubh.2025.1630351

Musicha, P., Morse, T., Cocker, D., Mugisha, L., Jewell, C. P., & Feasey, N. A. (2024). Time to define One Health approaches to tackling antimicrobial resistance. *Nature Communications*, *15*(1), 8782. DOI: 10.1038/s41467-024-53057-z

Mustafa, S. S., Batool, R., Kamran, M., Javed, H., & Jamil, N. (2022). Evaluating the role of wastewaters as reservoirs of antibiotic-resistant ESKAPEE bacteria using phenotypic and molecular methods. *Infection and Drug Resistance*, *15*, 5715–5728. DOI: 10.2147/IDR.S368886

Navarro-Muñoz, J. C., Selem-Mojica, N., Mullowney, M. W., Kautsar, S. A., Tryon, J. H., Parkinson, E. I., De Los Santos, E. L. C., Yeong, M., Cruz-Morales, P., Abubucker, S., Roeters, A., Lokhorst, W., Fernandez-Guerra, A., Cappelini, L. T. D., Goering, A. W., Thomson, R. J., Metcalf, W. W., Kelleher, N. L., Barona-Gomez, F., & Medema, M. H. (2020). A computational framework to explore large-scale biosynthetic diversity. *Nature Chemical Biology*, *16*(1), 60–68. DOI: 10.1038/s41589-019-0400-9

Nechaeva, A. M., Artyukhov, A. A., Svistunova, A. Y., Il'in, M., Loginova, T. P., Baranov, O. V., Shulgin, A. M., Komarova, L. G., Estifeeva, T. M., Kuzmin, P. S., Barmin, R. A., Gerasimov, V., Tsatsakis, A. M., Rezaee, R., Rudakovskaya, P. G., & Mezhuev, Y. O. (2026). Nanoparticles for doxorubicin delivery: Advances in carrier design and synergistic cancer therapy. *Topics in Current Chemistry (Cham)*, *384*(2), 9. Advance online publication. DOI: 10.1007/s41061-026-00547-x

Newman, M., Barabási, A.-L., & Watts, D. J. (2011). *The structure and dynamics of networks*. Princeton University Press., DOI: 10.1515/9781400841356

Nithyanand, P., Boya, B. R., Lee, J. H., & Lee, J. (2025). Polymicrobial biofilms: Interkingdom interactions, resistance and therapeutic strategies. *Microbial Biotechnology*, *18*(8), e70218. Advance online publication. DOI: 10.1111/1751-7915.70218

Nusrat, S., Aliyu, M., & Zohora, F. T. (2025). Mechanisms of antimicrobial resistance: From genetic evolution to clinical manifestations. *AIMS Microbiology*, *11*(4), 1007–1034. DOI: 10.3934/microbiol.2025045

Okkeh, M., Bloise, N., Restivo, E., De Vita, L., Pallavicini, P., & Visai, L. (2021). Gold nanoparticles: Can they be the next magic bullet for multidrug-resistant bacteria? *Nanomaterials (Basel, Switzerland)*, *11*(2), 312. DOI: 10.3390/nano11020312

Papo, D., & Buldú, J. M. (2026). A complex network perspective on brain disease. *Biological Reviews of the Cambridge Philosophical Society*, *101*(1), 364–399. DOI: 10.1111/brv.70086

Pathak, K., Sarma, M., Sahariah, M., Shankarishan, P., Sahariah, J. J., Deka, S., Das, A., Islam, M. A., Pramanik, P., Borthakur, P. P., Bharadwaj, A. R., & Bhaumik, R. (2026). Nanoparticles in the fight against antimicrobial challenges: A comprehensive review. *Next Nanotechnology*, *9*, 100420. Advance online publication. DOI: 10.1016/j.nxnano.2026.100420

Pati, R., Sahu, R., Panda, J., & Sonawane, A. (2016). Encapsulation of zinc-rifampicin complex into transferrin-conjugated silver quantum-dots improves its antimycobacterial activity and stability and facilitates drug delivery into macrophages. *Scientific Reports*, *6*, 24184. Advance online publication. DOI: 10.1038/srep24184

Patra, J. K., Das, G., Fraceto, L. F., Campos, E. V. R., Rodriguez-Torres, M. D. P., Acosta-Torres, L. S., Diaz-Torres, L. A., Grillo, R., Swamy, M. K., Sharma, S., Habtemariam, S., & Shin, H. S. (2018). Nano based drug delivery systems: Recent developments and future prospects. *Journal of Nanobiotechnology*, *16*(1), 71. Advance online publication. DOI: 10.1186/s12951-018-0392-8

Pattapulavar, V., S, S. K., Ramanujam, S., Subburaj, S., & John, G. C. (2026). Probiotic-based therapeutics for a One Health future: Redefining antibiotic dependency to combat antimicrobial resistance. *Frontiers in Microbiology, 17*, 1736436. https://doi.org/DOI: 10.3389/fmicb.2026.1736436

Pawar, S. S., & Rohane, S. H. (2021). Review on Discovery Studio: An important tool for molecular docking. *Asian Journal of Research in Chemistry*, *14*(1), 86–88. DOI: 10.52711/0974-4150.2021.00015

Pennisi, F., Pinto, A., Ricciardi, G. E., Signorelli, C., & Gianfredi, V. (2025). The role of artificial intelligence and machine learning models in antimicrobial stewardship in public health: A narrative review. *Antibiotics (Basel, Switzerland)*, *14*(2), 134. DOI: 10.3390/antibiotics14020134

Perini, H. F., Matos, B. S., de Oliveira, C. J. F., & da Silva, M. V. (2024). Biomimetic nanocarriers: Integrating natural functions for advanced therapeutic applications. *Beilstein Journal of Nanotechnology*, *15*, 1619–1626. DOI: 10.3762/bjnano.15.127

Pham, N.-P., Gingras, H., Godin, C., Feng, J., Groppi, A., Nikolski, M., Leprohon, P., & Ouellette, M. (2024). Holistic understanding of trimethoprim resistance in *Streptococcus pneumoniae* using an integrative approach of genome-wide association study, resistance reconstruction, and machine learning. *mBio*, *15*(9), e0136024. DOI: 10.1128/mbio.01360-24

Pochet, A., Bourguignon, T., Grandé, A., Godinez-Leon, J. A., Li, X., Fine, J., Sencio, V., Chatagnon, J., Dremierre, B., Vandeputte, A., Mascarau, R., Piveteau, C., Brodin, P., Machelart, A., & Gref, R. (2026). Selective accumulation of nanoparticles in infected cells for targeted drug delivery. *Journal of Controlled Release : Official Journal of the Controlled Release Society*, *389*, 114415. DOI: 10.1016/j.jconrel.2025.114415

Polianciuc, S. I., Gurzău, A. E., Kiss, B., Ştefan, M. G., & Loghin, F. (2020). Antibiotics in the environment: Causes and consequences. *Medicine and Pharmacy Reports*, *93*(3), 231–240. DOI: 10.15386/mpr-1742

Quan, W., Zhou, J., Wang, J., Huang, J., & Du, L. (2025). Machine learning-driven precision nutrition: A paradigm evolution in dietary assessment and intervention. *Nutrients*, *18*(1), 45. DOI: 10.3390/nu18010045

Rath, B. P., Kar, S., Pattnaik, S., & Mohanty, C. (2025). Computational methods to explore the therapeutic potential of Triphala, a polyherbal formulation, against LuxS protein of *Helicobacter pylori. Microbe*, *9*, 100559. DOI: 10.1016/j.microb.2025.100559

Redgrave, L. S., Sutton, S. B., Webber, M. A., & Piddock, L. J. (2014). Fluoroquinolone resistance: Mechanisms, impact on bacteria, and role in evolutionary success. *Trends in Microbiology*, *22*(8), 438–445. DOI: 10.1016/j.tim.2014.04.007

Reyed, M. R. (2019). Computational biotechnology: An approach in silico based modeling bioprocess. *International Journal of Research Studies in Microbiology and Biotechnology*, *5*(4), 115. DOI: 10.22259/2454-9428.0504001

Reyed, M. R. (2020). Investigation of antimicrobial resistance in wastewater based on public health: An integrated methodology for coronavirus disease 2019 (COVID-19) outbreak tracking schemes and real-time detection of reoccurrence is needed. *Journal of Infectious Diseases & Case Reports*, *1*(2), 115. DOI: 10.47363/JIDSCR/2020(1)115

Reyed, M. R. (2023). Simulate the dynamic interactions across nutritional supplements, the polybiome, and the biochemical approach dependent on patient-derived gastrointestinal knowledge. *Journal of Clinical Epidemiology & Toxicology*, *4*(4), 133. DOI: 10.47363/JCET/2023(4)133

Reyed, M. R. (2026a). Advancing diabetes reclassification using the NexDi AI system with multiomic insights: Egypt's living laboratory. In Reyed, R., Prabhakar, P., & Haghi, A. (Eds.), *Advancing Personalized Medicine With AI-Driven Microbiome and Nutrigenomics: Innovations and Future Perspectives* (pp. 1–92). IGI Global., DOI: 10.4018/979-8-3373-3121-8.ch001

Reyed, M. R. (2026c). Digital nanoplastic science (DNPS) paradigm: Computational intelligence and proteostasis disruptions. *Zenodo*. https://doi.org/DOI: 10.5281/zenodo.18435353

Reyed, M. R. (2026e). Polybiome systems medicine (PSM): A systems-level biodiversity conversion framework for precision nutrition and predictive health 2040. *Zenodo*. https://doi.org/DOI: 10.5281/zenodo.18727664

Reyed, M. R. (2026f). Polybiome systems medicine (PSM) – Volume III: AI nutrigenomics, precision fermentation, and genotype-aligned biomanufacturing. *Zenodo*. https://doi.org/DOI: 10.5281/zenodo.18871656

Reyed, M. R. (2026g). RPCF-AMR: From resistance surveillance to predictive control—A precision convergence framework for integrated antimicrobial resistance governance. *Zenodo*. https://doi.org/DOI: 10.5281/zenodo.19260312

Reyed, M. R., Patel, R., Patel, S., & Prajapati, B. G. (2026b). Application of synthetic biology in genetic disorder. In *Synthetic Biology and Its Consequences in Present Healthcare* (pp. 201–213). Elsevier., DOI: 10.1016/B978-0-443-33649-2.00001-0

Reyed, M. R., & Prabhakar, P. K. (2026d). Modulating neuroinflammation and autophagy in psychiatric disorders through mechanisms and therapies: Therapeutic strategies and pathways. In Mokhtari, T., & Uludag, K. (Eds.), *Autophagy and Inflammation in Neuropsychological Disorders* (pp. 61–124). IGI Global., DOI: 10.4018/979-8-3693-5908-2.ch003

Reyed, M. R., & Prabhakar, P. K. M. (2025). Polybiome derived therapies and fermentobiometrics: Novel strategies for blocking cancer pathways and progression. In *Cancer Etiology, Diagnosis and Treatments* (pp. 145–212). Nova Science Publishers. https://doi.org/DOI: 10.52305/JKMU3682

Rolff, J., Bonhoeffer, S., Kloft, C., Leistner, R., Regoes, R., & Hochberg, M. E. (2024). Forecasting antimicrobial resistance evolution. *Trends in Microbiology*, *32*(8), 736–745. DOI: 10.1016/j.tim.2023.12.009

Rusu, A. V., Trif, M., & Rocha, J. M. (2023). Microbial secondary metabolites via fermentation approaches for dietary supplementation formulations. *Molecules (Basel, Switzerland)*, *28*(16), 6020. DOI: 10.3390/molecules28166020

Sabih Ur Rehman, S., Nasar, M. I., Mesquita, C. S., Al Khodor, S., Notebaart, R. A., Ott, S., Mundra, S., Arasardanam, R. P., Muhammad, K., & Alam, M. T. (2025). Integrative systems biology approaches for analyzing microbiome dysbiosis and species interactions. *Briefings in Bioinformatics*, *26*(4), bbaf323. Advance online publication. DOI: 10.1093/bib/bbaf323

Sakagianni, A., Koufopoulou, C., Feretzakis, G., Kalles, D., Verykios, V. S., Myrianthefs, P., & Fildisis, G. (2023). Using machine learning to predict antimicrobial resistance—A literature review. *Antibiotics (Basel, Switzerland)*, *12*(3), 452. DOI: 10.3390/antibiotics12030452

Saleem, N., Kumar, N., & El-Omar, E.. (2026). Nanoantimicrobial peptides (nano-AMPs) to combat resistant gram-negative bacteria. *Drug Delivery and Translational Research*. Advance online publication. DOI: 10.1007/s13346-026-02085-x

Salehi, M., Khayami, R., Akbari, R., & Mirmozaffari, M. (2025). An integrated predictive impact-enhanced process mining framework for strategic oncology workflow optimization: Case study in Iran. *Bioengineering (Basel, Switzerland)*, *12*(12), 1288. DOI: 10.3390/bioengineering12121288

Salmani-Zarchi, H., Mousavi-Sagharchi, S. M. A., Sepahdoost, N., Ranjbar-Jamalabadi, M., Gross, J. D., Jooya, H., & Samadi, A. (2024). Antimicrobial feature of nanoparticles in the antibiotic resistance era: From mechanism to application. *Advanced Biomedical Research*, *13*, 113. DOI: 10.4103/abr.abr_92_24

Salo-Ahen, O. M. H., Alanko, I., Bhadane, R., Alexandre, A. M., Honorato, R. V., Hossain, S., Juffer, A. H., Kabedev, A., Lahtela-Kakkonen, M., Larsen, A. S., Lescrinier, E., Marimuthu, P., Mirza, M. U., Mustafa, G., Nunes-Alves, A., Pantsar, T., Saadabadi, A., Singaravelu, K., & Vanmeert, M. (2021). Molecular dynamics simulations in drug discovery and pharmaceutical development. *Processes (Basel, Switzerland)*, *9*(1), 71. DOI: 10.3390/pr9010071

Sambaza, S. S., & Naicker, N. (2023). Contribution of wastewater to antimicrobial resistance: A review article. *Journal of Global Antimicrobial Resistance*, *34*, 23–29. DOI: 10.1016/j.jgar.2023.05.010

Sarkar, D., Majumder, S., Giri, K., & Sabnam, N. (2023). In silico characterization, molecular docking, and dynamic simulation of a novel fungal cell-death suppressing effector, MoRlpA as potential cathepsin B-like cysteine protease inhibitor during rice blast infection. *Journal of Biomolecular Structure & Dynamics*, *41*(18), 9039–9056. DOI: 10.1080/07391102.2022.2139763

Sarker, I. H. (2021). Machine learning: Algorithms, real-world applications and research directions. *SN Computer Science*, *2*(3), 160. DOI: 10.1007/s42979-021-00592-x

Satheeskumar, R. (2025). Enhancing drug discovery with AI: Predictive modeling of pharmacokinetics using graph neural networks and ensemble learning. *Intelligent Pharmacy*, *3*(2), 127–140. DOI: 10.1016/j.ipha.2024.11.002

Scaglione, G., Mastroianni, N., Rizzo, A., Palomba, E., Carcione, D., Brigante, G., Principe, L., Colaneri, M., Gori, A., & Borgonovo, F. (2026). Integrating artificial intelligence with genome sequencing against antimicrobial resistance: A narrative review. *Frontiers in Public Health*, *14*, 1757161. DOI: 10.3389/fpubh.2026.1757161

Sharma, D., Gautam, S., Singh, S., Srivastava, N., Khan, A. M., & Bisht, D. (2025). Unveiling the nanoworld of antimicrobial resistance: Integrating nature and nanotechnology. *Frontiers in Microbiology*, *15*, 1391345. DOI: 10.3389/fmicb.2024.1391345

Shen, Y., Zhang, P., Luo, J., Chen, S., Gu, S., Lin, Z., & Tang, Z. (2026). Artificial intelligence drives advances in multiomics analysis and precision medicine for sepsis. *Biomedicines*, *14*(2), 261. DOI: 10.3390/biomedicines14020261

Shin, M. K., Lee, B., Kim, S. T., Yoo, J. S., & Sung, J. S. (2022). Designing a novel functional peptide with dual antimicrobial and anti-inflammatory activities *via* in silico methods. *Frontiers in Immunology*, *13*, 821070. DOI: 10.3389/fimmu.2022.821070

Shinu, P., Mouslem, A. K. A., Nair, A. B., Venugopala, K. N., Attimarad, M., Singh, V. A., Nagaraja, S., Alotaibi, G., & Deb, P. K. (2022). Progress report: Antimicrobial drug discovery in the resistance era. *Pharmaceuticals*, *15*(4), 413. DOI: 10.3390/ph15040413

Silva, P. M. A., & Queirós, O. (2025). Advances in nanodrug delivery for tumor microenvironment and drug resistance-insights from the special issue " Nanodrug delivery systems for targeting the tumor microenvironment and simultaneously overcoming drug resistance properties. *Pharmaceutics*, *17*(7), 942. DOI: 10.3390/pharmaceutics17070942

Silva, Y. P., Bernardi, A., & Frozza, R. L. (2020). The role of short-chain fatty acids from gut microbiota in gut-brain communication. *Frontiers in Endocrinology*, *11*, 25. DOI: 10.3389/fendo.2020.00025

Smyth, C., Leigh, R. J., Do, T. T., & Walsh, F. (2025). Communities of plasmids as strategies for antimicrobial resistance gene survival in wastewater treatment plant effluent. *npj Antimicrobials and Resistance*, *3*(1), 78. https://doi.org/DOI: 10.1038/s44259-025-00151-x

Sodagari, H. R., Ghasemi, M., Varga, C., & Habib, I. (2025). Machine learning prediction of multidrug resistance in swine-derived *Campylobacter* spp. using United States antimicrobial resistance surveillance data (2013-2023). *Veterinary Sciences*, *12*(10), 937. DOI: 10.3390/vetsci12100937

Spasic, A. M. (2018). Classification and methodology—finely dispersed systems. In Spasic, A. M. (Ed.), *Rheology of Emulsions* (Vol. 22, pp. 41–49). Elsevier., DOI: 10.1016/B978-0-12-813836-6.00003-9

Sraphet, S., & Javadi, B. (2022). Application of hierarchical clustering to analyze solvent-accessible surface area patterns in Amycolatopsis lipases. *Biology (Basel)*, *11*(5), 652. DOI: 10.3390/biology11050652

Srivastava, P., & Kowshik, M. (2013). Mechanisms of metal resistance and homeostasis in haloarchaea. *Archaea : an International Microbiological Journal, 732864*. Advance online publication. DOI: 10.1155/2013/732864

Sulimov, V. B., Kutov, D. C., & Sulimov, A. V. (2019). Advances in docking. *Current Medicinal Chemistry, 26*(42), 7555–7580. DOI: 10.2174/0929867325666180904115000

Sultana, A., Zare, M., Thomas, V., Kumar, T. S. S., & Ramakrishna, S. (2022). Nanobased drug delivery systems: Conventional drug delivery routes, recent developments and future prospects. *Medicine in Drug Discovery, 15*, 100134. DOI: 10.1016/j.medidd.2022.100134

Sum, D. K. C., Chong, Y. Y., & Tan, J. L. (2025). Comparative analyses suggest genome stability and plasticity in *Stenotrophomonas maltophilia. International Journal of Molecular Sciences, 26*(21), 10477. DOI: 10.3390/ijms262110477

Tabll, A. (2026). AI-Guided Monoclonal Antibody Design Integrating Polybiome and Nutrigenomic Insights: Integrating AI, Polybiome, and Nutrigenomics in Therapy. In Reyed, R., Prabhakar, P., & Haghi, A. (Eds.), *Advancing Personalized Medicine With AI-Driven Microbiome and Nutrigenomics: Innovations and Future Perspectives* (pp. 219–262). IGI Global Scientific Publishing., DOI: 10.4018/979-8-3373-3121-8.ch004

Tatonetti, N. P., Ye, P. P., Daneshjou, R., & Altman, R. B. (2012). Data-driven prediction of drug effects and interactions. *Science Translational Medicine, 4*(125), 125ra31. Advance online publication. DOI: 10.1126/scitranslmed.3003377

Tavares Luiz, M., Santos Rosa Viegas, J., Palma Abriata, J., Viegas, F., Testa Moura de Carvalho Vicentini, F., Lopes Badra Bentley, M. V., Chorilli, M., Maldonado Marchetti, J., & Tapia-Blácido, D. R. (2021). Design of experiments (DoE) to develop and to optimize nanoparticles as drug delivery systems. *European Journal of Pharmaceutics and Biopharmaceutics : Official Journal of Arbeitsgemeinschaft für Pharmazeutische Verfahrenstechnik e.V, 165*, 127–148. DOI: 10.1016/j.ejpb.2021.05.011

Thakur, B., Kaur, S., Rani, N., Kaur, R., Upadhyay, S. K., & Tripathi, M. (2025). Exploring microbial contributions to nutraceutical production: From natural to designed foods. *Molecular Biotechnology, 67*(8), 2946–2961. DOI: 10.1007/s12033-023-00937-2

Topol, E. J. (2019). High-performance medicine: The convergence of human and artificial intelligence. *Nature Medicine, 25*(1), 44–56. DOI: 10.1038/s41591-018-0300-7

Torchilin, V. P. (2006). Multifunctional nanocarriers. *Advanced Drug Delivery Reviews, 58*(14), 1532–1555. DOI: 10.1016/j.addr.2006.09.009

Törős, G., Gulyás, G., El-Ramady, H., Alibrahem, W., Muthu, A., Gangakhedkar, P., Atieh, R., & Prokisch, J. (2025). Sustainable nanotechnology strategies for modulating the human gut microbiota. *International Journal of Molecular Sciences, 26*(12), 5433. DOI: 10.3390/ijms26125433

Trucillo, P. (2021). Drug carriers: Classification, administration, release profiles, and industrial approach. *Processes (Basel, Switzerland), 9*(3), 470. DOI: 10.3390/pr9030470

van Pinxteren, D. J. M., & Jespers, W. (2025). Integrating machine learning into free energy perturbation workflows. *Journal of Chemical Information and Modeling, 65*(19), 9856–9864. DOI: 10.1021/acs.jcim.5c01449

Villanueva-Miranda, I., Xiao, G., & Xie, Y. (2025). Artificial intelligence in early warning systems for infectious disease surveillance: A systematic review. *Frontiers in Public Health, 13*, 1609615. DOI: 10.3389/fpubh.2025.1609615

Vivarelli, S., De Francesco, C., Paba, E., Giambò, F., & Fenga, C. (2026). The resistome bridge between livestock and workers: Novel frameworks for early detection and monitoring of antimicrobial resistance. *Frontiers in Public Health, 14*, 1746385. DOI: 10.3389/fpubh.2026.1746385

Voter, A. F. (1997). Hyperdynamics: Accelerated molecular dynamics of infrequent events. *Physical Review Letters, 78*(20), 3908–3911. DOI: 10.1103/PhysRevLett.78.3908

Wadan, A. S., Ali, M. H., & Ellakwa, D. E. (2026). Sustainable nanomaterials for precision dental medicine: Green synthesis, therapeutic applications, and future directions. *Journal of Nanobiotechnology, 24*(1), 190. DOI: 10.1186/s12951-025-04008-3

Wang, F., Harker, A., Edirisinghe, M., & Parhizkar, M. (2025). Advanced experimental design strategies for drug development. *Advanced Intelligent Discovery, 2025*, e202500087. DOI: 10.1002/aidi.202500087

Wang, K., Huang, Y., Wang, Y., You, Q., & Wang, L. (2024). Recent advances from computer-aided drug design to artificial intelligence drug design. *RSC Medicinal Chemistry, 15*(12), 3978–4000. DOI: 10.1039/d4md00522h

Waykar, R., & Kulkarni, Y. A. (2025). Informed consent: Navigating the bioethical complexities of clinical research. *Bioinformation, 21*(9), 2950–2955. DOI: 10.6026/973206300212950

Wegener Parfrey, L. (2015). Mock communities highlight the diversity of host-associated eukaryotes. *Molecular Ecology, 24*(17), 4337–4339. DOI: 10.1111/mec.13311

Wimalawansa, S. J. (2025). Reforming food, drug, and nutraceutical regulations to improve public health and reduce healthcare costs. *Foods, 14*(13), 2328. DOI: 10.3390/foods14132328

Wörheide, M. A., Krumsiek, J., Kastenmüller, G., & Arnold, M. (2021). Multiomics integration in biomedical research - A metabolomics-centric review. *Analytica Chimica Acta, 1141*, 144–162. DOI: 10.1016/j.aca.2020.10.038

Wu, J., & Koelzer, V. H. (2024). Toward generative digital twins in biomedical research. *Computational and Structural Biotechnology Journal, 23*, 3481–3488. DOI: 10.1016/j.csbj.2024.09.030

Wu, S., Zhou, H., Chen, D., Lu, Y., Li, Y., & Qiao, J. (2024). Multiomic analysis tools for microbial metabolites prediction. *Briefings in Bioinformatics, 25*(4), bbae264. Advance online publication. DOI: 10.1093/bib/bbae264

Wu, Z., Yao, T., & Shen, N. (2025). To face Disease X: Building resilient futures in the age of emergent threats. *BMJ Global Health, 10*(7), e020479. DOI: 10.1136/bmjgh-2025-020479

Xia, S., Chen, E., & Zhang, Y. (2023). Integrated molecular modeling and machine learning for drug design. *Journal of Chemical Theory and Computation, 19*(21), 7478–7495. DOI: 10.1021/acs.jctc.3c00814

Xiaolin, X., Xiaozhi, L., Guoping, H., Hongwei, L., Jinkuo, G., Xiyun, B., Zhen, T., Xiaofang, M., Yanxia, L., Na, X., Chunyan, Z., Rui, G., Kuan, W., Cheng, Z., Cuancuan, W., Mingyong, L., & Xinping, D. (2023). Overfit deep neural network for predicting drug-target interactions. *iScience, 26*(9), 107646. DOI: 10.1016/j.isci.2023.107646

Xie, J., Sun, L., & Zhao, Y. F. (2025). On the data quality and imbalance in machine learning-based design and manufacturing—A systematic review. *Engineering (Beijing), 45*(18), 105–131. DOI: 10.1016/j.eng.2024.04.024

Yang, S. Y., Han, S. M., Lee, J. Y., Kim, K. S., Lee, J. E., & Lee, D. W. (2025). Advancing gut microbiome research: The shift from metagenomics to multiomics and future perspectives. *Journal of Microbiology and Biotechnology, 35*, e2412001. DOI: 10.4014/jmb.2412.12001

Yoon, M. Y., & Yoon, S. S. (2018). Disruption of the gut ecosystem by antibiotics. *Yonsei Medical Journal, 59*(1), 4–12. DOI: 10.3349/ymj.2018.59.1.4

Yu, X., Gao, D., Gao, L., Lai, J., Zhang, C., Zhao, Y., Zhong, L., Jia, B., Wang, F., Chen, X., & Liu, Z. (2017). Inhibiting metastasis and preventing tumor relapse by triggering host immunity with tumor-targeted photodynamic therapy using photosensitizer-loaded functional nanographenes. *ACS Nano*, *11*(10), 10147–10158. DOI: 10.1021/acsnano.7b04736

Yu, Y., Mei, L., Shi, Y., Zhang, X., Cheng, K., Cao, F., Zhang, L., Xu, J., Li, X., & Xu, Z. (2020). Ag-conjugated graphene quantum dots with blue light-enhanced singlet oxygen generation for ternary-mode highly efficient antimicrobial therapy. *Journal of Materials Chemistry. B*, *8*(7), 1371–1382. DOI: 10.1039/c9tb02300c

Zdrazil, B. (2025). Fifteen years of ChEMBL and its role in cheminformatics and drug discovery. *Journal of Cheminformatics*, *17*(1), 32. DOI: 10.1186/s13321-025-00963-z

Zeng, L., An, L., & Wu, X. (2011). Modeling drug-carrier interaction in the drug release from nanocarriers. *Journal of Drug Delivery*, *370308*. Advance online publication. DOI: 10.1155/2011/370308

Zhang, C., Li, J., Tang, Q., Li, L., & Cao, D. (2025). Targeting proteostasis for cancer therapy: Current advances, challenges, and future perspectives. *Molecular Cancer*, *24*(1), 265. DOI: 10.1186/s12943-025-02472-x

Zhang, J., Lin, G., Xie, N., Ye, Y., Zhao, H., & Lu, J. (2026). Alternations along gut-brain axis in response to hypo- and hyperosmotic stress in yellowfin seabream (*Acanthopagrus latus*). *Aquaculture Reports*, *48*, 103533. DOI: 10.1016/j.aqrep.2026.103533

Zhang, T., Wang, J., Feng, Q., Xu, X., Zhu, W., Mao, S., & Liu, J. (2026). Age- and diet-driven assembly of the gut antibiotic resistome in humans and food-producing animals. *Gut Microbes*, *18*(1), 2610052. DOI: 10.1080/19490976.2025.2610052

Zhang, Y., Lin, S., Fu, J., Zhang, W., Shu, G., Lin, J., Li, H., Xu, F., Tang, H., Peng, G., Zhao, L., Chen, S., & Fu, H. (2022). Nanocarriers for combating biofilms: Advantages and challenges. *Journal of Applied Microbiology*, *133*(3), 1273–1287. DOI: 10.1111/jam.15640

Zhang, Z., Chen, K., Zhang, S., He, Y., Lei, G., Zhao, Y., & Liang, J. (2025). Identification of fecal microbiota and related metabolites associated with feed efficiency in DLY pigs. *Animals (Basel)*, *15*(20), 3026. DOI: 10.3390/ani15203026

Zhao, Y., Yin, J., Zhang, L., Zhang, Y., & Chen, X. (2024). Drug–drug interaction prediction: Databases, web servers and computational models. *Briefings in Bioinformatics*, *25*(1), bbad445. Advance online publication. DOI: 10.1093/bib/bbad445

Zhitlov, M. Y., Korshun, V. A., & Alferova, V. A. (2025). Microfluidics to meet antibiotic resistance: A growing research frontier. *Antibiotics (Basel, Switzerland), 14*(12), 1232. DOI: 10.3390/antibiotics14121232

Zhu, H. (2020). Big data and artificial intelligence modeling for drug discovery. *Annual Review of Pharmacology and Toxicology, 60*, 573–589. DOI: 10.1146/annurev-pharmtox-010919-023324

Zhu, Y., Tang, Y., Qi, X., & Zhu, X. (2026). Transformer models, graph networks, and generative AI in gut microbiome research: A narrative review. *Bioengineering (Basel, Switzerland), 13*(2), 144. DOI: 10.3390/bioengineering13020144

Zielińska, K., Udekwu, K. I., Rudnicki, W., Frolova, A., & Łabaj, P. P. (2025). Healthy microbiome-moving toward functional interpretation. *GigaScience, 14*, giaf015. Advance online publication. DOI: 10.1093/gigascience/giaf015

Zou, D., Du, X., Liu, Y., Fang, M., & Chu, M. (2026). Integrated DIA proteomics and transcriptomics unravel molecular mechanisms of uterine prolificacy in Yunshang Black goats during the estrous cycle. *BMC Genomics, 27*(1), 314. DOI: 10.1186/s12864-026-12634-w

Chapter 2
Beyond Antibiotics:
AI–Nanotech Synergy for a Sustainable Antimicrobial Future

Arun Agrawal
http://orcid.org/0000-0001-7233-6660
Institute of Technology and Management, Gwalior, India

Y. Dasaratha Rami Reddy
Chaitanya Bharathi Institute of Technology, India

K. Hussain
http://orcid.org/0000-0001-9744-4764
Sharad Institute of Technology College of Engineering, India

Srinivasan Nagaraj
Chaitanya Bharathi Institute of Technology, India

B. Venkateswarlu
Koneru Lakshmaiah Education Foundation, India

Rekha Gangula
http://orcid.org/0000-0001-8553-9176
Vaagdevi Engineering College, India

Deepak Gupta
http://orcid.org/0000-0003-3929-1362
Institute of Technology and Management, Gwalior, India

Deshdeepak Shrivastava
http://orcid.org/0009-0009-0909-3956
Institute of Technology and Management, Gwalior, India

ABSTRACT

The rising threat of antimicrobial resistance (AMR) has made conventional antibiotic discovery increasingly ineffective against drug-resistant pathogens. Beyond Antibiotics: AI–Nanotech Synergy for a Sustainable Antimicrobial Future explores how artificial intelligence and nanotechnology can transform antimicrobial innovation. AI accelerates drug discovery through predictive modeling and generative design, enabling rapid identification of novel candidates. Nanotechnology improves targeted delivery, stability, and therapeutic effectiveness. Integration of genomics and microbiome analysis supports personalized treatment strategies. The chapter

DOI: 10.4018/979-8-3373-5268-8.ch002

also highlights sustainability through lifecycle assessment, responsible data governance, and equitable access. It presents an interdisciplinary framework aligned with Sustainable Development Goals (SDGs) 3, 9, 12, and 17, while identifying key limitations, research gaps, and future directions for sustainable antimicrobial solutions.

1. INTRODUCTION

One of the most severe health challenges of the twenty-first century is antimicrobial resistance **(AMR)**. Over time, numerous bacteria have developed mechanisms that enable them to resist the effects of antibiotics, and infections that were previously manageable are becoming increasingly difficult to treat (Salam et al., 2023). Antibiotic-resistant common pathogens include Escherichia coli, Klebsiella pneumoniae, Staphylococcus aureus, and Pseudomonas aeruginosa, which resist antibiotics through enzyme production, efflux pumping, drug target mutations, and antibiotic degradation (Qin et al., 2022).

The effects of AMR are not confined to hospitals but profoundly impact healthcare systems, economies, and societies globally. Currently, more than 700,000 people die each year because of antimicrobial resistance, and this number may rise to almost 10 million deaths each year unless viable options are developed (Fong, 2023). Despite the significant contribution of antibiotics to human health, there is a decline in effectiveness due to misuse, overuse, and the slow pace of new drug development.

The history of antibiotics demonstrates that the process of discovery, extensive usage, and the emergence of resistance is a recurring cycle (Cook & Wright, 2022). As an illustration, the treatment of infectious diseases began to take a new dimension in 1928 when Alexander Fleming discovered penicillin (Chhabra et al., 2024). Subsequently, the treatment options were enhanced with the introduction of cephalosporins, tetracyclines, and fluoroquinolones (Nayeem et al., 2024). But over time, the bacteria also became resistant to these medications, with one of them being methicillin-resistant Staphylococcus aureus (MRSA) (Issa & Muhsin, 2023).

The process of creating the new antibiotics is also prolonged and costly as it may take 10-15 years and involve more than one billion dollars (Brown et al., 2022). Due to these difficulties, scientists consider new technologies like artificial intelligence (AI) and nanotechnology. Artificial intelligence assists in the analysis of massive biological and chemical volumes to determine new drug molecules and predict antimicrobial activity (Rahman et al., 2025). Nanotechnology enhances drug delivery by nanocarriers, which include liposomes, dendrimers, and polymeric nanoparticles, to increase drug stability and targeted delivery (Majumder & Minko, 2021).

Integration of nanotechnology and the artificial intelligence offers a synergetic system of forecasting molecular design and targeted drug delivery. AI models are capable of optimizing the nanocarrier design with the ability to predict the drug-carrier interactions and release kinetics, and experimental data in nanotechnology can enhance the accuracy of AI predictions based on accumulative feedback (Das et al., 2023). This closed-loop business model speed up the drug development process, enables highly individualized antimicrobial therapy using genomic and microbiome data, and is part of the sustainable and ethical medical system. Moreover, the discovery of new antibiotics, including halicin, and nanocarrier-mediated targeted delivery systems with the assistance of AI proves the feasibility of this combined approach. This interdisciplinary approach together with microfluidic and lab-on-chip validation systems provides a scalable, efficient, and sustainable solution to the creation of next-generation antimicrobial therapies.

Figure 1. AI with Nanotechnology

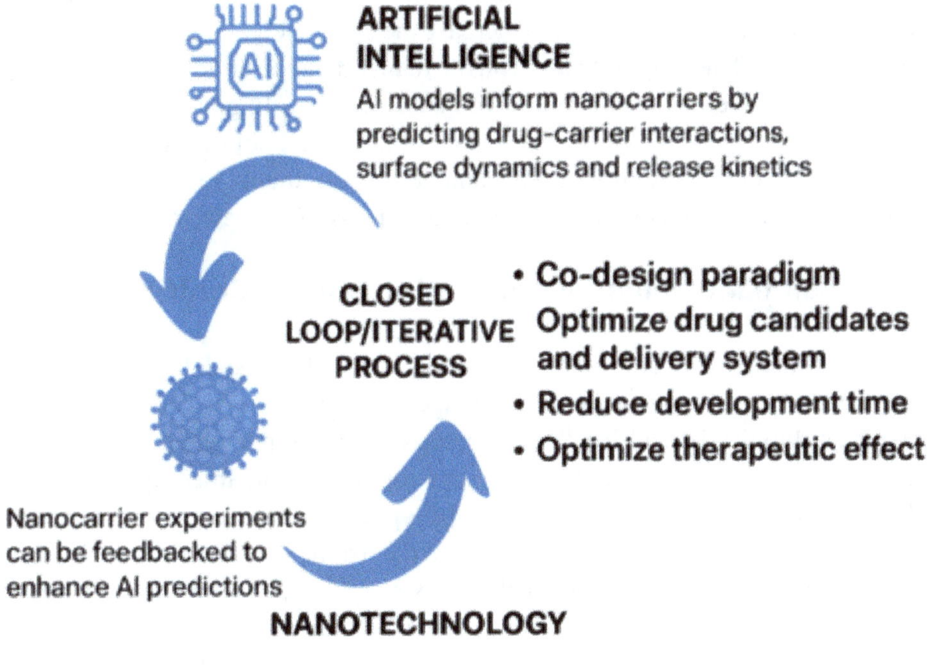

Although AI significantly accelerates antimicrobial discovery through predictive modeling and generative design, several challenges remain. The reliability of models requires high-quality datasets, which are usually not complete or biased to known compounds. Moreover, AI predictions have to be experimentally validated, which restricts their application in the real world. Moreover, regulatory unpredictability and unpublished frameworks of validation pose an obstacle to clinical adoption. It is these limitations that require hybrid methods using AI prediction alongside experimental microbiology and nanotechnology-directed delivery methods.

Despite significant progress, several critical research gaps remain. First, most current studies focus either on AI-based drug discovery or nanotechnology-based delivery independently, with limited integration of both approaches. Second, there has been less focus on the sustainability, lifecycle assessment and environmental impact of nanomaterials. Third, issues of quality of data, ethical application of AI, and clinical translation have not been examined thoroughly. Lastly, no extensive interdisciplinary models have been created to tie AI, nanotechnology, microbiome analysis, and personalized medicine.

This chapter presents the current status of artificial intelligence–based drug discovery and nanocarrier technologies and integrates these approaches to enable sustainable antimicrobial interventions.It offers an interdisciplinary approach where AI-based molecular design is used, nanotechnology-based targeted delivery, and personalized therapy are combined to increase the efficiency of treatment in antimicrobial resistance. The chapter focuses on the main practical, ethical, and regulatory considerations as well as puts emphasis on sustainability. In general, it provides a comprehensive roadmap of next-generation antimicrobial solutions in accordance with the global health systems and Sustainable Development Goals (SDGs), especially those regarding health, innovation, and responsible development.

1.1. Precision Medicine, Scope, and Scholarly Contribution

Precision medicine in antimicrobial therapy is based on the delivery of specific and personalized therapy instead of broad-spectrum antibiotics. The choice of the precision antibiotics depends on the characteristics of the pathogen, the resistance genes, the genetics of the patient, and the environment of the infection.

Artificial intelligence is also significant in this practice because it studies microbial genomes, resistance trends, and clinical data to determine useful antimicrobial compounds (Rahman et al., 2025). Nanotechnology also facilitates this plan as it allows the targeted drug delivery with the application of nanocarrier system and lowers the toxicity and enhances the success of the treatments (Majumder & Minko, 2021).

AI and nanotechnology combined allows discovering drugs more quickly and allows creating antimicrobial therapies that are personalized, which can be viewed as a promising method to address antimicrobial resistance.

This chapter analyzes how artificial intelligence, nanotechnology, and precision medicine can be used to deal with antimicrobial resistance. It puts into focus the recent breakthrough in AI-driven drug discovery and nanocarrier-based delivery platforms and explains how they can be combined to create next-generation antimicrobial drugs. Technical, ethical, and regulatory issues of these emerging technologies are also discussed in the chapter and they serve to help in coming up with sustainable and precision-based solutions in controlling how antimicrobial resistance is managed.

2. LITERATURE REVIEW

2.1 Global Burden of Antimicrobial Resistance

The burden of AMR is high and growing all over the world. Anti-resistance infections cause longer hospital stays, increased treatment expenses, and increased mortality, with the majority being predominantly high in low- and middle-income nations with low access to second-line agents (Otaigbe, 2025). Table 1 provides an overview of the burden by the chief six AMR pathogen categorized either critical or high-priority as per the WHO.

Table 1. Burden of Priority Antimicrobial-Resistant Pathogens Worldwide.

Pathogen	Resistance Mechanism	Annual Deaths Attributable	WHO Priority Category
Klebsiella pneumoniae (CR)	Carbapenemase production, efflux pumps	~215,000 (global est.)	Critical
Acinetobacter baumannii (CR)	OXA carbapenemases, outer membrane loss	~190,000 (global est.)	Critical
Pseudomonas aeruginosa (CR)	Efflux pumps, porin mutations, biofilm	~180,000 (global est.)	Critical
Staphylococcus aureus (MRSA)	mecA gene, altered PBP2a	~100,000 (global est.)	High
Escherichia coli (ESBL)	Beta-lactamase genes (blaCTX-M)	~130,000 (global est.)	High

Overuse of antibiotics in both clinical and agricultural practices, taking half of the courses, and insufficient infection control facilities are continuously added to the list of primary reasons of resistance diffusion (Endale et al., 2023). Natural product screening or optimization of synthetic chemistry discovery of classical antibiotics is intrinsically incompatible with the time scale of pathogen adaptation (Muteeb et al., 2023).

2.2 AI in Antimicrobial Drug Discovery

Machine learning, as well as deep learning, has already shown a great potential in changing antimicrobial discovery. Quantitative structure-activity relationship (QSAR) models are based on rapid prediction of antimicrobial activity using molecular structure, and can therefore pre-screen large chemical libraries without the need to synthesize them. The same can be applied to generative adversarial networks (GANs) and variational autoencoders (VAEs) that can be used to design new molecular scaffolds with specific pharmacophoric properties (Bian and Xie, 2021; Das, 2025; Macedo et al., 2024). One of the first landmark cases is the AI-aided discovery of the drug halicin, the compound with a broad-spectrum bactericidal action against the drug-resistant pathogens found in the screen of more than 100 million molecules: a task that cannot be done using conventional methods (Tiwari et al., 2023).

Regardless of these developments, there are a number of constraints that limit the existing AI-based methods. The model performance significantly relies on the quality of the dataset; numerous training libraries prefer the known chemical backbones and are not expected to be able to generalize to structurally novel antimicrobials. It is an expensive and time-intensive bottleneck to experimentally verify AI predictions. Moreover, the interpretability shortcomings of deep learning designs restrict the field of regulatory acceptability because the decision logic of black-box models cannot be easily audited (Qadri et al., 2025; Seal et al., 2025). These limitations are research gaps present at the substantive level, which are not fully completed in the existing literature (see Table 2).

Table 2. Informative Analysis of AI Methods in Antimicrobial Discovery: Capabilities and Research Gap.

AI Approach	Primary Function	Key Advantage	Limitation / Research Gap
Machine Learning (QSAR models)	Predict antimicrobial activity from structure	Fast screening of large compound libraries	Dependent on quality/ diversity of training data; poor generalization to novel scaffolds
Generative Adversarial Networks (GANs)	Design novel chemical scaffolds	Explores uncharted chemical space	Generated molecules require costly experimental validation; mode collapse risk
Variational Autoencoders (VAEs)	Latent-space molecular design and optimization	Controllable generation with pharmacophore constraints	Limited interpretability; difficulty encoding complex biological context

continued on following page

Table 2. Continued

AI Approach	Primary Function	Key Advantage	Limitation / Research Gap
Graph Neural Networks (GNNs)	Model molecular topology and drug-target interactions	Captures structural relationships invisible to fingerprints	Computationally intensive; limited AMR-specific datasets
Reinforcement Learning (RL)	Iteratively optimize molecules toward target properties	Aligns generation with multi-objective criteria (efficacy + safety)	Reward function design complex; sparse reward problem in drug contexts

2.3 Nanotechnology in Antimicrobial Drug Delivery

Nanotechnology fits complementary line of issues: as much as AI prompts the identification of promising compounds, nanocarrier capabilities surmount the pharmacodynamics and pharmacokinetics challenges preventing the effective compounds to reach the site of infection to exert their therapeutic effects. The key nanocarrier types that have been used in antimicrobial delivery are outlined in Table 3.

Table 3. The Nanocarrier Systems in Delivery of Antimicrobials: Properties, Strengths and Problems.

Nanocarrier Type	Composition	Antimicrobial Advantage	Drug Loading	Key Challenge
Liposomes	Phospholipid bilayer	Enhanced membrane penetration; co-delivery of hydrophilic and lipophilic agents	Moderate–High	Physical instability; leakage on storage
Polymeric Nanoparticles	PLGA, chitosan, PEG	Controlled release; biofilm penetration; pH-responsive delivery	High	Scalable synthesis uniformity
Dendrimers	Branched macromolecules	Multivalent surface for antibiotic conjugation; disrupts bacterial membranes	Moderate	Cytotoxicity at high concentrations
Solid Lipid Nanoparticles	Solid lipid matrix	Protects labile antibiotics; improves oral bioavailability	Moderate	Polymorphic transitions; drug expulsion
Biomimetic Carriers	Cell-membrane-coated nanoparticles	Immune evasion; pathogen-specific tropism; microbiome-compatible delivery	Variable	Complex fabrication; batch-to-batch variation

Functionalized nanocarriers can be utilized in pathogen-specific targeting by surface ligands, which bind bacterial surface markers and concentrate drug payloads at the sites of infection, reducing overall systemic exposure (Din et al., 2017). Spatial and temporal drug delivery is further optimized with the help of stimuli-responsive designs, responding to pH decreases, enzymatic activity, or redox gradients typical of an infected tissue (Mashele, 2025). Co-delivery systems, which simultaneously deliver two or more antimicrobial agents or adjuvants, provide a pharmacological basis for overcoming multidrug resistance.

However, there are still severe constraints. There is still no reliable technology to produce nanoparticles on a scale and reproducibly, along with the variability of manufacturing that compromises the translational reliability. The literature on long-term biocompatibility, immunogenicity, the environmental fate of nanomaterials, and especially ecotoxicology of these materials has not been characterized sufficiently (Ramos et al., 2022). These are gaps that are barriers of substantive clinical implementation to the implementation which have not been well dealt with by the available reviews.

2.4 Convergence of AI and Nanotechnology: Evidence and Gaps

Antimicrobial research is a relatively young branch where AI is becoming convergent with nanotechnology. To minimize nanoparticle physicochemical properties (size, surface charge, ligand density) via compatibility prediction of drugs/carrier and release kinetics via molecular descriptors, AI models have been utilized (Noury et al., 2025). On the other hand, the information of experimental nanocarriers has aided in enhancing the accuracy of AI models in the course of iterative training. The first microfluidics, lab-on-chip systems are also starting to have a level of physiologically relevant feedback loop in order to develop closed-loop AI-nanotechnology (Nunziata et al., 2025).

Nevertheless, there are gaps in the literature. Most published works are about AI and nanotechnology based discovery and delivery as independent of each other with little systematic integration. The studies which report integrated strategies do not possess standardized validation designs and hence comparing the studies across studies is problematic. Moreover, the available literature tends to involve microbiome data, patient genomics, and lifecycle sustainability metrics in the integrated design process infrequently. These gaps are conceptualizing the scholarly input and critical locating of the framework in the current chapter.

2.5 Critical Comparison of Existing Studies and Positioning of This Chapter

A critical comparison of the existing literature reveals three broad categories of prior reviews: (1) AI-focused reviews that emphasize computational drug discovery and machine learning models without addressing delivery challenges (e.g., Tiwari et al., 2023; Oyejide et al., 2025); (2) nanotechnology-focused reviews that examine nanocarrier systems and drug delivery optimization without integrating AI-driven molecular design (e.g., Sultana et al., 2022; Xu et al., 2025); and (3) a small number of hybrid reviews that mention both technologies but treat them as parallel rather than synergistic tools (e.g., Serrano et al., 2024; Noury et al., 2025). These three categories do not contain, particularly, a single operational model that integrates AI-driven molecular discovery, nanocarrier design optimization, and closed-loop experimental validation into one and the same, systematic model.

Furthermore, prior reviews consistently underaddress four critical research gaps: (i) the lack of open-access, pathogen-diverse AMR datasets needed to train generalizable AI models; (ii) the absence of standardized Good Manufacturing Practice (GMP) protocols for AI-designed nanocarriers; (iii) insufficient attention to the environmental lifecycle and ecotoxicological impact of engineered nanomaterials; and (iv) the absence of end-to-end discovery-to-clinic integration pipelines that link computational prediction with experimental validation and clinical translation. These gaps are addressed in this chapter with the suggestion of the Three-Pillar AI-Nanotechnology Synergy Framework that does not only bring together these technologies into practice but also explicitly outlines unaddressed challenges and is a structured research agenda. By doing this, it goes beyond the descriptive reviews to provide a critically positioned and actionable contribution to the field, which is sustainability conscious.

3. CONCEPTUAL FRAMEWORK: THE THREE-PILLAR AI–NANOTECHNOLOGY SYNERGY MODEL

The Three-Pillar AI Nanotechnology Synergy Framework that is suggested in this chapter offers a combined operational design of developing the next-generation antimicrobials. Each pillar is a very specific but interrelated stage: Discovery and Design, Delivery and Optimization, and Validation and Feedback. The framework goes further than the established methods in that AI and nanotechnology are seen as integrated, synergizing systems, not as sequential or independent instruments.

Figure 2. The Three-Pillar AI –Nanotechnology Synergy Framework to Next-Generation Antimicrobial Development.

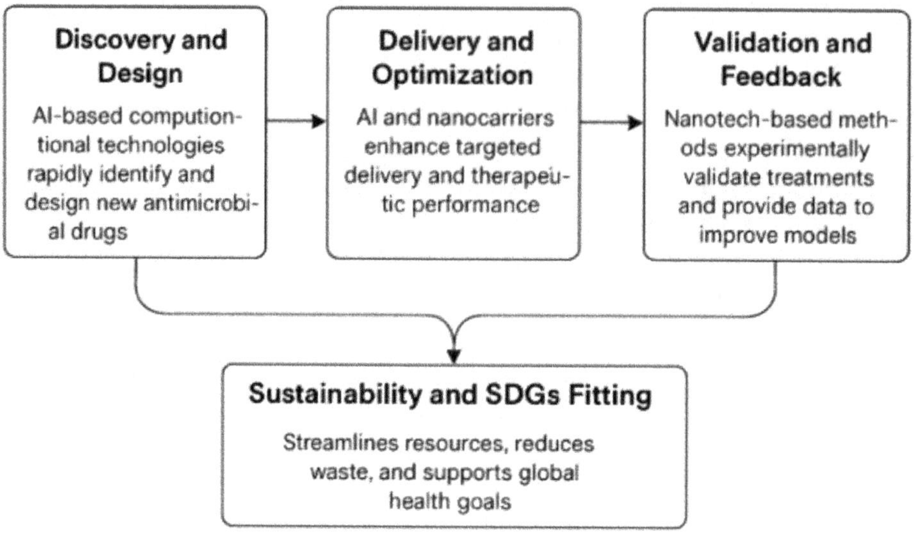

3.1 Pillar 1: Discovery and Design

The pillar one helps to identify and generate antimicrobial drugs in a short time and with the help of AI-based computer simulations and designs. Linear-based machine learning models (trained on chemical structure libraries, databases of microbial resistance and databases of pharmacokinetics) yield predictive rankings of candidate molecules when dependence is made on the predicted antimicrobial activity, toxicity, solubility, and target binding affinity. Additionally, generative models, such as GANs and VAEs, can be utilized to suggest completely new molecular scaffolds potentially containing entirely new mechanisms of action not found in current compound libraries, and in which manner this is one of the most essential gaps in the current antibiotic discovery pipeline.

Virtual screening and molecular docking further screen candidates before experimental synthesis, focusing laboratory resources on molecules that are most likely to be of clinical interest. At this point, genomic and microbiome information is included to assure that the discovered compounds do not cause negative exposure of the microbiome to patient-specific biologic conditions. Notably, proactive prediction of resistance is present in this pillar: AI models are engineered to evolve bacteria in presence of drugs and predict the emergence of molecular phenomena related to resistance, which can be used in pre-emptive design strategies.

3.2 Pillar 2: Delivery and Optimization

The second pillar brings the computationally discovered compounds into clinical implementable therapeutic systems by AI-mediated nanocarrier manufacturing. The existing formulation challenges, such as low aqueous solubility, metabolic and non-specific toxicity, and limited tissue penetration are overcome by rational design of nanocarriers based on AI predictions of the best physicochemical parameters.

The AI models anticipate nanocarrier properties such as particle size, surface charge, ligand density and release kinetics using input-response combinations in a prescribed drug-pathogen interaction profile, replacing the trial-and-error formulation of such models with predictive engineering. Selective accumulation at sites of infection is possible by pathogen-specific functionalization with ligands that recognize specific bacterial surface markers, and can be activated through stimuli-responsive release mechanisms. Nanocarrier-based co-delivery of two or more agents or antimicrobial adjuvants provides a mechanistic approach to surmounting multidrug resistance.

3.3 Pillar 3: Validation and Feedback

The third pillar defines an iterative loop of learning the AI predictions and designs of nanocarriers through the outcomes of the experiments. Microfluidic systems, organ-on-chip models as well as ex vivo models of tissues can offer physiologically significant performance information such as drug release kinetics, cellular uptake, antimicrobial performance, and toxicity information that cannot be predictable solely using in silico models. Such data are generated and fed back systematically into AI model training processes, and dynamic optimization of molecular scaffolds, nanocarrier parameters, and treatment protocols is performed.

This is the closing nature that sets the Three Pillar Framework apart of actual standalone approaches. With a unidirectional feedback signal between experimental reality and computational prediction, the architecture so that the decisions of development are founded on the iteratively validated data as opposed to the constant outputs of the model.

3.4 Comparative Positioning of the Three-Pillar Framework

Table 4 presents a tabular comparison of the Three Pillar Framework with AI-only and nano-only approaches across five key dimensions. This comparative analysis highlights the uniqueness of the integrated framework and clearly demonstrates how it differs from existing approaches.

Table 4. Comparative Analysis: AI-Only, Nano-Only and AI-Nano Integrated Framework.

Dimension	AI-Only Approach	Nano-Only Approach	Integrated AI-Nano Framework (This Chapter)
Discovery Speed	Fast computational screening	Slow: empirical formulation trials	Accelerated: AI pre-screens + nanocarrier optimization in silico
Delivery Precision	None: discovery only	High with functionalization	AI-guided ligand/surface design for pathogen-specific targeting
Resistance Prediction	Moderate: predictive models exist	None inherent	Proactive: AI anticipates resistance; nanocarrier adapts release
Personalization Capacity	Limited to genomic/ microbiome prediction	Limited to formulation adjustments	Full: patient genomics + microbiome inform both molecule selection and delivery
Feedback/Iterative Learning	Model retraining on new data	Empirical reformulation	Closed-loop: microfluidic/in vivo data continuously refines AI and nanocarrier design

The comparative analysis demonstrates that neither AI-only nor nano-only approaches can individually achieve the full range of capabilities required for next-generation antimicrobial therapy. The Discovery speed, delivery precision, resistance forecasting, personalization capacity, and iterative feedback are addressed individually by standalone solutions but are synergistically combined in the Three-Pillar Framework.

4. PROPOSED RESEARCH METHODOLOGY

4.1 Data Collection

Three-Pillar Framework is operationalized using a multi-modal strategy of data collection. The chemical diversity that forms the basis of the AI model training is the chemical compound libraries, both publicly accessible and proprietary antimicrobial databases. The microbial genomic repositories, annotated with both resistance gene profile and virulence factor information, can provide the biological context that can be used to select the resistance-aware compound (Dhaarani & Reddy, 2025; Qadri et al., 2025). The clinical isolate resistance profiles mirror the true landscapes of resistance and make sure that AI-high-priority candidates are applicable to the currently prevalent pathogens. Patient microbiome profilers can be used to design personalized treatment by providing the ability to select antimicrobial treatment that risk minimal risk of dysbiosis in individual patients.

The state of data is considered to be one of the limitations: partial, unbalanced, or shallowly sampled data directly compromises the generalizability of AI models. A solution to this weakness will involve open-access data-sharing systems, federated learning protocols, which maintain patient privacy, and standardized data curation pipelines. This is one of the most urgent gaps in research that have been identified in this chapter.

4.2 Computational Modeling

The computational core of the discovery and design pipeline is composed of machine learning algorithms such as gradient-boosted and graph neural network models, as well as transformer-based models trained with molecular representation of SMILES. Recent generative models, including GANs and VAEs, can search through a predefined space of chemicals not only by taking better advantage of the current library, but also by facilitating chemical synthesis. Virtual screening and molecular docking minimise the cost of resources in experiments by filtering out structurally most attractive candidates using common knowledge of their structure, before laboratory synthesis (Oyejide et al., 2025). Parallel computation of predictive toxicology models is performed to ensure optimization of efficacy, which does not affect the safety profile (Seal et al., 2025).

4.3 Nanocarrier Design

After candidate molecule identification through computational methods, AI-assisted nanocarrier design optimization proceeds to refine the delivery architecture for each compound–pathogen combination. Trained regression models are used to predict the physicochemical parameters of a particle system such as particle size (usually 50-300nm in the case of antimicrobial application), polymer coating, surface charge (zeta potential), and ligand selection. The features of the stimulus-responsive design are introduced depending on the eventualities of the anticipation of infection microenvironment. AI can optimize drug loading efficiency, release kinetics and specificity of the nanocarriers, resulting in a prioritizing list of nanocarrier designs to be fabricated in experiments (Noury et al., 2025).

4.4 Experimental Validation

Experiments at a tiered pipeline are used to validate computationally designed compound-nanocarrier systems. Primary in vitro evaluation of a standard minimum inhibitory concentration (MIC) assays and cytotoxicity screening offer baseline data of efficacy and safety. High-end validation makes use of microfluidic lab-on-chips which are used to in vivo environments of the tissues and drugs release profile, cellular uptake, antimicrobial action is measured under clinical conditions (Nunziata et al., 2025). Ex vivo and in vivo models give both the pharmacokinetic and immunological data to submit regulatory and translate to clinical use.

4.5 Iterative Feedback

These validation data are converted into AI model training systematically and generate the closed-loop learning system that is fundamental to the Third Pillar. This refinement enables the sustained evolution of molecular scaffolds and nanoparticle designs throughout development cycles. As clinical data are growing, it is possible to add resistance surveillance data in order to ensure the continued effectiveness of therapeutic use on the changing populations of pathogens. Regulatory transparency is also supported by the iterative feedback mechanism because an auditable record of model refinement history and experiment validation history is saved.

5. DISCUSSION

5.1 AI-Driven Molecular Discovery: Advances and Critical Limitations

Antimicrobial discovery is proven to have been accelerated with the help of artificial intelligence. The breakthrough with halicin is indicative of the capabilities of machine-learning-based screening: an anti-bacterial compound with a broad-spectrum and, crucially, some bactericidal activity, such as against both the mycobacterium tuberculosis and the carbapenem resistant enterobacteraceae, was found in a screen based on over 100 million molecular chances in just a few days (Tiwari et al., 2023). The capability of the generative models is extended with the creation of molecules whose structural scaffolds are novel and would be inaccessible with standard combinatorial chemistry.

Although these works are technically successful, as can be seen through a critical analysis, there are inherent substantive limitations that have been diminished in the reviews that were done up to now. To start with, AI model performance is dependent on the training data variety and quality. Published antimicrobial datasets heavily focus on Gram-positive pathogens, leaving a lot of gaps in AI effectiveness between Gram-negative and mycobacterial infections. Second, deep-learning architecture based on black-box architectures makes it harder to discover the relationship between structure and activity, sacrificing scientific interpretability and posing regulatory challenges. Third, the computational prediction-biological efficacy gap is still very wide; numerous AI-predicted compounds fail in preclinical experimental validation because of unexpected solubility, stability, or off-target action. To mitigate these limitations, the efforts will be necessary, not only to obtain bigger and more eclectic datasets but also to develop methodologies explainable AI and hybrid computational experimental validation procedures.

5.2 Nanocarrier-Based Delivery: Therapeutic Potential and Implementation Barriers

Nanocarrier systems eliminate important pharmacological drawbacks of newly identified antimicrobial agents namely; low aqueous solubility, high metabolic clearance and low biofilm penetrativity, otherwise rendering them inapplicable to clinical use. Pre-clinical models have shown functionalized nanoparticles are specifically targeted to sites of infection; stimulus-responsive systems have demonstrated significant opportunities for precision engineered drug delivery within the infected tissue microenvironment (Mashele, 2025; Xu et al., 2025). Multidrug-resistant biofilms have been reported to be synergistically effective in response to co-delivery of more than two antimicrobials in a single nanocarrier system (Ahsan et al., 2024).

Regardless of these developments, there are still a number of implementation barriers that are not addressed. Full-scale and reproducible nanoparticles production still poses considerable challenges to engineering: suspension-to-suspension size variations, surface chemistry, and ability to encapsulate continue to be problematic to well-established carrier systems such as PLGA nanoparticles and liposomes. The in vivo fate of new nanocarriers such as biodistribution, immune recognition, renal or hepatic retention and long-term biocompatibility should be systematically described, which is often lacking in the literature. In addition, another research gap that requires further research is the environmental ecotoxicology of nanomaterials released through pharmaceutical manufacturing or excreted by treated individuals, which has the potential to impact public health.

5.3 Microbiome Integration in Precision Antimicrobial Design

Human gut microbiome has far-reaching influence on the metabolism of antimicrobial drugs, treatment efficacy and resistance development. Alteration in microbiome habitats, termed in general as dysbiosis, following broad-spectrum antibiotic treatment is now understood to be a contributing factor to secondary infections such as colitis by Clostridioides difficile and systemic immune dysregulation (Ho et al., 2020). Accurate antimicrobial design using microbiome data has the potential to identify agents and delivery systems that are able to wipe out the target pathogens and maintain the health-promoting microbial communities.

An area of capability that is insufficiently explored in the literature is AI models that have the ability to combine microbiome sequencing data with pathogen genomics and host clinical data. Further functionalization of nanocarriers with surface biomarkers (e.g., exploiting gut microbiota transport systems) may enable site-specific antimicrobial delivery without causing collateral disruption of the microbiome. Nevertheless, the mathematical models of microbiome-conscious antimicrobial design are still at infancy and clinical base of microbiome-based prescribing is not well-developed. This is a priority gap in the research of the field.

5.4 Personalized and Precision Antimicrobial Therapy

The convergence of AI and nanotechnology enables a transition from population-level empirical prescribing to patient-specific precision antimicrobial therapy. Patient-level pathogen genomic sequencing can identify resistance genes, virulence factors, and mutation patterns relevant to antimicrobial susceptibility within hours of clinical presentation. AI models that integrate this genomic data with patient microbiome profiles, immune status, and pharmacogenomics information can generate treatment recommendations tailored to the biological characteristics of both the pathogen and the host (Serrano et al., 2024). Nanocarrier systems can then be configured to deliver selected agents with optimized pharmacokinetics for individual patient characteristics.

Personalized antimicrobial therapy on a large scale would entail a set of infrastructure requirements, such as the rapid signature sequencing, clinical decision support software, and specific pharmaceutical production, which are, as of now, not very accessible to clinical sectors with high incomes. Fair access to antimicrobial precision technologies should hence be tackled not just as a side-show feature of the entire design but as its fundamental attribute.

5.5 Critical Synthesis: Unresolved Gaps, Real-World Limitations, and Sustainability

Although most of the advancements are promising as evident in the Sections 5.1-5.4, a critical evaluation portrays that the discipline is hampered with numerous debilitating limitations that are interrelated and insufficiently addressed. To start with, the detection rate of translation transformation is very low: most of the AI-designed antimicrobial drugs fail preclinical validation, mostly because of the mismatch between in silico models and in vitro or in vivo biological reality. This sparsity is worsened by the use of the training data that excessively focuses on

Gram-positive organisms and commercially relevant classes of compounds, creating a poor generalizability of the developed models on new and unstudied pathogens and chemical scaffolds. Second, most AI-designed carrier architectures have not been solved in terms of scalability of nanocarrier systems between laboratory synthesis and pharmaceutical-grade manufacturing. Discrepancies between batches of size, charge at surface and encapsulation efficiency still inhibit translational reliability, especially to biomimetic and stimuli-responsive carriers.

Third, the level of regulatory environment has not been oriented to technological innovation. None of the significant jurisdictions has developed a clearly defined route of approval of combination products that include AI-generated molecules that are carried by engineered nanocarriers. This regulatory blank creates major delays and uncertainty to clinical translation timelines. Fourth, the AMR innovation literature does not consider the sustainability dimensions systematically. The energy usage of the massive training of AI models, and the ecotoxicological imprint of the discharge of nano-materials to the water and soil sources are real and amenable risks that are rarely documented in primary literature or been put into lifecycle consideration. Last but not least, global equity is a structural disparity: the systems needed to carry out accurate antimicrobial therapy (genomic sequencing, bioinformatics pipelines, and personalized production of nanocarriers) are located in the high-income countries, and there is no evident system of equal transfer of technology. It is important to note that it is crucial to discuss all these limitations together, not separately, and in order to promote the responsible and effective development of AI–nanotechnology antimicrobial solutions.

6. UNRESOLVED CHALLENGES, RESEARCH GAPS, AND FUTURE DIRECTIONS

6.1 Critical Research Gaps

Despite significant progress, the application of AI in nanotechnology to create antimicrobials has multiple substantive gaps in the field that have not been well identified in preceding reviews. Table 5 aligns these gaps with status quo and proposed future course of action.

Table 5. Outstanding Issues, Research Needs, and Future Research Opportunities in AI-Nanotechnology Antimicrobial Development.

Challenge Area	Current Status	Gap Identified	Recommended Future Direction
Dataset Quality for AI Training	Incomplete, biased toward known compounds	Limited open-access AMR-specific datasets	Open-data repositories; federated learning frameworks
Nanomaterial Scalability	Lab-scale synthesis inconsistency	No standardized GMP protocols for AI-designed nanocarriers	Microfluidic manufacturing; AI-assisted quality control
Regulatory Frameworks	Designed for small-molecule drugs	No adaptive pathway for AI-generated nanotherapeutics	Sandbox regulatory models; harmonized international guidelines
Environmental Impact of Nanomaterials	Largely unassessed lifecycle data	Absence of ecotoxicity standards for novel nanocarriers	Mandatory lifecycle assessment; green synthesis protocols
Clinical Translation Barrier	Computational predictions rarely reach Phase I trials	No end-to-end discovery-to-clinic integration pipeline exists	Collaborative industry-academia pipelines; adaptive trial designs

6.2 Real-World Implementation Limitations

When laboratory efficacy has been determined, there are a number of systemic factors limiting the real world use of AI-nanotechnology-based antimicrobial-based therapies. In none of the key jurisdictions is the regulatory approval pathway for AI-generated nanotherapeutics clearly defined. Current frameworks designed to assess small-molecule drugs do not provide a good fit to assess the combination products, where AI-designed molecules are capped with optimized nanocarrier systems. A lack of responsive regulatory processes introduces significant delays in the lab-to-patient distance.

Significant health system infrastructure requirements also present implementation barriers. Clinical decision support integration, rapid-turnaround sequencing, and bioinformatics pipelines are prerequisites for pathogen-genomics-based personalized antimicrobial therapy, yet such infrastructure is unavailable in most low- and middle-income country healthcare systems. AI-nanotechnology antimicrobial solutions may further fuel the current inequities in health inequalities around the globe without targeted equity policy.

Scalability in manufacturing is a third implementation challenge. Companies that need large-scale manufacturing of functionalized nanocarriers to the pharmaceutical level of quality have to use advanced technologies of manufacturing, such as microfluidic fabrication and real-time quality assurance, which are not accessible to the majority of companies yet. The affordability of personalized nanocarrier formulations can also be limited by the cost structures of the formulations in the resource-constrained environment.

6.3 Sustainability Considerations

The aspects of sustainability by addressing the overlap of artificial intelligence and nanotechnology in the development of antimicrobials are underexplored in existing materials. Computation with AI is often generally energy intensive: big screen molecular screening processes and generative model training all have a significant carbon footprint. Such emissions have not been yet tabulated or countered in published antimicrobial discovery studies. Therefore, it is suggested that the green chemistry principles and energy-efficient computing architectures can be incorporated into AI-driven discovery pipelines.

Environmental impact of engineered nanomaterials life-cycle is a unique sustainability issue. The nanoparticles discharged into the environment through pharmaceutical effluents or released by treated patients can still remain within the aquatic environment and have ecotoxical characteristics that are yet to be fully described. Development protocols need to include the mandatory life-cycle assessment and ecotoxicological profiling of novel nanocarriers, which will correspond to the principles of responsible innovation.

Both SDG 3 (Good Health and Well-Being) and SDG 9 (Industry, Innovation and Infrastructure) are also relevant to the model discussed in this paper. The realization of SDG 3 involves achieving not only the image of the effective design and production of antimicrobials but also ensuring the fair access of the world to them. SDG 9 lays stress on sustainable industrialization and infrastructure on innovation. The Three-Pillar Framework directly agrees with both of these objectives as it promotes open-data collaborative paradigms, green synthesis solutions, and fair implementation plans.

7. ETHICAL, REGULATORY, AND GOVERNANCE CONSIDERATIONS

Using patient genomic and microbiome data to discover drugs with the help of AI raises serious ethical concerns regarding their data management, data informed consent, and data privacy concerns. AI models that are based on clinical data should reveal that training data is also representative, the process of model curation is transparent, and model results can be understood and read by clinicians. Anonymous algorithmic decision-making in antimicrobial prescription settings, where wrong treatment could escalate the process of resistance formation, both carries an ethical and practice risk. Regulatory features should be viewed as explainable AI architecture and model audit protocols and not optional features (Qadri et al., 2025).

The regulatory environment needs to be modified in order to embrace AI-generated nanotherapeutics. Existing systems aimed at small-molecule drugs do not support the form of the combination products including AI-discovered molecular drug in the family with engineered nanoparticle delivery systems. To fill this gap, regulatory sandboxes, flexible approval mechanisms, and consensus guidance committees using researchers, regulating bodies and industry collaborators are needed. Global standardization of AI validation, nanoparticle characterization, and clinical translation standards would be of great value in cutting down the approval times without losing patient safety.

Different forms of equitable access to AI-nanotechnology antimicrobial therapies should be preemptively handled. The tethering of AI infrastructure, genomic sequencing capacity, and pharmaceutical manufacturing expertise to high-income nations creates structural barriers to equitable global access. Essential elements of an ethically-based innovation ecosystem are open-access archives of data, technology transfer arrangements, and capacity-building alliances.

8. FUTURE RESEARCH DIRECTIONS

The critical analysis of the chapter results in the following priority research directions. Both of them cover a certain gap or limitation that were recognized during the literature review, comparative analysis, or challenge mapping.

The future of precision antimicrobial discovery will be state-of-the-art personalized AI systems combining real-time clinical measurements, microbiome genome sequencing of patients and lab-on-chip experimental results. These models have to be able to dynamically respond to changing profiles of pathogen resistance and to the biological particularities of individual patients. These models need methodological innovation in multi-modal AI architectures and heavy investment in standardized, interoperable clinical data infrastructure (Yakobi & Nwodo, 2025).

The next-generation nanomaterials, especially biodegradable, biomimetic and stimuli-responsive designs should be tested not only regarding their therapeutic performance but also regarding their environmental safety in their entire lifecycle. The ecotoxicological profiling protocols and green synthesis methods must become part and parcel of the nanocarrier developmental pipelines. Designing environmentally benign nanocarriers, which maintain therapeutic efficacy using AI-assistance, is a research objective that can be approached with sufficient tractability and has direct sustainability impacts.

To achieve the translational potential of the Three-Pillar Framework end-to-end discovery-to-delivery pipelines linking AI molecular design, nanocarrier formulation and clinical validation would be necessary. By investing in microfluidic manufacturing platforms that have the capability to scale to clinical batches of nanocarrier production with automated quality monitoring systems connected to AI, the implementation gap between laboratory innovation and patient care will significantly decrease.

The creation of data variety and validation rigor to drive AI-nanotechnology antimicrobial cures will depend on international collaborative infrastructure, such as open-access AMR genomic databases, shared AI model libraries, such as coordinated cross-institutional research networks. Innovation in regulation should proceed similarly: Standardized AI-assisted drug discovery validation pathways, adaptive approval to AI-assisted drug discovery, and universal guidelines on nanotherapeutics characterization are all priorities of regulatory authorities.

9. CONCLUSION

Antimicrobial resistance has become existential to present-day medicine, and even the standard drug discovery models are on a structural footing not appropriately positioned to handle it. This chapter has provided a critical and integrative discussion of how, through operational synergy, artificial intelligence and nanotechnology under the Three-Pillar AI-Nanotechnology Synergy Framework, may provide a plausible and scientifically supported road-map to next-generation antimicrobial treatment.

The principal contribution of the framework is its integration of predictive molecular discovery, AI-guided nanocarrier delivery, and closed-loop experimental feedback within a single operational model. Comparative analysis demonstrates that this combined methodology addresses the limitations of both AI-only and nano-only approaches, enabling faster discovery, more precise delivery, improved resistance management, and more comprehensive personalized treatment than any standalone approach allows.

An orderly mapping of unresolved research gaps, such as the quality of data limitation, nanomanufacturing scalability challenges, regulatory preparedness gaps, sustainability pitfalls that need to be filled in order to move laboratory innovations into clinical delivery have also been established in the chapter. These items outline an actual research agenda of the field.

The realization of the antimicrobial development involving AI-nanotechnology will need not only scientific innovation, but also interdisciplinary work, ethical governance, and fair implementation design, as well as adaptive regulation. These technologies are now mature enough to play significant roles in the AMR crisis in the near future; the potential is possible to actualize upon the choices undertaken presently on data infrastructure, regulatory directions, and worldwide access. This chapter supplies the conceptual framework of such initiatives as well as the critical analysis that needs to be used so that the result of the choices is, in a way, clinically effective, environmentally responsible, and globally equitable.

REFERENCES

Ahsan, A., Thomas, N., Barnes, T. J., Subramaniam, S., Loh, T. C., Joyce, P., & Prestidge, C. A. (2024). Lipid nanocarriers-enabled delivery of antibiotics and antimicrobial adjuvants to overcome bacterial biofilms. *Pharmaceutics, 16*(3), 396. DOI: 10.3390/pharmaceutics16030396 PMID: 38543290

Bian, Y., & Xie, X. Q. (2021). Generative chemistry: Drug discovery with deep learning generative models. *Journal of Molecular Modeling, 27*(3), 71. DOI: 10.1007/s00894-021-04674-8 PMID: 33543405

Brown, D. G., Wobst, H. J., Kapoor, A., Kenna, L. A., & Southall, N. (2022). Clinical development times for innovative drugs. *Nature Reviews. Drug Discovery, 21*(11), 793–794. DOI: 10.1038/d41573-021-00190-9 PMID: 34759309

Chhabra, S., Taksande, A. B., & Munjewar, P. (2024). The penicillin pioneer: Alexander Fleming's journey to a medical breakthrough. *Cureus, 16*(7). Advance online publication. DOI: 10.7759/cureus.65179 PMID: 39176366

Cook, M. A., & Wright, G. D. (2022). The past, present, and future of antibiotics. *Science Translational Medicine, 14*(657), eabo7793. Advance online publication. DOI: 10.1126/scitranslmed.abo7793 PMID: 35947678

Das, K. P., & J, C. (2023). Nanoparticles and convergence of artificial intelligence for targeted drug delivery for cancer therapy: Current progress and challenges. *Frontiers in Medical Technology, 4*, 1067144. DOI: 10.3389/fmedt.2022.1067144 PMID: 36688144

Das, U. (2025). Generative AI for drug discovery and protein design: The next frontier in AI-driven molecular science. *Medicine in Drug Discovery, 100213*, 100213. Advance online publication. DOI: 10.1016/j.medidd.2025.100213

Dhaarani, R., & Reddy, M. K. (2025). Progressing microbial genomics: Artificial intelligence and deep learning driven advances in genome analysis and therapeutics. *Intelligence-Based Medicine, 11*, 100251. DOI: 10.1016/j.ibmed.2025.100251

Din, F. U., Aman, W., Ullah, I., Qureshi, O. S., Mustapha, O., Shafique, S., & Zeb, A. (2017). Effective use of nanocarriers as drug delivery systems for the treatment of selected tumors. *International Journal of Nanomedicine, 12*, 7291–7309. DOI: 10.2147/IJN.S146315 PMID: 29042776

Endale, H., Mathewos, M., & Abdeta, D. (2023). Potential causes of spread of antimicrobial resistance and preventive measures in one health perspective—A review. *Infection and Drug Resistance*, *16*, 7515–7545. DOI: 10.2147/IDR.S428837 PMID: 38089962

•Fong, I. W. (2023). Antimicrobial resistance: a crisis in the making. *New antimicrobials: for the present and the future*, 1-21.

Ho, D., Quake, S. R., McCabe, E. R., Chng, W. J., Chow, E. K., Ding, X., Gelb, B. D., Ginsburg, G. S., Hassenstab, J., Ho, C.-M., Mobley, W. C., Nolan, G. P., Rosen, S. T., Tan, P., Yen, Y., & Zarrinpar, A. (2020). Enabling technologies for personalized and precision medicine. *Trends in Biotechnology*, *38*(5), 497–518. DOI: 10.1016/j.tibtech.2019.12.021 PMID: 31980301

Issa, K. D., & Muhsin, M. D. (2023). Beta–Lactam drug resistance pattern in Staphylococcus aureus isolates: A review. *Eurasian Journal of Science and Engineering*, *9*(3), 141–153. DOI: 10.23918/eajse.v9i3p13

Macedo, B., Ribeiro Vaz, I., & Taveira Gomes, T. (2024). MedGAN: Optimized generative adversarial network with graph convolutional networks for novel molecule design. *Scientific Reports*, *14*(1), 1212. DOI: 10.1038/s41598-023-50834-6 PMID: 38216614

Majumder, J., & Minko, T. (2021). Multifunctional and stimuli-responsive nanocarriers for targeted therapeutic delivery. *Expert Opinion on Drug Delivery*, *18*(2), 205–227. DOI: 10.1080/17425247.2021.1828339 PMID: 32969740

Mashele, S. S. (2025). Stimuli-responsive, cell-mediated drug delivery systems: Engineering smart cellular vehicles for precision therapeutics. *Pharmaceutics*, *17*(8), 1082. DOI: 10.3390/pharmaceutics17081082 PMID: 40871101

Muteeb, G., Rehman, M. T., Shahwan, M., & Aatif, M. (2023). Origin of antibiotics and antibiotic resistance, and their impacts on drug development: A narrative review. *Pharmaceuticals (Basel, Switzerland)*, *16*(11), 1615. DOI: 10.3390/ph16111615 PMID: 38004480

Nayeem, A., Suresh, A. S., Vellapandian, C., Singh, S., Elossaily, G. M., & Prajapati, B. G. (2024). Comprehensive insights into cephalosporins: Spectrum, generations, and clinical applications. *Current Drug Therapy*. Advance online publication. DOI: 10.2174/0115748855339119241204071146

Noury, H., Rahdar, A., Ferreira, L. F. R., & Jamalpoor, Z. (2025). AI-driven innovations in smart multifunctional nanocarriers for drug and gene delivery: A mini-review. *Critical Reviews in Oncology/Hematology*, *210*, 104701. DOI: 10.1016/j.critrevonc.2025.104701 PMID: 40086770

Nunziata, G., Borroni, A., & Rossi, F. (2025). Advanced microfluidic strategies for core-shell nanoparticles: The next-generation of polymeric and lipid-based drug nanocarriers. *Chemical Engineering Journal Advances*, *22*, 100759. DOI: 10.1016/j.ceja.2025.100759

Otaigbe, I. I. (2025). Mitigating inequitable access to appropriate antibiotics in low-and middle-income countries. *JAC-Antimicrobial Resistance*, *7*(2), dlaf061. Advance online publication. DOI: 10.1093/jacamr/dlaf061 PMID: 40276025

Oyejide, A. J., Adekunle, Y. A., Abodunrin, O. D., & Atoyebi, E. O. (2025). Artificial intelligence, computational tools and robotics for drug discovery, development, and delivery. *Intelligent Pharmacy*, *3*(3), 207–224. DOI: 10.1016/j.ipha.2025.01.001

Qadri, Y. A., Shaikh, S., Ahmad, K., Choi, I., Kim, S. W., & Vasilakos, A. V. (2025). Explainable artificial intelligence: A perspective on drug discovery. *Pharmaceutics*, *17*(9), 1119. DOI: 10.3390/pharmaceutics17091119 PMID: 41012456

Qin, S., Xiao, W., Zhou, C., Pu, Q., Deng, X., Lan, L., Liang, H., Song, X., & Wu, M. (2022). Pseudomonas aeruginosa: Pathogenesis, virulence factors, antibiotic resistance, interaction with host, technology advances and emerging therapeutics. *Signal Transduction and Targeted Therapy*, *7*(1), 199. DOI: 10.1038/s41392-022-01056-1 PMID: 35752612

Rahman, M. U., Shah, J. A., Khan, M. N., Bilal, H., Zhu, D., Du, Z., & Mu, D. S. (2025). Innovative approaches to combat antimicrobial resistance: A review of emerging therapies and technologies. *Probiotics and Antimicrobial Proteins*, •••, 1–18. DOI: 10.1007/s12602-025-10676-2 PMID: 41026406

Ramos, T. I., Villacis-Aguirre, C. A., López-Aguilar, K. V., Santiago Padilla, L., Altamirano, C., Toledo, J. R., & Santiago Vispo, N. (2022). The Hitchhiker's guide to human therapeutic nanoparticle development. *Pharmaceutics*, *14*(2), 247. DOI: 10.3390/pharmaceutics14020247 PMID: 35213980

•Salam, M. A., Al-Amin, M. Y., Salam, M. T., Pawar, J. S., Akhter, N., Rabaan, A. A., & Alqumber, M. A. (2023, July). Antimicrobial resistance: a growing serious threat for global public health. In *Healthcare* (Vol. 11, No. 13, p. 1946). MDPI. https://doi.org/DOI: 10.3390/healthcare11131946

Seal, S., Mahale, M., García-Ortegón, M., Joshi, C. K., Hosseini-Gerami, L., Beatson, A., Greenig, M., Shekhar, M., Patra, A., Weis, C., Mehrjou, A., Badré, A., Paisley, B., Lowe, R., Singh, S., Shah, F., Johannesson, B., Williams, D., Rouquie, D., & Bender, A. (2025). Machine learning for toxicity prediction using chemical structures: Pillars for success in the real world. *Chemical Research in Toxicology, 38*(5), 759–807. DOI: 10.1021/acs.chemrestox.5c00033 PMID: 40314361

Serrano, D. R., Luciano, F. C., Anaya, B. J., Ongoren, B., Kara, A., Molina, G., Ramirez, B. I., Sánchez-Guirales, S. A., Simon, J. A., Tomietto, G., Rapti, C., Ruiz, H. K., Rawat, S., Kumar, D., & Lalatsa, A. (2024). Artificial intelligence (AI) applications in drug discovery and drug delivery: Revolutionizing personalized medicine. *Pharmaceutics, 16*(10), 1328. DOI: 10.3390/pharmaceutics16101328 PMID: 39458657

Sultana, A., Zare, M., Thomas, V., Kumar, T. S., & Ramakrishna, S. (2022). Nano-based drug delivery systems: Conventional drug delivery routes, recent developments and future prospects. *Medicine in Drug Discovery, 15*, 100134. DOI: 10.1016/j.medidd.2022.100134

Tiwari, P. C., Pal, R., Chaudhary, M. J., & Nath, R. (2023). Artificial intelligence revolutionizing drug development: Exploring opportunities and challenges. *Drug Development Research, 84*(8), 1652–1663. DOI: 10.1002/ddr.22115 PMID: 37712494

Xu, Z., Xie, Y., Chen, W., & Deng, W. (2025). Nanocarrier-based systems for targeted delivery: Current challenges and future directions. *MedComm, 6*(9), e70337. DOI: 10.1002/mco2.70337 PMID: 40859956

Yakobi, S. H., & Nwodo, U. U. (2025). *AI-driven modelling, antimicrobial discovery, and precision therapeutics for targeting bacterial persisters. In Silico Research in Biomedicine, 100062.*, DOI: 10.1016/j.insi.2025.100062

Yang, L., Pijuan-Galito, S., Rho, H. S., Vasilevich, A. S., Eren, A. D., Ge, L., Habibović, P., Alexander, M. R., de Boer, J., Carlier, A., van Rijn, P., & Zhou, Q. (2021). High-throughput methods in the discovery and study of biomaterials and materiobiology. *Chemical Reviews, 121*(8), 4561–4677. DOI: 10.1021/acs.chemrev.0c00752 PMID: 33705116

Chapter 3
Harnessing Microbiome and Genomics for Precision Antibiotics:
AI and Nanobiotechnology Innovations

Tarun Kumar Patel
http://orcid.org/0009-0003-3002-8100
Department of Biotechnology, Sant Guru Ghasidas Government P.G. College, Kurud, India

ABSTRACT

The chapter examines the AMR crisis, highlighting its global mortality and economic impacts, and proposes integrative solutions combining microbiome genomics, AI, and nanobiotechnology for precision antibiotics. The gut microbiome is identified as a key reservoir for ARGs, with broad-spectrum antibiotic-induced dysbiosis accelerating resistance via horizontal gene transfer. The purpose is to advocate sustainable, equitable strategies shifting to targeted therapies that minimize microbiome disruption. Content covers genomic personalization using multi-omics and pathogen profiling for microbiome-sparing interventions, including CRISPR editing; AI for resistance prediction, regimen optimization, and clinical de-escalation; and nanobiotechnology via stimuli-responsive nanocarriers for site-specific delivery and ARG silencing. Global collaboration and ethical data-sharing are emphasized to ensure accessibility, concluding that these advancements can reduce AMR burdens and sustain antibiotic efficacy through predictive, personalized stewardship.

DOI: 10.4018/979-8-3373-5268-8.ch003

INTRODUCTION

The antimicrobial resistance (AMR) crisis threatens modern medicine by rendering once-treatable infections deadly. In 2021, resistant bacteria directly caused 1.27 million global deaths and contributed to 4.95 million more, disproportionately burdening sub-Saharan Africa and South Asia (GBD 2021 Antimicrobial Resistance Collaborators, 2024). Without intervention, the WHO projects 10 million annual deaths by 2050 surpassing cancer and economic impacts comparable to the 2008 financial crisis (WHO, 2023). Overuse and misuse of antibiotics, with inappropriate prescriptions in some regions, fuel selective pressure for resistant strains (Laxminarayan et al., 2013; WHO, 2022).

The gut microbiome plays a central role as a reservoir for antibiotic resistance genes (ARGs). Broad-spectrum antibiotics induce dysbiosis loss of diversity, depletion of beneficial microbes, and overgrowth of opportunists creating niches for ARG transfer via conjugation, transformation, and transduction (Shen et al., 2025; Tan et al., 2025). Mobile elements like plasmids and integrons enable exchange across human, animal, and environmental bacteria, forming a One Health network (Dongre et al., 2025). Restoring homeostasis through probiotics, prebiotics, or fecal transplantation could mitigate ARG persistence.

Broad-spectrum antibiotics' indiscriminate action accelerates resistance and ecological harm, prompting a shift to precision antibiotics that target pathogens selectively while sparing the microbiome (Johnston & Badran, 2022). This strategy uses microbiome-genomics interplay for patient-specific stratification, reducing off-target effects (Diamantis et al., 2022). Precision therapeutics encompasses a range of innovative options designed to target pathogens selectively while minimizing disruption to the microbiome. These include repurposed narrow-spectrum agents derived from existing antibiotics, novel microbiome-sparing compounds that avoid broad ecological damage, and engineered biologics that exploit pathogen-specific virulence factors or metabolic vulnerabilities for instance, antivirulence inhibitors such as FimH antagonists in uropathogenic *E. coli*, which disrupt bacterial adhesion and prevent infection progression without exerting bactericidal pressure that could drive resistance.

Diagnostics like 16S rRNA sequencing for microbiome profiling and whole-genome sequencing for pathogen resistance guide tailored prescribing, avoiding depleting agents in vulnerable patients. Economically, this aligns incentives; curbs overuse, and extends drug lifespans (Bokhari et al., 2024). Artificial intelligence (AI) and nanobiotechnology drive transformation. Artificial intelligence integrates genomic, clinical, and surveillance data to predict resistance, personalize regimens, and identify transmission networks via machine learning (Branda & Scarpa, 2024; Bonazzetti et al., 2025; Sagar et al., 2025). Nanobiotechnology enables targeted delivery with stimuli-responsive nanoparticles (pH-, enzyme-, or quorum-triggered) for localized release, minimizing systemic exposure and dysbiosis (Patel 2025; Nazir et al., 2022). Advanced formulations (lipid-polymer hybrids, metal-organic frameworks) enhance antibiotic stability and penetration, while inorganic nanoantibiotics (e.g., silver nanoparticles) overcome resistance via multimodal mechanisms (Pisani et al., 2024; AlQurashi et al., 2025). Their synergy Artificial intelligence-guided prescribing with nano-delivery reduces selective pressure, preserves ecology, and prolongs antibiotic utility (Huang et al., 2024; Mondal et al., 2024).

Precision antibiotics, in this context, refer to tailored antimicrobial therapies that leverage patient-specific microbiome profiles, pathogen genomic sequences, and advanced delivery systems to selectively target resistant strains while preserving commensal microbial ecosystems. This approach minimizes dysbiosis and ARG dissemination by integrating multi-omics data (e.g., 16S rRNA and whole-genome sequencing) with targeted nanocarriers, enabling therapies like CRISPR-edited antivirulence agents or artificial intelligence-optimized narrow-spectrum dosing (Johnston & Badran, 2022; Diamantis et al., 2022). By focusing on pathogen vulnerabilities rather than broad-spectrum killing, precision antibiotics reduce selective pressure and support sustainable stewardship.

The chapter overviews integrating microbiome-genomics with artificial intelligence and nanobiotechnology for precision antibiotics, emphasizing sustainable global strategies: Artificial intelligence-driven discovery, smart nanocarriers, green nanotechnology, life-cycle assessments, and LMIC deployments via portable sequencing and artificial intelligence tools for tuberculosis. It bridges foundational AMR-microbiome insights to genomic personalization for targeted interventions.

MICROBIOME AND GENOMICS IN PRECISION ANTIBIOTICS

The human microbiome spanning gut, skin, and respiratory niches acts as a key reservoir and dissemination hub for antibiotic resistance genes (ARGs), fueled by dysbiosis and horizontal gene transfer between commensal and pathogenic bacteria (Dongre et al., 2025; Paul & Das, 2022). The gut microbiota, rich in density and mobile genetic elements like plasmids and integrons, drives ARG exchange, amplifying resistance in priority pathogens such as *Clostridioides difficile* and multidrug-resistant Enterobacteriaceae (MDRE), including carbapenem-resistant *Klebsiella pneumoniae* (Shen et al., 2025; Tan et al., 2025). *C. difficile* exploits antibiotic-induced dysbiosis for overgrowth, acquiring resistance via mutations and transposons in hospital outbreaks (Wickramage et al., 2021; Smits et al., 2025). MDRE persist post-treatment, often co-colonizing with *C. difficile* (Jain et al., 2020). Similar ARG harboring occurs in skin and respiratory microbiomes, exacerbated in ICU settings, while early infant colonization influenced by diet and antibiotics establishes lifelong resistance risk (Sakagianni et al., 2025; Theophilus & Taft, 2023; Ma et al., 2024).

The synergy between microbiome research and genomics lies in their complementary roles: microbiome profiling reveals community-level dysbiosis and ARG reservoirs, while genomic sequencing decodes pathogen-specific mechanisms, together enabling precision interventions. For instance, shotgun metagenomics identifies ARG abundance in gut niches (e.g., blaTEM in Enterobacteriaceae post-dysbiosis), which genomic analysis then maps to virulence factors like efflux pumps in *C. difficile* (Dongre et al., 2025; Shen et al., 2025). This integration informs targeted strategies, such as CRISPR-Cas9 editing of transposon-borne ARGs or pharmacogenomic-adjusted dosing to avoid microbiome-sparing failures in CYP2D6 poor metabolizers (Hossam Abdelmonem et al., 2024; Srivastav et al., 2025). In clinical practice, multi-omics pipelines (e.g., QIIME2 with transcriptomic overlays) predict treatment outcomes, as demonstrated in ICU cohorts where microbiome shifts correlated with genomic resistance evolution, guiding de-escalation and reducing recurrence (Ghosh et al., 2025). This linkage not only enhances efficacy but also addresses One Health dynamics, curbing ARG transfer across reservoirs.

Genomic personalization optimizes antibiotic efficacy by leveraging host-pathogen interactions. Host pharmacogenomics, especially cytochrome P450 variants (e.g., CYP2C9, CYP2D6), affects drug metabolism, plasma levels, and toxicity; poor metabolizers experience prolonged exposure, increasing dysbiosis and ARG selection (Hossam Abdelmonem et al., 2024). Integrating these variants into dosing algorithms enables genetically tailored regimens (Srivastav et al., 2025). Pathogen genomics identifies virulence factors adhesins, toxins, and efflux pumps targetable by CRISPR-Cas systems, which edit genes to render pathogens avirulent and susceptible, such as disrupting *C. difficile* toxins or MDRE carbapenemases (Echavarria Galindo & Lai, 2025). Artificial intelligence-enhanced CRISPR design, informed by whole-genome sequencing, minimizes off-target effects (Srivastav et al., 2025). Organ-on-chip models with patient-derived iPSCs simulate individualized responses in vitro (Palasantzas et al., 2023; Ingber, 2022). Portable platforms extend these approaches to resource-limited settings (Bah et al., 2018).

Multi-omics integration combining metagenomics, transcriptomics, and metabolomics builds comprehensive patient-specific AMR profiles for precise antibiotic selection (Babu & Snyder, 2023; Ghosh et al., 2025). Shotgun metagenomics detects taxa and ARGs across niches, while 16S rRNA sequencing maps community structure (Dongre et al., 2025). Transcriptomics uncovers upregulated resistance pathways (e.g., efflux pumps, beta-lactamases), and metabolomics identifies modulators like short-chain fatty acids influencing ARG transfer (Ma et al., 2024; Molla & Bitew, 2024). Standardized pipelines with tools like QIIME2 address batch effects (Zhou et al., 2023). Artificial intelligence, via graph neural networks, integrates layers to predict phenotypes, as seen in *C. difficile* outbreak analyses linking gene expression to treatment failure (Ghosh et al., 2025; Yetgin, 2025; Kan et al., 2025). This enables microbiome-sparing and CRISPR-guided therapies, shifting from reactive to predictive stewardship (Srivastav et al., 2025).

Despite promise, implementation faces hurdles: data heterogeneity from varying platforms and samples impedes reproducibility, especially in low- and middle-income countries (LMICs). Ethical issues include privacy risks from revealing susceptibilities, demanding strong consent and de-identification. Solutions involve federated learning, blockchain databases, and artificial intelligence harmonization to bridge silos, though gaps in infrastructure and expertise persist.

ARTIFICIAL INTELLIGENCE-DRIVEN DISCOVERY AND RATIONAL DESIGN

Machine Learning for Microbiome-Genomics Analysis

Machine learning (ML) revolutionizes microbiome-genomics analysis for precision antibiotics by extracting predictive insights from complex omics datasets. Supervised models, such as random forests and block forests, integrate metagenomic ARG profiles, host pharmacogenomics (e.g., CYP450 variants), and clinical metadata to forecast resistance patterns with promising accuracy, outperforming traditional susceptibility testing (Marcos-Zambrano et al., 2021; Hornung & Wright, 2019). These algorithms identify biomarker panels specific taxa, genes, or metabolites linked to *C. difficile* recurrence or MDRE colonization, guiding targeted interventions (Manandhar et al., 2021; Ghosh et al., 2025). Unsupervised clustering techniques, including multi-omics integration frameworks, delineate microbiome subtypes (enterotypes) and dysbiotic states, revealing patient strata with differential AMR risk (Vahabi & Michailidis, 2022; Muller et al., 2024). Multi-view learning harmonizes heterogeneous data layers, uncovering disease-associated microbial modules (Li et al., 2022). In real-world applications, ML predicts treatment responses in ICU settings and post-antibiotic recovery trajectories (Cao et al., 2025; Sakagianni et al., 2025).

Case Study: Artificial Intelligence Analysis of Microbiome Shifts Post-Antibiotic Exposure to Guide De-Escalation Protocols

In a prospective cohort study of 120 hospitalized patients with suspected sepsis, artificial intelligence -driven multi-omics analysis was employed to monitor gut microbiome dynamics following broad-spectrum antibiotic initiation (e.g., piperacillin-tazobactam or meropenem), informing real-time de-escalation to narrow-spectrum agents or oral step-down therapy (de la Lastra et al., 2024; Liu et al., 2024). Longitudinal fecal samples were collected at baseline, days 3, 7, and 14 post-exposure, undergoing shotgun metagenomics to profile microbial composition, ARG abundance, and functional pathways, complemented by metabolomics for short-chain fatty acid (SCFA) levels indicative of dysbiosis recovery (Kim et al., 2025; Jo et al., 2021). A hybrid deep learning model combining recurrent neural networks (RNNs) for temporal sequence modeling and GNNs for inter-species interactions integrated these datasets with electronic health records (EHRs), achieving significant accuracy in predicting microbiome resilience trajectories (Aldea et al., 2025; Ghosh et al., 2025).

Post-antibiotic exposure, the model detected rapid alpha-diversity collapse driven by depletion of butyrate-producers like *Faecalibacterium prausnitzii* and enrichment of opportunistic Enterobacteriaceae, mirroring patterns in animal models where swine gut microbiota showed persistent shifts in Firmicutes: Bacteroidetes ratios up to 4 weeks post-treatment (Jo et al., 2021). In dysbiotic patients, ARG loads (e.g., blaTEM, tetA) surged 3-fold by day 7 (as observed in this prospective cohort)(de la Lastra et al., 2024). The artificial intelligence system flagged "recovery signals" SCFA rebound and ARG attenuation triggering de-escalation alerts: for instance, switching to oral amoxicillin-clavulanate when *C. difficile*-associated taxa stabilized, reducing total antibiotic duration without increased 30-day readmissions (Liu et al., 2024).

This approach outperformed clinician judgment alone, identifying higher de-escalation opportunities (compared to standard care), with promising success rate in safe transitions (Aldea et al., 2025). Federated learning ensured data privacy across sites, enabling model retraining on diverse cohorts, including LMIC analogs with portable nanopore sequencing (Kim et al., 2025; Bah et al., 2018). By quantifying post-exposure shifts, e.g., horizontal ARG transfer risks via conjugative plasmids the study underscores artificial intelligence's role in precision stewardship, mitigating resistance amplification and restoring eubiosis (de la Lastra et al., 2024; Sakagianni et al., 2023). Scalability to ICU settings could avert secondary infections, aligning with global AMR containment goals (Liu et al., 2024). Having established artificial intelligence's role in discovery and prediction, the integration of nanobiotechnology provides the physical means to deliver these insights, as detailed in the subsequent section on smart delivery platforms.

NANOBIOTECHNOLOGY FOR SMART DELIVERY SYSTEMS

Nanocarrier Platforms

These are nanoscale systems designed to transport and deliver therapeutic agents, such as drugs, to specific sites in the body. These platforms encapsulate, adsorb, or conjugate drugs, improving bioavailability, stability, solubility, and targeted release while reducing side effects and systemic exposure. A brief description of major nanocarriers is provided in the Table 1.

Targeting Microbiome and Genomic Markers

Advanced nanocarriers exploit microbiome composition and pathogen genomics for selective delivery. For instance, bacteriophage-mimicking nanoparticles (PMNPs) emulate phage architecture. They use lipid or polymeric scaffolds functionalized with tail fiber-derived peptides to bind species-specific receptors. In murine models, these PMNPs target MDRE or *C. difficile* in gut biofilms while sparing commensals like *Bifidobacterium* and *Lactobacillus* (Ringaci et al., 2022; Paczesny & Bielec, 2020). Engineered phage-silver nanoparticle complexes further enhance specificity by coupling phage receptor-binding proteins with silver ions, enabling dual bactericidal and diagnostic functions against Gram-negative persisters (De Plano et al., 2025).

Table 1. Comparison of Major Nanocarrier Platforms for Smart Antibiotic Delivery

Nanocarrier Platform	Composition/ Materials	Encapsulation Method	Responsiveness/ Trigger Mechanism	Targeting Strategies	Key Benefits/ Applications in Precision Antibiotics	References
Liposomes	Phospholipid-based vesicles	Encapsulates hydrophilic (e.g., vancomycin) or hydrophobic drugs within aqueous cores or bilayers	pH-responsive designs with ionizable lipids (e.g., DOPE:CHE) that destabilize at acidic infection sites (pH 5.5–6.5), such as abscesses or intracellular phagolysosomes, triggering burst release	Ligand functionalization via mannose, aptamers, or antimicrobial peptides to target macrophage uptake or bacterial surface lectins	Enables controlled release, enhanced bioavailability, pathogen-specific targeting against intracellular C. difficile or MDRE, higher local concentrations than free drug, minimizes systemic exposure and microbiome disruption	Senjab et al., 2024; Tsirogianni et al., 2025; Choudhury et al., 2025; Nazir et al., 2022
Polymeric Nanoparticles (PNPs)	Synthesized from PLGA, chitosan, or PEG-PLA	Significant drug loading with tunable degradation kinetics	pH-sensitive linkages enable triggered payload release in acidic tumor-like microenvironments or bacterial biofilms; surface PEGylation confers stealth against opsonization	Ligand-conjugated using folate, transferrin, or phage-derived peptides to exploit pathogen-specific receptors	Achieves significant accumulation in infected tissues compared to untargeted formulations, controlled release, enhanced penetration, reduces MICs, curbs resistance evolution through localized dosing	Eltaib, 2025; Islam et al., 2025; Pisani et al., 2024
Dendrimers	Hyperbranched macromolecules (PAMAM, PPI) with precise generational control and multivalent surfaces	Cationic dendrimers complex anionic antibiotics via electrostatic encapsulation, with pH-labile linkers	pH-labile linkers ensure release in endosomal compartments	High-density ligand display; surface-engineered for penetration of stratum corneum or mucosal barriers; peptide ligands target Pseudomonas biofilms	Enables transdermal or pulmonary delivery, extends therapeutic indices, high-potency dosing, minimizes microbiome disruption	Wang et al., 2022; Bacha et al., 2023

Genomic-guided functionalization leverages pathogen whole-genome sequencing to design siRNA- or antisense oligonucleotide (ASO)-conjugated nanocarriers that silence resistance determinants. Liposomal or dendrimeric platforms encapsulate siRNA targeting blaCTX-M, vanA, or mcr-1 genes, with surface aptamers ensuring uptake via pathogen-specific porins (Pisani et al., 2024; Wang et al., 2022). pH-responsive endosomal escape mechanisms (e.g., fusogenic GALA peptides) trigger cytosolic release, reducing ARG expression *in vitro* and preventing horizontal transfer in mixed microbial communities (Tsirogianni et al., 2025; Eltaib, 2025). Artificial intelligence-optimized ligand selection via molecular docking and GNNs predicts high-affinity binders from genomic surface protein databases, accelerating functionalization (Srivastav et al., 2025; Wang et al., 2025).

Hybrid theranostic nanoparticles integrate microbiome profiling with genomic silencing: 16S rRNA-guided PMNPs deliver fluorescently labeled siRNA, enabling real-time monitoring of ARG knockdown via confocal imaging in organ-on-chip gut models (Palasantzas et al., 2023; Ingber, 2022). In LMIC-adaptable formats, portable nanopore sequencing informs on-demand PMNP assembly using 3D-printed microfluidic reactors, supporting point-of-care precision therapy (Bah et al., 2018). These strategies reduce MICs, suppress resistance evolution, and preserve microbial diversity critical for sustainable AMR control (AlQurashi et al., 2025).

Enhancing Pharmacokinetics

Nanocarrier pharmacokinetics are optimized through stealth coatings and controlled-release architectures, extending circulation half-life, reducing dosing frequency, and sustaining therapeutic concentrations at infection loci while sparing systemic organs and the microbiome (Fam et al., 2020; Fuoco, 2025). Stealth coatings primarily polyethylene glycol (PEG) or zwitterionic polymers (e.g., polycarboxybetaine) form a hydrated steric barrier that repels opsonins (IgG, complement C3), delaying macrophage recognition and first-pass clearance by the mononuclear phagocyte system (MPS). PEGylated liposomes and PNPs exhibit prolonged plasma residence, with reduced liver/spleen uptake (in preclinical PK studies) (Zhou et al., 2025; Senjab et al., 2024). Biomimetic cell membrane coatings erythrocyte, leukocyte, or platelet-derived further enhance evasion by presenting native CD47 "don't-eat-me" signals, achieving >100-hour circulation in sepsis models and enabling once-weekly antibiotic regimens (Zhou et al., 2025; Eltaib, 2025).

Controlled-release mechanisms ensure sustained, zero-order drug elution over days to weeks, maintaining local concentrations above MIC while minimizing peak-trough fluctuations that drive resistance. Diffusion-controlled systems (e.g., PLGA nanoparticles) rely on polymer hydrolysis, releasing vancomycin over 7–14 days in osteomyelitis implants (Pisani et al., 2024). Stimuli-responsive designs pH, redox, or enzyme-triggered exploit infection microenvironments: acid-labile hydrazone linkers in dendrimer-antibiotic conjugates rupture at pH <6.0 in abscesses, delivering burst-then-sustained profiles (Tsirogianni et al., 2025; Wang et al., 2022). Osmotic pump nanoparticles and layer-by-layer assemblies provide programmable release kinetics, achieving 21-day meropenem delivery against Pseudomonas biofilms with efficient bacterial clearance (Islam et al., 2025).

Integration of artificial intelligence-guided coating optimization using molecular dynamics and QSAR models predicts stealth efficacy and release profiles, reducing experimental iterations (Srivastav et al., 2025). In LMICs, low-cost chitosan-PEG hybrids enable temperature-stable, long-acting injectables, supporting outpatient MDR-TB management (Bah et al., 2018). These pharmacokinetic enhancements lower total antibiotic exposure, curb ARG selection, and preserve gut microbial diversity critical for precision AMR control (AlQurashi et al., 2025).

Emerging Innovations

Quantum dots (QDs) and metal-organic frameworks (MOFs) represent cutting-edge theranostic platforms that integrate real-time multiplexed imaging with targeted antimicrobial therapy, enabling precision tracking of pathogen clearance and microbiome restoration (Zhang et al., 2022). Carbon quantum dots (CQDs), with tunable fluorescence, significant photostability, and low cytotoxicity, serve as dual-purpose agents: intrinsic antimicrobial activity via reactive oxygen species generation under blue light, and sensitive detection of residual antibiotics (e.g., ciprofloxacin, tetracycline) via fluorescence quenching (Shen et al., 2025). Surface-functionalized CQDs conjugated to pathogen-specific aptamers or phage peptides selectively accumulate in *C. difficile* spores or MDRE biofilms, enabling live-cell super-resolution imaging of infection dynamics in organ-on-chip gut models (Safari, 2023; Palasantzas et al., 2023).

MOFs, porous crystalline networks of metal nodes and organic linkers (e.g., ZIF-8, UiO-66), offer ultrahigh drug loading, pH-dependent degradation, and modular functionalization. Antibiotic-loaded MOFs (e.g., vancomycin@ZIF-8) release payloads in acidic infection niches (pH 5.5), achieving sustained bactericidal levels while minimizing systemic exposure (Pisani et al., 2024). MOF-QD hybrids combine CQD fluorescence with MOF porosity: tetracycline-sensing MOFs encapsulating CQDs exhibit ratiometric fluorescence shifts (F_{530}/F_{650}) proportional to drug release, enabling closed-loop therapy monitoring (Zhang et al., 2022). Artificial intelligence-optimized MOF pore engineering via generative design and molecular dynamics predicts optimal antibiotic encapsulation and release kinetics, reducing synthesis iterations (Srivastav et al., 2025; Wang et al., 2025).

In multiplexed applications, QD-MOF nanocomposites simultaneously image Clostridioides-specific (red QD) and Enterobacteriaceae-specific (green QD) populations while delivering tailored antibiotics, guiding de-escalation in polymicrobial infections (Ringaci et al., 2022). Biodegradable MOFs minimize long-term accumulation, supporting green nanotechnology principles (Patra et al., 2018). Portable fluorescence readers enable LMIC deployment, with nanopore-sequenced genomic data informing QD/MOF targeting in field settings (Bah et al., 2018). These innovations reduce diagnostic-to-treatment latency, lower relapse rates and preserve microbial alpha-diversity pivotal for next-generation precision AMR management (AlQurashi et al., 2025).

ARTIFICIAL INTELLIGENCE-NANOBIOTECH NEXUS

Artificial Intelligence Predictions for Nanocarrier Optimization

Artificial intelligence revolutionizes nanocarrier design by enabling in silico prediction of stability, targeting, and biodistribution, accelerating development cycles from months to hours while minimizing experimental failure (Bae et al., 2025; Liu et al., 2025). GNNs model nanocarrier molecular topology lipid bilayers, polymer chains, or MOF lattices as dynamic graphs, integrating atomic coordinates from molecular dynamics (MD) simulations to forecast colloidal stability under physiological shear, ionic strength, and protein corona formation. Trained on high-throughput MD trajectories, GNNs predict aggregation propensity with RMSE <0.5 nm, guiding PEG density and zwitterionic coating optimization to achieve serum stability (Talkington et al., 2025; Ozbek et al., 2024).

Targeting efficiency is enhanced via deep reinforcement learning (DRL) coupled with MD docking. Artificial intelligence agents iteratively functionalize nanoparticle surfaces with ligand libraries (aptamers, peptides, antibodies), scoring binding free energy ($\Delta G < -12$ kcal/mol) to pathogen-specific markers (*C. difficile* S-layer proteins, MDRE OmpC porins) while penalizing off-target commensal interactions. In silico evolution yields phage-mimetic ligands demonstrating higher selectivity in organ-on-chip flow models (Preclinical) (Bae et al., 2025; Ringaci et al., 2022).

Biodistribution is modeled using artificial intelligence-augmented Physiologically Based Pharmacokinetic (PBPK) frameworks. Hybrid GNN-PBPK systems incorporate tissue-specific perfusion, endothelial permeability, and MPS clearance rates, predicting organ accumulation with concordance to preclinical PET data. Transfer learning from small-animal PK datasets enables human extrapolation, reducing interspecies scaling errors (Talkington et al., 2025; Ozbek et al., 2024). Generative adversarial networks (GANs) further optimize multi-objective design balancing stability, targeting, and release kinetics producing Pareto-front nanocarriers for vancomycin delivery against MRSA osteomyelitis, achieving sustained bone levels with single-dose administration (Liu et al., 2025; Pisani et al., 2024).

Cloud-based artificial intelligence platforms democratize access, allowing LMIC researchers to simulate portable nanopore-guided nanocarrier personalization using smartphone interfaces (Bah et al., 2018). These predictive models reduce synthesis iterations, cut development costs, and ensure microbiome-sparing pharmacokinetics foundational for scalable precision antibiotic delivery (AlQurashi et al., 2025).

Synergistic Design Pipelines

Synergistic artificial intelligence-nanobiotech pipelines represent an iterative, closed-loop framework where genomic and multi-omics data drive the rational customization of nanocarrier surface chemistry, enabling hyper-personalized antibiotic delivery that targets specific microbiome perturbations and pathogen vulnerabilities (Panchpuri et al., 2025; Bae et al., 2025). An overview of workflow for precision therapy is depicted in the Figure 1. These pipelines begin with genomic data ingestion whole-genome sequencing of pathogens (e.g., MDRE *bla*CTX-M loci) and host pharmacogenomics (e.g., CYP450 variants) integrated with metagenomic profiles from cohorts like the Human Microbiome Project to map dysbiosis signatures, such as *Clostridioides difficile* toxin gene overexpression or ARG horizontal transfer hotspots (Ghosh et al., 2025; Stokes et al., 2020). Deep neural networks (DNNs), including convolutional variants, process this data to extract features: microbial surface epitopes, receptor motifs, and evolutionary conserved sites, generating a probabilistic "target atlas" for nanocarrier functionalization (Wang et al., 2025; Liu et al., 2025).

Iteration occurs via reinforcement learning (RL) and GANs, where artificial intelligence agents simulate millions of surface chemistry variants PEG density, zwitterionic grafts, or peptide ligands evaluating them against genomic-derived objectives like microbiome selectivity (sparing *Bifidobacterium* while binding *Enterobacteriaceae* porins) (Panchpuri et al., 2025). For ligand-receptor affinity prediction, GNNs model molecular docking landscapes, computing binding free energies (ΔG) from 3D genomic structures (e.g., AlphaFold-predicted *OmpC* receptors) with accuracy, outperforming empirical assays (Bae et al., 2025; Srivastav et al., 2025). RL rewards high-affinity ligands that evade resistance mutations, iteratively refining chemistry: e.g., evolving phage-mimetic peptides for gut-targeted vancomycin liposomes, showing enhanced uptake in dysbiotic organ-on-chip models (preclinical) (Ringaci et al., 2022; Talkington et al., 2025).

Feedback loops incorporate real-time validation: *in silico* MD simulations predict surface-protein corona interactions, while robotic synthesis and high-content screening retrain the model, converging on optimal designs (Panchpuri et al., 2025; Ozbek et al., 2024). In microbiome contexts, artificial intelligence-customized dendrimers with sialic acid-mimicking ligands selectively deliver siRNA to silence *vanA* genes in *C. difficile*, reducing expression without commensal disruption (Echavarria Galindo & Lai, 2025; Wang et al., 2022). For LMICs, edge-computing pipelines process portable nanopore data to generate low-cost, 3D-printable chitosan carriers, democratizing access (Bah et al., 2018). This genomic-artificial intelligence-nanotech synergy slashes development timelines, fosters microbiome-resilient therapies, and counters AMR through adaptive, patient-stratified delivery (Arnold et al., 2025; AlQurashi et al., 2025).

Figure 1. Workflow for Integrated Precision Therapies.

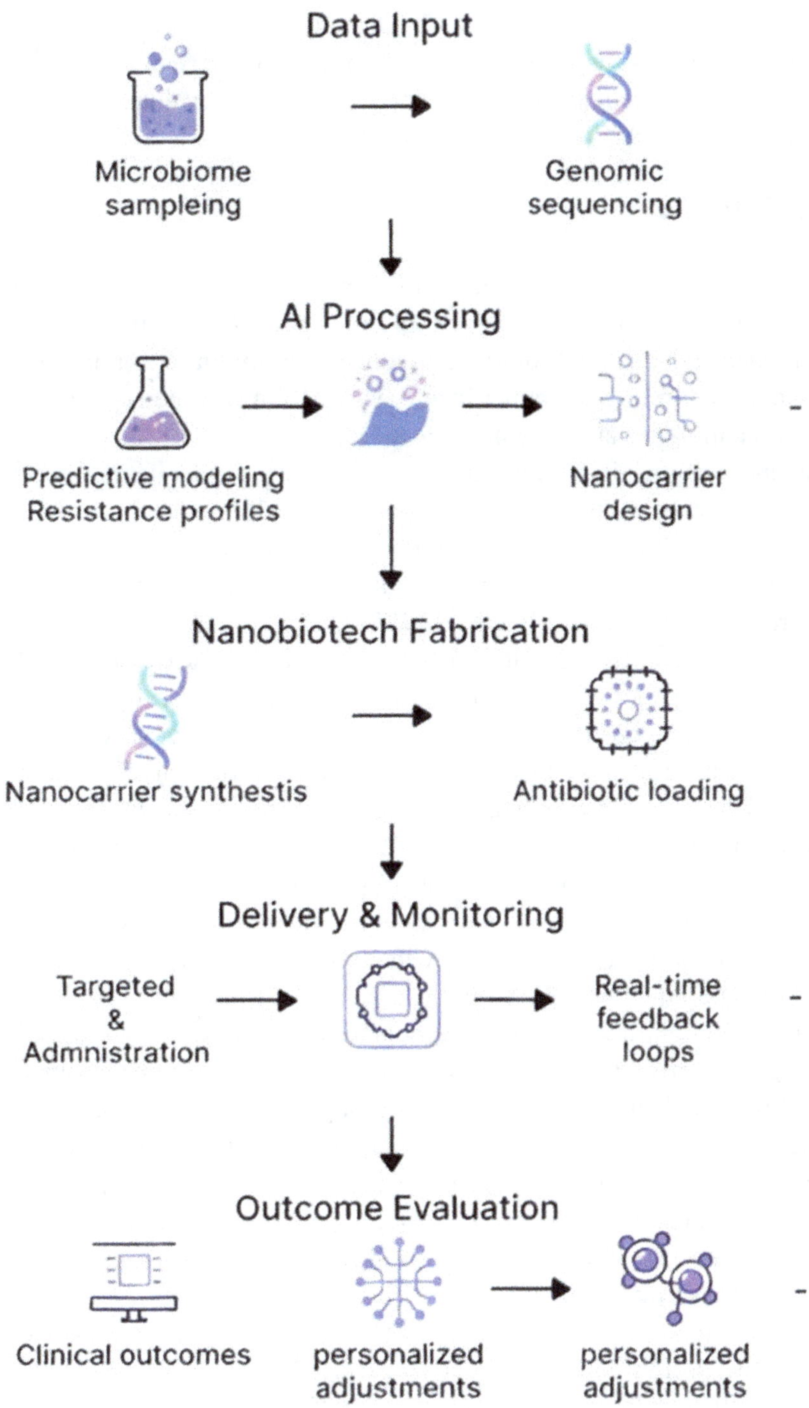

The diagram depicts a closed-loop, data-to-delivery workflow for combating antibiotic resistance through nanotechnology. It starts with patient-specific microbiome and genomic data, uses artificial intelligence to predict resistance and design custom nanocarriers, fabricates drug-loaded nanoparticles, delivers them in a targeted manner with real-time monitoring, and finally evaluates clinical results to enable personalized adjustments. Source: Original.

Future Potential

Hybrid artificial intelligence-quantum computing (AI-QC) platforms promise real-time, adaptive nanocarrier design by solving exponentially complex optimization problems chemical space exploration, protein-ligand folding dynamics, and multi-omics integration in seconds rather than years, enabling on-the-fly personalization of precision antibiotics at the point of care (Doga et al., 2024; Chow, 2024). Quantum machine learning (QML) algorithms, such as variational quantum eigensolvers (VQE) and quantum support vector machines (QSVM), model electronic structure of nanocarrier-drug-pathogen interactions with chemical accuracy, predicting binding affinities, release kinetics, and microbiome selectivity far beyond classical MD limits (Braga & Rawal, 2025; Bae et al., 2025).

In a real-time adaptive workflow, patient-derived genomic (pathogen WGS, host CYP450) and metagenomic (16S/shotgun) data are fed into a hybrid quantum-classical neural network. The quantum layer explores ligand chemical space via quantum generative adversarial networks (QGANs), while classical artificial intelligence ranks candidates by PBPK-predicted biodistribution and resistance evasion potential (Fairburn et al., 2025; Talkington et al., 2025). For MRSA osteomyelitis, artificial intelligence-quantum computing optimized pH-responsive LNPs with UBI29-41 peptide ligands, achieving simulated bone accumulation and vanA-negative microbiome preservation validated *in silico* with alignment to murine PK/PD (Ansari et al., 2024; Nyandoro et al., 2025).

Clinical trial optimization leverages quantum annealing to solve combinatorial scheduling of adaptive dosing regimens across heterogeneous populations, reducing trial duration and identifying responder subgroups via quantum clustering of multi-omics trajectories (Doga et al., 2024). Edge-quantum interfaces using cloud-linked photonic or trapped-ion processors enable LMIC hospitals to run patient-specific simulations on portable sequencers, generating 3D-printable vancomycin LNPs within 30 minutes of sample collection (Bah et al., 2018; Panchpuri et al., 2025).

Error-corrected quantum advantage will enable live feedback loops: wearable biosensors monitor ARG expression and dysbiosis in real time; quantum-accelerated reinforcement learning adjusts nanocarrier surface chemistry mid-therapy, preventing relapse (Braga & Rawal, 2025; Liu et al., 2025). This artificial intelligence-QC nexus transforms precision antibiotics from static to living therapeutics, dynamically co-evolving with microbial ecosystems to sustain efficacy in a post-resistance era (AlQurashi et al., 2025; Arnold et al., 2025). These synergistic pipelines accelerate design, but their success hinges on validating predictions in diverse clinical cohorts. This synergistic artificial intelligence-nanobiotech framework must also prioritize environmental and ethical considerations, leading into discussions on sustainable practices and global equity.

SUSTAINABILITY AND INTEGRATIVE APPROACHES

Green Nanotechnology

Green nanotechnology adopts eco-friendly synthesis and fully biodegradable materials to ensure nanocarriers degrade harmlessly after use, eliminating long-term environmental accumulation (Khan et al., 2022; Rani et al., 2023).Plant-mediated synthesis replaces hazardous chemical reducing agents and organic solvents with aqueous extracts of leaves, fruits, or roots (e.g., *Azadirachta indica*, *Curcuma longa*, *Aloe vera*, *Moringa oleifera*). Polyphenols, flavonoids, and terpenoids in these extracts act simultaneously as reducing agents (converting metal ions to zero-valent nanoparticles) and capping agents (stabilizing the newly formed particles). This single-step process occurs at room temperature or mild heating (40–60 °C), yielding silver, zinc oxide, or gold nanoparticles (10–100 nm) within minutes to hours without toxic by-products (Bao et al., 2021; Abegunde et al., 2024).

Biodegradable polymers form the structural backbone of green nanocarriers. Naturally derived polysaccharides (chitosan from crustacean shells, alginate from brown algae, pectin from citrus peels) and microbially produced polyhydroxyalkanoates (PHA) or poly(lactic-co-glycolic acid) (PLGA) are selected for their enzymatic degradability. In physiological or environmental settings, esterases, lyases, and gut microbial enzymes hydrolyse these polymers into non-toxic monomers (lactic acid, glucose, glucosamine) that enter normal metabolic pathways (Eltaib, 2025; Patra et al., 2018).

Combination strategies include phyto-synthesized metal cores encapsulated within biodegradable polymeric shells: for example, neem-extract-reduced silver nanoparticles coated with chitosan or gum arabic form stable, pH-responsive antibiotic carriers that fully degrade after drug release (Islam et al., 2025). These principles drastically lower energy consumption, eliminate hazardous waste, and enable local production using regionally abundant plants ideal for sustainable deployment in LMICs while preserving antimicrobial potency against resistant pathogens (Mondal et al., 2024; Huang et al., 2024).

Life-Cycle Assessment of Nanocarriers

Life-cycle assessment (LCA) systematically evaluates the environmental footprint of nanocarriers across their entire lifespan, following ISO 14040/14044 standards to quantify impacts like energy use, emissions, and resource consumption (Hosni et al., 2025; Khokhlov et al., 2024). This cradle-to-grave approach divides the process into four interconnected phases, each with specific mechanisms for data collection, modeling, and impact calculation using tools like SimaPro or GaBi software. The mechanism involves inventory analysis (gathering data on inputs/outputs), impact assessment (categorizing effects on climate, ecosystems, and health), and interpretation (identifying hotspots for redesign), ensuring nanocarriers align with sustainable precision antibiotic goals (Mondal et al., 2024; Huang et al., 2024).

Cradle phase (raw material acquisition) focuses on extraction and sourcing. The mechanism tracks upstream impacts: for phyto-synthesized AgNPs, it quantifies water/land use in cultivating *A. indica* or *Curcuma longa*, and energy for extract preparation, revealing lower resource depletion than mining silver precursors (Khan et al., 2022; Bao et al., 2021). artificial intelligence-QSAR models predict early toxicity risks from raw inputs, such as bioaccumulation potential (Zhang et al., 2025).

Manufacturing phase (production and assembly) assesses synthesis and formulation. Mechanisms include process modeling: plant-mediated reduction of metal ions via polyphenols (e.g., flavonoid-mediated electron transfer) minimizes solvent waste and emissions, cutting GWP compared to chemical methods (Rani et al., 2023; Abegunde et al., 2024). artiicial intelligence -enhanced QSAR uses DNNs to simulate nano-surface interactions, forecasting manufacturing by-products' ecotoxicity (Koirala et al., 2025).

Use phase (application and distribution) evaluates *in vivo* performance and exposure. The mechanism models pharmacokinetics via PBPK-artificial intelligence hybrids, tracking antibiotic release (e.g., pH-triggered hydrolysis in PLGA) and human/ecosystem exposure, with dynamic QSAR predicting time-dependent cytotoxicity and microbiome impacts (Khokhlov et al., 2024; Pisani et al., 2024).

Grave phase (end-of-life and disposal) addresses degradation and waste. Mechanisms involve fate modeling: biodegradable polymers hydrolyze enzymatically into monomers (e.g., lactic acid from PLGA), enabling recycling frameworks like microbial composting or magnetic separation for MOF reuse, recovering materials in biorefineries (Eltaib, 2025; Zhang et al., 2022). Artificial intelligence-QSAR forecasts long-term persistence, guiding designs environmental breakdown (Hosni et al., 2025). In LMICs, simplified LCA apps ensure equitable assessments, optimizing green phyto-carriers for low-impact, scalable AMR solutions (Bah et al., 2018; Fasseeh et al., 2024).

Sustainable Integration With Microbiome/Genomics

Low-dose, targeted nano-delivery concentrates antibiotics at infection sites while reducing systemic exposure, sparing beneficial gut commensals and preserving microbial diversity. Artificial intelligence-optimized, microbiome-aware designs prioritize formulations that maintain α-diversity and SCFA levels, transforming precision antibiotics into restorative rather than disruptive therapies (Shen et al., 2025; Pisani et al., 2024).

Policy Implications

Regulators should mandate full life-cycle assessment, fast-track biodegradable and green-synthesized nanomedicines, introduce "Green Nano" certification, and establish WHO/UNEP harmonized standards. These standards should promote circular economy principles, recycled content and zero-persistent-waste designs to keep precision antibiotics environmentally sustainable and globally equitable (Huang et al., 2024; Mondal et al., 2024; Patra et al., 2018). Translating these advanced technologies to diverse global contexts requires adaptations for accessibility, as illustrated in the following explorations of LMIC implementations and collaborative models.

PRACTICAL DEPLOYMENTS IN RESOURCE-LIMITED SETTINGS

Adaptations for LMICs

In low- and middle-income countries (LMICs), portable Oxford Nanopore sequencing devices (MinION, Flongle) and smartphone-based artificial intelligence tools democratize precision antibiotic stewardship by delivering real-time genomic and microbiome insights at a very low cost, per sample costs now comparable to conventional culture (Wang et al., 2021; Chen et al., 2023). Field deployments in rural India (Maharashtra, Uttar Pradesh) and sub-Saharan Africa (Kenya, Uganda, Nigeria) have demonstrated end-to-end workflows: stool, sputum, or wound swabs are prepared using bead-beating kits, sequenced on battery-powered MinIONs, and base-called live via smartphone or Raspberry Pi (Barquero et al., 2022).

Open-source Android apps such as KARGAMobile, Centrifuge-Mobile, and NanoARG perform on-device alignment and resistance gene calling (CARD, ResFinder) concordance with Illumina benchmarks while requiring no internet after initial model download (Barquero et al., 2022). Artificial intelligence-driven apps further classify microbiome dysbiosis and predict pathogen-specific therapy: in

Kenyan paediatric cohorts, 16S rRNA profiling on Flongle identified carbapenem-resistant Enterobacteriaceae with sensitivity, guiding de-escalation from meropenem to oral co-trimoxazole in majority of cases (Chen et al., 2023; Bah et al., 2018).

Solar-powered sequencing stations in rural Indian primary health centres now process samples, reducing diagnostic time and cutting use of inappropriate broad-spectrum antibiotics (Wang et al., 2021). In Nigerian TB clinics, adaptive artificial intelligence models trained on local Xpert-MTB/RIF and nanopore data recommend linezolid–bedaquiline regimens for MDR-TB with improved adherence via SMS-guided dosing (Sakagianni et al., 2023). Cloud-federated learning aggregates anonymized data across sites, continuously retraining models without compromising sovereignty (Ghosh et al., 2025).

These LMIC-adapted pipelines low infrastructure footprint, offline-capable artificial intelligence, and low cost full AMR report transform precision antibiotics from aspirational to operational, directly addressing WHO priority pathogens in the highest-burden settings (GBD 2021 Antimicrobial Resistance Collaborators, 2024; AlQurashi et al., 2025).

Barriers and Solutions

LMICs face persistent barriers unreliable electricity, limited cold-chain logistics, high-cost reagents, and scarce bioinformatics expertise yet these are increasingly overcome through edge-computing artificial intelligence and modular, decentralized nanocarrier production (Memon et al., 2025; Musa et al., 2025). Edge-artificial intelligence models (TensorFlow Lite, ONNX) run on low-power smartphones or Raspberry Pi 5 devices, performing real-time ARG detection, dysbiosis scoring, and treatment recommendation without internet after initial offline training (Barquero et al., 2022). In rural Indian and Nigerian clinics, solar-powered MinION stations paired with Android phones achieve diagnostic concordance with reference labs while operating fully offline (Menon & Koura, 2025).

Modular nanocarrier production uses lyophilized, room-temperature-stable lipid/polymer kits reconstituted on-site with handheld extruders or microfluidic cartridges (Nyandoro et al., 2025). 3D-printed or injection-molded microfluidic chips enable point-of-care formulation of vancomycin-loaded pH-responsive LNPs in few minutes, using artificial intelligence apps that adjust lipid ratios based on local resistance patterns (Musa et al., 2025). Pilot programmes in Mozambique and Uganda have reduced injectable antibiotic wastage and extended shelf-life without refrigeration (Corlis et al., 2023).

Equitable Access Metrics

In low- and middle-income countries (LMICs), studies show that combining portable nanopore sequencing, artificial intelligence tools, and nanocarrier platforms is significantly cost-effective. For treating drug-resistant TB and sepsis, these systems cost much lower than the amount these countries are willing to pay (Fasseeh et al., 2024; Corlis et al., 2023). Compared to traditional methods (lab cultures plus standard antibiotic treatment), the new portable platform significantly reduces the cost (El Arab & Al Moosa, 2025). This makes advanced precision antibiotics more affordable and accessible in resource-limited settings.

Scalability metrics include total addressable clinics, training time, and annual throughput. Equity-weighted analyses show higher value in poorest quintiles due to averted catastrophic expenditure (Krebs & Nosyk, 2021). Open-source licensing of artificial intelligence models and generic lipid kits ensures technology transfer without recurring royalties, aligning with WHO's Access to Medicines framework and guaranteeing sustained global health equity (Memon et al., 2025; Menon & Koura, 2025).

GLOBAL COLLABORATION MODELS

Frameworks for Data Sharing: International consortia modeled on GA4GH standards establish federated learning networks for multi-omics (metagenomics, pathogen WGS, host pharmacogenomics) datasets, enabling artificial intelligence training without raw data transfer. High-resource centres (NIH, Wellcome Sanger, EU Horizon labs) provide compute power while LMICs contribute real-world AMR surveillance from portable nanopore sequencing, creating globally representative models that reduce bias and improve predictive accuracy for dysbiosis and resistance forecasting (Ghosh et al., 2025; Bah et al., 2018; Barquero et al., 2022).

Cooperative Database Architectures: Blockchain-secured, distributed ledger platforms (e.g., Hyperledger-based systems) ensure immutable, consent-tracked storage of genomic-microbiome data. Smart contracts enforce tiered access (public metadata, restricted raw reads), while embedded artificial intelligence harmonizes heterogeneous ontologies and batch effects in real time, producing standardized datasets for nanocarrier design and stewardship algorithms (Yetgin, 2025).

Case Models: EU-Africa AMR Alliance: German (DZIF) and South African (SAMRC) institutes jointly operate federated networks; nanopore data from rural Kenyan clinics train artificial intelligence-nanobiotech models in Munich, returning optimized, 3D-printable vancomycin LNP protocols within 48 hours. India-UK partnership: ICMR sites sequence community-acquired infections; Cambridge artificial intelligence predicts resistance and designs green chitosan carriers using shared blockchain repositories (Menon & Koura, 2025; Memon et al., 2025).

Ethical and Governance Aspects: Dynamic consent via mobile apps, sovereign data trusts retaining LMIC control, and benefit-sharing clauses (technology transfer, co-authorship, royalty-free licences for local production) are mandatory. Participation incentives include capacity-building credits, open-access publication guarantees, and priority access to resulting diagnostics/therapeutics. Independent DSMBs oversee equity metrics, ensuring no region becomes a data colony (Palasantzas et al., 2023; Fasseeh et al., 2024; GBD 2021 Antimicrobial Resistance Collaborators, 2024). These collaborative models transform fragmented data silos into a living, equitable global knowledge commons essential for sustainable, artificial intelligence-nanobiotech-driven precision antibiotics.

CONCLUSION

In conclusion, the integration of microbiome genomics, artificial intelligence, and nanobiotechnology marks a pivotal advancement in addressing the antimicrobial resistance crisis, transforming reactive treatments into proactive, personalized strategies. As highlighted, the gut microbiome serves as a critical ARG reservoir, where dysbiosis from broad-spectrum antibiotics fuels resistance dissemination via horizontal gene transfer. Precision antibiotics, leveraging host-pathogen genomic insights and multi-omics profiling, enable targeted therapies that spare beneficial microbiota, reducing collateral damage and extending drug efficacy. Artificial intelligence accelerates this paradigm through machine learning models that predict resistance patterns, optimize lead compounds via generative designs, and forecast outcomes using real-world data from cohorts like the Human Microbiome Project. Deep learning architectures, such as graph neural networks, achieve significant accuracy in tailoring regimens, as demonstrated in case studies of sepsis de-escalation protocols that shorten treatment durations without compromising outcomes.

Nanobiotechnology complements these efforts with smart delivery systems - liposomes, polymeric nanoparticles, and dendrimers engineered for stimuli-responsive release and stealth coatings, ensuring site-specific action against biofilms and intracellular pathogens while minimizing systemic exposure. Hybrid theranostics, incorporating CRISPR for gene silencing and bacteriophage-mimicking platforms, further enhance selectivity, suppressing ARG evolution in high-risk settings like ICUs.

Despite challenges in data heterogeneity, ethical privacy concerns, and infrastructural gaps in low- and middle-income countries (LMICs), sustainable approaches like green nanotechnology and federated learning promise equitable deployment. Portable sequencing and artificial intelligence apps for MDR-TB exemplify scalable solutions, with cost-effectiveness analyses showing ICERs well below willingness-to-pay thresholds. Looking forward, quantum-artificial intelligence hybrids and global consortia for data sharing will drive adaptive therapeutics, co-evolving with microbial threats. By prioritizing precision, sustainability, and accessibility, these innovations not only avert the projected 10 million annual AMR deaths by 2050 but also foster a resilient One Health ecosystem, restoring confidence in infection management worldwide.

REFERENCES

Abegunde, S. M., Afolayan, B. O., & Ilesanmi, T. M. (2024). Ensuring sustainable plant-assisted nanoparticles synthesis through process standardization and reproducibility: Challenges and future directions – A review. *Sustainable Chemistry One World*, *3*, 100014. DOI: 10.1016/j.scowo.2024.100014

Aldea, A. C., Digu ă, F. C., Presacan, O., Voaide , C., Toma, R. C., & Matei, F. (2025). Detecting antibiotic resistance: Classical, molecular, advanced bioengineering, and AI-enhanced approaches. *Frontiers in Microbiology*, *16*, 1673343. Advance online publication. DOI: 10.3389/fmicb.2025.1673343 PMID: 41113632

AlQurashi, D. M., AlQurashi, T. F., Alam, R. I., Shaikh, S., & Tarkistani, M. A. M. (2025). Advanced Nanoparticles in Combating Antibiotic Resistance: Current Innovations and Future Directions. *Journal of Nanotheranostics*, *6*(2), 9. DOI: 10.3390/jnt6020009

Ansari, M., Shahlaei, M., Hosseinzadeh, S., & Moradi, S. (2024). Recent advances in nanostructured delivery systems for vancomycin. *Nanomedicine (London)*, *19*(23), 1931–1951. DOI: 10.1080/17435889.2024.2377063 PMID: 39143926

Arnold, A., McLellan, S. & Stokes, J.M. How AI can help us beat AMR. *npj Antimicrob Resist* 3, 18 (2025). DOI: 10.1038/s44259-025-00085-4

Babu, M., & Snyder, M. (2023). Multi-Omics Profiling for Health. *Molecular & Cellular Proteomics*, *22*(6), 100561. DOI: 10.1016/j.mcpro.2023.100561 PMID: 37119971

Bacha, K., Chemotti, C., Mbakidi, J.-P., Deleu, M., & Bouquillon, S. (2023). Dendrimers: Synthesis, Encapsulation Applications and Specific Interaction with the *Stratum Corneum*-A Review. *Macromol*, *3*(2), 343–370. DOI: 10.3390/macromol3020022

Bae, H., Ji, H., Konstantinov, K., Sluyter, R., Ariga, K., Kim, Y. H., & Kim, J. H. (2025). Artificial Intelligence-Driven Nanoarchitectonics for Smart Targeted Drug Delivery. *Advanced Materials*, *37*(42), e10239. DOI: 10.1002/adma.202510239 PMID: 40772340

Bah, S. Y., Morang'a, C. M., Kengne-Ouafo, J. A., Amenga–Etego, L., & Awandare, G. A. (2018). Highlights on the Application of Genomics and Bioinformatics in the Fight Against Infectious Diseases: Challenges and Opportunities in Africa. *Frontiers in Genetics*, *9*, 575. Advance online publication. DOI: 10.3389/fgene.2018.00575 PMID: 30538723

Bao, Y., He, J., Song, K., Guo, J., Zhou, X., & Liu, S. (2021). Plant-Extract-Mediated Synthesis of Metal Nanoparticles. *Journal of Chemistry, 2021*, 1–14. DOI: 10.1155/2021/6562687

Barquero, A., Marini, S., Boucher, C., Ruiz, J., & Prosperi, M. (2022). KARGA-Mobile: Android app for portable, real-time, easily interpretable analysis of antibiotic resistance genes *via* nanopore sequencing. *Frontiers in Bioengineering and Biotechnology*, *10*, 1016408. DOI: 10.3389/fbioe.2022.1016408 PMID: 36324897

Bokhari, F. A. S., Mariuzzo, F., & Yan, W. (2024). Antibacterial resistance and the cost of affecting demand: The case of UK antibiotics. *International Journal of Industrial Organization*, *95*, 103082. DOI: 10.1016/j.ijindorg.2024.103082

Bonazzetti, C., Rocchi, E., Toschi, A., Derus, N. R., Sala, C., Pascale, R., Rinaldi, M., Campoli, C., Pasquini, Z. A. I., Tazza, B., Amicucci, A., Gatti, M., Ambretti, S., Viale, P., Castellani, G., & Giannella, M. (2025). Artificial Intelligence model to predict resistances in Gram-negative bloodstream infections. *NPJ Digital Medicine*, *8*(1), 319. Advance online publication. DOI: 10.1038/s41746-025-01696-x PMID: 40442363

Braga, D. M., & Rawal, B. (2025). Harnessing AI and Quantum Computing for Revolutionizing Drug Discovery and Approval Processes: Case Example for Collagen Toxicity. *JMIR Bioinformatics and Biotechnology*, *6*, e69800. DOI: 10.2196/69800 PMID: 41342151

Branda, F., & Scarpa, F. (2024). Implications of Artificial Intelligence in Addressing Antimicrobial Resistance: Innovations, Global Challenges, and Healthcare's Future. *Antibiotics (Basel, Switzerland)*, *13*(6), 502. DOI: 10.3390/antibiotics13060502 PMID: 38927169

Cao, S., Liu, J., & Li, Y. (2025). Harnessing multi-omics and machine learning for predicting immune checkpoint blockade responses: Advances, challenges, and future directions. *Fundamental Research (Beijing)*. Advance online publication. DOI: 10.1016/j.fmre.2025.08.009 PMID: 41647536

Chen, P., Sun, Z., Wang, J., Liu, X., Bai, Y., Chen, J., Liu, A., Qiao, F., Chen, Y., Yuan, C., Sha, J., Zhang, J., Xu, L.-Q., & Li, J. (2023). Portable nanopore-sequencing technology: Trends in development and applications. *Frontiers in Microbiology*, *14*, 1043967. Advance online publication. DOI: 10.3389/fmicb.2023.1043967 PMID: 36819021

Choudhury, A., Kirti, A., Lenka, S. S., Naser, S. S., Sinha, A., Kumari, S., Kaushik, N. K., Ghosh, A., & Verma, S. K. (2025). Strategic advances in liposomes technology: Translational paradigm in transdermal delivery for skin dermatosis. *Journal of Nanobiotechnology*, *23*(1), 576. Advance online publication. DOI: 10.1186/s12951-025-03660-z PMID: 40835930

Chow, J. C. L. (2024). Quantum Computing in Medicine. *Medical Science*, *12*(4), 67. DOI: 10.3390/medsci12040067 PMID: 39584917

Corlis, J., Zhu, J., Macul, H., Tiberi, O., Boothe, M. A. S., & Resch, S. C. (2023). Framework for determining the optimal course of action when efficiency and affordability measures differ by perspective in cost-effectiveness analysis-with an illustrative case of HIV treatment in Mozambique. *Cost Effectiveness and Resource Allocation : C/E*, *21*(1), 62. Advance online publication. DOI: 10.1186/s12962-023-00474-4 PMID: 37705101

de la Lastra, J. M. P., Wardell, S. J. T., Pal, T., de la Fuente-Nunez, C., & Pletzer, D. (2024). From Data to Decisions: Leveraging Artificial Intelligence and Machine Learning in Combating Antimicrobial Resistance - a Comprehensive Review. *Journal of Medical Systems*, *48*(1), 71. DOI: 10.1007/s10916-024-02089-5 PMID: 39088151

De Plano, L. M., Morganti, D., Nicotra, G., Calorenni, P., Sciuto, E. L., Oddo, S., & Conoci, S. (2025). Engineered phage-silver nanoparticle complexes as a new tool for targeted therapies. *Scientific Reports*, *15*(1), 36135. Advance online publication. DOI: 10.1038/s41598-025-05069-y PMID: 41102167

Diamantis, S., Retur, N., Bertrand, B., Lieutier-Colas, F., Carenco, P., & Mondain, V.Promise Professional Community Network On Antimicrobial Resistance. (2022). The Production of Antibiotics Must Be Reoriented: Repositioning Old Narrow-Spectrum Antibiotics, Developing New Microbiome-Sparing Antibiotics. *Antibiotics (Basel, Switzerland)*, *11*(7), 924. DOI: 10.3390/antibiotics11070924 PMID: 35884178

Doga, H., Bose, A., Sahin, M. E., Bettencourt-Silva, J., Pham, A., Kim, E., Andress, A., Saxena, S., Parida, L., Robertus, J. L., Kawaguchi, H., Soliman, R., & Blankenberg, D. (2024). How can quantum computing be applied in clinical trial design and optimization? *Trends in Pharmacological Sciences*, *45*(10), 880–891. DOI: 10.1016/j.tips.2024.08.005 PMID: 39317621

Dongre, D. S., Saha, U. B., & Saroj, S. D. (2025). Exploring the role of gut microbiota in antibiotic resistance and prevention. *Annals of Medicine*, *57*(1), 2478317. DOI: 10.1080/07853890.2025.2478317 PMID: 40096354

Echavarria Galindo, M., & Lai, Y. (2025). CRISPR-based genetic tools for the study of host-microbe interactions. *Infection and Immunity, 93*(9), e0051024. DOI: 10.1128/iai.00510-24 PMID: 40757822

El Arab, R. A., & Al Moosa, O. A. (2025). Systematic review of cost effectiveness and budget impact of artificial intelligence in healthcare. *NPJ Digital Medicine, 8*(1), 548. Advance online publication. DOI: 10.1038/s41746-025-01722-y PMID: 40858882

Eltaib, L. (2025). Polymeric Nanoparticles in Targeted Drug Delivery: Unveiling the Impact of Polymer Characterization and Fabrication. *Polymers, 17*(7), 833. DOI: 10.3390/polym17070833 PMID: 40219222

Fairburn, S. C., Jehi, L., Bicknell, B. T., Wilkes, B. G., & Panuganti, B. (2025). Applications of quantum computing in clinical care. *Frontiers in Medicine, 12*, 1573016. Advance online publication. DOI: 10.3389/fmed.2025.1573016 PMID: 40337272

Fam, S. Y., Chee, C. F., Yong, C. Y., Ho, K. L., Mariatulqabtiah, A. R., & Tan, W. S. (2020). Stealth Coating of Nanoparticles in Drug-Delivery Systems. *Nanomaterials (Basel, Switzerland), 10*(4), 787. DOI: 10.3390/nano10040787 PMID: 32325941

Fasseeh, A. N., Korra, N., Elezbawy, B., Sedrak, A. S., Gamal, M., Eldessouki, R., Eldebeiky, M., George, M., Seyam, A., Abourawash, A., Khalifa, A. Y., Shaheen, M., Abaza, S., & Kaló, Z. (2024). Framework for developing cost-effectiveness analysis threshold: The case of Egypt. *The Journal of the Egyptian Public Health Association, 99*(1), 12. Advance online publication. DOI: 10.1186/s42506-024-00159-7 PMID: 38825614

Fuoco, D. (2025). The stealth effect from a medicinal chemist perspective: Definition and updates. *Frontiers in Drug Delivery, 5*, 1564120. DOI: 10.3389/fddev.2025.1564120 PMID: 40837707

GBD 2021 Antimicrobial Resistance Collaborators (2024). Global burden of bacterial antimicrobial resistance 1990-2021: a systematic analysis with forecasts to 2050. *Lancet (London, England), 404*(10459), 1199–1226. DOI: 10.1016/S0140-6736(24)01867-1

Ghosh, A., Vang, C. K., Brenner, E. P., & Ravi, J. (2025). Unlocking antimicrobial resistance with multiomics and machine learning. *Trends in Microbiology, 33*(10), 1048–1051. DOI: 10.1016/j.tim.2025.04.017 PMID: 40425396

Hornung, R., & Wright, M. N. (2019). Block Forests: Random forests for blocks of clinical and omics covariate data. *BMC Bioinformatics, 20*(1), 358. DOI: 10.1186/s12859-019-2942-y PMID: 31248362

Hosni, Z., Achour, S., Saadi, F., Chen, Y., & Al Qaraghuli, M. (2025). Machine learning-driven nanoparticle toxicity. *Ecotoxicology and Environmental Safety, 299*, 118340. DOI: 10.1016/j.ecoenv.2025.118340 PMID: 40393320

Hossam Abdelmonem, B., Abdelaal, N. M., Anwer, E. K. E., Rashwan, A. A., Hussein, M. A., Ahmed, Y. F., Khashana, R., Hanna, M. M., & Abdelnaser, A. (2024). Decoding the Role of CYP450 Enzymes in Metabolism and Disease: A Comprehensive Review. *Biomedicines, 12*(7), 1467. DOI: 10.3390/biomedicines12071467 PMID: 39062040

Huang, Y., Guo, X., Wu, Y., Chen, X., Feng, L., Xie, N., & Shen, G. (2024). Nanotechnology's frontier in combatting infectious and inflammatory diseases: Prevention and treatment. *Signal Transduction and Targeted Therapy, 9*(1), 34. Advance online publication. DOI: 10.1038/s41392-024-01745-z PMID: 38378653

Ingber, D. E. (2022). Human organs-on-chips for disease modelling, drug development and personalized medicine. *Nature Reviews. Genetics, 23*(8), 467–491. DOI: 10.1038/s41576-022-00466-9 PMID: 35338360

Islam, S., Ahmed, M. M. S., Islam, M. A., Hossain, N., & Chowdhury, M. A. (2025). Advances in nanoparticles in targeted drug delivery–A review. *Results in Surfaces and Interfaces, 19*, 100529. DOI: 10.1016/j.rsurfi.2025.100529

Jain, E., Shin, J. H., Wells, K. M., Pannone, A. F., Hornsby, P. P., & Warren, C. A. (2020). Association between Clostridioides difficile infection and multidrug-resistant organism colonization or infection among hospitalized adults: A case-control study. *American Journal of Infection Control, 48*(10), 1276–1278. DOI: 10.1016/j.ajic.2020.01.008 PMID: 32145992

Jo, H. E., Kwon, M. S., Whon, T. W., Kim, D. W., Yun, M., Lee, J., Shin, M. Y., Kim, S. H., & Choi, H. J. (2021). Alteration of Gut Microbiota After Antibiotic Exposure in Finishing Swine. *Frontiers in Microbiology, 12*, 596002. DOI: 10.3389/fmicb.2021.596002 PMID: 33643231

Johnston, C. W., & Badran, A. H. (2022). Natural and engineered precision antibiotics in the context of resistance. *Current Opinion in Chemical Biology, 69*, 102160. DOI: 10.1016/j.cbpa.2022.102160 PMID: 35660248

Kan, Z., Wen, J., Bonato, V., Webster, J., Yang, W., Ivanov, V., Kim, K. H., Roh, W., Liu, C., Mu, X. J., Lapira-Miller, J., Oyer, J., VanArsdale, T., Rejto, P. A., & Bienkowska, J. (2025). Real-world clinical multi-omics analyses reveal bifurcation of ER-independent and ER-dependent drug resistance to CDK4/6 inhibitors. *Nature Communications, 16*(1), 932. Advance online publication. DOI: 10.1038/s41467-025-55914-x PMID: 39843429

Khan, F., Shariq, M., Asif, M., Siddiqui, M. A., Malan, P., & Ahmad, F. (2022). Green Nanotechnology: Plant-Mediated Nanoparticle Synthesis and Application. *Nanomaterials (Basel, Switzerland)*, *12*(4), 673. DOI: 10.3390/nano12040673 PMID: 35215000

Khokhlov, I., Legashev, L., Bolodurina, I., Shukhman, A., Shoshin, D., & Kolesnik, S. (2024). Prediction of Dynamic Toxicity of Nanoparticles Using Machine Learning. *Toxics*, *12*(10), 750. DOI: 10.3390/toxics12100750 PMID: 39453170

Kim, M., Gim, D., Kim, S., Park, S., Eom, T. P., Seol, J., Yeo, J., Jo, C., Seo, G., Ku, H., & Kim, J. H. (2025). From Dysbiosis to Prediction: AI-Powered Microbiome Insights into IBD and CRC. *Gastroenterology Insights*, *16*(3), 34. DOI: 10.3390/gastroent16030034

Koirala, M., Yan, L., Mohamed, Z., & DiPaola, M. (2025). AI-Integrated QSAR Modeling for Enhanced Drug Discovery: From Classical Approaches to Deep Learning and Structural Insight. *International Journal of Molecular Sciences*, *26*(19), 9384. DOI: 10.3390/ijms26199384 PMID: 41096653

Krebs, E., & Nosyk, B. (2021). Cost-Effectiveness Analysis in Implementation Science: A Research Agenda and Call for Wider Application. *Current HIV/AIDS Reports*, *18*(3), 176–185. DOI: 10.1007/s11904-021-00550-5 PMID: 33743138

Laxminarayan, R., Duse, A., Wattal, C., Zaidi, A. K., Wertheim, H. F., Sumpradit, N., Vlieghe, E., Hara, G. L., Gould, I. M., Goossens, H., Greko, C., So, A. D., Bigdeli, M., Tomson, G., Woodhouse, W., Ombaka, E., Peralta, A. Q., Qamar, F. N., Mir, F., & Cars, O. (2013). Antibiotic resistance-the need for global solutions. *The Lancet. Infectious Diseases*, *13*(12), 1057–1098. DOI: 10.1016/S1473-3099(13)70318-9 PMID: 24252483

Li, P., Luo, H., Ji, B., & Nielsen, J. (2022). Machine learning for data integration in human gut microbiome. *Microbial Cell Factories*, *21*(1), 241. Advance online publication. DOI: 10.1186/s12934-022-01973-4 PMID: 36419034

Liu, G.-Y., Yu, D., Fan, M.-M., Zhang, X., Jin, Z.-Y., Tang, C., & Liu, X.-F. (2024). Antimicrobial resistance crisis: Could artificial intelligence be the solution? *Military Medical Research*, *11*(1), 7. Advance online publication. DOI: 10.1186/s40779-024-00510-1 PMID: 38254241

Liu, Y., Zhang, L., Jiang, Z., Tian, X., Li, P., Wu, P., Du, W., Yuan, B., Xie, C., Bu, G., Zhong, L., Yang, Y., Li, T., Zeng, M., & Sun, C. (2025). Applications of Artificial Intelligence in Biotech Drug Discovery and Product Development. *MedComm*, 6(8). Portico. DOI: 10.1002/mco2.70317

Ma, Z., Zuo, T., Frey, N., & Rangrez, A. Y. (2024). A systematic framework for understanding the microbiome in human health and disease: From basic principles to clinical translation. *Signal Transduction and Targeted Therapy*, *9*(1), 237. Advance online publication. DOI: 10.1038/s41392-024-01946-6 PMID: 39307902

Manandhar, I., Alimadadi, A., Aryal, S., Munroe, P. B., Joe, B., & Cheng, X. (2021). Gut microbiome-based supervised machine learning for clinical diagnosis of inflammatory bowel diseases. *American Journal of Physiology. Gastrointestinal and Liver Physiology*, *320*(3), G328–G337. DOI: 10.1152/ajpgi.00360.2020 PMID: 33439104

Marcos-Zambrano, L. J., Karaduzovic-Hadziabdic, K., Loncar Turukalo, T., Przymus, P., Trajkovik, V., Aasmets, O., Berland, M., Gruca, A., Hasic, J., Hron, K., Klammsteiner, T., Kolev, M., Lahti, L., Lopes, M. B., Moreno, V., Naskinova, I., Org, E., Paciência, I., Papoutsoglou, G., & Truu, J. (2021). Applications of Machine Learning in Human Microbiome Studies: A Review on Feature Selection, Biomarker Identification, Disease Prediction and Treatment. *Frontiers in Microbiology*, *12*, 634511. DOI: 10.3389/fmicb.2021.634511 PMID: 33737920

Memon, S., Bibi, S., & He, G. (2025). Integration of AI and ML in Tuberculosis (TB) Management: From Diagnosis to Drug Discovery. *Diseases (Basel, Switzerland)*, *13*(6), 184. DOI: 10.3390/diseases13060184 PMID: 40558595

Menon, S., & Koura, K. G. (2025). Artificial intelligence for tuberculosis control: A scoping review of applications in public health. *Journal of Global Health*, *15*, 04192. DOI: 10.7189/jogh.15.04192 PMID: 40709592

Molla, G., & Bitew, M. (2024). Revolutionizing Personalized Medicine: Synergy with Multi-Omics Data Generation, Main Hurdles, and Future Perspectives. *Biomedicines*, *12*(12), 2750. DOI: 10.3390/biomedicines12122750 PMID: 39767657

Mondal, S. K., Chakraborty, S., Manna, S., & Mandal, S. M. (2024). Antimicrobial nanoparticles: Current landscape and future challenges. *RSC Pharmaceutics*, *1*(3), 388–402. DOI: 10.1039/D4PM00032C

Muller, E., Shiryan, I., & Borenstein, E. (2024). Multi-omic integration of microbiome data for identifying disease-associated modules. *Nature Communications*, *15*(1), 2621. Advance online publication. DOI: 10.1038/s41467-024-46888-3 PMID: 38521774

Musa, S. S., Ibrahim, A. M., Alhassan, M. Y., Musa, A. H., Jibo, A. G., Auwal, A. R., Okesanya, O. J., Othman, Z. K., Abubakar, M. S., Ahmed, M. M., Barroso, C. J. V., Sium, A. F., Garcia, M. B., Flores, J. B., Maikifi, A. S., Kouwenhoven, M. B. N., & Lucero-Prisno, D. E. (2025). Nanotechnology and machine learning: A promising confluence for the advancement of precision medicine. *Intelligence-Based Medicine*, *12*, 100267. DOI: 10.1016/j.ibmed.2025.100267

Nazir, F., Tabish, T. A., Tariq, F., Iftikhar, S., Wasim, R., & Shahnaz, G. (2022). Stimuli-sensitive drug delivery systems for site-specific antibiotic release. *Drug Discovery Today*, *27*(6), 1698–1705. DOI: 10.1016/j.drudis.2022.02.014 PMID: 35219858

Nyandoro, V. O., Ismail, E. A., Tageldin, A., Gafar, M. A., Peters, X. Q., Mautsoe, R., Omolo, C. A., & Govender, T. (2025). Potential of nanocarrier-mediated delivery of vancomycin for MRSA infections. *Expert Opinion on Drug Delivery*, *22*(3), 347–365. DOI: 10.1080/17425247.2025.2459756 PMID: 39949087

Ozbek, O., Genc, D. E., & Ulgen, O., K. (. (2024). Advances in Physiologically Based Pharmacokinetic (PBPK) Modeling of Nanomaterials. *ACS Pharmacology &amp. Translational Science*, *7*(8), 2251–2279. DOI: 10.1021/acsptsci.4c00250 PMID: 39144562

Paczesny, J., & Bielec, K. (2020). Application of Bacteriophages in Nanotechnology. *Nanomaterials (Basel, Switzerland)*, *10*(10), 1944. DOI: 10.3390/nano10101944 PMID: 33003494

Palasantzas, V. E. J. M., Tamargo-Rubio, I., Le, K., Slager, J., Wijmenga, C., Jonkers, I. H., Kumar, V., Fu, J., & Withoff, S. (2023). iPSC-derived organ-on-a-chip models for personalized human genetics and pharmacogenomics studies. *Trends in Genetics*, *39*(4), 268–284. DOI: 10.1016/j.tig.2023.01.002 PMID: 36746737

Panchpuri, M., Painuli, R., & Kumar, C. (2025). Artificial intelligence in smart drug delivery systems: A step toward personalized medicine. *RSC Pharmaceutics*, *2*(5), 882–914. DOI: 10.1039/D5PM00089K

Patel, T. K. (2025). Nanotechnology in Herbal-Based Wound Care. Applications of Herbal Excipients and Polymers in Wound Healing, 247–280. DOI: 10.4018/979-8-3373-3977-1.ch010

Patra, J. K., Das, G., Fraceto, L. F., Campos, E. V. R., Rodriguez-Torres, M. D. P., Acosta-Torres, L. S., Diaz-Torres, L. A., Grillo, R., Swamy, M. K., Sharma, S., Habtemariam, S., & Shin, H. S. (2018). Nano based drug delivery systems: Recent developments and future prospects. *Journal of Nanobiotechnology*, *16*(1), 71. DOI: 10.1186/s12951-018-0392-8 PMID: 30231877

Paul, D., & Das, B. (2022). Gut microbiome in the emergence of antibiotic-resistant bacterial pathogens. *Human Microbiome in Health and Disease - Part B*, 1–31. DOI: 10.1016/bs.pmbts.2022.07.009

Pisani, S., Tufail, S., Rosalia, M., Dorati, R., Genta, I., Chiesa, E., & Conti, B. (2024). Antibiotic-Loaded Nano-Sized Delivery Systems: An Insight into Gentamicin and Vancomycin. *Journal of Functional Biomaterials, 15*(7), 194. DOI: 10.3390/jfb15070194 PMID: 39057315

Rani, N., Singh, P., Kumar, S., Kumar, P., Bhankar, V., & Kumar, K. (2023). Plant-mediated synthesis of nanoparticles and their applications: A review. *Materials Research Bulletin, 163*, 112233. DOI: 10.1016/j.materresbull.2023.112233

Ringaci, A., Shevchenko, K. G., Zelepukin, I. V., Popova, A. V., & Nikitin, M. P. (2022). Phage-mimicking nanoagents for rapid depolymerase specificity screening against multidrug resistant bacteria. *Biosensors & Bioelectronics, 213*, 114444. DOI: 10.1016/j.bios.2022.114444 PMID: 35691082

Safari, M. (2023). *Recent Advances in Quantum Dots-Based Biosensors*. Quantum Dots - Recent Advances, New Perspectives and Contemporary Applications., DOI: 10.5772/intechopen.108205

Sagar, A., Kolluru, V., Jaiswal, U., Kumavat, G., Hole, S. R., & Kumar, A. (2025). Machine Learning and Artificial Intelligence for Predictive Modeling in Antimicrobial Resistance Data Sets, Challenges, and Future Directions. 2025 3rd International Conference on Smart Systems for Applications in Electrical Sciences (ICSSES), 1–6. DOI: 10.1109/ICSSES64899.2025.11009316

Sakagianni, A., Koufopoulou, C., Feretzakis, G., Kalles, D., Verykios, V. S., Myrianthefs, P., & Fildisis, G. (2023). Using Machine Learning to Predict Antimicrobial Resistance—A Literature Review. *Antibiotics (Basel, Switzerland), 12*(3), 452. DOI: 10.3390/antibiotics12030452 PMID: 36978319

Sakagianni, A., Koufopoulou, C., Koufopoulos, P., Feretzakis, G., Anastasiou, A., Theodorakis, N., & Myrianthefs, P. (2025). Influence of Microbiome Interactions on Antibiotic Resistance Development in the ICU Environment: Insights and Opportunities with Machine Learning. *Acta Microbiologica Hellenica, 70*(2), 14. DOI: 10.3390/amh70020014

Senjab, R. M., AlSawaftah, N., Abuwatfa, W. H., & Husseini, G. A. (2024). Advances in liposomal nanotechnology: From concept to clinics. *RSC Pharmaceutics, 1*(5), 928–948. DOI: 10.1039/D4PM00176A

Shen, F., Lu, Z., Yan, K., Luo, K., Pei, S., & Xiang, P. (2025). Synthesis and properties of carbon quantum dots as an antimicrobial agent and detection of ciprofloxacin. *Scientific Reports, 15*(1), 28535. Advance online publication. DOI: 10.1038/s41598-025-14383-4 PMID: 40764737

Shen, Y., Fan, N., Ma, S. X., Cheng, X., Yang, X., & Wang, G. (2025). Gut Microbiota Dysbiosis: Pathogenesis, Diseases, Prevention, and Therapy. *MedComm, 6*(5), e70168. DOI: 10.1002/mco2.70168 PMID: 40255918

Smits, W. K., Garey, K. W., Riley, T. V., & Johnson, S. (2025). Clostridioides difficile is a bacterial priority pathogen. *Anaerobe, 93*, 102965. DOI: 10.1016/j.anaerobe.2025.102965 PMID: 40288746

Srivastav, A. K., Mishra, M. K., Lillard, J. W.Jr, & Singh, R. (2025). Transforming Pharmacogenomics and CRISPR Gene Editing with the Power of Artificial Intelligence for Precision Medicine. *Pharmaceutics, 17*(5), 555. DOI: 10.3390/pharmaceutics17050555 PMID: 40430848

Stokes, J. M., Yang, K., Swanson, K., Jin, W., Cubillos-Ruiz, A., Donghia, N. M., MacNair, C. R., French, S., Carfrae, L. A., Bloom-Ackermann, Z., Tran, V. M., Chiappino-Pepe, A., Badran, A. H., Andrews, I. W., Chory, E. J., Church, G. M., Brown, E. D., Jaakkola, T. S., Barzilay, R., & Collins, J. J. (2020). A Deep Learning Approach to Antibiotic Discovery. *Cell, 180*(4), 688–702.e13. DOI: 10.1016/j.cell.2020.01.021 PMID: 32084340

Talkington, A. M., Cao, Y., Kearsley, A. J., & Lai, S. K. (2025). Opportunities for machine learning and artificial intelligence in physiologically-based pharmacokinetic (PBPK) modeling. *Advanced Drug Delivery Reviews, 227*, 115716. Advance online publication. DOI: 10.1016/j.addr.2025.115716 PMID: 41093240

Tan, R., Jin, M., Li, J., & Yang, D. (2025). The dissemination, health risks, and mitigation approaches of antibiotic resistance genes in the gut microbiome. *Journal of Hazardous Materials Advances, 17*, 100634. DOI: 10.1016/j.hazadv.2025.100634

Theophilus, R. J., & Taft, D. H. (2023). Antimicrobial Resistance Genes (ARGs), the Gut Microbiome, and Infant Nutrition. *Nutrients, 15*(14), 3177. DOI: 10.3390/nu15143177 PMID: 37513595

Tsirogianni, A. G., Chountoulesi, M., & Pippa, N. (2025). Recent Advances in pH-Responsive Liposomes: Lessons Learnt and New Directions in Nanomedicine Development. *Materials (Basel), 18*(18), 4295. DOI: 10.3390/ma18184295 PMID: 41010139

Vahabi, N., & Michailidis, G. (2022). Unsupervised Multi-Omics Data Integration Methods: A Comprehensive Review. *Frontiers in Genetics, 13*, 854752. Advance online publication. DOI: 10.3389/fgene.2022.854752 PMID: 35391796

Wang, J., Li, B., Qiu, L., Qiao, X., & Yang, H. (2022). Dendrimer-based drug delivery systems: History, challenges, and latest developments. *Journal of Biological Engineering*, *16*(1), 18. Advance online publication. DOI: 10.1186/s13036-022-00298-5 PMID: 35879774

Wang, Q., Sun, B., Yi, Y., Velkov, T., Shen, J., Dai, C., & Jiang, H. (2025). Progress of AI-Driven Drug–Target Interaction Prediction and Lead Optimization. *International Journal of Molecular Sciences*, *26*(20), 10037. DOI: 10.3390/ijms262010037 PMID: 41155330

Wang, Y., Zhao, Y., Bollas, A., Wang, Y., & Au, K. F. (2021). Nanopore sequencing technology, bioinformatics and applications. *Nature Biotechnology*, *39*(11), 1348–1365. DOI: 10.1038/s41587-021-01108-x PMID: 34750572

Wickramage, I., Spigaglia, P., & Sun, X. (2021). Mechanisms of antibiotic resistance of Clostridioides difficile. *The Journal of Antimicrobial Chemotherapy*, *76*(12), 3077–3090. DOI: 10.1093/jac/dkab231 PMID: 34297842

World Health Organization. (2022). Global Antimicrobial Resistance and Use Surveillance System (GLASS) report: Antibiotic use data for 2022. https://www.who.int/publications/i/item/9789240108127

World Health Organization. (2023). Antimicrobial resistance. https://www.who.int/news-room/fact-sheets/detail/antimicrobial-resistance

Yetgin, A. (2025). Revolutionizing multi-omics analysis with artificial intelligence and data processing. *Quantitative Biology*, 13(3). Portico. DOI: 10.1002/qub2.70002

Zhang, J., Li, H., Zhang, Y., Huang, J., Ren, L., Zhang, C., Zou, Q., & Zhang, Y. (2025). Computational toxicology in drug discovery: Applications of artificial intelligence in ADMET and toxicity prediction. *Briefings in Bioinformatics*, *26*(5), bbaf533. Advance online publication. DOI: 10.1093/bib/bbaf533 PMID: 41052279

Zhang, L., Chen, J., Zhang, F., Xu, J., Bi, N., Gou, J., & Jia, L. (2022). Silicon quantum dots and MOFs hybrid multicolor fluorescent nanosensor for ultrasensitive and visual intelligent sensing of tetracycline. *Colloids and Surfaces. A, Physicochemical and Engineering Aspects*, *652*, 129853. DOI: 10.1016/j.colsurfa.2022.129853

Zhou, R., Ng, S. K., Sung, J. J. Y., Goh, W. W. B., & Wong, S. H. (2023). Data pre-processing for analyzing microbiome data – A mini review. *Computational and Structural Biotechnology Journal*, *21*, 4804–4815. DOI: 10.1016/j.csbj.2023.10.001 PMID: 37841330

Zhou, Y., Wang, X., Tian, X., Zhang, D., Cui, H., Du, W., Yang, Z., Li, J., Li, W., Xu, J., Duanmu, Y., Yu, T., Cai, F., Li, W., Jin, Z., Wu, W., & Huang, H. (2025). Stealth missiles with precision guidance: A novel multifunctional nano-drug delivery system based on biomimetic cell membrane coating technology. *Materials Today. Bio*, *33*, 101922. DOI: 10.1016/j.mtbio.2025.101922 PMID: 40528836

Chapter 4

AI-Enabled Nanobiotechnological Platforms for Precision Antibiotic Discovery and Targeted Antimicrobial Delivery

Eanest Jebasingh
http://orcid.org/0000-0003-1827-3464
SRM Madurai College for Engineering and Technology, India

G. Ashwin Prabhu
St. Joseph's College of Engineering, India

S. Karuppaswamy
Meenakshi College of Engineering, India

K. M. Kumar
http://orcid.org/0000-0002-4268-6309
St. Joseph's College of Engineering, India

Senthil Nehru
P.T. Lee CRPT College, India

Ravi Kumar Panthangi
CMR College of Engineering and Technology, India

Anin Joe B.
St. Joseph's College of Engineering, India

Aalan Jayaraj J.
St. Joseph's College of Engineering, India

DOI: 10.4018/979-8-3373-5268-8.ch004

ABSTRACT

An emerging convergence of artificial intelligence and nanobiotechnology is transforming precision antibiotic discovery and targeted antimicrobial delivery. This chapter per the authors analyzes how machine learning models and bioinformatics enable data-driven drug discovery through pathogen genomics and molecular targeting strategies. The discussion further highlights the use of nanocarriers for controlled drug release and smart drug delivery, supporting improved therapeutic optimization while minimizing off-target effects and toxicity. Emphasis is placed on resistance mitigation through predictive modeling and adaptive delivery mechanisms. By integrating computational intelligence with systems biology and translational medicine principles, the chapter presents an interdisciplinary framework demonstrating how AI-enabled nanobiotechnological platforms can enhance the accuracy, effectiveness, and sustainability of next-generation antimicrobial therapies.

1. INTRODUCTION

Antimicrobial resistance (AMR) has emerged as one of the most critical global public health threats of the twenty-first century, undermining decades of progress in infectious disease control. The increasing prevalence of multidrug-resistant pathogens has significantly reduced the effectiveness of conventional antibiotics, leading to prolonged illnesses, higher healthcare costs, and increased mortality worldwide (World Health Organization [WHO], 2023). Factors such as the overuse and misuse of antibiotics, limited surveillance systems, and slow innovation in drug development have accelerated resistance evolution, creating an urgent need for alternative therapeutic strategies. Traditional antibiotic discovery pipelines, which rely heavily on empirical screening and incremental molecular modification, are increasingly inadequate to address the complexity and adaptability of modern pathogens. Conventional antibiotic discovery is further constrained by long development timelines, high attrition rates, and escalating research and development costs. Many promising compounds fail during late-stage clinical trials due to toxicity, poor bioavailability, or limited pathogen specificity. Moreover, classical delivery methods often result in suboptimal drug concentrations at infection sites, contributing to off-target effects and selective pressure that accelerates resistance development. These limitations highlight the necessity for precision-driven approaches that can enhance both drug discovery efficiency and therapeutic effectiveness (Ventola, 2015).

Recent advances in artificial intelligence (AI) have introduced transformative opportunities in biomedical research, particularly in the domain of antimicrobial drug discovery. The transformative impact of AI has been demonstrated in diverse domains, from personalized education systems enabling adaptive learning and student support (Ruba et al., 2026) to precision agriculture through AI-powered multipurpose robots capable of adaptive crop health management and predictive analytics (Dhanushya et al., 2025). As discussed in Ruba et al. (2026), where smart pedagogies and intelligent campuses illustrate AI's role in reimagining higher education. Similarly, IoT-integrated solutions for smart vertical crop cultivation (Gowthaman et al., 2025) highlight how AI and IoT converge to optimize resource use and enhance sustainability in agriculture. Such examples underscore AI's potential to process complex datasets, predict outcomes, and optimize interventions—capabilities directly translatable to antimicrobial drug discovery. Machine learning models and bioinformatics tools enable the analysis of vast biological datasets, including pathogen genomics, protein structures, and molecular interaction networks. By leveraging data-driven methodologies, AI facilitates rapid identification of novel antimicrobial candidates, prediction of molecular targeting efficacy, and optimization of drug properties prior to experimental validation. These computational approaches significantly reduce discovery timelines and enable rational design strategies tailored to pathogen-specific characteristics (Stokes et al., 2020). Parallel to these developments, nanobiotechnology has emerged as a powerful enabler of advanced drug delivery and therapeutic optimization. Nanocarriers such as polymeric nanoparticles, liposomes, and bio-inspired nanostructures offer enhanced control over drug stability, solubility, and release kinetics. At the nanoscale, antimicrobial agents can be protected from premature degradation and delivered selectively to infection sites, improving therapeutic index while minimizing systemic toxicity. Controlled drug release mechanisms further allow sustained antimicrobial exposure, reducing dosing frequency and mitigating resistance development associated with fluctuating drug concentrations. Similar to autonomous AI-controlled systems in engineering applications, such as real-time rover navigation platforms. Comparable innovations include IoT-enabled smart traffic control systems for hill roads with emergency vehicle priority (Poovithaa et al., 2025), which demonstrate AI's adaptability in complex, real-world environments. Nano-biotechnological delivery systems can dynamically adapt release profiles in response to changing biological conditions, demonstrating the potential of AI to enable real-time adaptive therapeutic interventions. The convergence of AI and nanobiotechnology represents a paradigm shift in precision antibiotic development. AI-driven drug discovery pipelines can be seamlessly integrated with nanocarrier-based delivery systems to create smart antimicrobial platforms capable of adaptive response and targeted intervention. Through molecular targeting and pathogen-specific modeling, these platforms support personalized and context-aware therapeutic strategies. Such integration aligns with systems biology principles, enabling holistic optimization across drug design, delivery, and clinical application. Importantly, these intelligent platforms provide novel avenues for resistance mitigation by minimizing unnecessary antimicrobial exposure and enabling predictive treatment adjustment. From a translational medicine perspective, AI-enabled nanobiotechnological platforms hold significant promise for bridging the gap between laboratory innovation and clinical implementation. These technologies facilitate scalable design, improved reproducibility, and data-driven decision-making throughout the therapeutic development lifecycle. By incorporating real-time data analytics and adaptive modeling, antimicrobial interventions can be continuously refined to address evolving resistance patterns and diverse patient populations. This translational potential is particularly relevant for global health settings, where resource constraints demand efficient and sustainable therapeutic solutions. The objective of this chapter is to critically examine the role of AI-enabled nanobiotechnological platforms in advancing precision antibiotic discovery and targeted antimicrobial delivery. The chapter explores how machine learning models, bioinformatics, and pathogen genomics contribute to data-driven drug discovery, while nanocarriers and smart drug delivery systems enhance therapeutic optimization. It further analyzes integrated strategies for resistance mitigation and discusses translational and clinical implications within modern healthcare ecosystems. By synthesizing interdisciplinary perspectives, this chapter contributes to a comprehensive understanding of next-generation antimicrobial interventions. In positioning AI-enabled nanobiotechnological platforms as a transformative solution, this chapter underscores their potential to redefine antimicrobial therapeutics in an era of escalating resistance. Through intelligent integration of computational and nanoscale technologies, precision-based approaches offer sustainable pathways for improving treatment efficacy, reducing resistance emergence, and advancing global infectious disease management. As antimicrobial resistance continues to challenge conventional paradigms, such innovative platforms are poised to play a central role in the future of precision antibiotics.

2. BACKGROUND AND CONCEPTUAL FOUNDATIONS

The growing global burden of antimicrobial resistance (AMR) has exposed fundamental limitations in conventional antibiotic discovery and delivery paradigms. Traditional drug discovery pipelines rely heavily on empirical screening, incremental chemical modification, and prolonged trial-and-error experimentation. These approaches are resource intensive, time consuming, and increasingly ineffective against rapidly evolving pathogens. The emergence of multidrug-resistant bacteria, biofilm-associated infections, and host–pathogen heterogeneity has intensified the need for precision-driven strategies capable of addressing microbial complexity at molecular and systems levels (World Health Organization [WHO], 2023). One of the primary challenges in conventional antibiotic discovery is the declining rate of novel antimicrobial targets. Many existing antibiotics act on a narrow set of bacterial processes, such as cell wall synthesis or protein translation, leading to predictable resistance mechanisms. Furthermore, traditional in vitro screening methods often fail to capture the dynamic biological environments encountered in vivo, resulting in poor translational success. These challenges are compounded by the genetic plasticity of pathogens, horizontal gene transfer, and adaptive resistance mechanisms that rapidly diminish therapeutic efficacy (Ventola, 2015). Emerging pathogen complexity necessitates a paradigm shift from single-target approaches toward system-level understanding of microbial behavior. Pathogen genomics has revealed extensive diversity within and across bacterial species, highlighting strain-specific virulence factors, metabolic pathways, and resistance genes. This complexity requires analytical frameworks capable of integrating large-scale biological data and identifying meaningful therapeutic patterns. As a result, computational approaches have become central to modern biomedicine, enabling the analysis of genomic, proteomic, and metabolomic datasets with unprecedented precision.

Bioinformatics serves as a foundational pillar in this transition by providing tools and methodologies for managing and interpreting biological data. Through sequence alignment, functional annotation, and comparative genomics, bioinformatics enables the identification of potential drug targets and resistance markers. When combined with machine learning models, these tools facilitate predictive insights into pathogen susceptibility and drug–target interactions. Such data-driven approaches significantly enhance the efficiency and accuracy of antibiotic discovery by reducing reliance on exhaustive laboratory experimentation (Chen et al., 2018). Closely aligned with bioinformatics is systems biology, which emphasizes holistic modeling of biological systems rather than isolated components. Systems biology integrates molecular interactions, regulatory networks, and environmental influences to understand emergent behaviors within biological systems. In the context of antimicrobial research, systems-level models enable the simulation of host–pathogen interactions, prediction of resistance evolution, and evaluation of therapeutic interventions under diverse conditions. This integrative perspective is essential for designing antibiotics that remain effective in complex biological environments. While computational intelligence advances discovery, effective therapeutic outcomes also depend on efficient drug delivery mechanisms. Conventional drug delivery systems typically rely on passive diffusion and systemic circulation, often resulting in suboptimal drug concentrations at infection sites and unintended off-target effects. These limitations can contribute to toxicity, reduced efficacy, and accelerated resistance development. Consequently, innovative delivery strategies are required to ensure precise spatial and temporal control of antimicrobial agents.

Nanobiotechnology introduces transformative solutions to these challenges through nanoscale engineering of drug delivery platforms. Nanocarriers, including polymeric nanoparticles, liposomes, and bio-inspired nanostructures, are designed to encapsulate therapeutic agents and protect them from degradation. Their tunable physicochemical properties allow for enhanced bioavailability, prolonged circulation, and targeted interaction with microbial cells. By operating at the nanoscale, these systems can interact more effectively with biological barriers and cellular components (Peer et al., 2020). A key conceptual advancement enabled by nanocarrier design is controlled drug release. Unlike conventional delivery systems, nanobiotechnological platforms can be engineered to release antimicrobial agents in response to specific stimuli, such as pH changes, enzymatic activity, or microbial metabolites. This controlled release capability ensures that therapeutic agents are delivered at optimal concentrations precisely where and when they are needed, thereby improving therapeutic optimization and reducing systemic toxicity. The distinction between conventional and smart drug delivery systems lies in the integration of computational intelligence with nanoscale engineering. Smart drug delivery systems leverage AI-driven models to predict infection dynamics, optimize release profiles, and adapt therapeutic strategies in real time. By combining machine learning insights with nanocarrier functionality, these systems enable personalized and adaptive antimicrobial interventions. This integration represents a critical conceptual shift from static treatment regimens to intelligent, data-driven therapeutic platforms. Parallel research on AI-driven healthcare equity highlights the importance of designing adaptive and inclusive systems during crises, reinforcing the need for antimicrobial platforms that prioritize accessibility and resilience (Ponkarthika et al., 2026). Collectively, the convergence of bioinformatics, systems biology, and nanobiotechnology establishes a robust conceptual foundation for AI-enabled precision antibiotics. These interdisciplinary frameworks support not only efficient drug discovery but also targeted antimicrobial delivery and resistance mitigation. Understanding these foundational concepts is essential for advancing next-generation antimicrobial strategies that are effective, sustainable, and clinically translatable.

3. ROLE OF ARTIFICIAL INTELLIGENCE IN PRECISION ANTIBIOTIC DISCOVERY

Artificial intelligence (AI) has emerged as a transformative force in precision antibiotic discovery by addressing long-standing limitations associated with conventional trial-and-error–based drug development. Traditional antimicrobial discovery pipelines are often time-intensive, costly, and constrained by limited success rates, particularly in the context of rapidly evolving antimicrobial resistance. AI-driven computational approaches enable data-centric exploration of biological complexity, allowing for faster, more accurate identification and optimization of antimicrobial candidates through advanced learning and prediction mechanisms.

3.1 Machine Learning Models in Drug Discovery

Machine learning models play a central role in enhancing drug discovery efficiency by learning complex patterns from large-scale biological and chemical datasets. AI's ability to recognize complex patterns in heterogeneous datasets has been demonstrated in both organizational contexts (Ruba et al., 2026) and cybersecurity, where machine learning models analyze high-dimensional features to detect malicious activity (Krithika et al., 2025). Such approaches parallel antimicrobial discovery workflows, where predictive pattern recognition can accelerate identification of candidate molecules and resistance mechanisms. Supervised, unsupervised, and deep learning algorithms are increasingly applied to predict antimicrobial activity, toxicity profiles, and pharmacokinetic properties. These models can process multidimensional data derived from chemical structures, biological assays, and genomic information, enabling early-stage screening of vast compound libraries with significantly reduced experimental burden (Chen et al., 2018). Deep learning architectures, such as convolutional and graph neural networks, have demonstrated superior performance in modeling molecular interactions and predicting structure–activity relationships. By learning hierarchical representations of chemical features, these models facilitate accurate assessment of candidate efficacy before laboratory validation, thereby reducing development timelines and improving success rates.

3.2 Pathogen Genomics and Bioinformatics Integration

The integration of pathogen genomics with AI-driven bioinformatics has revolutionized precision antibiotic discovery. High-throughput sequencing technologies generate extensive genomic datasets that capture pathogen variability, virulence factors, and resistance determinants. AI algorithms analyze these datasets to identify conserved genomic targets and strain-specific vulnerabilities, supporting precision molecular targeting strategies. Machine learning–based genomic analysis enables the prediction of resistance mechanisms and evolutionary trajectories, which is critical for designing antibiotics with sustained effectiveness. By correlating genomic features with phenotypic resistance data, AI systems facilitate the selection of drug targets less prone to resistance development. This genomics-informed approach enhances the specificity and clinical relevance of antimicrobial.

3.3 Molecular Targeting and Target Identification

AI enhances molecular targeting by enabling precise identification of biologically relevant targets within complex pathogen systems. Computational models analyze protein structures, metabolic pathways, and interaction networks to prioritize targets essential for pathogen survival. Network-based and systems biology approaches supported by AI uncover critical nodes within cellular pathways, allowing for rational target selection. Predictive modeling techniques assess target druggability and binding affinity, guiding the design of molecules with improved specificity. This approach minimizes off-target effects and reduces toxicity risks, aligning with precision medicine objectives. AI-driven target prioritization also supports the repurposing of existing compounds by identifying novel target interactions, accelerating therapeutic development.

3.4 Structure–Activity Relationship Modeling

Structure–activity relationship (SAR) modeling is a cornerstone of AI-enabled antibiotic discovery. Machine learning algorithms analyze the relationship between molecular structure and biological activity, enabling systematic optimization of antimicrobial compounds. Quantitative SAR models leverage chemical descriptors and molecular fingerprints to predict efficacy, solubility, and toxicity profiles. AI-enhanced SAR modeling allows rapid iteration and refinement of candidate molecules, supporting rational design rather than empirical testing. This capability significantly reduces the number of experimental cycles required to achieve optimal therapeutic properties. Advanced deep learning–based SAR models further improve predictive accuracy by capturing non-linear relationships within chemical datasets (Stokes et al., 2020).

3.5 In Silico Screening of Antimicrobial Compounds

In silico screening represents one of the most impactful applications of AI in antibiotic discovery. Virtual screening platforms powered by machine learning can evaluate millions of compounds against multiple biological targets in a fraction of the time required for experimental assays. These platforms integrate docking simulations, predictive scoring functions, and learning-based ranking systems to identify high-potential candidates. AI-driven screening not only accelerates candidate identification but also enables the exploration of chemical spaces beyond traditional antibiotic scaffolds. This capability is particularly valuable for discovering structurally novel compounds capable of overcoming existing resistance mechanisms.

By prioritizing candidates with favorable safety and efficacy profiles, AI reduces downstream attrition rates.

3.6 Predictive Analytics for Resistance Mitigation

Resistance mitigation is a critical focus of AI-enabled antibiotic discovery. Predictive analytics models analyze historical resistance patterns, genomic mutations, and treatment outcomes to forecast resistance emergence. These models support the design of antibiotics with enhanced durability by identifying features associated with rapid resistance development. Adaptive modeling techniques allow continuous learning from new data, enabling real-time refinement of predictive frameworks. This dynamic approach supports proactive therapeutic strategies that anticipate resistance trends rather than reacting to clinical failures. By integrating resistance prediction with molecular design and delivery optimization, AI contributes to sustainable antimicrobial development. Comparable predictive analytics frameworks have been applied in energy systems, such as AI-enhanced wind turbine maintenance, demonstrating how intelligent monitoring can anticipate failures and optimize interventions—principles directly translatable to antimicrobial resistance mitigation (Cloudin et al., 2026).

3.7 AI-Driven Bioinformatics Workflows

AI-driven bioinformatics workflows integrate data preprocessing, feature extraction, model training, and validation into cohesive pipelines that streamline antibiotic discovery. These workflows enable seamless analysis of heterogeneous datasets, including genomic sequences, chemical structures, and experimental results. Automation of these processes reduces human bias and enhances reproducibility. The interoperability of AI tools with high-throughput experimental platforms further accelerates translational progress. By bridging computational predictions with laboratory validation, AI-enabled workflows shorten discovery cycles and improve decision-making accuracy. This integration supports the rapid transition from computational insights to clinically viable antimicrobial candidates.

3.8 Implications for Precision and Translational Medicine

The application of AI in precision antibiotic discovery has significant implications for translational medicine. By enabling patient- and pathogen-specific therapeutic design, AI supports personalized antimicrobial strategies that improve clinical outcomes. The alignment of genomic insights, molecular targeting, and predictive analytics enhances therapeutic optimization while minimizing adverse effects. AI-driven discovery frameworks also facilitate scalability and adaptability, making them suitable for global resistance surveillance and rapid response to emerging pathogens. As computational models continue to evolve, their integration with nanobiotechnological delivery systems further strengthens the precision and effectiveness of next-generation antimicrobial therapies.

4. NANOBIOTECHNOLOGICAL PLATFORMS AND NANOCARRIER SYSTEMS

Nanobiotechnological platforms have emerged as a transformative approach for addressing long-standing limitations in antimicrobial therapy, particularly those related to drug stability, bioavailability, and targeted delivery. By operating at the nanoscale, these platforms enable precise control over physicochemical properties, biological interactions, and therapeutic performance. In the context of precision antibiotic discovery and delivery, nanocarrier systems play a critical role in enhancing antimicrobial efficacy while minimizing systemic toxicity and resistance development.

4.1 Design Principles of Nanobiotechnological Platforms

The design of nanobiotechnological platforms is guided by interdisciplinary principles encompassing materials science, molecular biology, and biomedical engineering. Key design considerations include particle size, surface charge, morphology, and functionalization, all of which directly influence circulation time, cellular uptake, and targeting efficiency. Nanoscale dimensions allow carriers to traverse biological barriers, such as cellular membranes and biofilms, which are often impermeable to conventional antibiotic formulations. Surface modification strategies, including ligand attachment and polymer coating, are commonly employed to enhance biocompatibility and specificity. Functionalization with targeting moieties enables selective interaction with pathogenic cells while reducing off-target accumulation in healthy tissues. Such design flexibility makes nanobiotechnological platforms highly adaptable to pathogen-specific therapeutic requirements.

4.2 Classification of Nanocarrier Systems

Nanocarriers used in antimicrobial delivery can be broadly classified based on their material composition and structural characteristics. Among the most widely studied systems are polymeric nanoparticles, liposomes, and bio-inspired nanostructures.

- Polymeric nanoparticles are constructed from biodegradable polymers such as poly(lactic-co-glycolic acid) (PLGA) and chitosan. These carriers offer high structural stability and tunable drug release profiles. Their ability to encapsulate both hydrophilic and hydrophobic antimicrobial agents makes them particularly suitable for precision drug delivery applications. Additionally, polymeric nanoparticles can be engineered to respond to environmental stimuli, such as pH or enzymatic activity, enabling controlled release at infection sites (Kamaly et al., 2016).
- Liposomes are lipid-based vesicles that mimic biological membranes, offering exceptional biocompatibility and reduced immunogenicity. Their bilayer structure allows for efficient encapsulation of antibiotics and protects active compounds from enzymatic degradation. Liposomes have demonstrated effectiveness in enhancing drug solubility and prolonging circulation time, thereby improving therapeutic outcomes in antimicrobial treatments (Allen & Cullis, 2013).
- Bio-inspired nanostructures, including peptide-based and biomimetic nanoparticles, represent an emerging class of nanocarriers. These systems are designed to replicate natural biological functions, such as membrane fusion or receptor recognition. Their inherent compatibility with biological systems enables efficient cellular interaction and targeted antimicrobial delivery. Such platforms are particularly valuable in overcoming microbial defense mechanisms, including biofilm formation.

4.3 Functional Mechanisms in Antimicrobial Delivery

Nanocarrier systems function through multiple mechanisms that collectively enhance antimicrobial performance. One primary mechanism is the protection of active compounds from premature degradation. Antibiotics encapsulated within nanocarriers are shielded from enzymatic breakdown and unfavorable physiological conditions, thereby preserving their therapeutic activity until reaching the target site. Another critical mechanism involves enhanced cellular uptake. Due to their nanoscale size, carriers are readily internalized by infected cells through endocytosis, ensuring localized drug accumulation. This targeted delivery reduces the required dosage and minimizes systemic exposure, contributing to improved safety profiles. Nanocarriers also facilitate controlled drug release, allowing sustained therapeutic concentrations over extended periods. Controlled release mechanisms reduce dosing frequency and prevent peak plasma concentrations that may cause toxicity. Similar IoT-enabled wearable health monitoring systems demonstrate how real-time adaptive feedback can enhance patient outcomes, offering insights into how smart antimicrobial delivery platforms may integrate continuous monitoring for precision therapy

(Sibiya et al., 2025). This feature is particularly relevant in antimicrobial therapy, where maintaining effective drug levels is essential for eradicating pathogens and preventing resistance development.

4.4 Role of Nanoscale Engineering in Improving Bioavailability

Nanoscale engineering significantly improves drug bioavailability by addressing solubility and permeability challenges associated with many antibiotics. Poor aqueous solubility is a major limitation for numerous antimicrobial agents, leading to reduced absorption and therapeutic inefficiency. Nanocarriers enhance solubility by encapsulating hydrophobic drugs within their core or lipid bilayer, facilitating improved dispersion in biological fluids (Huh & Kwon, 2011). Furthermore, nanoscale surface engineering enables prolonged circulation by reducing recognition and clearance by the immune system. Polyethylene glycol (PEG) coating, for instance, is commonly used to enhance stealth properties and extend systemic residence time. These advancements ensure that a higher proportion of the administered drug reaches the intended infection site.

4.5 Targeting Efficiency and Precision Therapeutics

Targeting efficiency is a defining advantage of nanobiotechnological platforms in antimicrobial delivery. Active targeting strategies utilize surface ligands that bind to pathogen-specific receptors or infection-associated biomarkers. This precision approach aligns with the principles of personalized and targeted therapy, enabling selective pathogen eradication while preserving beneficial microbiota. Passive targeting mechanisms, such as enhanced permeability and retention effects at inflamed or infected tissues, further contribute to delivery efficiency. By combining these strategies, nanocarriers achieve high localization accuracy, which is essential for effective antimicrobial intervention and resistance mitigation.

4.6 Challenges and Future Optimization

Despite their advantages, nanobiotechnological platforms face challenges related to scalability, reproducibility, and regulatory approval. Manufacturing consistency and long-term safety evaluation remain critical concerns for clinical translation. Ongoing research focuses on optimizing material selection, improving formulation stability, and integrating computational modeling to enhance design predictability. Future developments are expected to leverage artificial intelligence for rational nanocarrier design, enabling data-driven optimization of size, composition, and release kinetics. Such integration will further strengthen the role of nanobiotechnological platforms in next-generation precision antimicrobial therapies.

5. CONTROLLED DRUG RELEASE AND SMART ANTIMICROBIAL DELIVERY

Controlled drug release and smart antimicrobial delivery represent critical advancements in addressing the limitations of conventional antibiotic therapies, particularly issues related to systemic toxicity, poor bioavailability, and suboptimal therapeutic concentrations at infection sites. Nanobiotechnological platforms enable precise control over drug release kinetics and spatial targeting, thereby improving therapeutic optimization and minimizing off-target effects. When integrated with artificial intelligence (AI), these platforms offer adaptive, data-driven delivery strategies that respond dynamically to biological and environmental cues.

5.1 Principles of Controlled Drug Release

Controlled drug release refers to the regulated delivery of therapeutic agents over a predetermined time frame to maintain optimal drug concentrations at the target site. Nanocarriers play a central role in this process by encapsulating antimicrobial agents and protecting them from premature degradation. Controlled release profiles reduce dosing frequency, enhance patient compliance, and limit systemic exposure, which is particularly important in antimicrobial therapy where toxicity and resistance development are major concerns (Peer et al., 2020). Nanocarrier-based systems enable sustained, pulsatile, or delayed release depending on their physicochemical properties, including size, surface charge, and material composition. These properties can be engineered to align with pharmacokinetic and pharmacodynamic requirements of specific antimicrobial agents. AI-driven modeling supports the optimization of these parameters by predicting release behavior under physiological conditions, allowing for rational design of delivery systems tailored to pathogen-specific and patient-specific needs.

5.2 Stimuli-Responsive Nanocarriers

Stimuli-responsive nanocarriers represent a significant advancement in smart drug delivery. These systems release their therapeutic payload in response to specific internal or external stimuli, such as pH changes, enzymatic activity, temperature variations, or redox conditions. In infectious microenvironments, localized physiological changes—such as acidic pH or elevated enzyme concentrations—can trigger targeted antimicrobial release, enhancing efficacy while sparing healthy tissues (Mura et al., 2013). AI-enabled bioinformatics and machine learning models enhance the design of stimuli-responsive systems by analyzing complex biological data to identify optimal trigger conditions. Predictive modeling enables the selection of appropriate stimuli-response mechanisms based on pathogen characteristics and infection site microenvironments. This integration supports precision antimicrobial delivery, reducing the likelihood of drug degradation and improving site-specific action.

5.3 Site-Specific Targeting and Smart Delivery

Site-specific targeting is a cornerstone of smart antimicrobial delivery. In other domains, AI has enabled precise and adaptive operations, such as ROV-based inspections for infrastructure monitoring and adaptive assessment frameworks in education systems (Santhini et al., 2025). These examples highlight AI's ability to achieve site-specific and context-aware interventions, analogous to targeted antimicrobial delivery systems. Nanocarriers can be functionalized with targeting ligands, such as peptides, antibodies, or small molecules, that selectively bind to pathogen-associated markers or infected tissues. This molecular targeting approach ensures higher drug accumulation at infection sites, thereby improving antimicrobial efficacy and reducing systemic toxicity. AI-driven molecular targeting strategies further enhance specificity by identifying optimal ligand–receptor interactions through large-scale genomic and proteomic analyses. Pathogen genomics data enable the identification of conserved surface markers, which can be exploited for targeted delivery. Machine learning models assist in predicting binding affinity, stability, and delivery efficiency, thereby accelerating the development of smart drug delivery systems.

5.4 Adaptive Release Profiles and Therapeutic Optimization

Adaptive release profiles allow nanocarriers to modulate drug release rates in response to changing biological conditions. This capability is particularly valuable in antimicrobial therapy, where infection severity and pathogen load may vary over time. AI-enabled adaptive systems can adjust release kinetics based on real-time data, such as biomarker levels or therapeutic response indicators, enabling dynamic therapeutic optimization. Machine learning algorithms analyze historical and real-time data to predict optimal dosing strategies, reducing the risk of underdosing or overdosing. Such adaptive approaches contribute to resistance mitigation by maintaining effective drug concentrations that suppress pathogen survival while minimizing selective pressure for resistant strains. The integration of AI with controlled release technologies thus supports more sustainable antimicrobial treatment paradigms.

5.5 Minimizing Toxicity and Off-Target Effects

One of the primary advantages of smart antimicrobial delivery systems is their ability to minimize toxicity and off-target effects. Conventional antibiotics often distribute non-specifically throughout the body, leading to adverse effects and disruption of beneficial microbiota. Nanocarrier-mediated delivery confines drug activity to infected regions, reducing collateral damage to healthy tissues. AI-driven safety modeling enhances this process by predicting toxicity profiles and off-target interactions during the design phase. Computational simulations assess how nanocarriers interact with biological systems, allowing researchers to optimize formulations before clinical application. This predictive capability reduces trial-and-error experimentation and improves the overall safety profile of antimicrobial therapies.

5.6 Clinical Implications and Future Potential

The integration of controlled drug release, smart delivery systems, and AI-driven optimization holds significant promise for clinical translation. These technologies support personalized antimicrobial therapy by accounting for patient-specific factors, pathogen variability, and disease progression. While challenges remain—such as scalability, regulatory approval, and long-term safety evaluation—the convergence of AI and nanobiotechnology offers a robust framework for next-generation antimicrobial interventions. Continued advances in machine learning, systems biology, and nanocarrier engineering are expected to further enhance delivery precision and therapeutic outcomes. As antimicrobial resistance continues to pose a global health threat, smart antimicrobial delivery systems represent a critical component of future precision medicine strategies.

6. INTEGRATING AI AND NANOBIOTECHNOLOGY FOR RESISTANCE MITIGATION

Antimicrobial resistance (AMR) has emerged as one of the most pressing global health threats, driven by the overuse and misuse of antibiotics, slow drug development pipelines, and the adaptive evolution of pathogenic microorganisms. Conventional antimicrobial strategies often fail to account for the dynamic and systems-level nature of resistance development, resulting in diminished therapeutic efficacy over time. In this context, the integration of artificial intelligence (AI) and nanobiotechnology offers a promising paradigm for resistance mitigation by enabling predictive modeling, adaptive therapeutic design, and intelligent drug delivery systems. AI-enabled resistance mitigation begins with predictive resistance modeling, which leverages machine learning algorithms and pathogen genomics to anticipate resistance patterns before they emerge clinically. By analyzing large-scale genomic datasets, AI systems can identify resistance-conferring mutations, gene transfer mechanisms, and evolutionary trajectories across bacterial populations. Such predictive capabilities allow researchers to design antibiotics that are less susceptible to known resistance pathways and to anticipate potential escape mechanisms. This data-driven foresight represents a significant advancement over traditional trial-and-error approaches in antimicrobial development.

Nanobiotechnological platforms complement these predictive models by providing precise and adaptable drug delivery mechanisms. Nanocarriers can be engineered to enhance intracellular penetration, bypass bacterial defense mechanisms, and deliver therapeutics directly to infection sites. When integrated with AI-driven insights, these nanoscale systems can be optimized for molecular targeting, dosage control, and release kinetics, thereby reducing selective pressure that often accelerates resistance development (Huh & Kwon, 2021). The convergence of AI and nanotechnology thus enables a proactive rather than reactive approach to resistance mitigation. Adaptive therapeutic strategies represent another critical dimension of this integrated framework. AI models continuously analyze patient-specific data, pathogen response profiles, and treatment outcomes to dynamically adjust therapeutic regimens. This adaptability allows for real-time optimization of antibiotic combinations, dosing schedules, and delivery modes. Nanocarriers further enhance this adaptability by enabling controlled drug release and stimuli-responsive behavior, such as pH- or enzyme-triggered activation at infection sites. Together, these technologies minimize unnecessary antibiotic exposure, which is a key driver of resistance emergence (Topol, 2019).

Systems-level optimization is central to the success of AI-enabled nanobiotechnological resistance mitigation. Biological systems are inherently complex, involving interactions between host immunity, microbial communities, and pharmacological agents. Systems biology approaches, supported by AI-based modeling, enable holistic analysis of these interactions to identify leverage points for intervention. By integrating multi-omics data—including genomics, proteomics, and metabolomics—AI platforms can predict how pathogens respond to therapeutic pressure at both individual and population levels. Nanobiotechnology translates these insights into physical delivery systems capable of precise intervention, creating a closed-loop optimization framework. An important advantage of this integrated approach lies in its ability to slow the evolutionary pace of resistance. Intelligent drug delivery systems reduce off-target exposure and maintain therapeutic concentrations within optimal windows, limiting opportunities for resistant strains to emerge. Additionally, AI-guided molecular targeting ensures that antibiotics engage specific bacterial pathways with minimal collateral impact on beneficial microbiota. This targeted precision reduces ecological disruption, which is increasingly recognized as a contributor to resistance proliferation. The integration of AI and nanobiotechnology also supports combination therapies, which are widely regarded as effective resistance mitigation strategies. AI models can identify synergistic drug combinations that disrupt multiple bacterial processes simultaneously, making resistance development statistically less likely.

Nanocarriers enable co-delivery of these agents in controlled ratios, ensuring synchronized action at the infection site. Such coordinated interventions extend antibiotic lifespan and improve long-term treatment outcomes. Despite its promise, this integrated framework faces challenges related to data quality, model interpretability, and translational scalability. AI systems depend on high-quality, representative datasets, which are often limited in clinical microbiology. Similarly, the regulatory pathways for AI-driven nanomedicine remain underdeveloped, posing barriers to widespread adoption. Addressing these challenges requires interdisciplinary collaboration among data scientists, microbiologists, clinicians, and regulatory bodies to ensure ethical, safe, and effective implementation. In summary, integrating AI and nanobiotechnology offers a transformative approach to antimicrobial resistance mitigation. Predictive resistance modeling, adaptive therapeutic strategies, and systems-level optimization collectively enable precise, responsive, and sustainable antimicrobial interventions. By combining pathogen genomics, molecular targeting, and intelligent delivery systems, this interdisciplinary framework not only slows resistance emergence but also extends the clinical utility of existing and future antibiotics. As resistance continues to threaten global health, AI-enabled nanobiotechnological platforms represent a critical pathway toward resilient and next-generation antimicrobial solutions.

7. TRANSLATIONAL AND CLINICAL IMPLICATIONS

The translation of AI-enabled nanobiotechnological platforms from laboratory research to clinical application represents a critical step in advancing precision antibiotic discovery and targeted antimicrobial delivery. Translational medicine bridges computational innovation, nanoscale engineering, and clinical practice, ensuring that intelligent antimicrobial solutions are not only technologically robust but also clinically feasible, safe, and scalable. As antimicrobial resistance continues to compromise conventional therapies, these integrated platforms offer promising pathways for personalized, efficient, and sustainable therapeutic interventions. One of the primary translational considerations involves scalability and manufacturability of nanocarrier-based antimicrobial systems. While nanoscale drug delivery platforms demonstrate high precision and controlled release capabilities in preclinical studies, large-scale production must meet stringent quality, reproducibility, and cost-effectiveness requirements. Variability in nanoparticle size, surface chemistry, and drug-loading efficiency can significantly affect therapeutic outcomes. AI-driven optimization models are increasingly used to standardize nanocarrier synthesis and predict performance variability, supporting scalable manufacturing pipelines suitable for clinical deployment.

Regulatory and safety evaluation frameworks represent one of the most critical hurdles in translating AI-driven nanobiotechnological antimicrobial platforms from research into clinical practice. Regulatory authorities, including agencies such as the U.S. Food and Drug Administration (FDA) and the European Medicines Agency (EMA), require a rigorous and multidimensional assessment to ensure patient safety. For nanomaterial-based therapeutics, this includes thorough characterization of nanomaterial toxicity, biodistribution, pharmacokinetics, and long-term biocompatibility. The unique physicochemical properties of nanoparticles—such as high surface area, surface charge, and size-dependent cellular interactions—necessitate specialized testing protocols to evaluate potential off-target effects, immunogenicity, and cumulative toxicity over extended periods. The integration of artificial intelligence into these platforms introduces additional complexity for regulatory assessment. Unlike conventional therapeutics, AI models are dynamic systems capable of learning and adapting based on incoming data. Regulatory agencies thus require transparency in algorithm design, reproducibility of results, provenance of training datasets, and interpretability of predictive outputs. Explainable AI (XAI) techniques become crucial for demonstrating clinical reliability, as they provide interpretable decision pathways that allow regulators and clinicians to understand why a particular therapeutic candidate or dosage recommendation is suggested. Ethics-by-design principles and responsible AI frameworks complement these requirements by embedding accountability, patient safety, and transparency into every stage of development, from algorithm selection to deployment, thereby facilitating regulatory approval and fostering trust in clinical applications (Topol, 2019).

From a clinical feasibility perspective, AI-driven nanobiotechnological platforms hold significant promise in advancing precision antimicrobial therapy. By integrating pathogen genomics, molecular target information, and patient-specific infection profiles, machine learning models can identify optimal therapeutic candidates tailored to individual infections. Nanocarriers, when designed for targeted delivery and controlled drug release, minimize systemic exposure, reduce off-target toxicity, and enhance local drug concentrations at infection sites. This precision-based approach is particularly valuable in combating multidrug-resistant pathogens, where conventional broad-spectrum antibiotics often fail, contribute to resistance evolution, or cause collateral damage to the host microbiome. The combination of AI-guided candidate selection with nanotechnology-mediated delivery thus represents a convergence of computational and material innovation with high translational potential. Despite these advances, practical implementation faces significant barriers related to data integration and interoperability. Effective translational deployment requires seamless aggregation of heterogeneous data sources, including genomic sequences, electronic health records, pharmacological profiles, and real-time patient response metrics. AI-driven bioinformatics platforms must implement robust data standardization, ensure interoperability across disparate healthcare systems, and maintain secure data sharing protocols. High-performance computational architectures—such as pre-computation optimized memory systems (Kameswaran et al., 2017) and FPGA-based adaptive signal processing platforms (Shanmugaraja et al., 2023)—provide the technological backbone for accelerated AI model training, real-time predictive analytics, and scalable deployment. However, the absence of harmonized data infrastructures can limit model generalizability, reduce clinical trust, and hinder widespread adoption. Addressing these challenges necessitates robust data governance policies, cross-institutional collaboration among clinicians, data scientists, and biomedical engineers, and continual validation to ensure both efficacy and patient safety in clinical settings.

Real-world deployment challenges also include clinician acceptance, infrastructure readiness, and ethical considerations. The successful adoption of AI-driven antimicrobial platforms depends on clinician trust in algorithmic recommendations and delivery mechanisms. Training healthcare professionals to interpret AI-assisted insights and understand nanocarrier-based therapies is essential for clinical integration. Furthermore, ethical concerns related to data privacy, algorithmic bias, and equitable access must be proactively addressed to prevent disparities in healthcare delivery. Translational strategies must therefore emphasize human-centered design and inclusive healthcare innovation. Despite these challenges, interdisciplinary collaboration offers significant opportunities for accelerating clinical adoption. Collaborative ecosystems involving academia, industry, regulatory bodies, and healthcare institutions are critical for translating AI-enabled nanobiotechnological innovations into clinical practice. Public–private partnerships and translational research consortia can facilitate pilot trials, real-world validation, and regulatory alignment. Such collaborations also support iterative refinement of AI models and nanocarrier systems based on clinical feedback, enhancing long-term therapeutic effectiveness. Looking forward, future clinical adoption of AI-driven nanobiotechnological antimicrobial platforms is expected to advance toward personalized and adaptive treatment strategies. Continuous learning algorithms capable of updating therapeutic recommendations based on emerging resistance patterns and patient outcomes will play a pivotal role. The integration of real-time clinical data with intelligent delivery systems may enable dynamic dose adjustment and adaptive release profiles, further improving treatment precision. As translational research matures, these platforms have the potential to redefine antimicrobial therapy by aligning computational intelligence with nanoscale biomedical innovation. In summary, the translational and clinical implications of AI-enabled nanobiotechnological platforms extend beyond technological advancement to encompass regulatory readiness, clinical feasibility, ethical responsibility, and collaborative innovation. Addressing scalability, safety, and data integration challenges is essential for real-world deployment. Through interdisciplinary collaboration and responsible design, these intelligent platforms can significantly enhance the effectiveness and sustainability of next-generation antimicrobial therapies.

8. CONCLUSION AND FUTURE RESEARCH DIRECTIONS

The convergence of artificial intelligence and nanobiotechnology represents a transformative paradigm for addressing the persistent and evolving challenges associated with antimicrobial therapy. Throughout this chapter, the integration of data-driven drug discovery, molecular targeting, and smart antimicrobial delivery has been examined as a comprehensive strategy for improving the precision, efficacy, and sustainability of next-generation antibiotics. By leveraging machine learning models, bioinformatics, and pathogen genomics, AI-enabled platforms enable the rapid identification and optimization of antimicrobial candidates, significantly reducing development timelines while enhancing therapeutic specificity. When combined with nanobiotechnological delivery systems, these computational advances facilitate controlled drug release and site-specific action, thereby minimizing toxicity and improving clinical outcomes. A central insight emerging from this discussion is the role of nanocarriers as critical enablers of precision antimicrobial delivery. Nanoscale systems enhance drug stability, protect active compounds from premature degradation, and enable targeted interactions with pathogenic cells. The incorporation of smart drug delivery mechanisms, including stimuli-responsive and adaptive release systems, further strengthens the therapeutic potential of antibiotics by aligning drug activity with pathogen-specific environments. Such approaches not only improve treatment efficacy but also contribute to resistance mitigation by reducing unnecessary exposure and sub-therapeutic dosing. Equally significant is the contribution of artificial intelligence to resistance mitigation strategies. Predictive analytics and systems-level modeling allow for early detection of resistance patterns and enable the design of adaptive therapeutic interventions. The integration of AI-driven insights with nanobiotechnological platforms creates a feedback-driven ecosystem in which antimicrobial therapies can be continuously refined based on real-time biological data. This interdisciplinary synergy underscores the importance of combining computational intelligence with nanoscale engineering to extend the lifespan of existing antibiotics and guide the development of novel agents. Looking forward, several future research directions are essential for advancing AI-enabled nanobiotechnological platforms. One key area involves the deeper integration of advanced machine learning techniques, such as deep learning and reinforcement learning, to improve predictive accuracy in drug discovery and delivery optimization. These approaches can enhance molecular targeting, optimize nanocarrier design, and enable dynamic adjustment of therapeutic strategies in response to evolving pathogen profiles. Improved data interoperability and standardized bioinformatics frameworks will also be critical for maximizing the effectiveness of AI-driven systems. Personalized antimicrobial strategies represent another promising direction for future investigation. By incorporating patient-specific data, microbial profiles, and pharmacokinetic parameters, AI-enabled platforms can support individualized treatment regimens that maximize efficacy while minimizing adverse effects. The combination of personalized analytics with smart drug delivery systems offers the potential to tailor antimicrobial interventions to specific infections, thereby improving treatment outcomes and reducing the risk of resistance development.

Global resistance surveillance systems constitute a further area of strategic importance. The integration of AI-driven analytics with large-scale genomic and epidemiological data can support real-time monitoring of resistance trends across regions and healthcare settings. Such systems can inform public health decision-making, guide antibiotic stewardship programs, and facilitate the rapid deployment of targeted therapies. When linked with nanobiotechnological delivery platforms, global surveillance insights can be translated into adaptive therapeutic responses that address both local and global resistance challenges. In conclusion, AI-enabled nanobiotechnological platforms offer a robust and forward-looking framework for precision antibiotic discovery and targeted antimicrobial delivery. The interdisciplinary integration of machine learning models, bioinformatics, nanocarriers, and smart drug delivery systems provides a pathway toward sustainable antimicrobial development in an era of escalating resistance. Continued research, collaboration, and translational efforts will be essential to fully realize the long-term potential of intelligent, precision-based approaches in reshaping the future of antimicrobial therapy.

REFERENCES

Allen, T. M., & Cullis, P. R. (2013). Liposomal drug delivery systems: From concept to clinical applications. *Advanced Drug Delivery Reviews*, *65*(1), 36–48. DOI: 10.1016/j.addr.2012.09.037 PMID: 23036225

Chen, H., Engkvist, O., Wang, Y., Olivecrona, M., & Blaschke, T. (2018). The rise of deep learning in drug discovery. *Drug Discovery Today*, *23*(6), 1241–1250. DOI: 10.1016/j.drudis.2018.01.039 PMID: 29366762

Cloudin, S., Priya, R. S., Prabhu, G. A., Ruba, M., Joel, R., & Vanitha, R. (2026). AI-Enhanced Predictive Analytics and Reliability Management for Wind Turbine Maintenance. In AI-Powered Analysis, Modeling, and Monitoring of Wind Energy Systems (pp. 191-218). IGI Global Scientific Publishing.

Dhanushya, J., Ruba, M., Ponkarthika, M., Soundarya, B., Sakthi Priya, V., Dhanushri, S., & Harikrishnan, S. (2025, April). AI-Powered Multipurpose Agrobot for Precision Farming: Crop Health Management, Fertilizer Recommendation and Yield Prediction. In *2025 International Conference on Recent Advances in Electrical, Electronics, Ubiquitous Communication, and Computational Intelligence (RAEEUCCI)* (pp. 1-6). IEEE.

Gowthaman, N., Ruba, M., Aswath, S., Hemavarshni, S. N., Bragadeshwaran, R., & Nithish, S. J. (2025, November). Iot Integrated Solution For Smart Vertical Crop Cultivation. In 2025 IEEE 7th International Conference on Computing, Communication and Automation (ICCCA) (pp. 1-6). IEEE

Huh, A. J., & Kwon, Y. J. (2011). "Nanoantibiotics": A new paradigm for treating infectious diseases using nanomaterials in the antibiotics resistant era. *Journal of Controlled Release*, *156*(2), 128–145. DOI: 10.1016/j.jconrel.2011.07.002 PMID: 21763369

Kamaly, N., Yameen, B., Wu, J., & Farokhzad, O. C. (2016). Degradable controlled-release polymers and polymeric nanoparticles: Mechanisms of controlling drug release. *Chemical Reviews*, *116*(4), 2602–2663. DOI: 10.1021/acs.chemrev.5b00346 PMID: 26854975

Kameswaran, A., Nanthakumar, K., & Ruba, M. (2017). High performance precomputation based self-controlled precharge-free content-addressable memory. *Journal of Electronics and Communication Systems*, *2*, 1–9.

Krithika, S., Jayachandran, T., Raghapriya, N. R., Ruba, M., Janani, C., & Sathiyavathi, S. (2025, April). An Effective Denial of Service Texture Features from Malware Grayscale Image. In 2025 3rd International Conference on Artificial Intelligence and Machine Learning Applications Theme: Healthcare and Internet of Things (AIMLA) (pp. 1-7). IEEE. DOI: 10.1109/AIMLA63829.2025.11041501

Mura, S., Nicolas, J., & Couvreur, P. (2013). Stimuli-responsive nanocarriers for drug delivery. *Nature Materials*, *12*(11), 991–1003. DOI: 10.1038/nmat3776 PMID: 24150417

Peer, D., Karp, J. M., Hong, S., Farokhzad, O. C., Margalit, R., & Langer, R. (2020). Nanocarriers as an emerging platform for cancer therapy. Nano-enabled medical applications, 61-91.

Ponkarthika, M., Prabhu, A., Ruba, M., Kishore, Y. R., & GV, M. (2026). Bridging Health Equity in Times of Crisis: Ensuring Inclusive Care During Emergencies and Disasters. In Advancing Equity in Global Healthcare Systems (pp. 27-54). IGI Global Scientific Publishing.

Poovithaa, M., Ruba, M., Ponkarthika, M., Ragul, N., Ragul, V., & Sangeetha, K. (2025, November). IoT Smart Traffic Control System for Hill Roads with Emergency Vehicle Priority. In 2025 IEEE 7th International Conference on Computing, Communication and Automation (ICCCA) (pp. 1-6). IEEE.

Ruba, M., Ashwin Prabhu, G., Robinson Joel, M., Vishnu Priya, K., Dharshani, L., & Dharani, R. (2026). Integrating emerging technologies for organizational agility: An interdisciplinary approach to digital transformation. In Roache, D. (Ed.), *High performance leadership for organizational excellence: Innovation, technology, and resilience* (pp. 63–92). IGI Global Scientific Publishing.

Santhini, C., Ruba, M., Sakthi Priya, V., & Hemalatha, R. (2025). Intercultural communication and assessment practices in the integration of AI with education. In *Generative AI Approaches to Sustainable Development in Higher Education*. IGI Global. DOI: 10.4018/979-8-3693-5623-4.ch010

Shanmugaraja, T., Ruba, M., Priya, V. S., Venkatesh, T., Supriya, M., & Nallamuthu, M. (2023, September). ECG noise elimination using an FPGA version of a modified error normalized LMS adaptive filter. In AIP Conference Proceedings (Vol. 2764, No. 1, p. 060033). AIP Publishing LLC. DOI: 10.1063/5.0144367

Sibiya, S., Ruba, M., Ponkarthika, M., Vijayasabariswari, M., Sheba, S., Vignesh, S., . . . Sankar, S. S. (2025, October). Integrated Wearable Health Monitoring and Emergency Response System With IOT Sensors, AED and Mobile App for Enhanced Patient Outcomes. In 2025 IEEE 2nd International Conference on Green Industrial Electronics and Sustainable Technologies (GIEST) (pp. 1-6). IEEE.

Stokes, J. M., Yang, K., Swanson, K., Jin, W., Cubillos-Ruiz, A., Donghia, N. M., MacNair, C. R., French, S., Carfrae, L. A., Bloom-Ackermann, Z., Tran, V. M., Chiappino-Pepe, A., Badran, A. H., Andrews, I. W., Chory, E. J., Church, G. M., Brown, E. D., Jaakkola, T. S., Barzilay, R., & Collins, J. J. (2020). A deep learning approach to antibiotic discovery. *Cell*, *180*(4), 688–702. DOI: 10.1016/j.cell.2020.01.021 PMID: 32084340

Topol, E. J. (2019). High-performance medicine: The convergence of human and artificial intelligence. *Nature Medicine*, *25*(1), 44–56. DOI: 10.1038/s41591-018-0300-7 PMID: 30617339

Ventola, C. L. (2015). The antibiotic resistance crisis: part 1: causes and threats. *P&T : a Peer-Reviewed Journal for Formulary Management*, *40*(4), 277. PMID: 25859123

World Health Organization. (2023). *GLASS manual for antimicrobial resistance surveillance in common bacteria causing human infection*. World Health Organization.

Chapter 5
Integrating Artificial Intelligence and Nanobiotechnology for Precision Antibiotic Delivery and Targeted Antimicrobial Therapy

Dhanya K. S.
http://orcid.org/0009-0006-7017-354X
Stella Maris College, University of Madras, Chennai, India

Dhanushiya R.
http://orcid.org/0009-0004-8413-1821
Stella Maris College, University of Madras, Chennai, India

Veena Gayathri K.
http://orcid.org/0009-0004-9889-1573
Stella Maris College, University of Madras, Chennai, India

ABSTRACT

Antibiotics have been essential in treating bacterial infections and assisting with life-saving medical procedures; however, the development of antibiotic-resistant bacteria as a result of overuse of antibiotics presents serious problems. Artificial intelligence (AI) is revolutionizing precision medicine and drug discovery through quick compound screening, biological interaction modeling, and drug repurposing. Nanocarriers serve as advanced delivery platforms that address limitations of con-

DOI: 10.4018/979-8-3373-5268-8.ch005

ventional therapies such as poor bioavailability, systemic toxicity, and inadequate penetration into complex infection sites like biofilms. By incorporating patient data, AI improves diagnostic capabilities and allows for customized treatment plans. The fusion of AI and nanobiotechnology promise enormous scope for tailoring antibiotic therapy to individuals and microbial types. To use these potentials to their full potential in the improvement of global health, scientists, physicians and policymakers must work and cooperate together in the future.

INTRODUCTION

Antibiotics are the significant discovery of the medical field of the twentieth century and serve as a pivotal defense against bacterial infections. Antibiotics were historically discovered from soil microbes by screening secondary metabolites having antibacterial activities (Hutchings et al., 2019). Discovering new antibiotics to treat infections were becoming tedious because of identical compounds being isolated (Katz & Baltz, 2016). The drug discovery and development is time consuming and requires a lot of intensive procedures to synthesize them involving thousands of chemicals (Liu et al., 2024). The study of pharmacokinetic and pharmacodynamic (PKPD) principles has significantly advanced our knowledge of the interactions among the antibiotic, the pathogen, and the infected patient over the last few decades (Mohsen et al., 2020; Blondeau, 2009). The pharmaceutical industry has become much more digitalized, which has created difficulties in gathering and analyzing knowledge to handle complicated clinical problems. Due to its ability to handle massive amounts of data through improved automation, using cutting-edge tools and networks that mimic human intellect while maintaining the necessity for human intervention, artificial intelligence has gained more attention. Artificial Intelligence (AI) uses software and systems to analyze input data and make decisions on its own that are targeted towards particular goals (Ghaderzadeh et al., 2024). Integrating advanced technologies like artificial intelligence (AI) provides promising opportunities in drug discovery, precision medicines and diagnostics. However, the rapid emergence of antibiotic resistance among microorganisms has intensely undermined the treatment plans leading to a health crisis. Human activities and industrial livestock production have led to the rise in drug-resistant microorganisms, which have impacted the environment and concerns in public health (Yarahmadi et al., 2025). According to Global research on antimicrobial resistance (GRAM) study, 67.5% increase in annual deaths is due to bacterial AMR rising from 1.14 million in 2021 to an estimated 1.91 million in 2050. Deaths linked to AMR might also increase by 74% from 4.71 million in 2021, to 8.22 million deaths in 2050. Antimicrobial resistance (AMR) represents an integral and accelerating global health crisis, with projections indicating up to 10 million fatalities annually by 2050. The reducing effects of antibiotics are leading to increased morbidity, mortality, and substantial economic burden (Gargate et al., 2025). The quest for novel antibiotics and antimicrobials is essential since resistance is viewed as unavoidable (Nwobodo et al., 2022). Antimicrobial resistance is compelling innovative strategies to improve antibiotic efficacy. Despite the pressing demand for novel antimicrobial agents, progress in their development is significantly constrained by complex scientific challenges, high research and development costs, and stringent regulatory frameworks factors that have driven numerous major pharmaceutical companies to disengage from antibiotic innovation (Gargate et al., 2025). The difficulties of using therapeutic agents can be overcome by using nano-based drug delivery systems (DDSs), which enhance drug efficiency and reduce side effects. But due to their complexities and time-consuming nature, artificial intelligence (AI) tools are integrated with DDSs for personalized drug delivery, optimization of drugs, and drug property predictions. Precision medicine, a medical model that enables treatment to individual patients' response to disease, is very promising to improve the treatment of the illness (Watkins, 2022). Utilizing precision medicine and AI in DDSs, safer therapeutic outcomes can lead to improving patient's health and advancing pharmaceutical sciences (Heydari et al., 2024).

Recent research has looked at how NPs can get past biological barriers and improve precision with more individualized treatment approaches by using patient data, including genetic, environmental, and historical factors, as well as biomarker information, to create customized treatment regimens. Nanoparticles can improve drug specificity, optimize drug dose delivery, and improve response to precision medicine therapies (Das & J C, 2023; Mitchell et al., 2021). In this context, drug delivery could be revolutionized by computational pharmaceutics, a field that combines big data, artificial intelligence, and multi-scale modelling techniques. It offers multi-scale insights across pre-formulation studies, formulation screening, in vivo prediction, and clinic precision medicine (Wang et al., 2021). The design and optimization of drug delivery systems based on nanoparticles have further improved by advances in AI, increasing the effectiveness and versatility of targeted therapies (Kapoor et al., 2024). A critical factor in effective antibiotic therapy is understanding the action of a drug inside the pathogen and the corresponding reaction of the pathogen. In the clinic nowadays, genomic-based assays such as a molecular fast diagnostic test can identify genes that confer resistance to microorganisms significantly reducing the time required to select appropriate antibiotic treatments and thereby lowering infection-related mortality (Bulman et al., 2022; Timbrook et al., 2017). However, substantial gaps persist in mapping bacterial genotype to phenotype, particularly for resistance mechanisms including rare resistance genes, efflux pumps, and genes responsible for antibiotic uptake or penetration (Bulman et al., 2022; Huang Y et al., 2021b; Huang Y et al., 2022;). Addressing these gaps and integrating high-resolution diagnostic information with NP-based delivery, AI-supported design, and model-informed dosing is central to the development of precision antibiotics and infection-focused drug delivery systems aimed at mitigating AMR.

AI-DRIVEN PRECISION ANTIBIOTIC DISCOVERY AND NANODELIVERY

Feynman was the first physicist to introduce nanotechnology in 1959 (Afzal et al., 2022; Baig et al., 2021). In the 1960s, studies on liposomes and controlled drug release systems made substantial progress in developing drug delivery nanocarriers (Gao_et al., 2024). In the late 1970s, the relation of nanoparticles to biomedicine was exhibited and over 10,000 publications have referred to this with the term "nanomedicine" (Afzal et al., 2022; Bonifácio et al., 2014). The emergence of Artificial Intelligence (AI) can be traced back to the mid 20th century, where the innovators Alan Turing and John McCarthy introduced the idea of machines capable of simulating human reasoning and learning. In the 1950s–1970s symbolic AI and rule-based expert systems were characterized which relied on manually encoded logic. Neural networks and statistical learning were introduced in the 1980s and 1990s, which promoted data-driven computation. The deep learning revolution, which culminated in multilayer neural architectures for high-dimensional pattern recognition, was sparked by the growth of digital data and computing power in the 2010s. The field has recently made great strides thanks to transformer-based and generative models, which improve performance in natural language processing, drug development, and medical imaging (Trasca et al., 2025; Hirani et al., 2024; Avanzo et al., 2024; Gou et al., 2024). In recent years developments in AI and ML have led to a major shift in the rational design of nanomedicine formulations for drug delivery (Chou et al., 2025). Together, these AI-driven approaches seek to improve patient outcomes, reduce the time and expense of experimental testing, and increase the discovery of clinically feasible nanomaterials (Han et al., 2025; Saha, 2025).

However, due to inadequate effectiveness in drug delivery nanocarriers and a lack of knowledge about how nanocarriers interact with biological systems, clinical translation of nanocarriers is limited and requires interdisciplinary cooperation. To enhance nanocarrier design, scientists employ data-driven and physics-based models; yet, integrating these models proves difficult in terms of standardizing methodologies and guaranteeing model validation across biological systems As illustrated in figure 1, using AI, computational advancements, and machine learning, as well as integrating different modelling methodologies and making sure that models are validated, can improve data infrastructure, therapeutic outcomes, and nanocarrier systems (Gao_et al., 2024). (Figure 1)

Figure 1. Applications of AI in Drug development

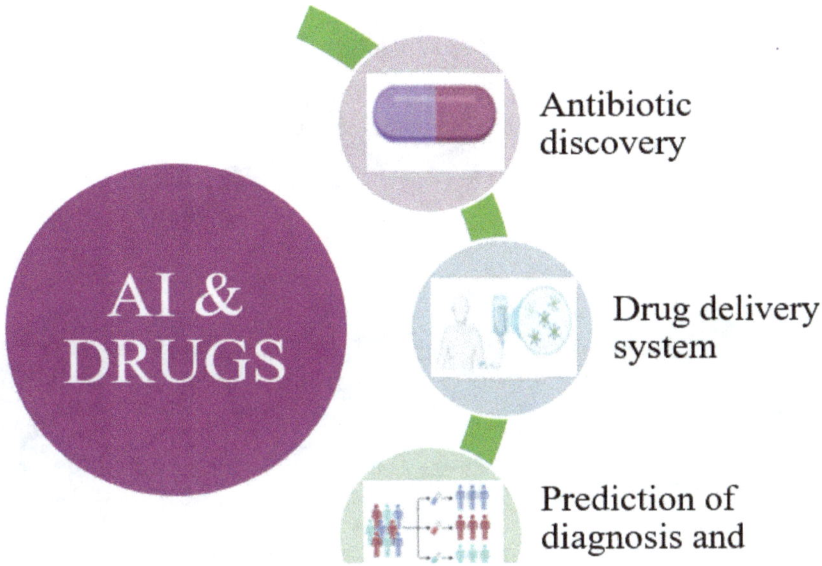

Source: Created by the authors. The figure illustrates the application of AI in antibiotic discovery, drug delivery systems and in the prediction of diagnosis and treatment.

Artificial intelligence techniques like machine learning and deep learning approaches like neural networks have significantly advanced nanoparticle design and drug discovery as illustrated in figure 2. These techniques enable rapid parameter screening, optimization of nanoparticles configurations, and improved drug loading and targeting specificity. By integrating predictive modelling with experimental workflows, AI accelerates drug development and supports the emergence of personalized therapeutic strategies (Kapoor et al., 2024; Popa et al., 2022). Novel approaches to antibiotic discovery can yield faster and more coherent results when machine learning is combined with Raman spectroscopy and quantitative structure-activity connection methodologies. As a result, AI is a pivotal tool for scientists and medical professionals fighting bacterial resistance (Popa et al., 2022) (Figure 2).

Figure 2. Machine Learning concepts

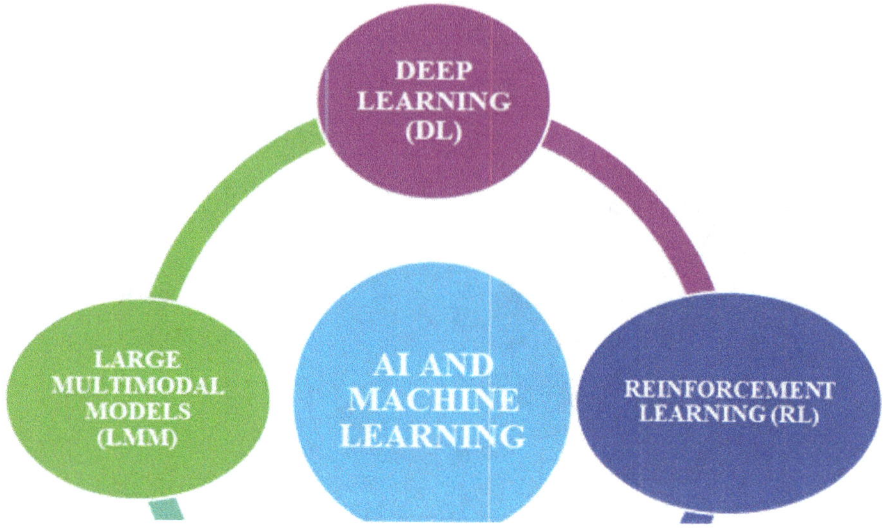

Source: Created by the authors. The figure illustrates the central role of AI and machine learning, highlighting key approaches such as large language models, deep learning, graph neural networks, and reinforcement learning in advanced biomedical applications.

AI And Algorithms in Drug Discovery and Development

Graph neural networks (GNNs) have surfaced as powerful models for learning from graph- structured data, which is well suited for representing molecular structures, drug discovery and drug repurposing (Corso et al., 2024). In drug discovery, their capability to capture spatial and relational dependencies enables more accurate drug- target interaction (DTI) prediction compared to traditional descriptor based or sequence based machine learning approaches. As a result, the newly enhanced graph neural network has been used to speed up forecasting DTIs and has shown promise in repositioning medications and hastening drug discovery (Zhang et al., 2022). Recent studies show that GNN can effectively encode molecular information and conserve key structural features thereby strengthening predictive accuracy (Wang et al., 2025; Sun et al., 2020). For instance, eXplainable Graph-based Drug Response Prediction (XGDP) technique, provides accurate drug response prediction by linking molecular features with gene - level interactions, addressing a key limitation of many deep learning approaches (Wang et al., 2025). However, despite these positives, GNNs are highly dependent on large, high- quality datasets and are computationally intensive, which may reduce their scalability and real - world clinical translation.

Reinforcement Learning has shown promise in ReLeaSE (Reinforcement Learning for Structural Evolution) with the help of a simplified molecular-input line-entry system (SMILES) which makes simple representations of molecules. Generative models produce chemically feasible SMILES strings by training with the stack-augmented memory network. Predictive models are designed to forecast the desired properties of the de novo- generated compounds. RL technique trains both models together to bias the creation of novel chemical structures in favor of those with the required biological and /or physical characteristics (Popova et al., 2018). Compared to GNNs, which focus on prediction tasks, RL is more efficient in exploring chemical space and generating new candidates. However, RL models often encounter training instability and may produce chemically infeasible or suboptimal molecules which highlights the need for hybrid approaches.

Large Language Models (LLMs) in Drug Development

Large language models (LLMs), like GPT-4, extend AI techniques beyond molecular modelling into knowledge extraction, automation and decision support (Zheng et al., 2025; Lu et al., 2025). LLM provides substantial chances for early discovery in AI-driven quantitative structure-activity relationship improvement. Additionally, Natural Language Processing (NLP) has effectively converted quantitative structure–activity relationship (QSAR) data into suitable vectors (Lu et al., 2025; Bouhedjar et al., 2020). As computer algorithms are controlled by advances in robotic automation, LLM can be used in compound synthesis and can be used for predicting diseases, diagnosis and is also used for assessing the targets in therapeutic medicine. Few examples are determining aging - related diseases and guidelines for treating cancer patients using the magnetic resonance imaging radionics (Saleh et al., 2025; Singhal et al., 2023; Cascella et al., 2023; Jo et al., 2023). The multi-LLM-based intelligent agent created can automate intricate scientific tasks like document and online searches, code development, and manipulating robotic platforms (Boiko et al., 2023). LLMs are being used more often in target safety evaluations in sectors like chemicals and pharmaceuticals, where they analyze data to determine environmental dangers, predict risks, and detect toxicological risks (Lu et al., 2025). Thus, LLMs are becoming indispensable in drug development and are being modified to interpret scientific data, such as DNA, proteins, and chemical structures. They can predict efficacy and safety, design new drug molecules, propose drug targets, expedite clinical trials, and pinpoint the biological causes of disease. LLMs have the potential to revolutionize future therapy development by automating these procedures, which could drastically cut down on the time and expense of introducing new medications to the market. However, the ethical issues of misuse, accountability, privacy, and fairness confront LLM-driven drug discovery. LLM also have the tendency to hallucinate which results in responses that are inaccurate or irrelevant. Hallucinations may have detrimental effects on diagnosis and data interpretation in clinical settings. Techniques like plugin integration, parameter tuning, and knowledge editing are used to improve factual accuracy. Enhancing output relevance is achieved by integrating retrieval-augmented generation with external documents into large language models (LLMs). Increasing alignment with real data through decoding algorithm refinements, such as factuality enhanced decoding, strengthens the dependability of LLM-driven drug discovery (Zheng et al., 2025).

Next - Generation Antibiotics: Innovations Shaping the Future of Drug Delivery and Treatment

AI improves drug research through adverse effect identification, formulation optimization, and by predicting chemical activity. By using data analysis to customize treatments based on genetic profiles, it promotes personalized medicine and improves therapeutic outcomes. AI integration in biotechnology helps in developing precision medicine and increases the effectiveness of research and ultimately improving the healthcare industry (Bhati et al., 2024; Sharma et al., 2024). To predict the host and pathogen protein interactions, a novel computational whole-genome approach has been created that scans genomes for comparable proteins it checks the probable interactions based on the available structure, and filters using biological context such as expression levels. These models demonstrate strong enrichment of biologically relevant interactions, highlighting their utility in guiding antimicrobial target discovery (Davis et al., 2007). Such insights into host–pathogen dynamics have accelerated interest in alternative therapeutic strategies, particularly Antimicrobial peptides (AMPs) which are emerging as viable solutions to the urgent need to produce new antibacterial drugs due to the global expansion of multi-drug-resistant bacteria. In multicellular organisms, these naturally occurring chemicals act as a first line of defense. The majority of current AMP prediction techniques rely on 2D characteristics such as amino acid composition and peptide charge, despite the existence of a small number of experimental 3D structures. The development of activity prediction models for AMPs has been aided by the application of predicted protein models in structure-based drug discovery made possible by developments in structural modelling (Liu et al., 2018). The development of AMPs as therapeutic agents is anticipated to accelerate in the upcoming years due to recent advancements in characterizing their mechanism-of-action, clever formulation strategies improving the chemical and metabolic stability, and sophisticated chemical synthesis protocols lowering manufacturing costs, despite acknowledged challenges in bringing the nonclinical candidate AMPs into clinical use (Mahlapuu et al., 2020).

Along with these developments, deep learning frameworks have been introduced to generalise the prediction of drug - target interaction (DTI). However, bioinformaticians and computer scientists face difficulties in deep learning and biomedical domain respectively. Models such as DeepPurpose provide flexible architectures for DTI prediction, integrating multiple encoders for chemical and protein features. By enabling customizable and scalable modelling, such platforms bridge computational prediction with practical drug development workflows (Huang et al., 2021a).

Modelling Resistance Evolution of the Antibiotics

An important public health concern is the development of antibiotic resistance, which is an example of quick adaptation through natural selection. Mathematical models provide insight into the mechanisms of resistance both inside hosts, addressing how resistance changes during treatment, and between hosts, such as in communities where hospitalized people or farm animals receive antibiotic therapy. Strategies to stop the spread of resistance through reduced transmission and customized antibiotic prescriptions are informed by these models (Blanquart, 2019). The accuracy of forecasting resistance patterns is improved by recent developments in computational modelling, especially using artificial intelligence and hybrid statistical models. By combining traditional models like seasonal autoregressive integrated moving average (SARIMA) with machine learning approaches like long short-term memory (LSTM) networks, it is possible to predict resistance trajectories and make evidence-based policy decisions (Wang et al., 2024).

Using the data from 49,872 infections in Stanford's medical records, the researchers created machine learning models known as "personalized antibiograms" that predicted resistance to five important antibiotics. Prescriptions, comorbidities, past resistance, and demographics were all included in the models using LightGBM. Prior resistance and prescriptions were identified as important features by the models, which demonstrated discriminative abilities with AUROCs ranging from 0.74 to 0.78. These results show how machine learning can help with safer empirical prescribing and better antibiotic stewardship (Haredasht et al., 2025). Although the coexistence of drug-sensitive and resistant strains in the larger community is difficult to duplicate, fitting these models to hospital data can forecast intervention effects. In line with experimental results, within-host models show that resistance risk peaks at moderate antibiotic dosages. Irrespective of the global awareness of AMR, the studies have focused on single pathogens and have included a few antibiotic classes and failed to understand the evolution of AMR worldwide (Shepherd et al., 2024). For improved epidemiological interpretation and resistance evolution prediction, more sophisticated models taking into account intricate host interactions and genetic factors are required (Blanquart, 2019).

Optimizing Drug Formulation and Drug Delivery Strategies

A recent study focused on the use of gold nanoparticles and mesoporous silica nanoparticles with unease enzymes as nanomotors. These radiolabelled nanomotors were used for in vivo imaging. Furthermore, PET was used to quantitatively track nanomotors, enhancing real-time imaging and tracking of active swarming dynamics and opening the door for diagnostic drug delivery applications (Hortelao et al., 2021; Das & C, 2023). The physical and chemical properties of SLNPs (Soluble Lipid nanoparticles) was examined which was encapsulated with progesterone utilizing an emulsification ultrasonication technique and ANN for increased prediction accuracy. The ANN (Artificial neural network) model showed the best particle size, polydispersity index, and zeta potential for prediction accuracy when compared to the response surface methods (Suriyaamporn et al., 2024). When loaded with limonene, the PG-SLNPs (SLNP encapsulated with Progesterone) formulation showed improved drug administration through the skin, and the ANN model accelerated production by lowering resource and time requirements (Kapoor et al., 2024).

AI in Existing Workflows

Halicin Discovery (MIT) As Proof Of AI-Enabled Antibiotic Identification

A deep neural network was instructed to anticipate the antibacterial compounds from the Drug Repurposing Hub. They discovered a compound halicin, a structurally distinct antibiotic that efficiently treats a variety of illnesses and pathogens. In addition, the model demonstrated the potential of deep learning in antibiotic discovery by identifying eight structurally distant antimicrobial compounds from a total of over 107 million molecules in the ZINC15 database (Stokes et al., 2020). Technion (The Israel Institute of Technology) researchers have created a deep learning model that can be applied to the optimization of current compounds and the design of new antibiotics. Certain bacteria can be targeted by the model, stopping bad bacteria from eradicating beneficial bacteria. In the field of antibiotic discovery, this groundbreaking work signifies a paradigm shift (Trafton, 2020).

Early - Stage Applications in Drug Delivery System Design

Precise targeting of cells implicated in disease progression is required, especially for life-threatening illnesses, due to recent discoveries in molecular pharmacology and disease processes. By reducing systemic exposure, this accuracy helps lessen the negative effects of medicinal drugs. Modern drug delivery systems (DDS) use cutting-edge technology to improve delivery to specific locations, increasing treatment efficacy and lowering side effects. These novel DDS, which frequently make use of nanoparticles or smaller biocompatible and biodegradable devices with multifunctional capabilities and longer circulation periods, exhibit higher performance, automation, precision, and efficacy when compared to conventional systems (Ezike et al., 2023). However, these modern drug delivery systems face challenges like scalability, stability for long - term usage and toxicity. Extensive characterization, accessing the formulations at different parameters and safety measures are required for using these drug delivery systems in real life.

AI-GUIDED NANOCARRIERS: SMART PLATFORMS FOR PRECISION DELIVERY

Nanocarriers for Precision Antibiotic Delivery

Nanocarriers, which are essentially colloidal nanoparticles, are frequently used to deliver therapeutic agents or any other substances to a target site. Nanocarrier diameters range from 1 to 100nm which is also suitable for therapeutic applications because the body's microcapillaries are 200 nm in diameter. The primary forms of nanocarriers are polymeric nanoparticles, liposomes, dendrimers, and micelles. Liposomes are spherical, lipid-based vesicles with an aqueous centre that can transport bioactive substances such as drugs. To enable controlled, on-demand release, gold nanoparticles and verteporfin can be incorporated into X-ray-triggerable liposomes, which release the encapsulated drug when exposed to X-rays, as the nanoparticles absorb the radiation and trigger liposome break down (Afreen et al., 2022). A dendrimer is made up of many arms that originate from the core and are made of sugar, nucleotides, and amino acids. Dendrimers have been extensively investigated in cancer therapy, because of its capability to reduce the amount of bioavailable copper in tumours. Since the traditional technique of dendrimer synthesis is a time-consuming process, an efficient approach for producing poly (acyl thiourea) and polythiourea dendrimers was developed (Chamundeeswari et al., 2019). Micelles are colloidal aggregates formed by surfactants in aqueous or nonpolar media; with hydrophilic and hydrophobic regions whose size and shape are influenced by temperature, pH, ionic strength, and surfactant type (Bose et al., 2021). Polymeric nanoparticles are colloidal particles made of biodegradable polymers that come in two different shapes: nanospheres and nanocapsules. Nanosphere is made by encapsulating the drug in a polymeric matrix and Nanocapsule is made by dissolving or dispersing a drug in a liquid core and then encasing it in a polymeric membrane. (Crintea et al., 2022). In the context of precision antibiotic therapy, these nanocarrier platforms are increasingly investigated for infection-site targeting, enhanced penetration of bacterial biofilms, and localized antibiotic delivery, thereby minimizing off-target exposure and reducing the emergence of antimicrobial resistance.

Advantages of Nanocarriers in Combating Antimicrobial Resistance

Nanocarriers have emerged as a revolutionary platform in the field of precision antibiotic delivery, addressing several limitations of conventional antimicrobial therapy. Conventional antibiotics mostly suffer from poor stability, limited bioavailability and inadequate penetration into complex infection sites such as bacterial biofilms. The advent of nanocarrier systems including liposomes, polymeric nanoparticles, micelles, and dendrimers has given several benefits combining encapsulation, sustained release and biofilm penetration to enhance therapeutic efficacy. The major benefits of nanocarriers include:

- **Enhanced retention and targeted delivery:** Polymeric nano and microparticulate delivery systems can significantly improve the therapeutic efficiency of antibiotics against biofilms by preserving drug activity, enhancing retention at the infection site, and preventing systemic dispersion. This targeted delivery approach allows higher concentrations of antibiotics to reach biofilm embedded bacteria while minimizing systemic exposure and associated side effects (Birk et al., 2021).
- **Sustained release:** Stimuli-responsive nanocarrier systems have demonstrated significant promise in this area, offering precise drug release triggered by infection-specific cues such as pH gradients and bacterial enzymes. For instance, enzyme-responsive nanoplatforms engineered to degrade in the presence of bacterial proteases can selectively release antibiotics directly within biofilm matrices, maximizing antibacterial action while minimizing off-target toxicity. Similarly, pH-responsive nanocarriers use the acidic microenvironment of infection sites to trigger antibiotic liberation where the pathogens are present. (Ding et al., 2022).
- **Biofilm penetration:** Biofilms are structured bacterial communities embedded in a self-produced extracellular matrix, notoriously resistant to conven-

tional antibiotics due to poor drug penetration and altered metabolic states of resident bacteria. Protein-based nanocarriers, such as α-lactalbumin nanospheres and nanotubes, have demonstrated superior permeability into biofilms, allowing antibiotic payloads to diffuse more efficiently and exert bactericidal effects in otherwise refractory infections (Yan et al., 2024). This improved permeability and targeted release make these nanocarrier systems particularly promising for managing biofilm-associated infections where standard antibiotic therapy struggles to succeed.

- **Enhanced lipid based delivery:** Lipid-based nanocarriers (such as liposomes, solid lipid nanoparticles, and nanostructured lipid carriers) have been shown to improve antibiotic stability, enable controlled release, and deliver antibiotics together with antimicrobial adjuvants into biofilms, thereby increasing local drug concentration and overcoming biofilm-based resistance (Ahsan et al., 2024).

Pathogen-Specific Drug Delivery Needs

The increasing prevalence of drug-resistant microorganisms like Methicillin-resistant *Staphylococcus aureus* (MRSA) has shown the importance for targeted and efficient drug delivery systems. Vancomycin is one of the treatments for MRSA despite a variety of antibacterial choices which raises concerns about the medical expenses, mortality, and extended hospital stays. New delivery methods, such as liposome entrapment enabling higher drug dose and lowers toxicity, are being explored to fight antibiotic resistance and prolong antibiotic efficacy. Furthermore, liposomes can be modified to have better drug-release mechanisms and stealth qualities to avoid the immune system and prolong circulation (Mat Rani et al., 2021). Hence, liposome- based delivery systems have emerged as a promising approach due to their ability to enhance drug bioavailability and reduce systemic toxicity. However, conventional liposome formulations often face challenges related to stability, targeting efficiency, and large-scale reproducibility. To overcome these limitations, recent studies have explored the integration of AI into liposome design and optimization like attachment of targeting ligands or surface modifications (Sharma et al., 2024). Compared to traditional trial-and-error approaches, AI-driven design significantly accelerates the development of customized delivery systems. AI accelerates the development of customized liposome-based therapies by modelling liposome interactions with target cells to find efficient ligands and surface changes that improve targeting specificity and drug delivery performance (Sharma et al., 2024; Kaur & Singh, 2021).

Bacterial biofilms are resistant to traditional treatments and pose a challenge to drug delivery using nanovehicles, especially liposomes. To address this, novel liposomal functionalization strategies have been developed. For example, Poly[2-(methacryloyloxy)ethyl phosphorylcholine] (pMPC) is a modified liposome that optimizes drug delivery to biofilms and targets *Pseudomonas aeruginosa*. When compared to conventional PEGylation techniques, pMPC-liposomes show better cellular absorption, cargo transport, and biofilm removal. This advancement is explained by molecular interactions between bacterial lipopolysaccharides and pMPC's zwitterionic groups that are made possible by divalent ions. According to the results, pMPCylation may transform liposomal drug delivery methods for the treatment of infections linked to biofilms (Kluzek et al., 2022). However, their performance may vary depending on biofilm composition and in vivo conditions, highlighting the need for further validation.

Smart & Responsive Nanocarriers: Stimuli-Responsive Systems

Concerns over bacterial infections have increased with the use of implanted biomaterials, particularly because of high recurrence rates and antibiotic resistance. Due to the shortcomings of conventional therapies, interest in stimuli-responsive biomaterials that can modify medication release rates has grown. Developments in bacteria-responsive drug delivery systems, which are categorized by trigger modes such as physical stimuli, virulence-factor responsiveness, and host-immune response, and are intended to release antibacterial agents autonomously at infection sites (Yang et al., 2022). Stimuli-responsive systems can be broadly classified into endogenous (pH, temperature, enzymes, and redox) and extracorporeal (light, magnetic fields, and ultrasound) factors. Compared to conventional delivery systems, these approaches improve therapeutic efficacy while minimizing off-target effects (Wang et al., 2023).

After surgery, controlling peri-implant infections requires the release of antibacterial drugs under infection control. Alginate and poly-l-lysine are conjugated with ciprofloxacin using a click reaction to generate polymers with antibiotics linked via cleavable linkers utilizing nanogels. These stable nanogel dispersions create films that are stable in water, but when enzymes are present, they break down and release ciprofloxacin (Bourgat et al., 2021). Lactic acid builds up and the extracellular pH surrounding tumours is lowered (pH 7.2-7.4) as a result of the Warburg effect, a glycolytic metabolism of cancer cells that increases glucose consumption and lactate generation. The release of chemicals from acid-sensitive components in tumour regions may be triggered by this acidic environment. Several pH-responsive nanocarriers, such as polymersomes, liposomes, micelles, dendrimers, and nanoparticles, have been developed to deliver Pt-based antitumor agents that remain stable in circulation and release their payload in the acidic tumour environment in order to take advantage of this endogenous pH stimulus (Wang et al., 2023; Li et al., 2019; Gabison, 2002).

In addition to biochemical triggers, externally controlled systems such as photothermal nanocarriers provide precise spatial and temporal control over drug release. A protocol was designed to combine CuFeS2 chalcopyrite nanoparticles with a thermoresponsive polymer based on poly (ethylene glycol methacrylate) so that heat - mediated drug delivery can happen along with photothermal heat damage. These nanoparticles when exposed to a laser beam (in the range of 0.5 and 1.5 W/cm^2) display a high heating performance along with a rise in solution temperature. The nanoparticles can also act as nanocarriers delivering doxorubicin, a chemotherapeutic agent which can be designed to be released by exposing the nanoparticles to a laser beam (where hyperthermia temperature above 42°C could be reached) (Conteh et al., 2023). Despite their potential, challenges related to tissue penetration of external stimuli and safety considerations remain.

AI's Role in Nanocarrier Design

In precision antibiotic development, AI-guided design strategies enable the optimization of antibiotic loading, prediction of pathogen–nanocarrier interactions and tailoring of nanocarrier delivery systems to resistance profiles and infection microenvironments. Artificial Intelligence-Driven Nanoarchitectonics for Smart Targeted Drug Delivery highlights that AI enables high-throughput formulation screening using algorithms such as random forests (RF) and support vector machines (SVMs) for prediction, classification, and optimization. It also incorporates advanced models including deep neural networks (DNNs), convolutional neural networks (CNNs), and generative models that capture complex structure function relationships. These tools support an interactive workflow in which diverse nanocarrier formulations are computationally generated, evaluated, and experimentally refined, thereby accelerating discovery and reducing reliance on empirical trial-and-error methods (Bae et al., 2025).

An integrated AI workflow to accelerate formulation development was applied in the CXCR4-targeted dual-drug nanocarrier study. And those AI techniques were used to predict optimal drug synergy identifying berberine and paclitaxel as an effective combination and to computationally screen peptide ligands for strong CXCR4 binding. Thus, these AI-driven selections informed the nanocarrier's material composition, targeting strategy, and drug-loading profile, producing a core-shell system with enhanced cellular uptake, sustained release, and marked antitumor activity (Jang et al., 2025). Collectively, these works show that AI can guide material selection, targeting design, and performance optimization, establishing a strong foundation for next-generation intelligent nanocarrier development as illustrated in figure 3. (Figure 3).

Figure 3. AI-Guided Nanocarrier Design & Clinical Application.

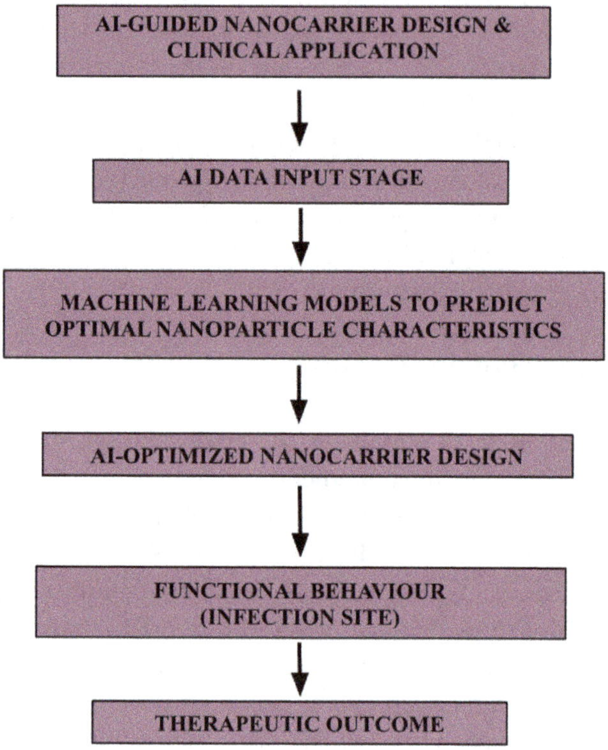

Source: Created by the authors. This figure illustrates the workflow of AI-assisted nanocarrier development, where data-driven models optimize nanoparticle design and predict functional behaviour at infection sites to improve therapeutic outcomes.

Building on these AI-enabled advances in nanocarrier design, it is important to consider how such approaches compare with traditional antibiotic strategies in clinical practice. Traditional antibiotic therapies typically rely on empirical or guideline-based selection and fixed dosing regimens within well-established clinical workflows. In contrast, AI-guided approaches function primarily as clinical decision-support systems that incorporate resistance data to optimize antibiotic selection and dosing. While these AI-supported strategies offer the potential to improve antimicrobial stewardship and reduce inappropriate antibiotic use, they also introduce additional validation, safety, and implementation challenges and remain at an early translational stage compared with conventional therapies (AlGain et al., 2025; Hua et al., 2018). The data is summarised in Table 1. (Table 1).

Table 1. Comparison of Traditional Antibiotic Strategies and AI-Guided Approaches in Antimicrobial Therapy.

Aspect	**Traditional Antibiotic Strategies (Hua et al., 2018)**	**AI-Guided Strategies (AlGain et al., 2025)**
Antibiotic selection	Empirical or guideline-based	Data-driven clinical decision support
Dosing adaptability	Fixed dosing regimens	Adaptive dosing recommendations
Resistance pressure	Higher due to broad antimicrobial exposure	Potentially reduced through optimized antibiotic use
Clinical workflow maturity	Well-established clinical workflows	Early clinical adoption
Validation requirements	Standard clinical validation pathways	Extensive validation with clinician oversight
Safety considerations	Well-characterized safety profiles	Ongoing evaluation due to decision-support risks
Scalability	High and established	Currently limited by implementation and validation barriers

Source: Adapted from Hua et al., (2018) and AlGain et al., (2025). This table highlights key differences between traditional and AI-guided approaches in terms of decision-making, adaptability, validation requirements, and clinical implementation, illustrating the transition toward data-driven antimicrobial therapy.

Predictive Models of Antibiotics

Infection Microenvironment Behaviour

A major limitation in antibiotic development lies in accurately modelling the infection microenvironment, where complex host–pathogen interactions govern therapeutic outcomes. Traditional two-dimensional (2-D) monolayers are unable to replicate the intricate aspects of the native microenvironment that are essential for comprehending infections, such as multicellular interactions, three-dimensional (3-D) structure, and physiologically significant forces. Lately researchers on infectious diseases have widely accepted to make 3-D models to study the mechanism of pathogenesis, for biomarkers and for screening drug candidates. Creating dynamic 3-D models that effectively depict these settings is a major challenge in tissue engineering for research on infectious diseases. Technologies such as Organ-on-a-chip (OAC) models, extracellular matrix (ECM)-embedded/organoid models and 3-D organotypic cultures in rotating wall vessel (RWV) bioreactors provide a more relevant framework for investigating infectious disease processes and antimicrobial therapeutics (Barrila et al., 2018). However, challenges remain in standardization, scalability, and reproducibility across different experimental setups.

Effective non-pharmaceutical interventions depend on a better understanding of environmental transmission dynamics, yet these processes are still not properly understood. To investigate pathogen dispersal and degradation mechanisms, a modelling framework was created. By performing experiments on *Campylobacter jejuni* transmission in broilers, which produced helpful estimates for three important parameters: pathogen inactivation, diffusion coefficient, and transmission rate the model's validation and calibration was done. The framework can provide important mechanistic insights into environmental pathogen transmission, as evidenced by the model's time and distance-dependent transmission patterns matching experimental observations (Gamża et al., 2023).

In addition to population-level modelling, microscale systems are being developed to understand localized infection environments, such as granuloma formation in tuberculosis. In order to investigate the signalling between mycobacterial infections and their surroundings, the study presented a unique in vitro model that makes use of a suspended microfluidic substrate called Stacks. The model combines a mycobacterial strain with human immune cells, making it simple to adjust to different microenvironmental settings. Validation shows that the model can produce 3D granuloma aggregates and microscale infection cultures with increased proinflammatory components. Later, under various culture conditions, a coculture with an angiogenesis model demonstrated notable morphological alterations. The modular granuloma model, by improving the in vivo research, provides more opportunities to scrutinize the interaction of granuloma with their surroundings (Berry et al., 2020). Compared to traditional in vitro models, such systems provide greater flexibility and physiological relevance; however, their complexity and technical requirements may limit widespread adoption.

AI-Integrated Biosensor Feedback Loops for Real-Time Drug Release

By fusing microfluidic medication delivery with AI-driven biosensors to enable real-time monitoring and therapy, the Smart Cancer Patch revolutionizes the field of medicines. It uses AI to evaluate cancer risk and detects cancer biomarkers in saliva or sweat, delivering medication only when required. This non-invasive method has the potential to improve patient outcomes globally by enhancing cancer identification and treatment (Jinan et al., 2025).

Biosensor-integrated drug delivery systems are cutting-edge medical technologies that allow continuous drug administration and monitoring. They mainly target chronic ailments including diabetes, cancer, and cardiovascular diseases, as well as advancements in regenerative medicine, using smart polymers, bioMEMS (biomedical Micro-Electro-Mechanical systems) and electrochemical sensors. The system releases medications in response to particular signals after continuously analyzing biological substances using sensors (Cicha et al., 2022). BioMEMS are significant in biomedical engineering for biomolecular analysis and sensing since they are made for biomedical applications. By transforming different signals into electrical impulses that enable drug release based on sensor feedback once implanted in the human body, they provide advantages including short response time, scalability, and sensitivity (Cicha et al., 2022; Coffell & Nuxoll, 2018; Gray et al., 2018).

Artificial intelligence (AI) and wearables have recently captured attention due to AI's capacity to handle high-dimensional data from wearable biosensing networks. By revealing hidden patterns in biosensing data, this movement advances customized treatment through effective, economical, and precise point-of-care diagnostics. Smartphone-based biosensing systems have been created as a result of the use of wearables for sensor readout, wireless data transfer, data processing and storage, result presentation, and cloud server connectivity (Zhang et al., 2023) as illustrated in figure 4. (Figure 4).

Figure 4. AI - Assisted Wearable Biosensors

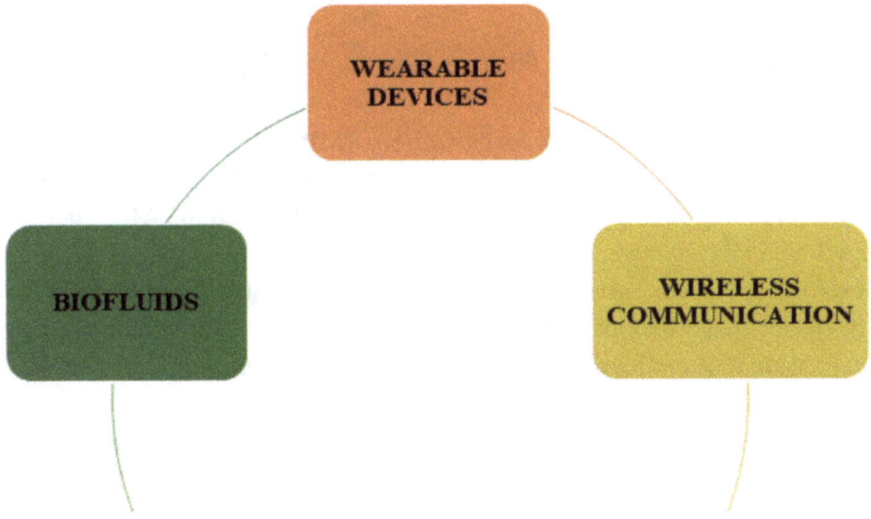

Source: Created by the authors. The figure depicts different AI- assisted wearable biosensors which enables monitoring and intelligent data - driven clinical insights.

The biofluids from the body surface can be collected by the wearable biosensors to identify health related biomarker levels (Zhang et al., 2023; Kim et al., 2019;

Yetisen et al., 2018). The data collected by biosensors from the wireless communication devices are transferred to the personal smart readout device (Zhang et al., 2023; Bandodkar and Wang, 2014; Jin et al., 2020). The raw data are then evaluated and stored in cloud servers or local devices which can help in diagnosing diseases using machine learning algorithms (Zhang et al., 2023; Pathinarupothi et al., 2019; Jeong et al., 2021).

Model interpretability, dependability, flexibility, wearable power constraints, and data security are some of the issues AI- assisted wearable biosensors must deal with. Notwithstanding these obstacles, these biosensors seek to minimize latency and boost dependability while delivering accurate health information for clinical diagnosis and individual health profiles. AI - assisted wearable biosensors provides an alternative for disease diagnosis and health monitoring through the integration of AI, wearable biosensors, and IoT. It has the ability to change the medical system into a personalized remote healthcare model that encourages healthy lifestyles (Zhang et al., 2023).

APPLICATIONS OF AI-GUIDED NANOCARRIERS IN AMR MANAGEMENT

Clinical Case Studies in Infectious Diseases

Smart Nanoparticles for Multidrug-Resistant MRSA

Methicillin-resistant *Staphylococcus aureus* (MRSA) remains to pose clinical challenges due to its resistance to multiple antibiotics and its ability to form biofilms that protect against conventional therapies. Building on the AI-guided design strategies described above, recent advancements in nanotechnology have led to the development of smart nanoparticles designed to combat these resistant strains effectively. Targeted nanoparticle systems have emerged as an effective approach for the treatment of methicillin-resistant *Staphylococcus aureus* (MRSA), particularly due to their ability to concentrate therapeutic agents at the infection site while minimizing cytotoxic effects on healthy tissues. Several types of nanoparticles, including metallic, polymeric, and lipid-based formulations are being engineered to improve bacterial specificity and reduce off-target toxicity. By enhancing drug stability, improving penetration into bacterial structures, and enabling controlled release profiles, these platforms hold promise for increasing therapeutic efficacy while potentially minimizing systemic exposure and resistance pressure (Andrade et al., 2023).

Recent innovations have also explored the use of red phosphorus nanoparticles (RPNPs) combined with low-temperature photothermal treatment (45°C) substantially sensitised methicillin-resistant *Staphylococcus aureus* (MRSA) to aminoglycoside antibiotics. Their mechanistic study showed that the heat generated by RPNPs inhibited the activity of the aminoglycoside-modifying enzyme APH (2″), thereby restoring antibiotic efficacy against resistant MRSA in vivo. This strategy indicates on the exogenous physical activation of nanomaterials in enhancing the performance of conventional antibiotics against resistant pathogens (Tan et al., 2020)

Moreover, dual-membrane bioinspired nanocarriers for targeted delivery in MRSA infections describes the development of rifampicin-loaded nanoparticles cloaked with two different cell membranes (neutrophil and lung epithelial A549) that inherit the targeting and cell-interaction properties of each. The nanoparticles had an average diameter of ~191 nm and a near-neutral zeta potential (−2.7 mV). They exhibited rapid uptake by lung epithelial cells (within five minutes) and selectively accumulated in inflamed lung tissue in a murine MRSA-induced acute lung injury (ALI) and bacteraemia model. The study found enhanced efficacy against MRSA in its planktonic, biofilm, and intracellular forms, and significantly improved lung targeting with reduced overall tissue damage compared to controls (Lin et al., 2025). While these studies demonstrate promising preclinical and early translational outcomes, large-scale clinical trials of AI-guided antibiotic nanocarriers remain limited, highlighting the need for rigorous clinical validation.

Enzyme-Triggered Polymeric Nanoparticles for Lung Infections

Enzyme-responsive polymeric nanoparticles have emerged as an innovative and highly promising approach for targeted drug delivery in the treatment of lung infections, especially those caused by multidrug-resistant pathogens. These nanocarriers are engineered to release their therapeutic payloads selectively in response to specific enzymatic activities within the infected lung tissue, thereby improving drug efficacy while minimizing systemic toxicity.

Recent advancements have shown that pH- and enzyme-responsive nanocarriers can effectively target inflamed or infected pulmonary tissues. In one study, pH-sensitive nanoparticles coated with anti-ICAM-1 antibodies were designed to deliver the anti-inflammatory drug TPCA-1 specifically to inflamed lung endothelia. These nanoparticles recognized adhesion molecules upregulated during inflammation and released their payload in the acidic microenvironment, significantly alleviating acute lung injury (Zhang et al., 2019). Enzyme-responsive nanocarriers have also been engineered to enable drug release in infection-associated enzyme-rich environments, improving therapeutic precision (Lim et al., 2016).

Collectively, these findings underscore the potential of enzyme-triggered polymeric nanoparticles as a next-generation therapeutic platform for combating severe lung infections. By harnessing the biochemical cues unique to infected tissues, these smart nanocarriers can ensure precise, localized drug delivery enhancing therapeutic efficiency while minimizing off-target effects. The data is summarised in Table 2. (Table 2).

Table 2. Advanced Nanocarrier Strategies for MRSA and Lung Infection Therapy.

Technology	Mechanism	Key Outcome	Reference
Dual-membrane nanocarriers	Neutrophil + A549 membrane coating	Rapid lung targeting, kills planktonic + biofilm MRSA	Lin et al., 2025
Red phosphorus Nanoparticles	Low-temp photothermal effect inhibits resistance enzyme	Restores aminoglycoside activity	Tan et al., 2020
Enzyme-responsive polymeric Nanoparticles	Release triggered by infection enzymes	Precise drug delivery to inflamed lung	Zhang et al., 2019; Lim et al., 2016

Source: Compiled from Lin et al., (2025), Tan et al. (2020), Zhang et al., (2019), and Lim et al., (2016). This table summarizes emerging nanocarrier platforms and their mechanisms, highlighting their potential to enhance targeting, overcome resistance mechanisms, and improve therapeutic efficacy in complex infections such as MRSA and lung diseases.

Applications in Low- and Middle-Income Countries

AI-Localized Resistance Prediction

In low- and middle-income countries, the successful deployment of AI-guided antibiotic nanotechnologies depends not only on technical performance but also on affordability, infrastructure compatibility, and scalability. Antimicrobial resistance (AMR) represents one of the most pressing global health challenges, and its burden in such regions contributes to delays in effective diagnosis and treatment. In these contexts, AI-guided resistance prediction approaches can be applied to identify localized resistance patterns, thereby enabling precision antimicrobial stewardship. Recent advances use ML to support antimicrobial resistance (AMR) surveillance and therapeutic decision-making in low- and middle-income countries (LMICs).

Machine learning models, including support vector machines, random forests, logistic regression, and light gradient boosting, have been applied to whole-genome sequencing data from *Escherichia coli* isolates in multi-country African settings, achieving high predictive accuracy for resistance to ciprofloxacin, cefotaxime and ampicillin, and identifying key genomic mutations linked with resistance (Nsubuga et al., 2024).

Similar frameworks have been developed for *Klebsiella pneumoniae*, aiming to support clinical therapy selection and optimize antibiotic prescriptions based on local resistance profiles. These predictive tools hold promise for tailoring antibiotic prescriptions to local resistance profiles, reducing reliance on empirical broad-spectrum therapy and improving AMR surveillance in resource-limited settings. However, significant challenges remain including data quality, standardization across sites, model validation in real-world clinical workflows and integration into frontline care in underserved regions (Condorelli et al., 2024).

Low-Cost NPs (Alginate, Chitosan) Tailored With AI for Region-Specific Pathogens

In low- and middle-income countries (LMICs), the escalating threat of antimicrobial resistance (AMR) necessitates innovative approaches to enhance treatment efficacy while minimizing costs. Polysaccharide-based nanoparticles, particularly those derived from alginate and chitosan, have garnered attention for their biocompatibility, biodegradability, and ability to be engineered for targeted drug delivery. These nanoparticles can be synthesized using cost-effective methods and tailored to address the specific pathogens prevalent in regional settings. For instance, chitosan-alginate nanoparticles exhibited direct antimicrobial and anti-inflammatory activity against *Propionibacterium acnes* in vitro and enhanced delivery of benzoyl peroxide with improved efficacy and reduced eukaryotic toxicity (Friedman et al., 2012). Although this study focused on skin infections rather than systemic antibiotic delivery, it highlights the potential of polysaccharide-nanoparticle platforms for broader therapeutic adaptation in resource-limited settings. Incorporating AI into the development of low-cost, polysaccharide-based nanoparticles represents a promising avenue for combating AMR in LMICs. By leveraging local data and tailoring treatments to the specific needs of the population, this approach can improve patient outcomes, reduce healthcare costs, and contribute to the global effort to manage antimicrobial resistance.

CHALLENGES

One of the major obstacles in translating nano-enabled therapeutics from research to clinical reality is the economic and manufacturing burden associated with nanomedicine platforms.

The clinical translation of nanoparticulate nanomedicines (NNMs) is widely recognized as an expensive and time-consuming process, compounded by the technical complexity and precision required in nanoparticle fabrication. Many nanocarrier systems involve complex and laborious synthesis procedures, which introduce fundamental limitations hindering clinical adoption. These include high material and manufacturing costs, low production yields, and insufficient batch-to-batch reproducibility, consistency, and storage stability of the final product. Collectively, these factors show that, despite promising therapeutic outcomes, advanced nanocarriers often face significant financial, scalability, and quality-control barriers that prevent progression beyond the laboratory stage (Hua et al., 2018).

Even when manufacturing and scale-up challenges are addressed, AI-guided antibiotic nanocarrier systems face additional technical, safety, and deployment-related barriers that limit real-world translation. Nanocarrier-based therapeutics are associated with variability in toxicity, immune interactions, and off-target accumulation, which can contribute to discrepancies between preclinical efficacy and clinical outcomes. Such safety concerns, along with biological complexity and reproducibility challenges, have been identified as major contributors to the failure of nanomedicine platforms during clinical translation (Metselaar & Lammers, 2020). From an artificial intelligence perspective, clinical deployment is further complicated by issues related to data quality, algorithmic bias, model drift, and insufficient external validation, emphasizing the need for rigorous validation, continuous performance monitoring, and clinician oversight to ensure safe implementation of AI-supported therapeutic systems (Kelly et al., 2019).

These deployment and validation challenges are increasingly evident in clinical evaluations of AI-assisted antimicrobial decision-making. A recent systematic review reported that machine learning-based clinical decision support systems can improve antimicrobial resistance prediction and optimize antibiotic use, whereas large language models currently show higher error rates and raise patient safety concerns. The authors emphasize that careful clinical validation, regular model updating, and continued clinician oversight are necessary for the safe integration of AI into antibiotic decision-making, which is particularly relevant for the clinical adoption of AI-guided antibiotic nanotechnologies (AlGain et al., 2025).

ETHICAL AND GOVERNANCE CHALLENGES OF AI-GUIDED ANTIBIOTIC NANOTECHNOLOGIES

Global health organizations have emphasized the need for ethical governance of AI in healthcare, including transparency, data privacy, and accountability, particularly for high-risk clinical applications (World Health Organization [WHO], 2021).

At the regulatory level, agencies such as the FDA, EMA, and CDSCO require established proof of safety and efficacy, which can conflict with the adaptive nature of AI algorithms. Key issues include the validation of these algorithms, model version control, and the evaluation of both nanoparticle pharmacology and AI model robustness. Disparities in AI regulation across jurisdictions further complicate global deployment, necessitating new models for rigorous safety assessment alongside continuous monitoring (Kataria, 2025).

Ethical challenges, including algorithmic bias and data privacy, also arise from AI-enhanced nanoparticles, highlighting the need for careful testing. Effective implementation of AI-guided nanomedicines therefore demands collaboration among regulators, researchers, and industry, combined with ethical safeguards to ensure patient trust and equitable access to treatments (Elabiad & Salaheldin, 2025).

FUTURE DIRECTIONS

Building on these challenges in manufacturing, regulation, and clinical adoption, new approaches to nanomedicine production have been explored. The NANOSPRESSO platform, for instance, shifts production from large industrial facilities to hospital-pharmacy settings. Using modular cartridges and microfluidics, the system enables point-of-care preparation of personalized nucleic acid nanomedicines, allowing hospital pharmacists to assemble tailored therapeutic formulations directly at the bedside. As highlighted by the authors, this model provides high-quality, personalized nucleic acid therapeutics manufactured conveniently at local healthcare facilities (Estapé Senti et al., 2025). It also highlights a scalable path for future nanomedicine deployment, especially in settings where rapid, patient-specific formulation and on-site manufacturing will be essential.

CONCLUSION

The integration of artificial intelligence with nanobiotechnology heralds a new era in the fight against antimicrobial resistance. AI-guided nanocarriers enable precision antibiotics by delivering drugs specifically to infection sites, optimizing dosage, and minimizing off-target exposure, thereby reducing selective pressure that drives resistance. Clinical and computational strategies now allow for patient- and pathogen-specific interventions, ensuring that therapies are both effective and safer.

By incorporating predictive modelling, real-time biosensor feedback, and advanced nanocarrier design, these technologies provide a framework for stewardship-driven antibiotic delivery, supporting responsible use and extending the clinical utility of existing antimicrobials. In low- and middle-income settings, AI-tailored, cost-effective nanocarrier solutions can further improve access while mitigating the global AMR burden.

Looking forward, the continued convergence of AI, nanotechnology, and precision medicine will be pivotal in developing next-generation antimicrobial strategies. To fully realize this potential, collaborative efforts among researchers, clinicians, and policymakers are essential for ensuring regulatory alignment, ethical deployment, and sustainable stewardship, ultimately transforming how infectious diseases are prevented and treated worldwide.

REFERENCES

Afreen, S., Zare, M., Thomas, V., Kumar, T. S. S., & Ramakrishna, S. (2022). Nano-based drug delivery systems: Conventional drug delivery routes, recent developments and future prospects. *Medicine in Drug Discovery*, *15*, 100134. DOI: 10.1016/j.medidd.2022.100134

Afzal, O., Altamimi, A. S. A., Nadeem, M. S., Alzarea, S. I., Almalki, W. H., Tariq, A., Mubeen, B., Murtaza, B. N., Iftikhar, S., Riaz, N., & Kazmi, I. (2022). Nanoparticles in drug delivery: From history to therapeutic applications. *Nanomaterials (Basel, Switzerland)*, *12*(24), 4494. DOI: 10.3390/nano12244494

Ahsan, A., Thomas, N., Barnes, T. J., Subramaniam, S., Loh, T. C., Joyce, P., & Prestidge, C. A. (2024). Lipid nanocarriers-enabled delivery of antibiotics and antimicrobial adjuvants to overcome bacterial biofilms. *Pharmaceutics*, *16*(3), 396. DOI: 10.3390/pharmaceutics16030396

AlGain, S., Marra, A. R., Kobayashi, T., Marra, P. S., Celeghini, P. D., Hsieh, M. K., Shatari, M. A., Althagafi, S., Alayed, M., Ranavaya, J. I., Boodhoo, N. A., Meade, N. O., Fu, D., Sampson, M. M., Rodriguez-Nava, G., Zimmet, A. N., Ha, D., Alsuhaibani, M., Huddleston, B. S., & Salinas, J. L. (2025). Can we rely on artificial intelligence to guide antimicrobial therapy? A systematic literature review. *Antimicrobial Stewardship & Healthcare Epidemiology : ASHE*, *5*(1), e90. DOI: 10.1017/ash.2025.47

Andrade, S., Ramalho, M. J., Santos, S. B., Melo, L. D. R., Santos, R. S., Guimarães, N., Azevedo, N. F., Loureiro, J. A., & Pereira, M. C. (2023). Fighting methicillin-resistant *Staphylococcus aureus* with targeted nanoparticles. *International Journal of Molecular Sciences*, *24*(10), 9030. DOI: 10.3390/ijms24109030

Avanzo, M., Stancanello, J., Pirrone, G., Drigo, A., & Retico, A. (2024). The evolution of artificial intelligence in medical imaging: From computer science to machine and deep learning. *Cancers (Basel)*, *16*(21), 3702. DOI: 10.3390/cancers16213702

Bae, H., Ji, H., Konstantinov, K., Sluyter, R., Ariga, K., Kim, Y. H., & Kim, J. H. (2025). Artificial intelligence-driven nanoarchitectonics for smart targeted drug delivery. *Advanced Materials*, *37*(42), e10239. DOI: 10.1002/adma.202510239

Baig, N., Kammakakam, I., & Falath, W. (2021). Nanomaterials: A review of synthesis methods, properties, recent progress, and challenges. *Materials Advances*, *2*(6), 1821–1871. DOI: 10.1039/D0MA00807A

Bandodkar, A. J., & Wang, J. (2014). Non-invasive wearable electrochemical sensors: A review. *Trends in Analytical Chemistry*, *76*, 173–181. DOI: 10.1016/j.trac.2015.07.005

Barrila, J., Crabbé, A., Yang, J., Franco, K., Nydam, S. D., Forsyth, R. J., Davis, R. R., Gangaraju, S., Ott, C. M., Coyne, C. B., Bissell, M. J., & Nickerson, C. A. (2018). Modeling host–pathogen interactions in the context of the microenvironment: Three-dimensional cell culture comes of age. *Infection and Immunity*, *86*(9), e00282–e18. DOI: 10.1128/IAI.00282-18

Berry, S. B., Gower, M. S., Su, X., Seshadri, C., & Theberge, A. B. (2020). A modular microscale granuloma model for immune–microenvironment signaling studies in vitro. *Frontiers in Bioengineering and Biotechnology*, *8*, 551913. DOI: 10.3389/fbioe.2020.00931

Bhati, D., Shukla, P., Shukla, P., Devi, B. N., Sarakula, P., Chopade, P. D., Kundu, H., & Bhatt, P. (2024). Fused and substituted pyrimidine derivatives as potent anticancer agents. *Biochemical and Cellular Archives*, *24*(1). Advance online publication. DOI: 10.51470/bca.2024.24.1.749

Birk, S. E., Boisen, A., & Nielsen, L. H. (2021). Polymeric nano- and microparticulate drug delivery systems for treatment of biofilms. *Advanced Drug Delivery Reviews*, *174*, 30–52. DOI: 10.1016/j.addr.2021.04.005

Blanquart, F. (2019). Evolutionary epidemiology models to predict the dynamics of antibiotic resistance. *Evolutionary Applications*, *12*(3), 365–383. DOI: 10.1111/eva.12753

Blondeau, J. M. (2009). What have we learned about antimicrobial use and the risks for *Clostridium difficile*-associated diarrhoea? *The Journal of Antimicrobial Chemotherapy*, *63*(2), 238–242. DOI: 10.1093/jac/dkn477

Boiko, D. A., MacKnight, R., Kline, B., & Gomes, G. (2023). Autonomous chemical research with large language models. *Nature*, *624*(7992), 570–578. DOI: 10.1038/s41586-023-06792-0

Bonifácio, B. V., da Silva, P. B., dos Santos Ramos, M. A., Negri, K. M. S., Bauab, T. M., & Chorilli, M. (2014). Nanotechnology-based drug delivery systems and herbal medicines: A review. *International Journal of Nanomedicine*, *9*, 1–15. DOI: 10.2147/IJN.S57357

Bose, A., Roy Burman, D., Sikdar, B., & Patra, P. (2021). Nanomicelles: Types, properties and applications in drug delivery. *IET Nanobiotechnology / IET*, *15*(1), 19–27. DOI: 10.1049/nbt2.12018

Bouhedjar, K., Boukelia, A., Khorief Nacereddine, A., Boucheham, A., Belaidi, A., & Djerourou, A. (2020). A natural language processing approach based on embedding deep learning from heterogeneous compounds for quantitative structure–activity relationship modeling. *Chemical Biology & Drug Design*, *96*(3), 961–972. DOI: 10.1111/cbdd.13742

Bourgat, Y., Mikolai, C., Stiesch, M., Klahn, P., & Menzel, H. (2021). Enzyme-responsive nanoparticles and coatings made from alginate/peptide ciprofloxacin conjugates as drug release system. *Antibiotics (Basel, Switzerland)*, *10*(6), 653. DOI: 10.3390/antibiotics10060653

Bulman, Z. P., Wicha, S. G., Nielsen, E. I., Lenhard, J. R., Nation, R. L., Theuretzbacher, U., Derendorf, H., Tängdén, T., Zeitlinger, M., Landersdorfer, C. B., Bulitta, J. B., Friberg, L. E., Li, J., & Tsuji, B. T. (2022). Research priorities towards precision antibiotic therapy to improve patient care. *The Lancet Microbe*, *3*(10), e795–e802. DOI: 10.1016/S2666-5247(22)00121-5

Cascella, M., Montomoli, J., Bellini, V., & Bignami, E. (2023). Evaluating the feasibility of ChatGPT in healthcare: An analysis of multiple clinical and research scenarios. *Journal of Medical Systems*, *47*(1), 33. DOI: 10.1007/s10916-023-01925-4

Chamundeeswari, M., Jeslin, J., & Verma, M. L. (2019). Nanocarriers for drug delivery applications. *Environmental Chemistry Letters*, *17*(2), 849–865. DOI: 10.1007/s10311-018-00841-1

Chou, W. C., Canchola, A., Zhang, F., & Lin, Z. (2025). Machine learning and artificial intelligence in nanomedicine. *Wiley Interdisciplinary Reviews. Nanomedicine and Nanobiotechnology*, *17*(4), e70027. DOI: 10.1002/wnan.70027

Cicha, I., Priefer, R., Severino, P., Souto, E. B., & Jain, S. (2022). Biosensor-integrated drug delivery systems as new materials for biomedical applications. *Biomolecules*, *12*(9), 1198. DOI: 10.3390/biom12091198

Coffel, J., & Nuxoll, E. (2018). BioMEMS for biosensors and closed-loop drug delivery. *International Journal of Pharmaceutics*, *544*(2), 335–349. DOI: 10.1016/j.ijpharm.2018.01.030

Condorelli, C., Nicitra, E., Musso, N., Bongiorno, D., Stefani, S., Gambuzza, L. V., Carchiolo, V., & Frasca, M. (2024). Prediction of antimicrobial resistance of *Klebsiella pneumoniae* from genomic data through machine learning. *PLoS One*, *19*(9), e0309333. DOI: 10.1371/journal.pone.0309333

Conteh, J. S., Nucci, G. E., Fernandez Cabada, T., Mai, B. T., Soni, N., De Donato, F., Pasquale, L., Catalano, F., Prato, M., Manna, L., & Pellegrino, T. (2023). $CuFeS_2$ nanoparticles functionalized with a thermoresponsive polymer for photothermia and externally controlled drug delivery. *ACS Applied Materials & Interfaces*, *15*(19), 22999–23011. DOI: 10.1021/acsami.3c03902

Corso, G., Stark, H., Jegelka, S., Jaakkola, T., & Barzilay, R. (2024). Graph neural networks. *Nature Reviews. Methods Primers*, *4*(1), 1–23. DOI: 10.1038/s43586-024-00294-7

Crintea, A., Dutu, A. G., Sovrea, A., Constantin, A.-M., Samasca, G., Masalar, A. L., Ifju, B., Linga, E., Neamti, L., Tranca, R. A., Fekete, Z., Silaghi, C. N., & Craciun, A. M. (2022). Nanocarriers for drug delivery: An overview with emphasis on vitamin D and K transportation. *Nanomaterials (Basel, Switzerland)*, *12*(8), 1376. DOI: 10.3390/nano12081376

Das, K. P., & J, C. (2023). Nanoparticles and convergence of artificial intelligence for targeted drug delivery for cancer therapy: Current progress and challenges. *Frontiers in Medical Technology*, *4*, 1067144. DOI: 10.3389/fmedt.2022.1067144

Davis, F. P., Barkan, D. T., Eswar, N., McKerrow, J. H., & Sali, A. (2007). Host–pathogen protein interactions predicted by comparative modeling. *Protein Science : a Publication of the Protein Society*, *16*(12), 2597–2608. DOI: 10.1110/ps.073228407

Ding, M., Zhao, W., Song, L. J., & Luan, S. F. (2022). Stimuli-responsive nanocarriers for bacterial biofilm treatment. *Rare Metals*, *41*(2), 482–498. DOI: 10.1007/s12598-021-01802-4

Elabiad, M., & Salaheldin, A. (2025). Artificial intelligence in nanotechnology: Transforming innovation at the atomic scale. SSRN *Electronic Journal*. DOI: 10.2139/ssrn.5108426

Estapé Senti, M., Ceccaldi, A., Luciani, M., Saber, N., Schurmann, P. J. L., Geerlings, M. W., Holig, P., de Beer, J., Hannus, M., Campbell, F., Winter, M., van Hoeve, W., Westerik, M., Dorst-Mooiman, K. D., Dekker, J., Fuchs, S., Wolbers, F., Segers, T., IJzerman, M., & Schiffelers, R. M. (2025). NANOSPRESSO: Toward personalized, locally produced nucleic acid nanomedicines. *Frontiers in Science*, *3*, 1458636. DOI: 10.3389/fsci.2025.1458636

Ezike, T. C., Okpala, U. S., Onoja, U. L., Nwike, C. P., Ezeako, E. C., Okpara, O. J., Okoroafor, C. C., Eze, S. C., Kalu, O. L., Odoh, E. C., Nwadike, U. G., Ogbodo, J. O., Umeh, B. U., Ossai, E. C., & Nwanguma, B. C. (2023). Advances in drug delivery systems, challenges and future directions. *Heliyon*, *9*(6), e17488. DOI: 10.1016/j.heliyon.2023.e17488

Friedman, A., Phan, J., Schairer, D., Champer, J., Qin, M., Pirouz, A., Blecher-Paz, K., Oren, A., Liu, P., Modlin, R., & Kim, J. (2012). Antimicrobial and anti-inflammatory activity of chitosan–alginate nanoparticles: A targeted therapy for cutaneous pathogens. *The Journal of Investigative Dermatology*, *132*(4), 1234–1245. DOI: 10.1038/jid.2012.399

Gamża, A. M., Hagenaars, T. J., Koene, M. G. J., & de Jong, M. C. M. (2023). Combining a parsimonious mathematical model with infection data from tailor-made experiments to understand environmental transmission. *Scientific Reports*, *13*(1), 38817. DOI: 10.1038/s41598-023-38817-z

Gao, X. J., Ciura, K., Ma, Y., Mikolajczyk, A., Jagiello, K., Wan, Y., Gao, Y., Zheng, J., Zhong, S., Puzyn, T., & Gao, X. (2024). Toward the integration of machine learning and molecular modeling for designing drug delivery nanocarriers. *Advanced Materials*, *36*(45), 2407793. DOI: 10.1002/adma.202407793

. Gargate, N., Laws, M., & Rahman, K. M. (2025). Current economic and regulatory challenges in developing antibiotics for Gram-negative bacteria. *npj Antimicrobial Resistance,* 3._DOI: 10.1038/s44259-025-00123-1

Ghaderzadeh, M., Shalchian, A., Irajian, G., Sadeghsalehi, H., Zahedi Bialvaei, A., & Sabet, B. (2024). Artificial intelligence in drug discovery and development against antimicrobial resistance: A narrative review. *Majallah-i Mikrub/Shinasi-i Pizishki-i Iran*, *18*(3), 135–147. DOI: 10.30699/ijmm.18.3.135

Gou, F., Liu, J., Xiao, C., & Wu, J. (2024). Research on artificial-intelligence-assisted medicine: A survey on medical artificial intelligence. *Diagnostics (Basel)*, *14*(14), 1472. DOI: 10.3390/diagnostics14141472

Gray, M., Meehan, J., Ward, C., Langdon, S. P., Kunkler, I. H., Murray, A., & Argyle, D. (2018). Implantable biosensors and their contribution to the future of precision medicine. *Veterinary Journal (London, England)*, *239*, 21–29. DOI: 10.1016/j.tvjl.2018.07.011

Han, Y., Kim, D. H., & Pack, S. P. (2025). Nanomaterials in drug delivery: Leveraging artificial intelligence and big data for predictive design. *International Journal of Molecular Sciences*, *26*(22), 11121. DOI: 10.3390/ijms262211121

Haredasht, F. N., Maddali, M. V., Ma, S. P., Chang, A., Kim, G. Y. E., Banaei, N., Deresinski, S., Goldstein, M. K., Asch, S. M., & Chen, J. H. (2025). Enhancing antibiotic stewardship: A machine learning approach to predicting antibiotic resistance in inpatient care. *AMIA Annual Symposium Proceedings,* 857–864.

Heydari, S., Masoumi, N., Esmaeeli, E., Ayyoubzadeh, S. M., Ghorbani-Bidkorpeh, F., & Ahmadi, M. (2024). Artificial intelligence in nanotechnology for treatment of diseases. *Journal of Drug Targeting*, *32*(10), 1247–1266. DOI: 10.1080/1061186X.2024.2393417

Hirani, R., Noruzi, K., Khuram, H., Hussaini, A. S., Aifuwa, E. I., Ely, K. E., Lewis, J. M., Gabr, A. E., Smiley, A., Tiwari, R. K., & Etienne, M. (2024). Artificial intelligence and healthcare: A journey through history, present innovations, and future possibilities. *Life (Basel, Switzerland)*, *14*(5), 557. DOI: 10.3390/life14050557

Hortelao, A. C., Simó, C., Guix, M., Garrido, S. G., Julián, E., Vilela, D., Rejc, L., Ramos-Cabrer, P., Cossío, U., Gómez-Vallejo, V., & Patiño, T. (2021). Swarming behavior and in vivo monitoring of enzymatic nanomotors within the bladder. *Science Robotics*, *6*(52), eabd2823. Advance online publication. DOI: 10.1126/scirobotics.abd2823

Hua, S., de Matos, M. B. C., Metselaar, J. M., & Storm, G. (2018). Current trends and challenges in the clinical translation of nanoparticulate nanomedicines: Pathways for translational development and commercialization. *Frontiers in Pharmacology*, *9*, 790. DOI: 10.3389/fphar.2018.00790

Huang, K., Fu, T., Glass, L. M., Zitnik, M., Xiao, C., & Sun, J. (2021a). DeepPurpose: A deep learning library for drug–target interaction prediction. *Bioinformatics (Oxford, England)*, *36*(22–23), 5545–5547. DOI: 10.1093/bioinformatics/btaa1005

Huang, Y., Rana, A. P., Wenzler, E., Ozer, E. A., Krapp, F., Bulitta, J. B., Hauser, A. R., & Bulman, Z. P. (2022). Aminoglycoside-resistance gene signatures are predictive of aminoglycoside MICs for carbapenem-resistant *Klebsiella pneumoniae. The Journal of Antimicrobial Chemotherapy*, *77*(2), 356–363. DOI: 10.1093/jac/dkab381

Huang, Y., Sokolowski, K., Rana, A., Singh, N., Wang, J., Chen, K., Lang, Y., Zhou, J., Kadiyala, N., Krapp, F., Ozer, E. A., Hauser, A. R., Li, J., Bulitta, J. B., & Bulman, Z. P. (2021b). Generating genotype-specific aminoglycoside combinations with ceftazidime/avibactam for KPC-producing *Klebsiella pneumoniae. Antimicrobial Agents and Chemotherapy*, *65*(8), e00692–e21. DOI: 10.1128/AAC.00692-21

Hutchings, M. I., Truman, A. W., & Wilkinson, B. (2019). Antibiotics: Past, present and future. *Current Opinion in Microbiology*, *51*, 72–80. DOI: 10.1016/j.mib.2019.10.008

Jang, Y., Babu, A., Chahal, S., Vasukutty, A., Moon, J. J., Park, I.-K., & Park, H. (2025). AI-guided design of a CXCR4-targeted core–shell nanocarrier for co-delivery of berberine/paclitaxel in cancer therapy. *Journal of Nanobiotechnology*, *23*, 350. DOI: 10.1186/s12951-025-03850-9

Jeong, H., Lee, J. Y., Lee, K., Kang, Y. J., Kim, J. T., Avila, R., Tzavelis, A., Kim, J., Ryu, H., Kwak, S. S., Kim, J. U., Banks, A., Jang, H., Chang, J.-K., Li, S., Mummidisetty, C. K., Park, Y., Nappi, S., Chun, K. S., & Rogers, J. A. (2021). Differential cardiopulmonary monitoring system for artifact-canceled physiological tracking of athletes, workers, and COVID-19 patients. *Science Advances*, *7*(23), eabg3092. Advance online publication. DOI: 10.1126/sciadv.abg3092

Jin, S., Wang, Y., Deng, W., & Li, J. (2020). Wearable sensors for human health monitoring: From materials to devices. *Biosensors & Bioelectronics*, *165*, 112413. DOI: 10.1016/j.bios.2020.112413

Jinan, A. F., Fazeeha, M., & Meera, K. A. (2025). Smart cancer patch: AI-integrated biosensor with adaptive drug release for universal cancer detection and treatment. *International Journal for Research in Applied Science and Engineering Technology*, *13*(6), 438–442. Advance online publication. DOI: 10.22214/ijraset.2025.73593

Jo, E., Epstein, D. A., Jung, H., & Kim, Y. H. (2023). Understanding the benefits and challenges of deploying conversational AI leveraging large language models for public health intervention. In *Proceedings of the CHI Conference on Human Factors in Computing Systems* (Article 16). DOI: 10.1145/3544548.3581503

Kapoor, D. U., Sharma, J. B., Gandhi, S. M., Prajapati, B. G., Thanawuth, K., Limmatvapirat, S., & Sriamornsak, P. (2024). AI-driven design and optimization of nanoparticle-based drug delivery systems. *Science. Engineering and Health Studies*, *18*, 24010003. Advance online publication. DOI: 10.69598/sehs.18.24010003

Kataria, S. (2025). AI-guided design of personalized nanomedicine: A review of data-driven approaches in nanoparticle formulation. *International Journal for Multidisciplinary Research*, *7*(4), 53283. Advance online publication. DOI: 10.36948/ijfmr.2025.v07i04.53283

Katz, L., & Baltz, R. H. (2016). Natural product discovery: Past, present, and future. *Journal of Industrial Microbiology & Biotechnology*, *43*(2–3), 155–176. DOI: 10.1007/s10295-015-1723-5

Kaur, T., & Singh, S. (2021). Controlled release of bi-layered malvidin tablets using 3D printing techniques. *Journal of Pharmaceutical Research International*, *32*(40), 70–78. DOI: 10.9734/jpri/2020/v32i4031034

Kelly, C. J., Karthikesalingam, A., Suleyman, M., Corrado, G., & King, D. (2019). Key challenges for delivering clinical impact with artificial intelligence. *BMC Medicine*, *17*(1), 195. Advance online publication. DOI: 10.1186/s12916-019-1426-2

Kim, J., Campbell, A. S., de Ávila, B. E.-F., & Wang, J. (2019). Wearable biosensors for healthcare monitoring. *Nature Biotechnology*, *37*(4), 389–406. DOI: 10.1038/s41587-019-0045-y

Kluzek, M., Oppenheimer-Shaanan, Y., Dadosh, T., Morandi, M. I., Avinoam, O., Raanan, C., Goldsmith, M., Goldberg, R., & Klein, J. (2022). Designer liposomic nanocarriers are effective biofilm eradicators. *ACS Nano*, *16*(8), 12451–12468. DOI: 10.1021/acsnano.2c04232

Li, M., Du, C., Guo, N., Teng, Y., Meng, X., Sun, H., Li, S., Yu, P., & Galons, H. (2019). Composition design and medical application of liposomes. *European Journal of Medicinal Chemistry*, *164*, 640–653. DOI: 10.1016/j.ejmech.2019.01.007

Lim, Y. H., Tiemann, K. M., Hunstad, D. A., Elsabahy, M., & Wooley, K. L. (2016). Polymeric nanoparticles in development for treatment of pulmonary infectious diseases. *Wiley Interdisciplinary Reviews. Nanomedicine and Nanobiotechnology*, *8*(6), 842–871. DOI: 10.1002/wnan.1401

Lin, C.-Y., Chang, Y.-T., Chung, Y.-K., Alalaiwe, A., Yu, H.-P., & Fang, J.-Y. (2025). Dual-membrane bioinspired nanocarriers for targeted therapy of MRSA-induced acute lung injury and bacteremia. *Journal of Materials Chemistry. B*, *13*(19), 5686–5704. DOI: 10.1039/D4TB02742F

Liu, G. Y., Yu, D., Fan, M. M., Zhang, X., Jin, Z. Y., Tang, C., & Liu, X. F. (2024). Antimicrobial resistance crisis: Could artificial intelligence be the solution? *Military Medical Research*, *11*(1), 7. Advance online publication. DOI: 10.1186/s40779-024-00510-1

Liu, S., Bao, J., Lao, X., & Zheng, H. (2018). Novel 3D structure-based model for activity prediction and design of antimicrobial peptides. *Scientific Reports*, *8*(1), 11189. Advance online publication. DOI: 10.1038/s41598-018-29566-5

Lu, J., Choi, K., Eremeev, M., Gobburu, J., Goswami, S., Liu, Q., Mo, G., Musante, C. J., & Shahin, M. H. (2025). Large language models and their applications in drug discovery and development: A primer. *Clinical and Translational Science*, *18*(4), e70205. Advance online publication. DOI: 10.1111/cts.70205

Mahlapuu, M., Björn, C., & Ekblom, J. (2020). Antimicrobial peptides as therapeutic agents: Opportunities and challenges. *Critical Reviews in Biotechnology*, *40*(7), 978–992. DOI: 10.1080/07388551.2020.1796576

Mat Rani, N. N. I., Hussein, Z. M., Mustapa, F., Azhari, H., Sekar, M., Yi, C. X., & Mohd Amin, M. C. I. (2021). Exploring the possible targeting strategies of liposomes against methicillin-resistant Staphylococcus aureus (MRSA). *European Journal of Pharmaceutics and Biopharmaceutics*, *165*, 84–105. DOI: 10.1016/j.ejpb.2021.04.021

Metselaar, J. M., & Lammers, T. (2020). Challenges in nanomedicine clinical translation. *Drug Delivery and Translational Research*, *10*(3), 721–725. DOI: 10.1007/s13346-020-00740-5

Mitchell, M. J., Billingsley, M. M., Haley, R. M., Wechsler, M. E., Peppas, N. A., & Langer, R. (2021). Engineering precision nanoparticles for drug delivery. *Nature Reviews. Drug Discovery*, *20*(2), 101–124. DOI: 10.1038/s41573-020-0090-8

Mohsen, S., Dickinson, J. A., & Somayaji, R. (2020). Update on the adverse effects of antimicrobial therapies in community practice. *Canadian Family Physician Médecin de Famille Canadien*, *66*(9), 651–659.

Nsubuga, M., Galiwango, R., Jjingo, D., & Mboowa, G. (2024). Generalizability of machine learning in predicting antimicrobial resistance in *E. coli*: A multi-country case study in Africa. *BMC Genomics*, *25*(1), 287. Advance online publication. DOI: 10.1186/s12864-024-10214-4

Nwobodo, D. C., Ugwu, M. C., Anie, C. O., Al-Ouqaili, M. T. S., Ikem, C. J., Uchenna, V. C., & Saki, M. (2022). Antibiotic resistance: The challenges and some emerging strategies for tackling a global menace. *Journal of Clinical Laboratory Analysis*, *36*(9), e24655. Advance online publication. DOI: 10.1002/jcla.24655

Pathinarupothi, R. K., Durga, P., & Rangan, E. S. (2019). IoT-based smart edge for global health: Remote monitoring with severity detection and alerts transmission. *IEEE Internet of Things Journal*, *6*(2), 2449–2462. DOI: 10.1109/JIOT.2018.2870068

Popa, S. L., Pop, C., Dita, M. O., Brata, V. D., Bolchis, R., Czako, Z., Saadani, M. M., Ismaiel, A., Dumitrascu, D. I., Grad, S., David, L., Cismaru, G., & Padureanu, A. M. (2022). Deep learning and antibiotic resistance. *Antibiotics (Basel, Switzerland)*, *11*(11), 1674. Advance online publication. DOI: 10.3390/antibiotics11111674

Popova, M., Isayev, O., & Tropsha, A. (2018). Deep reinforcement learning for de novo drug design. *Science Advances*, *4*(7), eaap7885. Advance online publication. DOI: 10.1126/sciadv.aap7885

Saha, S. (2025). How nanoinformatics could pave the way to safer design of engineered nanomaterials? *Frontiers in Nanotechnology*, *7*, 1559053. DOI: 10.3389/fnano.2025.1559053

Saleh, Y., Abu Talib, M., Nasir, Q., & Dakalbab, F. (2025). Evaluating large language models: A systematic review of efficiency, applications, and future directions. *Frontiers of Computer Science*, *7*, 1523699. Advance online publication. DOI: 10.3389/fcomp.2025.1523699

. Sharma, A., Gupta, K., Lakshmi, Dixit, R., Tyagi, M., Khurana, A., & Garg, S. (2024). AI-driven engineering of liposomes for targeted drug delivery of marine bioactive peptides. *African Journal of Biological Sciences,* 6(S.i3), 2971–2987. DOI: 10.48047/AFJBS.6.Si3.2024.2971-2987

Shepherd, M. J., Fu, T., Harrington, N. E., Kottara, A., Cagney, K., Chalmers, J. D., Paterson, S., Fothergill, J. L., & Brockhurst, M. A. (2024). Ecological and evolutionary mechanisms driving within-patient emergence of antimicrobial resistance. *Nature Reviews. Microbiology*, *22*(10), 650–665. DOI: 10.1038/s41579-024-01041-1

Singhal, K., Azizi, S., Tu, T., Mahdavi, S. S., Wei, J., Chung, H. W., Scales, N., Tanwani, A., Cole-Lewis, H., Pfohl, S., Payne, P., Seneviratne, M., Gamble, P., Kelly, C., Babiker, A., Schärli, N., Chowdhery, A., Mansfield, P., Demner-Fushman, D., & Natarajan, V. (2023). Large language models encode clinical knowledge. *Nature*, *620*(7972), 172–180. DOI: 10.1038/s41586-023-06291-2

Stokes, J. M., Yang, K., Swanson, K., Jin, W., Cubillos-Ruiz, A., Donghia, N. M., MacNair, C. R., French, S., Carfrae, L. A., Bloom-Ackermann, Z., Tran, V. M., Chiappino-Pepe, A., Badran, A. H., Andrews, I. W., Chory, E. J., Church, G. M., Brown, E. D., Jaakkola, T. S., Barzilay, R., & Collins, J. J. (2020). A deep learning approach to antibiotic discovery. *Cell*, *180*(4), 688–702.e13. DOI: 10.1016/j.cell.2020.01.021

Sun, M., Zhao, S., Gilvary, C., Elemento, O., Zhou, J., & Wang, F. (2020). Graph convolutional networks for computational drug development and discovery. *Briefings in Bioinformatics*, *21*(3), 919–935. DOI: 10.1093/bib/bbz042

Suriyaamporn, P., Pamornpathomkul, B., Wongprayoon, P., Rojanarata, T., Ngawhirunpat, T., & Opanasopit, P. (2024). The artificial intelligence and design of experiment assisted in the development of progesterone-loaded solid-lipid nanoparticles for transdermal drug delivery. *Farmatsiia (Sofia)*, *71*, 1–12. DOI: 10.3897/pharmacia.71.e123549

Tan, L., Zhou, Z., Liu, X., Li, J., Zheng, Y., Cui, Z., Yang, X., Liang, Y., Li, Z., Feng, X., Zhu, S., Yeung, K. W. K., Yang, C., Wang, X., & Wu, S. (2020). Overcoming multidrug-resistant MRSA using conventional aminoglycoside antibiotics. *Advanced Science (Weinheim, Baden-Wurttemberg, Germany)*, *7*(14), 1902070. Advance online publication. DOI: 10.1002/advs.201902070

Timbrook, T. T., Morton, J. B., McConeghy, K. W., Caffrey, A. R., Mylonakis, E., & LaPlante, K. L. (2017). The effect of molecular rapid diagnostic testing on clinical outcomes in bloodstream infections: A systematic review and meta-analysis. *Clinical Infectious Diseases*, *64*(1), 15–23. DOI: 10.1093/cid/ciw649

Trafton, A. (2020). Artificial intelligence yields new antibiotic. *MIT News*. https://news.mit.edu/2020/artificial-intelligence-identifies-new-antibiotic-0220

Trasca, D.-M., Dorin, P. I., Carmen, S., Varut, R.-M., Singer, C. E., Radivojevic, K., & Stoica, G. A. (2025). Artificial intelligence in biomedicine: A systematic review from nanomedicine to neurology and hepatology. *Pharmaceutics*, *17*(12), 1564. DOI: 10.3390/pharmaceutics17121564

Wang, C., Kumar, G. A., & Rajapakse, J. C. (2025). Drug discovery and mechanism prediction with explainable graph neural networks. *Scientific Reports*, *15*(1), 179. Advance online publication. DOI: 10.1038/s41598-024-83090-3

Wang, T., Wu, C., Hu, Y., Zhang, Y., & Ma, J. (2023). Stimuli-responsive nanocarrier delivery systems for Pt-based antitumor complexes: A review. *RSC Advances*, *13*(24), 16488–16511. DOI: 10.1039/D3RA00866E

Wang, W., Ye, Z., Gao, H., & Ouyang, D. (2021). Computational pharmaceutics—A new paradigm of drug delivery. *Journal of Controlled Release*, *338*, 119–136. DOI: 10.1016/j.jconrel.2021.08.030

Wang, Z., Wang, Y., Zhang, S., Wang, S., Xu, Z., & Feng, Z. J. (2024). Trend analysis and prediction of gonorrhea in mainland China based on a hybrid time series model. *BMC Infectious Diseases*, *24*(1), 113. Advance online publication. DOI: 10.1186/s12879-023-08969-4

Watkins, R. R. (2022). Antibiotic stewardship in the era of precision medicine. *JAC-Antimicrobial Resistance*, *4*(3), dlac066. Advance online publication. DOI: 10.1093/jacamr/dlac066

World Health Organization. (2021). *Ethics and governance of artificial intelligence for health*. https://www.who.int/publications/i/item/9789240029200

Yan, H., Wen, P., Tian, S., Zhang, H., Han, B., Khan, J., Xue, Y., Chen, X., Li, X., & Li, Y. (2024). Enhancing biofilm penetration and antibiofilm efficacy with protein nanocarriers against pathogenic biofilms. *International Journal of Biological Macromolecules*, *256*(Pt 1), 128300. Advance online publication. DOI: 10.1016/j.ijbiomac.2023.128300

Yang, Y., Jiang, X., Lai, H., & Zhang, X. (2022). Smart bacteria-responsive drug delivery systems in medical implants. *Journal of Functional Biomaterials, 13*(4), 173. Advance online publication. DOI: 10.3390/jfb13040173

Yarahmadi, A., Najafiyan, H., Yousefi, M. H., Khosravi, E., Shabani, E., Afkhami, H., & Aghaei, S. S. (2025). Beyond antibiotics: Exploring multifaceted approaches to combat bacterial resistance in the modern era: A comprehensive review. *Frontiers in Cellular and Infection Microbiology, 15*, 1493915. Advance online publication. DOI: 10.3389/fcimb.2025.1493915

Yetisen, A. K., Martinez-Hurtado, J. L., Unal, B., Khademhosseini, A., & Butt, H. (2018). Wearables in medicine. *Advanced Materials, 30*(8), 1706910. Advance online publication. DOI: 10.1002/adma.201706910

Zhang, C. Y., Lin, W., Gao, J., Shi, X., Davaritouchaee, M., Nielsen, A. E., Mancini, R. J., & Wang, Z. (2019). pH-responsive nanoparticles targeted to lungs for improved therapy of acute lung inflammation/injury. *ACS Applied Materials & Interfaces, 11*(18), 16380–16390. DOI: 10.1021/acsami.9b04051

Zhang, Y., Hu, Y., Jiang, N., & Yetisen, A. K. (2023). Wearable artificial intelligence biosensor networks. *Biosensors & Bioelectronics, 219*, 114825. Advance online publication. DOI: 10.1016/j.bios.2022.114825

Zhang, Z., Chen, L., Zhong, F., Wang, D., Jiang, J., Zhang, S., Jiang, H., Zheng, M., & Li, X. (2022). Graph neural network approaches for drug target interactions. *Current Opinion in Structural Biology, 73*, 102327. Advance online publication. DOI: 10.1016/j.sbi.2021.102327

Zheng, Y., Koh, H. Y., Ju, J., Yang, M., May, L. T., Webb, G. I., Li, L., Pan, S., & Church, G. (2025). Large language models for drug discovery and development. *Patterns (New York, N.Y.), 6*(10), 101346. Advance online publication. DOI: 10.1016/j.patter.2025.101346

ADDITIONAL READING

Adir, O., Poley, M., Chen, G., Froim, S., Krinsky, N., Shklover, J., & Schroeder, A. (2020). Integrating artificial intelligence and nanotechnology for precision cancer medicine. *Advanced Materials, 32*(13), 1901989. DOI: 10.1002/adma.201901989

de Villiers, M. M., Aramwit, P., & Kwon, G. S. (Eds.). (2008). *Nanotechnology in drug delivery*. Springer.

Ibrahim, H. K. (2024). From nanotech to AI: The cutting-edge technologies shaping the future of medicine. *African Journal of Advanced Pure and Applied Sciences*, 410–427.

Lorkowski, J., Kolaszyńska, O., & Pokorski, M. (2021). Artificial intelligence and precision medicine: A perspective. In Pokorski, M. (Ed.), *Integrative Clinical Research* (pp. 1–11). Springer International Publishing., DOI: 10.1007/5584_2021_652

Nandini, R., Sharma, B., & Gabrani, R. (2025). Nanotechnology in precision medicine. In *Nanotechnologies for Drug Delivery and Biopharmaceutical Development* (pp. 533–551). Springer Nature Switzerland. DOI: 10.1007/978-3-032-00796-4_17

National Academies of Sciences, Engineering, and Medicine. (2019). *Artificial intelligence in health care: The hope, the hype, the promise, the peril.* National Academies Press.

Orozco, J. (2023). Nanoscience, nanotechnology, and disruptive technologies in the context of precision medicine. *Revista de la Academia Colombiana de Ciencias Exactas, Físicas y Naturales*, *47*(183), 221–241. DOI: 10.18257/raccefyn.1895

Samin, S., Khan, N. A., Bangash, S. A., & Riaz, S. (2025). Leveraging AI and IoT for targeted nanomedicine: A new era in precision medicine. *Journal of Neonatal Surgery*, *14*(28S, 28s), 137–148. DOI: 10.63682/jns.v14i28S.6569

Tropsha, A., Isayev, O., Varnek, A., Schneider, G., & Cherkasov, A. (2023). Integrating QSAR modelling and deep learning in drug discovery: The emergence of deep QSAR. *Nature Reviews. Drug Discovery*, *23*(2), 141–155. DOI: 10.1038/s41573-023-00832-0

Wang, B., Siahaan, T. J., & Soltero, R. A. (2016). *Drug delivery: Principles and applications*. Wiley. DOI: 10.1002/9781118833322

KEY TERMS AND DEFINITIONS

AI-Guided Nanocarrier Design: The use of artificial intelligence techniques to optimize nanocarrier composition, predict biological interactions, and enhance targeted delivery performance.

Antimicrobial Peptides: Antimicrobial peptides are small, usually positively charged peptides (10-50 amino acids long) that can kill or inhibit the growth of microorganisms, primarily by disrupting their cell membranes or interfering with essential cellular functions.

Antimicrobial Resistance (AMR): The ability of microorganisms to survive exposure to antimicrobial agents, rendering standard treatments ineffective and increasing the risk of disease spread and mortality.

Artificial Intelligence (AI) in Drug Discovery: The use of computational algorithms, including machine learning and deep learning, to analyze biological data, predict drug–target interactions, and accelerate the identification of novel therapeutics.

Biofilm: A structured community of microorganisms enclosed within a self-produced extracellular matrix that adheres to surfaces and exhibits increased resistance to antibiotics.

Nanobiotechnology: The application of nanotechnology principles in biological and medical systems to develop advanced diagnostic tools, drug delivery systems, and therapeutic strategies.

Nanocarriers: Nanoscale delivery systems (1–100 nm) designed to transport therapeutic agents to specific biological targets, improving drug stability, bioavailability, and efficacy.

Precision Antibiotic Therapy: An approach that tailors antibiotic selection and dosing based on patient-specific data, pathogen characteristics, and resistance profiles to improve treatment outcomes.

Targeted Drug Delivery: A therapeutic approach in which drugs are directed specifically to diseased tissues or cells, minimizing systemic exposure and reducing adverse effects.

APPENDIX

Abbrevations

AI – Artificial Intelligence
ML – Machine Learning
DL – Deep Learning
RL – Reinforcement Learning
LLM – Large Language Models
AMR – Antimicrobial Resistance
NP – Nanoparticles
AMP – Antimicrobial Peptides
FMT – Fecal Microbiota Transplantation
TA – Toxin-Antitoxin
ANN – Artificial Neural Network
GNN – Graph Neural Network
NNMs – Nanoparticulate Nanomedicines
LMICs – Low- and Middle-Income Countries
MRSA – Methicillin-Resistant Staphylococcus aureus
RPNPs – Red Phosphorus Nanoparticles
MIPD – Model-Informed Precision Dosing
TDM – Therapeutic Drug Monitoring
GCN – Graph Convolutional Networks
GRNs – Graph Recurrent Networks
DTI – Drug-Target Interaction
XGDP – eXplainable Graph-based Drug Response Prediction
ReLeaSE – Reinforcement Learning for Structural Evolution
SMILES – Simplified Molecular-Input Line-Entry System
NLP – Natural Language Processing
QSAR – Quantitative Structure–Activity Relationship
DDS – Drug Delivery Systems
2-D – Two-Dimensional
3-D – Three-Dimensional
OAC – Organ-on-a-Chip
ECM – Extracellular Matrix
RWV – Rotating Wall Vessel
TB – Tuberculosis
ALI – Acute Lung Injury
RF – Random Forests

SVMs – Support Vector Machines
DNNs – Deep Neural Networks
CNNs – Convolutional Neural Networks

Chapter 6
AI-Guided Stimuli-Responsive Nanocarriers for Precision Antimicrobial Therapy

Mona M. Khalifa
http://orcid.org/0009-0000-6721-9558
Faculty of Science. Cairo University. Egypt

Ahmed Ibrahim Mohammed
Faculty of Biotechnology, Badr University in Cairo, Egypt

Halah Abdulrahman Hafiz
Faculty of Applied Medical Sciences, Umm Al-Qura University, Saudi Arabia

Mariam Sayed A. Ahmed
Faculty of Medicine, Cairo University, Egypt

Mohamed Ihab Al Refaei
Faculty of Education, Ain Shams University, Egypt

Eman Ibrahim Mohammed
Faculty of Biotechnology, Badr University in Cairo, Egypt

Hagar Sayed Abdelgwad
Faculty of Medicine, Banha University, Egypt

ABSTRACT

Antimicrobial resistance (AMR) poses an escalating threat to world health by diminishing the efficacy of common antibiotics. Antimicrobial timing, use, and placement must be carefully planned to combat AMR. These are crucial to advanced infection management. This chapter discusses clever nanocarriers that distribute antibiotics only when needed. These devices deliver medicine when infection is detected, improving treatment and saving waste. We examine how AI and advanced computer models are transforming smart nanocarrier design and utilization. These

DOI: 10.4018/979-8-3373-5268-8.ch006

technologies help uncover better ways to distribute drugs that fight diseases, protect good bacteria, and speed up drug development. We do not randomly administer antibiotics and hope for efficacy. AI can help fight major diseases and resistance, according to the chapter. Novel nanocarriers composed of polymers, lipids, metals, or a combination of these materials can detect infection signals like pH, enzymes, and biofilms and deliver medication exactly to the targeted spot.

1. INTRODUCTION

Antimicrobial resistance (AMR) is recognized as one of the most critical global health challenges of the twenty-first century. Recent global analyses indicate that drug-resistant infections are responsible for millions of deaths each year and impose an increasing economic burden on healthcare systems worldwide (Craig et al., 2024). The rapid emergence and global spread of resistant pathogens now threaten many of the medical advances achieved over the past decades, potentially undermining essential clinical procedures including surgery, cancer chemotherapy, and organ transplantation (Darby et al., 2023). The declining effectiveness of conventional antibiotics is largely driven by the remarkable adaptability of microorganisms. Resistance mechanisms encompass enzymatic degradation of antimicrobial agents, structural modification of drug targets, activation of efflux pumps, and the formation of protective biofilms that serve as physical and biochemical barriers to treatment. (Powers, 2024).

The notion of context-specific antimicrobial therapy has arisen as a viable approach to enhance therapeutic efficacy while reducing systemic toxicity. Targeted antimicrobial systems are engineered to administer therapeutic drugs directly to the infection site, facilitating the selective eradication of harmful pathogens while safeguarding beneficial microbiota. By diminishing off-target drug exposure, these delivery methods also mitigate the selective pressures that foster the development of antimicrobial resistance. Among targeted delivery platforms, nanobiotechnology has emerged as a potent foundation for next-generation antimicrobial systems, offering regulated drug release, enhanced stability, improved pharmacokinetics, and superior tissue penetration compared with conventional formulations. Among these systems, stimuli-responsive smart nanocarriers have attracted significant scientific attention due to their ability to respond dynamically to infection-associated microenvironmental signals including acidic pH conditions, elevated levels of reactive oxygen species (ROS), pathogen-derived enzymes, and biofilm-associated molecular signals (Chaudhary et al., 2025).

By responding to these stimuli, nanocarriers release antimicrobial agents selectively at infection sites, improving therapeutic efficacy while reducing systemic side effects. The integration of artificial intelligence (AI) with nanomedicine further advances this paradigm: machine learning algorithms enable researchers to predict optimal nanocarrier architectures, improve targeting strategies, and design stimuli-responsive drug release profiles from large biological datasets, overcoming the limitations of iterative experimental optimization (Kapoor et al., 2024; Fagbemi et al., 2025).

The integration of nanobiotechnology with artificial intelligence has created new opportunities for the rational design of precision antimicrobial therapies. AI-driven algorithms leverage microbial genomic, infection microenvironment, and pharmacokinetic data to optimize nanocarrier systems; nanocarriers can be engineered to respond to infection-associated signals such as pH gradients, enzymatic activity, and redox imbalances; and stimuli-responsive release delivers antimicrobials selectively within infected tissues, preserving beneficial microbiota and avoiding off-target damage. This chapter proposes a five-stage integrated framework that maps this logic explicitly: (1) AI-driven molecular design of both antibiotic payload and carrier architecture, (2) stimuli-responsive nanocarrier engineering tuned to infection-specific microenvironmental signals,(3) targeted in vivo delivery concentrating therapeutic activity at the infection site, (4) controlled, microbiota-sparing drug release that preserves host commensal communities, and (5) translational advancement through regulatory-aligned safety validation and scalable manufacturing. Artificial intelligence operates across all five stages from generative design in Stage 1 to patient stratification algorithms in Stage 5. To integrate the interdisciplinary components of precision antimicrobial therapy, a conceptual framework is proposed to illustrate the role of artificial intelligence in guiding the design, optimization, and clinical translation of stimuli-responsive nanocarriers Figure 1.

This chapter investigates the evolution of sophisticated infection-responsive nanomaterial systems, analyzing their functional processes and prospective clinical applications. Significant focus is on artificial intelligence-assisted design methodologies, stimuli-responsive nanocarriers, and their contributions to augmenting antibacterial efficacy, reducing systemic toxicity, maintaining the host microbiota, and promoting clinical translation. This chapter demonstrates how interdisciplinary innovation, integrating microbiology, nanotechnology, pharmacology, and artificial intelligence, is advancing the development of precision antimicrobial therapies.

Figure 1. Conceptual framework of AI-guided stimuli-responsive nanocarriers for targeted antimicrobial therapy.

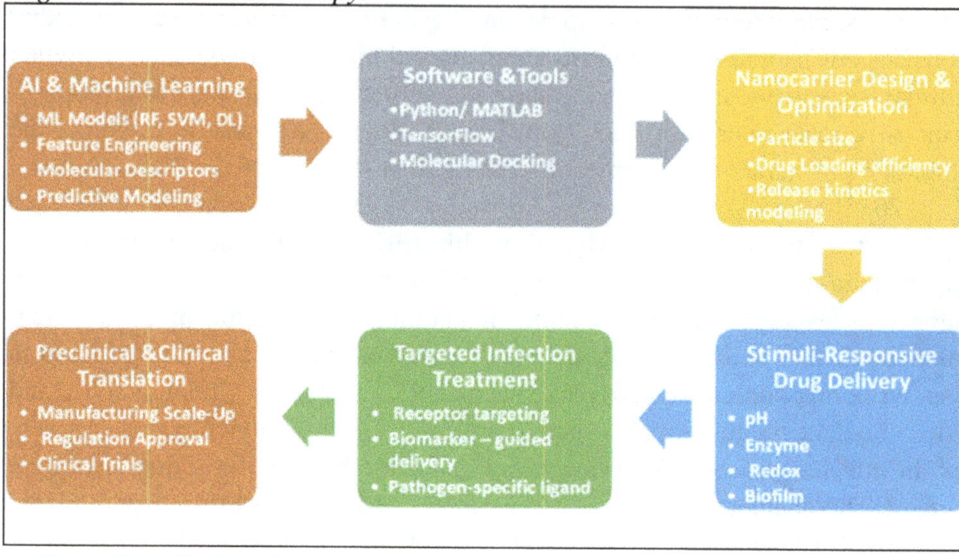

Source: Created by the authors.

Artificial intelligence and machine learning tools enable data-driven design and optimization of nanocarriers, which are engineered for targeted infection treatment through stimuli-responsive drug release mechanisms (pH, enzyme, redox, and biofilm-responsive). The framework further highlights the pathway from computational design to preclinical and clinical translation.

2. MECHANISMS OF ANTIMICROBIAL RESISTANCE AND LIMITATIONS OF CONVENTIONAL ANTIBIOTICS

2.1 Global Landscape and Drivers of Antimicrobial Resistance

The widespread misuse and overuse of antibiotics in clinical medicine, agriculture, and environmental settings have transformed antimicrobial resistance (AMR) into a major global health emergency. Excessive prescribing of broad-spectrum antibiotics, patient self-medication, incomplete treatment courses, and inadequate infection control practices have collectively accelerated the emergence of multidrug-resistant (MDR) pathogens. Global surveillance data now indicate that antibiotic-resistant infections contribute to a growing number of deaths worldwide, with the burden falling disproportionately on low- and middle-income countries where antimicrobial stewardship programs remain limited (Craig et al., 2024).

Horizontal gene transfer (HGT) represents one of the most important mechanisms by which resistance spreads across bacterial populations. Through conjugation, transformation, and transduction, resistance genes can be exchanged between bacterial species. Mobile genetic elements, including plasmids, transposons, and integrons, play a critical role in disseminating resistance determinants both within clinical environments and in broader ecological reservoirs (Tokuda & Shintani, 2024). In addition, environmental sources such as pharmaceutical waste, hospital effluents, and agricultural runoff serve as reservoirs of antimicrobial residues and resistant microorganisms. These interconnected pathways demonstrate that AMR is not solely a clinical issue but a One Health challenge involving human, animal, and environmental health systems (Tzavellas et al., 2026).

2.2 Molecular and Cellular Mechanisms of Antibiotic Resistance

Bacteria employ a wide range of biological strategies to neutralize antimicrobial agents. These resistance mechanisms may arise through intrinsic traits or acquired genetic adaptations, and multiple mechanisms frequently coexist within a single organism, resulting in highly resistant phenotypes. Antimicrobial resistance arises through multiple complex mechanisms that limit antibiotic efficacy. These include enzymatic degradation, target-site modification, reduced permeability, and biofilm formation, as illustrated in Figure 2.

2.2.1 Enzymatic Destruction of Antimicrobial Agents

One of the most prevalent mechanisms of antibiotic resistance involves the production of enzymes that modify or degrade antibiotics, preventing them from reaching their intended cellular targets. A prominent example of this mechanism is the β-lactamase family of enzymes. These enzymes hydrolyze the β-lactam ring structure found in antibiotics such as penicillins and cephalosporins, effectively neutralizing their antimicrobial efficacy. More sophisticated forms of these enzymes, including extended-spectrum β-lactamases (ESBLs) and carbapenemases, provide resistance to a wide range of antibiotics, including those considered as last-resort treatments. This resistance strategy has become a major challenge in the management of hospital-acquired infections. (Belay et al., 2024).

2.2.2 Target-Site Modification

Bacteria can acquire resistance by modifying the molecular targets of antibiotics, which include essential components such as penicillin-binding proteins (PBPs), ribosomal subunits, and DNA gyrase. Genetic mutations or enzymatic modifications can diminish the binding affinity of antibiotics while maintaining the normal biological function of these targets. For instance, changes in PBPs are responsible for β-lactam resistance in methicillin-resistant Staphylococcus aureus (MRSA), while modifications in ribosomal subunits hinder the effectiveness of macrolide antibiotics. Additionally, enzymatic processes like rRNA methylation further reduce antibiotic susceptibility and are frequently transmitted between bacterial populations via horizontal gene transfer (Xie et al., 2025).

2.2.3 Efflux Pumps and Decreased Membrane Permeability

Resistance in Gram-negative bacteria is commonly associated with two major mechanisms: active efflux pump systems and decreased membrane permeability. Efflux pumps actively expel antibiotic molecules from the bacterial cell, reducing intracellular drug concentrations below the minimum inhibitory threshold. Concurrently, downregulation or structural alteration of outer membrane porin channels restricts antibiotic influx, further limiting intracellular drug accumulation. The combination of these two mechanisms significantly diminishes antibiotic efficacy against many Gram-negative pathogens (Gaurav et al., 2023)

2.2.4 The Formation of Biofilm

Biofilms are structured communities of bacteria encased within a self-produced extracellular polymeric matrix (EPM) that adheres to biological surfaces. The extracellular polymeric matrix significantly limits antibiotic penetration into the biofilm structure, while simultaneously shielding resident bacteria from host immune defense. Within biofilm communities, a substantial proportion of bacteria exist in a slow-growing or metabolically dormant state, rendering them inherently less susceptible to antibiotics that target actively dividing cells. Biofilm formation is consequently a major contributor to the persistence and recurrence of healthcare-associated infections (Bamford et al., 2023). Together, these resistance strategies enzymatic drug inactivation, target-site modification, efflux pump activation, and biofilm formation frequently co-occur within a single organism, enabling the highly resistant phenotypes characteristic of multidrug-resistant pathogens (Rezasoltani et al., 2020).

The increasing complexity of how bacteria resist antibiotics highlights the limitations of traditional treatments. Standard antibiotic treatments spread throughout the entire body, affecting both good and harmful bacteria, which raises the risk of resistance developing. As a result, there has been growing interest in targeted delivery systems that release antibiotics only at the site of infection. The next section will explore how nanotechnology-based platforms, specifically stimuli-responsive nanocarriers, can apply insights from resistance biology to create more targeted approaches for treating infections.

Figure 2. Major mechanisms of antimicrobial resistance in bacterial pathogens. These include (1) enzymatic drug degradation (e.g., β-lactamases), (2) target-site modification such as ribosomal methylation and penicillin-binding protein mutations, (3) reduced permeability and active efflux pump systems, and (4) biofilm formation that limits drug penetration and promotes bacterial persistence.

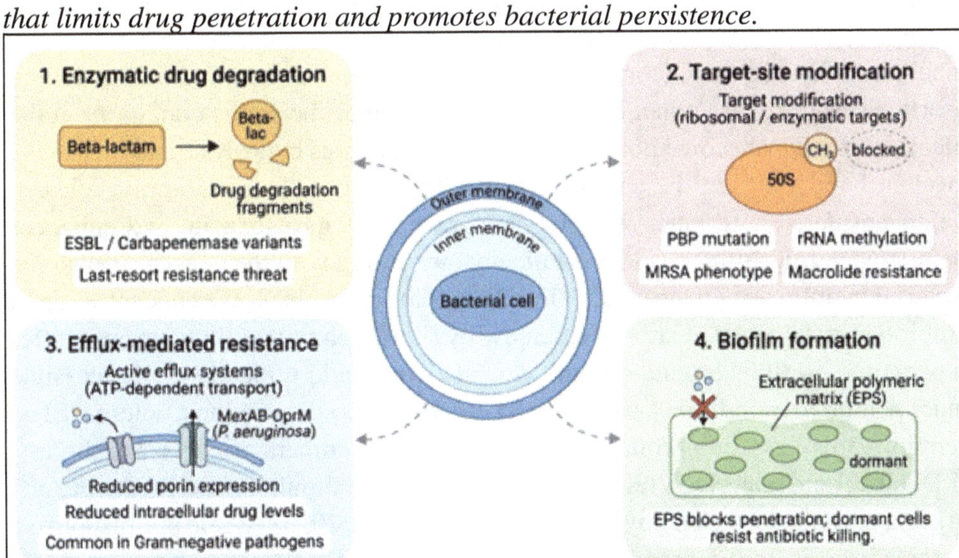

Source: Created by the authors.

Figure 2 illustrates This resistance mechanisms collectively highlight the limitations of conventional antibiotics and underscore the need for precision antimicrobial strategies, including stimuli-responsive nanocarriers capable of bypassing or overcoming these barriers. The overlap between the biology of resistance and the design of responsive carriers is not coincidental: each trigger exploited in Section 3 is directly derived from a resistance or virulence mechanism described in this section.

3. STIMULI-RESPONSIVE ANTIBIOTIC RELEASE SYSTEMS

Stimuli-responsive nanocarriers improve antibiotic delivery by identifying infection-specific triggers pH, enzymes, or redox conditions and initiating drug release exclusively at the target location. This precision diminishes off-target effects, decreases systemic toxicity, and alleviates antibiotic resistance pressure by concentrating therapeutic activity where it is needed. Their versatility renders them effective against complex infections including biofilms and intracellular pathogens, and their compatibility with imaging modalities enables real-time monitoring of therapeutic progress (Zhang et al., 2023).

3.1. pH-Triggered Release

Bacterial fermentation of food (such as lactic acid production), tissue inflammation, and low oxygen levels all contribute to the creation of an acidic environment at infection sites (pH 4.5–6.5). Stimuli-responsive nanocarriers take advantage of this pH difference by incorporating acid-sensitive bonds into structures like liposomes, micelles, or hydrogels. Common acid-sensitive bonds include hydrazone, acetal, or orthoester linkages. When these linkages encounter the acidic conditions at the infection site, they become protonated, which accelerates hydrolysis and breaks the covalent bond, causing the carrier structure to collapse. This disruption releases the encapsulated antibiotic, such as vancomycin or ciprofloxacin, and enhances membrane permeability by electrostatically interacting with the negatively charged bacterial membrane (Zhang et al., 2023; Mi, 2020). As a result, the concentration of antibiotics in the affected area can increase by five to ten times, allowing antibiotics to penetrate biofilms around chronic wounds that would typically be inaccessible under neutral pH conditions. pH-sensitive mesoporous silica nanoparticles (MSNs) exemplify this class: upon contact with the acidic environment of a bacterial abscess, MSNs swell and selectively release their payload while simultaneously reducing off-target consequences in healthy tissue (Konwar & Kim, 2025). While pH-responsive systems benefit from a broadly applicable trigger and relatively mature synthetic chemistry, their principal limitation is selectivity: tumor microenvironments and intracellular lysosomes in healthy tissue are also acidic, raising off-target release risk in patients with concurrent malignancy or chronic inflammation. This context-dependence motivates the development of more specific enzymatic and multi-trigger approaches described in the sections below.

3.2. Enzyme-Responsive Mechanisms

During infection, numerous bacterial species synthesize enzymes that facilitate tissue penetration and biofilm development. These enzymes phospholipases, hyaluronidases, and β-lactamases function as precise biochemical catalysts for targeted drug release. Enzyme-responsive nanocarriers utilize peptide or ester linkers that are specifically cleaved by pathogen-associated enzymes. Upon recognition and cleavage of the linker sequence by the enzyme, the nanocarrier structure becomes destabilized, resulting in the release of the antibiotic payload directly into the contaminated microenvironment (Jiang & Loo, 2021).

This approach's specificity derives from the structural compatibility between the enzyme's active site and the designed linker sequence. By engineering linkers that are selectively identified by pathogen-derived enzymes instead of host enzymes, the system can attain highly localized drug activation. This molecular selectivity provides a benefit over traditional antibiotics, which generally disseminate throughout the body indiscriminately affecting both infected and healthy tissues. Experimental investigations have yielded encouraging outcomes. Enzyme-responsive nanocarriers that target β-lactamase activity have significant antibacterial efficacy against multidrug-resistant Pseudomonas aeruginosa in vitro, while minimizing damage to adjacent host cells (Zhang et al., 2023).

Another benefit of this technology is the potential for cascade activation. In certain designs, the initial enzyme-mediated cleavage reveals supplementary drug reservoirs or secondary triggers, resulting in enhanced antimicrobial activity. Due to the elevated enzymatic activity in biofilms associated with numerous chronic infections, enzyme-responsive nanocarriers are seen as especially promising for the treatment of persistent biofilm-related infections. (Zhang et al., 2023).

3.3. Redox Conditions as Triggers

Tissue infection induces an oxidative stress response marked by increased reactive oxygen species (ROS), including H_2O_2, which can reach concentrations of up to 1 mM, generated during the neutrophil oxidative burst. Redox-responsive nanocarriers utilize their surroundings via disulfide bonds or thioketal connections in their polymer backbone, which are cleaved under oxidative circumstances. The molecular mechanism underlying this rupture involves the two-electron oxidation of thiol groups by H_2O_2 or peroxynitrite ($ONOO^-$), resulting in the transformation of thioether or cysteine residues into chemically unstable sulfenic acid intermediates; these intermediates subsequently engage in rapid condensation reactions that cleave the polymer chain and liberate the encapsulated antibiotic payload (Zhang et al., 2023).

Disulfide-crosslinked carriers are notably susceptible to reductive breakage by increased intracellular glutathione in infected macrophages or neutrophils. The bidirectional redox sensitivity renders disulfide-based nanocarriers particularly effective for infections characterized by both external oxidative bursts and internal bacterial reservoirs, such as Staphylococcus aureus and Mycobacterium tuberculosis. Dual redox-pH responsive liposomes exhibit 80% efficacy against E. coli biofilms, as diminished intracellular glutathione synergizes with external oxidation to yield enhanced antibacterial results (Konwar & Kim, 2025). Redox gradients in the periplasm of Gram-negative bacteria can activate these systems intracellularly, improving their efficacy against persistent infections.

3.4. Toxins and Inflammation Triggers

Pathogen-derived toxins and host inflammatory mediators represent a further category of infection-specific signals exploitable for triggered drug release. Matrix metalloproteases (MMPs), upregulated in response to bacterial toxins such as α-hemolysin and pro-inflammatory cytokines, are endopeptidases that degrade extracellular matrix components during active infection. Nanocarriers incorporating peptide-crosslinked hydrogels exploit this mechanism: elevated MMP activity promotes progressive hydrolysis of the hydrogel matrix, enabling spatially controlled antibiotic release directly at the infection site. Hyaluronidase, an enzyme produced by numerous bacterial pathogens, similarly degrades hyaluronic acid-based nanocarrier scaffolds, releasing cationic antimicrobial payloads that disrupt biofilm integrity and impair bacterial adhesion. Hyaluronic acid-based hydrogels additionally sequester inflammatory cytokines through electrostatic interactions (Qin et al., 2025), conferring dual antibacterial and anti-inflammatory activity.

These systems exemplify the concept of sense-and-respond therapeutics, in which infection-associated biochemical signals directly trigger drug release. In certain formulations, inflammatory signaling induces the generation of reactive nitrogen species, including nitric oxide and peroxynitrite ($ONOO^-$) derived from L-arginine intermediates — which synergizes with co-released antibiotics. This synergy has been shown to achieve bacterial kill rates of up to 97% even at antibiotic concentrations below the minimum inhibitory concentration (Zhang et al., 2023), offering a compelling strategy for managing infections where conventional dosing is insufficient.

3.5. Integrated Site-Specific Delivery

Multifunctional nano systems advance infection management. These technologies combine medication delivery, toxin neutralization, and stimuli-responsive activation in a nanostructure.

pH- and enzyme-responsive hybrid devices are promising for penetrating dense biofilms and activating antibiotics specifically in contaminated microenvironments. Antimicrobial nanoparticles like silver nanoparticles (AgNPs) damage bacterial membranes, generate reactive oxygen species, and interfere with microbial metabolic pathways to boost therapeutic efficiency.

Recent research has investigated multimodal techniques that combine biological specificity with nanomaterial responsiveness. Bacteriophage–nanomaterial hybrids attack multidrug-resistant infections using stimuli-responsive drug delivery systems (Singh et al., 2025).

External physical cues can boost endogenous signaling. Light, magnetic fields, or ultrasound activate nanocarriers in deep-tissue infections, controlling drug release spatially and temporally. These methods reduce dose frequency, handle infection microenvironment heterogeneity, and restore antibiotic activity against multidrug-resistant bacteria. However, preclinical research continues to address large-scale synthesis, in vivo stability, and regulatory review (Nazir et al., 2022).

3.6 Critical Comparison of Stimuli-Responsive Systems.

pH-responsive systems are the most extensively studied among stimuli-responsive nanocarriers. Bacterial fermentation frequently generates acidic microenvironments at infection sites, providing a broadly applicable biological trigger. One major advantage of this approach is its synthetic accessibility, as acid-labile linkages are chemically well defined and relatively easy to incorporate into nanocarrier systems. However, selectivity remains a limitation because acidic environments are also present in tumor tissues and intracellular lysosomes, which may lead to unintended drug release in non-infected sites. Despite this challenge, pH-responsive platforms remain the most advanced in terms of clinical translation. Lipid nanoparticle systems based on similar pH-sensitive mechanisms have already entered oncology trials, providing an important regulatory framework, although no nano-antibiotic formulation has yet progressed beyond Phase II clinical testing (Dinya et al., 2026).

Enzyme-responsive nanocarriers offer improved biological selectivity because pathogen-specific enzyme expression is generally more restricted to infected tissues than pH changes alone. Many bacterial pathogens produce proteases, lipases, or other hydrolytic enzymes that can cleave specifically designed peptide or ester linkers within the nanocarrier structure. This enzymatic activation significantly reduces off-target drug exposure and is particularly advantageous in polymicrobial infections characterized by high enzymatic activity within biofilms. However, an important limitation is the variability of enzyme expression across pathogens and patients. Linker cleavage efficiency may vary depending on peptide sequence, tertiary structure, and the local enzyme concentration within the infection microenvironment. Although enzyme-responsive systems have shown promising results in preclinical studies, they have not yet advanced to antimicrobial clinical trials (Li et al., 2020).

Redox-responsive platforms exploit the elevated levels of reactive oxygen species (ROS) generated during the neutrophil oxidative burst that accompanies many bacterial infections. These systems often incorporate disulfide or thioketal linkages that respond to oxidative or reductive conditions. A key advantage of this strategy is its dual-compartment sensitivity: disulfide-linked carriers can respond to the intracellular reducing environment characterized by high glutathione concentrations, while thioketal-linked systems are activated by extracellular oxidative stress. Consequently, redox-responsive nanocarriers are particularly suitable for targeting intracellular pathogens such as Mycobacterium tuberculosis or intracellular Staphylococcus aureus. However, ROS gradients can be transient and spatially heterogeneous, especially in chronic inflammatory diseases. In such cases, premature carrier activation may occur, particularly in patients with metabolic disorders or persistent inflammatory conditions. (Zhang et al., 2023).

Multi-trigger systems represent the most sophisticated category of stimuli-responsive nanocarriers. These systems combine two or more signals such as pH, enzymatic activity, or oxidative stress to improve activation specificity and therapeutic robustness. While multi-trigger platforms offer enhanced selectivity, their increased formulation complexity introduces challenges in reproducibility, large-scale synthesis, and regulatory evaluation. Table 1 provides a comparative overview of the four major stimuli-responsive nanocarrier classes evaluated in this chapter, summarizing their activation mechanisms, selectivity profiles, biofilm penetration capacity, current clinical stage, and principal limitations.

Table 1. Comparison of Stimuli-Responsive Nanocarrier Systems for Antimicrobial Applications. Source: Compiled by the authors based on reviewed literature.

Stimuli	Mechanism	Selectivity	Biofilm Penetration	Clinical Stage	Key Limitation
pH-Responsive	Acidic pH (4.5–6.5)	Moderate	Good (protonation aids membrane disruption)	Most advanced; Phase I oncology analogues	Off-target release in healthy acidic compartments
Enzyme-Responsive	β-lactamase, hyaluronidase, phospholipase	High pathogen-specific	Excellent for biofilm-associated enzymes	Preclinical, in vivo animal studies	Linker cleavage variability across patients
Redox-Responsive	ROS (H_2O_2), glutathione gradient	High for intracellular pathogens	Moderate requires intracellular access	Early preclinical	Premature activation in chronic oxidative disease
Toxin/ Inflammation	α-hemolysin, cytokines, MMPs	Very high pathogen-derived signal	Moderate context-dependent	Proof-of-concept only	Trigger concentration varies with infection stage
Multi-Trigger	Dual pH/ enzyme, ROS	Highest orthogonal signals	Best combines membrane and enzymatic action	Early preclinical	High formulation complexity; regulatory uncertainty

Collectively, Table 1 illustrates that no single stimuli-responsive platform is universally optimal: pH-responsive systems offer the broadest applicability and the most mature translational basis, making them the preferred choice for proof-of-concept studies in acute infections. Enzyme-responsive carriers provide superior selectivity for chronic or biofilm-associated infections driven by a defined pathogen, while redox-responsive platforms are most compelling for intracellular bacterial reservoirs such as Mycobacterium tuberculosis and intracellular Staphylococcus aureus. Multi-trigger systems represent the most promising future direction, combining the highest activation specificity with the best biofilm penetration, though their increased formulation complexity and regulatory uncertainty remain significant barriers to clinical translation.

Responsive nanocarriers provide conditional activation mechanisms, but their design requires multidimensional optimization of trigger sensitivity, carrier geometry, surface chemistry, payload loading, and release kinetics for each clinical context. Iterative experimentation is impossible for multi-trigger systems due to their structural complexity. Artificial intelligence alters the area as a new design paradigm, not an incremental accelerator. Section 4 shows how AI and machine learning turn nanocarrier engineering into a computational problem, condensing years of bench labor into targeted experiments.

4. AI-GUIDED DESIGN AND OPTIMIZATION OF NANOCARRIERS

Artificial intelligence (AI) and machine learning (ML) are increasingly used to support the development of nanocarriers for antimicrobial therapy. By analyzing large and complex datasets, these tools help researchers predict nanocarrier behavior and reduce the reliance on time-consuming trial-and-error experimentation. These technologies facilitate the design of stimuli-responsive systems that optimize targeted drug delivery against MDR bacteria while improving both efficacy and safety. By integrating data-driven insights across molecular, cellular, and physiological scales, they enable the development of adaptive, personalized nanotherapeutics that represent a fundamental advance in nanomedicine (Esmaeilpour et al., 2026).

4.1 Role of AI and ML in Nanomedicine

AI-based models accelerate nanocarrier design by integrating physicochemical, genetic, and pharmacokinetic data to simulate drug–carrier interactions and predict formulation performance, compressing development timelines from years to months. Supervised machine learning predicts nanocarrier stability and bioavailability, while generative models such as GANs construct novel lipid or polymer compositions specifically suited to antibiotic payloads (Han et al., 2025). In antimicrobial applications, AI leverages resistance pattern data to optimize therapeutic payloads, as demonstrated by the ML-based design of mRNA-lipid nanoparticles (LNPs) with enhanced delivery characteristics (Abughalia et al., 2025).

Neural networks trained on combined simulation and experimental data identify non-intuitive formulation optima such as optimal surface charge that improve biofilm penetration and bacterial targeting. Surface charge is particularly non-intuitive because the optimal zeta potential is not a fixed value, it is a function of the target pathogen's biofilm composition. Pseudomonas aeruginosa biofilms dominated by negatively charged alginate favor cationic carriers; Staphylococcus aureus biofilms containing teichoic acids shift the optimum toward neutral or zwitterionic surfaces; polymicrobial wound biofilms present a charge landscape that varies with infection stage and prior antibiotic exposure. No single empirical experiment can efficiently navigate this multidimensional space, which is precisely why neural network models integrating pathogen genomics, biofilm proteomics, and nanoparticle characterization data provide a decisive design advantage. (Bannigan et al., 2023). To further illustrate the role of artificial intelligence in nanocarrier development, a detailed data-driven workflow is presented in Figure 3.

Figure 3. AI-guided workflow for the design and optimization of stimuli-responsive nanocarriers. The framework integrates physicochemical characterization, machine learning–based prediction, and iterative experimental validation through a feedback-driven optimization loop.

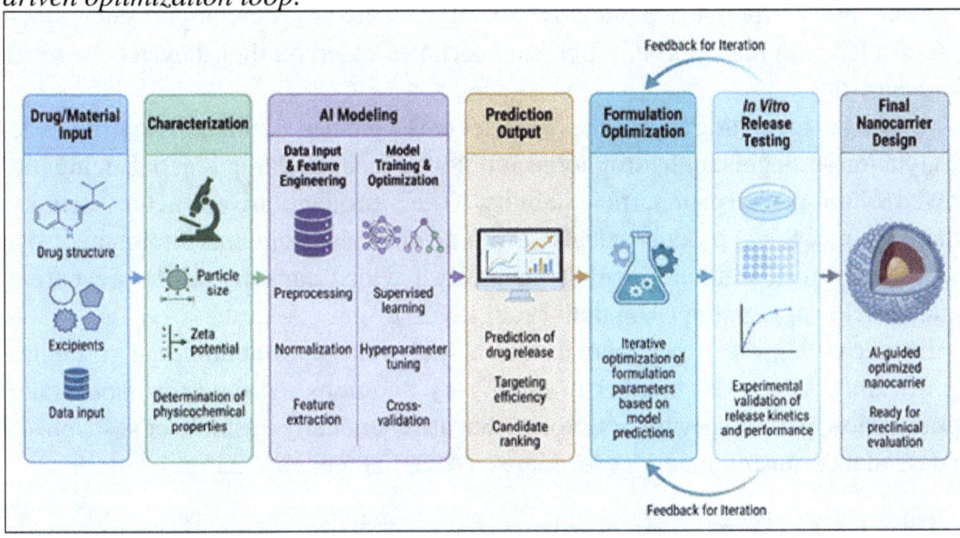

Source: Created by the authors.

Figure 3 highlights how machine learning models bridge experimental data and formulation design by enabling predictive optimization, thereby reducing experimental burden and accelerating nanocarrier development. Drug and material inputs are first characterized based on physicochemical properties such as particle size and zeta potential. These data are processed through machine learning pipelines, including feature engineering, normalization, and supervised learning with hyperparameter optimization and cross-validation. The trained models generate predictive outputs related to drug release, targeting efficiency, and candidate ranking. These predictions guide iterative formulation optimization, followed by in vitro validation of release kinetics and performance. The workflow ultimately enables the development of AI-guided, optimized nanocarriers suitable for preclinical evaluation.

4.2. Computational Modeling of Drug Loading

Computational modelling approaches combining molecular dynamics (MD) simulations with machine learning have become valuable tools for predicting drug encapsulation efficiency in nanocarrier systems. MD simulations provide atomistic insights into drug–nanoparticle interactions, including ligand orientation, solvent effects, and carrier stability. Machine learning models can then analyze these datasets to identify formulation parameters that maximize drug loading while maintaining structural stability. Graph neural networks (GNNs), in particular, are well suited to this task because they represent molecules as graph structures in which atoms serve as nodes and chemical bonds as edges. Using this architecture, GNN models have achieved high predictive accuracy ($R^2 > 0.9$) in estimating vancomycin encapsulation efficiency in mesoporous silica nanoparticles based on their physicochemical characteristics.

Additional modelling approaches include coarse-grained simulations that optimize PEGylation strategies for hydrophobic antibiotics such as rifampicin, balancing improved solubility with nanoparticle stability. AI-driven quantitative structure–activity relationship (QSAR) models can also evaluate multiple excipients simultaneously, helping identify combinations that maximize payload capacity while preventing nanoparticle aggregation (Abughalia et al., 2025).

By narrowing the experimental search space, these computational strategies significantly reduce the number of laboratory iterations required for nanocarrier optimization, particularly for antibiotics that are chemically unstable or susceptible to degradation during loading procedures. (Abughalia et al., 2025).

4.3. Optimizing Targeting Efficiency

Machine learning models improve drug delivery precision by forecasting receptor ligand interactions and nanoparticle circulation times. Models incorporating aptamers or pathogen-specific antibodies enable targeted delivery with enhanced specificity and reduced off-target effects. Aptamers offer advantages over antibodies in terms of production cost, stability, and binding affinity, making them well-suited to therapeutic applications; molecular dynamics simulations accelerate aptamer design by modelling binding dynamics computationally. Reinforcement learning optimizes biodistribution by modelling particle size (<100 nm) and zeta potential for specific pathogens such as S. aureus. Convolutional networks applied to imaging data optimize surface coatings such as mannose conjugation for macrophage-targeted intracellular delivery and ensemble methods integrating omics data predict efficacy against patient-specific resistomes. (Abughalia et al., 2025).

4.4. AI Workflows for Targeted Antimicrobial Therapy

AMR requires a strategic approach that emphasizes early-stage prevention through screening in clinical, environmental, farming, and livestock settings to identify nonvulnerable antimicrobial agents and the associated genes. Conventional AMR diagnostics, like antibiotic susceptibility testing, possess drawbacks, including high costs, time-consuming processes, and significant manpower requirements, underscoring the need for intelligent, prompt, and on-site diagnostic techniques. Nanoenabled artificial intelligence (AI)-supported smart optical biosensors present a potential solution by facilitating rapid point-of-care AMR detection with real-time, sensitive, and portable capabilities. These AI pipelines predict resistance trajectories and support the design of adaptive carriers incorporating machine learning (ML)-guided phage integration. (Taha et al., 2024). AI pipelines start with data curation from databases like NanoBank, followed by feature engineering (descriptors, fingerprints). Pretrained transformers encode structures; ML trains on labeled outcomes like MIC reduction. (Bess et al., 2022). While the technological foundations of AI-guided stimuli-responsive nanocarriers are increasingly robust, their real-world impact depends on successful clinical translation. Beyond laboratory validation, issues of regulatory approval, safety profiling, manufacturing scalability, and ethical governance must be carefully addressed. Table 2 summarizes the principal artificial intelligence (AI) and machine learning (ML) architectures currently applied to nanocarrier optimization, highlighting their specific applications, key advantages, and representative examples in antimicrobial drug delivery. (Abughalia et al., 2025). The following section discusses these translational dimensions in detail.

Table 2. AI tools accelerating nanocarrier optimization.

AI Tool/Model	Application in Nanocarriers	Key Advantages	Examples in Antimicrobials	References
Graph Neural Networks (GNNs)	Drug loading & stability prediction	Handles molecular graphs; high accuracy (R^2>0.9)	Vancomycin MSN encapsulation	Han et al.,2025
Recurrent Neural Networks (RNNs/ LSTMs)	Release kinetics modeling	Captures time-series data	pH-responsive liposomes	Kantesaria & Panda, 2026
Generative Adversarial Networks (GANs)	Novel polymer design	De novo compositions	Lipid NPs for mRNA antibiotics	Han et al.,2025
Bayesian Optimization	Hyperparameter tuning	Efficient global search	Redox-triggered colistin release	Kantesaria & Panda, 2026
Reinforcement Learning	Biodistribution/targeting	Adaptive to environments	Biofilm-penetrating aptamers	Abughalia et al.,2025

As shown in Table 2, each AI architecture addresses a distinct phase of the nanocarrier design pipeline: graph neural networks (GNNs) maximize drug-loading predictive accuracy through molecular graph representations; recurrent neural networks (RNNs) and long short-term memory (LSTM) networks capture the temporal dynamics of drug release kinetics; generative adversarial networks (GANs) enable de novo polymer and lipid composition generation; Bayesian optimization efficiently navigates high-dimensional formulation parameter spaces; and reinforcement learning dynamically adapts targeting strategies to heterogeneous infection microenvironments. Together, these tools provide comprehensive computational coverage across the full design-to-delivery workflow, substantially improving design efficiency, predictive accuracy, and targeting effectiveness compared to traditional trial-and-error approaches.

4.5. Critical Evaluation of AI-Driven Nanocarrier Design

While artificial intelligence may enhance the efficiency nanocarrier design, several methodological limitations must be addressed:

1. Dataset Limitations: The dependence on small, diverse datasets restricts the generalizability of AI models across various bacterial strains and infection contexts. Although these models demonstrate high predictive accuracy in silico, they frequently do not translate reliably into biological performance in vivo. This gap highlights the necessity for larger, standardized, and clinically representative datasets to enhance model robustness and clinical utility in managing bacterial infections. (Sanapalli et al., 2026)
2. Mechanistic Validation: Although AI is proficient in forecasting the physical and chemical characteristics of nanoparticles as dimensions, charge, and drug-loading efficacy, it frequently lacks the biological context essential for modeling intricate in vivo activities. Conventional empirical methods provide profound molecular understanding by elucidating dynamic interactions such as immunological responses, protein corona formation, and microbiota-induced metabolic variability, which are essential for practical efficacy. Combining these methodologies could harness AI's rapidity with empirical precision, enhancing predictive accuracy and clinical applicability. (Shen et al., 2026).
3. Scalability Concerns: AI-engineered nanocarriers demonstrate potential in improving targeting accuracy and reducing systemic toxicity, utilizing computational modeling for enhanced delivery optimization. Nonetheless, scalability continues to pose a challenge owing to complex production processes and ambiguous regulatory frameworks. Notwithstanding progress in AI-driven design, converting laboratory-scale achievements to industrial production necessitates standardized methodologies and validation frameworks. Ensuring reproducibility, batch uniformity, and long-term safety is essential for clinical

implementation. Environmental and biocompatibility issues further hamper extensive implementation. (Samathoti et al., 2025).

4. Model Interpretability: Black-box algorithms in healthcare may compromise clinical trust and regulatory compliance because of their limited interpretability. Lack of transparency hinders clinicians' ability to verify decision-making processes, thereby elevating the risk of biased or erroneous outcomes. The lack of transparency may jeopardize patient safety and impede adoption, particularly when AI impacts essential diagnoses or treatments. Ethical issues emerge, as patients possess the right to comprehend the decision-making processes that impact their care. To address these challenges, it is essential to develop explainable AI frameworks that balance performance and clarity, thereby ensuring accountability and fostering confidence among both providers and patients. (Cross et al., 2024).
5. Resistance Forecasting Limitations: AI-driven models proficiently simulate the evolution of microbial resistance through the analysis of genetic mutations and horizontal gene transfer. Nonetheless, real-world adaptation is influenced by intricate, dynamic factors such as host immunological responses, environmental conditions, and microbial community interactions elements that are frequently challenging to quantify or anticipate. The unexpected variables constrain the resistance aware intervention predictions, particularly in varied clinical and ecological contexts. Integrating real-time surveillance data with artificial intelligence may enhance predictive capabilities; nonetheless, intrinsic biological complexity continues to be a substantial obstacle. (Wang et al., 2026).

These advances illustrate how AI can guide early formulation decisions. The next stage of development focuses on predicting how efficiently antibiotics can be incorporated into nanocarrier systems

5. TRANSLATIONAL POTENTIAL, CLINICAL CHALLENGES, AND SAFETY CONSIDERATIONS

Stimuli-responsive nanocarriers possess significant potential to address AMR by facilitating targeted, microbiota-preserving drug administration, thereby diminishing resistance selection pressure. Translating these systems to clinical practice requires overcoming toxicological, regulatory, and manufacturing challenges for safe and scalable implementation (Mouzakis et al., 2025).

5.1. AMR Mitigation and Microbiota Preservation

Nano-delivery methods augment antibiotic effectiveness while markedly reducing collateral damage to the microbiota the commensal bacterial communities essential for immune regulation and metabolic health. By focusing on specific infection sites, these systems reduce off-target effects on commensal bacteria, and regulated release prevents subtherapeutic drug concentrations that would otherwise foster resistance selection. Preclinical research indicates a 70–90% decrease in microbiome disruption relative to conventional antibiotics, highlighting their capacity to improve treatment outcomes while mitigating long-term resistance threats (Singh et al., 2025).

These methods show a targeted antibacterial approach that breaks down infection cycles by blocking toxins and virulence factors and specifically killing pathogens like C. difficile while leaving behind beneficial microbiota. pH-responsive carriers allow exact distribution in the gastrointestinal environment, which helps keep probiotic populations high and eubiosis back to normal. This approach improves antimicrobial stewardship by reducing harm to the gut microbiome, protecting the resilience of microbes, and making existing antibiotics work longer while new antibiotics aren't being made as much (Bratkovič et al., 2024).

5.2. Safety, Toxicity, and Biocompatibility

Physicochemical properties such as nanoparticle size, surface charge, and composition play a major role in determining cytotoxicity and biodistribution. Size, shape, surface area, and chemical composition are the key physicochemical determinants of nanoparticle toxicity, and interactions with proteins, enzymes, and DNA further modulate their biological behavior (Xuan et al., 2023). Metallic nanoparticles induce cellular stress primarily through ROS generation, leading to oxidative damage in lipids, proteins, and DNA, with downstream risks of inflammation and apoptosis (Hassan et al., 2022). In contrast, biodegradable polymers such as PLGA produce minimal ROS, offering lower cytotoxicity and improved biocompatibility through degradation to non-toxic byproducts (Kuperkar et al., 2024).

Smaller nanoparticles enhance tissue and cellular penetration, enabling targeted delivery and organelle localization, but may trigger heightened immune responses (Kumar et al., 2025). Positively charged particles improve cellular uptake via electrostatic interactions yet risk hemolysis and inflammation (Desai et al., 2025). These adverse effects are often dose-dependent and may be reduced through surface modifications such as PEGylation, which provides nanoparticles with stealth properties that limit immune recognition. Stimuli-responsive lipid nanoparticles mitigate persistent organ accumulation by facilitating prompt clearance upon drug release, and AI-optimized surface designs reduce opsonization by approximately

50%, improving circulation duration and reducing immune recognition (Desai et al., 2025). A multidisciplinary approach encompassing in vitro testing, in vivo studies, advanced imaging, computational modelling, and regulatory compliance is essential for comprehensive nanoparticle safety evaluation in medical contexts (Khalifa, 2024).

5.3. Regulatory and Manufacturing Challenges

Regulatory agencies including the FDA and EMA classify nano-antimicrobials as combination products, requiring demonstration not only of nanomaterial novelty but also of equivalence in antimicrobial mechanism relative to conventional small-molecule antibiotics. ISO technical standards mandate batch-to-batch physicochemical consistency, creating a significant challenge in transitioning from laboratory-scale synthesis to GMP-compliant manufacturing, since variations in nanoparticle size, surface charge, and colloidal stability directly affect both safety and therapeutic efficacy (Hock et al., 2020). AI-assisted quality-by-design (QbD) approaches address this gap by predicting formulation variability and enabling manufacturing processes with reproducibility exceeding 95% across batches (Yang et al., 2025). Regulatory evaluation of nanomedicines encompasses safety profiling, pharmacokinetics, long-term toxicity, and demonstration of appropriate manufacturing controls. Both the EU and US are progressively updating their regulatory frameworks to address nanomaterial-specific considerations; however, regulatory evolution continues to lag behind the pace of scientific innovation (Desai et al., 2025; Vyas et al., 2026).

To facilitate regulatory alignment, researchers are encouraged to integrate nanoparticle-specific quality attributes, including trigger-response characterization (pH, ROS), reproducible release kinetics, off-target accumulation risk, and immunogenicity profiling, early in the design process. Adoption of standardized characterization methods applicable across nanomaterial classes is essential to ensure reproducible and scalable precision antimicrobial therapy.

5.4. Clinical Translation Barriers

Translating nano-antimicrobials from laboratory to clinic is fundamentally limited by the inability to simultaneously maximize infection-site drug accumulation and minimize systemic toxicity within the complex heterogeneity of human physiology (Younis et al., 2022). The degree of vascular permeability varies between infected sites within a single patient and between patients, depending on fibrous tissue content and immune cell accumulation; preclinical murine models do not adequately replicate the architecture and inflammatory dynamics of human infection (Cai et al., 2022). Specific barriers include:

- Differences in nanoparticle biodistribution between animal models and humans, particularly regarding vascular permeability and tissue architecture.
- Heterogeneity of infection microenvironments, where pH, oxygen tension, enzyme activity, and biofilm presence all vary spatially and temporally.
- Variable host immune responses in patients with comorbidities such as diabetes, cancer, or immunosuppression, which complicate trigger activation and safety profiles.

The enhanced permeability and retention (EPR) effect is well characterized in rodent models but incompletely validated in humans, potentially leading to overestimation of nanocarrier accumulation and therapeutic efficacy in preclinical studies (Vagena et al., 2025). A new generation of translational strategies is being developed to address these challenges, including: human organoid and micro physiological infection models that better replicate human tissue architecture; Phase 0 microdosing clinical trials enabling early pharmacokinetic assessment in humans; adaptive and platform-based trial designs improving efficiency and patient stratification; and AI-assisted patient stratification integrating genomic, microbiome, and clinical datasets to match patients with optimal nano-antimicrobial therapies (Shandilya et al., 2020).

5.5. Economic Feasibility and Global Health Access

Nano-enabled antimicrobial platforms may be limited to high-income markets, the opposite of where AMR is prevalent, due to high development and production costs. Complex multistep nanoparticle manufacturing, severe quality control, regulatory approval expenses, and cold-chain storage requirements boost costs. Modular nanoparticle platforms that can be used with multiple antibiotics, public–private partnerships (PPPs) that share development risks, global AMR financing mechanisms like pooled procurement and advanced market commitments, and scalable generic nanocarrier frameworks for efficient large-scale production are strategic solutions under development. AMR mitigation requires equitable access to precise nano-antimicrobials, which corresponds with One Health policies linking human, animal, and environmental health (Kariuki, 2024).

These scientific, regulatory, and economic challenges require technical innovation and a shared strategic vision for how precision nano-antimicrobials will reach the most vulnerable patients. The following concluding section outlines a concrete, time-horizoned roadmap.

5.5.1. Summary of Translational Challenges and Strategic Solutions

Table 3 summarizes the principal translational barriers facing stimuli-responsive nano-antimicrobials and the strategic solutions currently under development

Table 3. Summary of Translational Challenges and Strategic Solutions

Challenge	Description	Potential Solutions	Refences
Toxicity/Safety	ROS, accumulation, off-target effects	Biodegradable materials, PEGylation, dose optimization via AI	Wang et al., 2022
Scale-up/ Manufacturing	Batch variability, high costs	Continuous flow reactors, QbD with ML predictions	Wang et al., 2022
Regulatory Approval	Novelty classification, standardization	Harmonized ISO/FDA guidelines, PK/ PD modeling	Mouzakis et al., 2025
Translation Gaps	Animal-human discordance	Human organoids, Phase 0 microdosing, patient stratification	Zhang, 2023
Cost/Access	High development expenses	Generic platforms, public-private partnerships	Mouzakis et al., 2025

Table 3 reveals that AI-assisted quality-by-design (QbD) addresses at least three of the five major translational barriers simultaneously toxicity prediction, manufacturing variability, and patient stratification, underscoring its central role in clinical translation. The remaining barriers regulatory harmonization and cost access equity require policy intervention and international cooperation alongside scientific innovation. To clarify how these scientific and translational challenges may be addressed in practice, the following roadmap summarizes the key milestones required for the development of precision nano-antimicrobials, from early standardization efforts to fully programmable therapeutic systems.

6. CONCLUSION: A STRATEGIC ROADMAP FOR PRECISION NANO-ANTIMICROBIALS

Antimicrobial resistance has become one of the major challenges facing modern medicine. Traditional antibiotics are losing effectiveness against many pathogenic microorganisms, which makes the search for new therapeutic approaches increasingly important. The integration of artificial intelligence with stimuli-responsive nanocarrier systems represents a promising new strategy for addressing the growing global challenge of antimicrobial resistance. This chapter emphasizes that advancements in nanocarrier engineering, data-driven design, and targeted drug delivery are revolutionizing the development and application of antimicrobial therapies.

Experimental research has demonstrated the feasibility of pH-, enzyme-, and redox-responsive nanocarriers, while AI-assisted modeling is improving formulation design and optimization. Despite this progress, much effort remains necessary before these technologies can be reliably integrated into clinical practice. Rectifying this discrepancy requires coordinated scientific, regulatory, and translational efforts. This roadmap outlines a strategic framework to promote the development of precision nano-antimicrobials in the coming years.

In the near term, research efforts are expected to focus on strengthening the experimental and computational foundations of AI-guided nanomedicine. This includes generating larger and well-characterized datasets linking nanoparticle physicochemical properties with antimicrobial activity, improving the reliability and interpretability of AI models, and refining stimuli-responsive nanocarrier design for enhanced selectivity and controlled drug release. At the same time, standardized characterization protocols and harmonized regulatory frameworks will be essential to support the safe evaluation of nano-antibiotic candidates in early translational studies.

In the intermediate stage, the focus is likely to shift toward validating the clinical relevance of these systems. This will require rigorous preclinical studies, improved scalability of nanomaterial synthesis, and the development of manufacturing strategies capable of producing reproducible nanoparticle formulations at industrial scale. Addressing regulatory requirements, ensuring long-term safety, and demonstrating microbiota-sparing antimicrobial activity will be key steps for advancing nano-antibiotic systems toward clinical implementation.

Looking further ahead, advances in machine learning, biosensing technologies, and nanomaterial engineering may enable the development of adaptive antimicrobial platforms capable of responding dynamically to infection-specific microenvironments. AI-assisted diagnostic tools, integrated with programmable nanocarriers, could support more precise and personalized antimicrobial therapies. Such systems may eventually allow real-time optimization of antimicrobial treatment based on pathogen identity, resistance mechanisms, and host physiological conditions.

This strategic perspective highlights the growing potential of AI-enabled nanomedicine in addressing the global antimicrobial resistance crisis. Continued interdisciplinary collaboration between microbiologists, materials scientists, clinicians, and data scientists will be essential to translate these emerging technologies into safe, effective, and clinically accessible therapeutic solutions.

CONFLICT OF INTEREST

The author declares no conflict of interest.

ETHICAL STATEMENT

This chapter is based exclusively on previously published literature and does not involve new human or animal subjects.

AI-assisted tools were used only for language editing and formatting. All scientific content, analysis, and conclusions are the original work of the authors.

AUTHOR CONTRIBUTIONS

Mona M. Khalifa conceived the chapter concept, designed the framework, supervised the work, and wrote the main manuscript, Ahmed I. Mohamed contributed to literature collection, data organization, and manuscript drafting, Halah Abdulrahman Hafiz assisted in literature review and contributed to writing specific sections of the chapter, Mariam S. Ahmed participated in data compilation, reference formatting, and manuscript editing, Mohamed Ihab A. Al-Rifai contributed to scientific revision and critical review of the manuscript, Eman I. Mohamed assisted with data interpretation and language editing, Hagar Sayed Abdelgwad contributed to literature analysis and final proofreading of the chapter.

REFERENCES

Abughalia, A., Flynn, M., Clarke, P. F. A., Fayne, D., & Gobbo, O. L. (2025). The Use of Computational Approaches to Design Nano delivery Systems. *Nanomaterials (Basel, Switzerland)*, *15*(17), 1354. DOI: 10.3390/nano15171354 PMID: 40938032

Bamford, N. C., MacPhee, C. E., & Stanley-Wall, N. R. (2023). Microbial Primer: An introduction to biofilms - what they are, why they form and their impact on built and natural environments. *Microbiology (Reading, England)*, *169*(8), 001338. DOI: 10.1099/mic.0.001338 PMID: 37526065

Bannigan, P., Bao, Z., Hickman, R. J., Aldeghi, M., Häse, F., Aspuru-Guzik, A., & Allen, C. (2023). Machine learning models to accelerate the design of polymeric long-acting injectables. *Nature Communications*, *14*(1), 35. DOI: 10.1038/s41467-022-35343-w PMID: 36627280

Belay, W. Y., Getachew, M., Tegegne, B. A., Teffera, Z. H., Dagne, A., Zeleke, T. K., Abebe, R. B., Gedif, A. A., Fenta, A., Yirdaw, G., Tilahun, A., & Aschale, Y. (2024). Mechanism of antibacterial resistance, strategies and next-generation antimicrobials to contain antimicrobial resistance: A review. *Frontiers in Pharmacology*, *15*, 1444781. DOI: 10.3389/fphar.2024.1444781 PMID: 39221153

Bess, A., Berglind, F., Mukhopadhyay, S., Brylinski, M., Griggs, N., Cho, T., Galliano, C., & Wasan, K. M. (2022). Artificial intelligence for the discovery of novel antimicrobial agents for emerging infectious diseases. *Drug Discovery Today*, *27*(4), 1099–1107. DOI: 10.1016/j.drudis.2021.10.022 PMID: 34748992

Bratkovič, T., Zahirović, A., Bizjak, M., Rupnik, M., Štrukelj, B., & Berlec, A. (2024). New treatment approaches for *Clostridioides difficile* infections: Alternatives to antibiotics and fecal microbiota transplantation. *Gut Microbes*, *16*(1), 2337312. DOI: 10.1080/19490976.2024.2337312 PMID: 38591915

Cai, D., Gao, W., Li, Z., Zhang, Y., Xiao, L., & Xiao, Y. (2022). Current Development of Nano-Drug Delivery to Target Macrophages. *Biomedicines*, *10*(5), 1203. DOI: 10.3390/biomedicines10051203 PMID: 35625939

Chaudhary, A., Krishnarth, N., Tripathi, P. N., Chaudhary, A. K., Rejeeth, C., & Sharma, A. (2025). Smart Materials Design for Antibacterial Application. *Pharmaceutical Nanotechnology, 13*(3), 411-431. https://doi.org/DOI: 10.2174/01221 17385274576240209054652

Craig, M., Jernigan, D., Laserson, K., McBride, S., Fairbanks, J., Sievert, D., Armstrong, P. A., Ewing Ogle, H., & Zucker, H. (2024). Antimicrobial resistance at a crossroads: The cost of inaction. *Lancet, 404*(10458), 1083–1085. DOI: 10.1016/S0140-6736(24)01705-7 PMID: 39265589

Cross, J. L., Choma, M. A., & Onofrey, J. A. (2024). Bias in medical AI: Implications for clinical decision-making. *PLOS Digital Health, 3*(11), e0000651. DOI: 10.1371/journal.pdig.0000651 PMID: 39509461

Darby, E. M., Trampari, E., Siasat, P., Gaya, M. S., Alav, I., Webber, M. A., & Blair, J. M. A. (2023). Molecular mechanisms of antibiotic resistance revisited. *Nature Reviews. Microbiology, 21*(5), 280–295. DOI: 10.1038/s41579-022-00820-y PMID: 36411397

Desai, N., Rana, D., Patel, M., Bajwa, N., Prasad, R., & Vora, L. K. (2025). Nanoparticle Therapeutics in Clinical Perspective: Classification, Marketed Products, and Regulatory Landscape. *Small (Weinheim an der Bergstrasse, Germany), 21*(29), e2502315. https://doi.org/DOI: 10.1002/smll.202502315

Dinya, M., Dinya, E., & Mórotz, G. M. (2026). From Pathology to Formulation: Designing Biodegradable Polymers for Personalized Drug Delivery. *Pharmaceutics, 18*(3), 330. https://doi.org/DOI: 10.3390/pharmaceutics18030330

Esmaeilpour, D., Hamblin, M. R., Cheng, J., Khosravi, A., Liu, J., Zarepour, A., Zarrabi, A., Sillanpää, M., Nazarzadeh Zare, E., Shen, J., & Karimi-Maleh, H. (2026). Artificial intelligence driven protein design and sustainable nanomedicine for advanced theranostics. *Bioactive Materials, 60*, 425–455. DOI: 10.1016/j.bioactmat.2026.01.036 PMID: 41674557

Fagbemi, O. A., Essuah, M., Ugwu, D. C., Ajibola, A. B., Julius, S. O., Onyeyili, I. N., & Ojo-omoniyi, D. S. (2025). Nanobiotechnology-driven innovations for tackling antimicrobial resistance. *World J. Biol. Pharm. Health Sci, 21*(02), 300-328. https://doi.org/DOI: 10.30574/wjbphs.2025.21.2.0093

Gaurav, A., Bakht, P., Saini, M., Pandey, S., & Pathania, R. (2023). Role of bacterial efflux pumps in antibiotic resistance, virulence, and strategies to discover novel efflux pump inhibitors. *Microbiology (Reading, England), 169*(5), 001333. DOI: 10.1099/mic.0.001333 PMID: 37224055

Han, Y., Kim, D. H., & Pack, S. P. (2025). Nanomaterials in Drug Delivery: Leveraging Artificial Intelligence and Big Data for Predictive Design. *International Journal of Molecular Sciences, 26*(22), 11121. DOI: 10.3390/ijms262211121 PMID: 41303604

Hassan, W., Noreen, H., Rehman, S., Kamal, M. A., & da Rocha, J. B. T. (2022). Association of Oxidative Stress with Neurological Disorders. *Current Neuropharmacology, 20*(6), 1046–1072. DOI: 10.2174/1570159X19666211111141246 PMID: 34781871

Hock, S. C., Kian, S. M., & Wah, C. L. (2020). Global challenges in the manufacture, regulation and international harmonization of GMP and quality standards for biopharmaceuticals. *Generics and Biosimilars Initiative Journal, 9*(2), 52-64. https://doi.org/DOI: 10.5639/gabij.2020.0902.010

Jiang, L., & Loo, S. C. J. (2021). Intelligent Nanoparticle-Based Dressings for Bacterial Wound Infections. *ACS Applied Bio Materials*, *4*(5), 3849–3862. DOI: 10.1021/acsabm.0c01168 PMID: 34056562

Kantesaria, R., & Panda, H. S. (2026, March 09). (2026). A Review on AI-Based Data-Driven Models for Optimization of Nanocarriers as Drug Delivery Systems. *ACS Biomaterials Science & Engineering*, *12*(3), 1397–1418. Advance online publication. DOI: 10.1021/acsbiomaterials.5c01998 PMID: 41668335

Kapoor, D. U., Sharma, J. B., Gandhi, S. M., Prajapati, B. G., Thanawuth, K., Limmatvapirat, S., & Sriamornsak, P. (2024). AI-driven design and optimization of nanoparticle-based drug delivery systems. *Science, Engineering and Health Studies*, 24010003-24010003. https://doi.org/DOI: 10.69598/sehs.18.24010003

Kariuki, S. (2024). Global burden of antimicrobial resistance and forecasts to 2050. *Lancet*, *404*(10459), 1172–1173. DOI: 10.1016/S0140-6736(24)01885-3 PMID: 39299259

Khalifa, M. M. A. H. (2024). Safety of nanoparticles in medicine. In *Nanocarriers in neurodegenerative disorders* (pp. 94–100). CRC Press. DOI: 10.1201/9781003383376-7

Konwar, B., & Kim, K. S. (2025). Recent advances in bacterial hyaluronidase-responsive nano-drug delivery systems for targeting antibiotic-resistant Staphylococcus aureus. Chemical Engineering Journal, 165827. https://doi.org/DOI: 10.1016/j.cej.2025.165827

Kumar, R., Kumar, A., Bhardwaj, S., Sikarwar, M., Sriwastaw, S., Sharma, G., & Gupta, M. (2025). Nanotoxicity unveiled: Evaluating exposure risks and assessing the impact of nanoparticles on human health. *Journal of Trace Elements and Minerals, 13*, 100252. https://doi.org/.DOI: 10.1016/j.jtemin.2025.100252

Kuperkar, K., Atanase, L. I., Bahadur, A., Crivei, I. C., & Bahadur, P. (2024). Degradable Polymeric Bio(nano)materials and Their Biomedical Applications: A Comprehensive Overview and Recent Updates. *Polymers, 16*(2), 206. DOI: 10.3390/polym16020206 PMID: 38257005

Li, M., Zhao, G., Su, W. K., & Shuai, Q. (2020). Enzyme-responsive nanoparticles for anti-tumor drug delivery. *Frontiers in chemistry, 8*, 647. https://doi.org/DOI: 10.3389/fchem.2020.00647

Mi, P. (2020). Stimuli-responsive nanocarriers for drug delivery, tumor imaging, therapy and theranostics. *Theranostics, 10*(10), 4557–4588. DOI: 10.7150/thno.38069 PMID: 32292515

Mouzakis, A., Panagopoulos, P., Papazoglou, D., & Petrakis, V. (2025). A Comprehensive Review of Nanoparticles in the Fight Against Antimicrobial Resistance. *Pathogens (Basel, Switzerland), 14*(11), 1090. DOI: 10.3390/pathogens14111090 PMID: 41305328

Nazir, F., Tabish, T. A., Tariq, F., Iftikhar, S., Wasim, R., & Shahnaz, G. (2022). Stimuli-sensitive drug delivery systems for site-specific antibiotic release. *Drug Discovery Today, 27*(6), 1698–1705. DOI: 10.1016/j.drudis.2022.02.014 PMID: 35219858

Powers, J. H.III. (2024). Antimicrobial stewardship. *BMJ (Clinical Research Ed.), 385*, q1170. DOI: 10.1136/bmj.q1170 PMID: 38839082

Qin, H., Li, Z., Li, S., Huang, J., Ren, J., & Wu, X. (2025). Advances in hydrogels for capturing and neutralizing inflammatory cytokines. *Journal of Tissue Engineering, 16*, 20417314251342175. DOI: 10.1177/20417314251342175 PMID: 40575355

Rezasoltani, S., Yadegar, A., Hatami, B., Asadzadeh Aghdaei, H., & Zali, M. R. (2020). Antimicrobial Resistance as a Hidden Menace Lurking Behind the COVID-19 Outbreak: The Global Impacts of Too Much Hygiene on AMR. *Frontiers in Microbiology, 11*, 590683. DOI: 10.3389/fmicb.2020.590683 PMID: 33384670

Samathoti, P., Kumarachari, R. K., Bukke, S. P. N., Rajasekhar, E. S. K., Jaiswal, A. A., & Eftekhari, Z. (2025). The role of nanomedicine and artificial intelligence in cancer health care: Individual applications and emerging integrations-a narrative review. *Discover. Oncology, 16*(1), 697. DOI: 10.1007/s12672-025-02469-4 PMID: 40338421

Sanapalli, B. K. R., Palit, S., Deshpande, A., Tokala, R., Sigalapalli, D. K., & Sanapalli, V. (2026). Artificial Intelligence and the Discovery of Antibiotics: Reinventing with Opportunities, Challenges, and Clinical Translation. *Antibiotics (Basel, Switzerland)*, *15*(2), 233. DOI: 10.3390/antibiotics15020233 PMID: 41750530

Shandilya, R., Pathak, N., Lohiya, N. K., Sharma, R. S., & Mishra, P. K. (2020). Nanotechnology in reproductive medicine: Opportunities for clinical translation. *Clinical and Experimental Reproductive Medicine*, *47*(4), 245–262. DOI: 10.5653/cerm.2020.03650 PMID: 33227186

Shen, C., Zhang, M., Lu, M., Chang, E., Gao, Z., Ban, W., Liu, Q., Zuo, Z., & Jiang, C. (2026). Machine learning empowered formulation design, optimization and characterization of nanoparticulate drug delivery systems: Current applications, challenges, and future perspectives. *Acta Pharmaceutica Sinica. B*, *16*(2), 665–685. DOI: 10.1016/j.apsb.2025.12.011 PMID: 41685160

Singh, H., Balusamy, S. R., Sukweenadhi, J., Saravanan, M., Aruchamy, M., Mijakovic, I., & Singh, P. (2025). Smart hybrid nanomaterials for chronic infections: Microbiome-responsive and sustainable therapeutic platforms. *Journal of Nanobiotechnology*, *23*(1), 698. DOI: 10.1186/s12951-025-03712-4 PMID: 41184906

Taha, B. A., Ahmed, N. M., Talreja, R. K., Haider, A. J., Al Mashhadany, Y., Al-Jubouri, Q., Huddin, A. B., Mokhtar, M. H. H., Rustagi, S., Kaushik, A., Chaudhary, V., & Arsad, N. (2024). Synergizing Nanomaterials and Artificial Intelligence in Advanced Optical Biosensors for Precision Antimicrobial Resistance Diagnosis. *ACS Synthetic Biology*, *13*(6), 1600–1620. DOI: 10.1021/acssynbio.4c00070 PMID: 38842483

Tokuda, M., & Shintani, M. (2024). Microbial evolution through horizontal gene transfer by mobile genetic elements. *Microbial Biotechnology, 17*(1), e14408. https://doi.org/DOI: 10.1111/1751-7915.14408

Tzavellas, N. P., Atzemoglou, N., Bozidis, P., & Gartzonika, K. (2026). The Role of Environmental and Climatic Factors in Accelerating Antibiotic Resistance in the Mediterranean Region. *Acta Microbiologica Hellenica*, *71*(1), 1. DOI: 10.3390/amh71010001

Vagena, I.-A., Malapani, C., Gatou, M.-A., Lagopati, N., & Pavlatou, E. A. (2025). Enhancement of EPR Effect for Passive Tumor Targeting: Current Status and Future Perspectives. *Applied Sciences (Basel, Switzerland)*, *15*(6), 3189. DOI: 10.3390/app15063189

Vyas, H., Bhardwaj, M., Das, A., Agarwal, J., & Tiwari, A. K. (2026). Safety and regulatory considerations in nano-bio-medicine. In *Nano-Bio-Medicine for Gastrointestinal Diseases* (pp. 311–328). Academic Press., DOI: 10.1016/B978-0-443-34039-0.00002-0

Wang, X., Shan, M., Zhang, S., Chen, X., Liu, W., Chen, J., & Liu, X. (2022). Stimuli-responsive antibacterial materials: molecular structures, design principles, and biomedical applications. *Advanced Science, 9*(13), 2104843. https://doi.org/ DOI: 10.1002/advs.202104843

Wang, X. W., Wang, T., & Liu, Y. Y. (2026). Artificial intelligence for microbiology and microbiome research. *Cell Systems, 17*(2). https://doi.org/DOI: 10.1016/j.cels.2026.101531

Xie, Y., Lu, H., Liu, Y., Hu, G., Lian, S., Liu, J., Pang, S., Zhu, G., & Ding, X. (2025). Unveiling the Mechanisms of Bacterial Resistance and Countermeasures. *Pathogens (Basel, Switzerland), 14*(11), 1085. DOI: 10.3390/pathogens14111085 PMID: 41305322

Xuan, L., Ju, Z., Skonieczna, M., Zhou, P. K., & Huang, R. (2023). Nanoparticles-induced potential toxicity on human health: Applications, toxicity mechanisms, and evaluation models. *MedComm, 4*(4), e327. DOI: 10.1002/mco2.327 PMID: 37457660

Yang, J., des Rieux, A., & Malfanti, A. (2025). Stimuli-Responsive Nanomedicines for the Treatment of Non-cancer Related Inflammatory Diseases. *ACS Nano, 19*(16), 15189–15219. DOI: 10.1021/acsnano.5c00700 PMID: 40249331

Yang, S., Hu, X., Zhu, J., Zheng, B., Bi, W., Wang, X., Wu, J., Mi, Z., & Wu, Y. (2025). Aspects and Implementation of Pharmaceutical Quality by Design from Conceptual Frameworks to Industrial Applications. *Pharmaceutics, 17*(5), 623. DOI: 10.3390/pharmaceutics17050623 PMID: 40430914

Younis, M. A., Tawfeek, H. M., Abdellatif, A. A. H., Abdel-Aleem, J. A., & Harashima, H. (2022). Clinical translation of nanomedicines: Challenges, opportunities, and keys. *Advanced Drug Delivery Reviews, 181*, 114083. DOI: 10.1016/j.addr.2021.114083 PMID: 34929251

Zhang, J., Tang, W., Zhang, X., Song, Z., & Tong, T. (2023). An Overview of Stimuli-Responsive Intelligent Antibacterial Nanomaterials. *Pharmaceutics, 15*(8), 2113. DOI: 10.3390/pharmaceutics15082113 PMID: 37631327

Zhang N. (2023). Promoting the bench-to-bedside translation of nanomedicines. *Medical review (2021), 3*(1), 1–3. https://doi.org/DOI: 10.1515/mr-2023-0007

Chapter 7
Smart Nanocarriers and AI-Driven Release Systems for Targeted Antibiotic Delivery

Rabia Yasa Kostas

Gumushane Vocational School, Department of Computer Technologies, Gumushane University, Turkey

Pinar Bozbeyoglu

http://orcid.org/0000-0002-3704-2701

Biomedical Device Technology Program, Electronics and Automation Department, Gümüşhane University, Turkey

Elif Esra Altuner

http://orcid.org/0000-0001-7663-6898

Kocaeli Health and Tecnology University, Turkey

Muzaffer Elmas

Kocaeli Health and Technology University, Turkey

ABSTRACT

The antimicrobial resistance (AMR) crisis is not merely a consequence of the emergence of resistant pathogens. Furthermore, the remaining shortcomings of traditional antibiotic production and development processes have become a profound health threat, exacerbated by their pharmaceutical and pharmacokinetic limitations. This book chapter redefines the AMR crisis as a problem of drug and treatment discovery as well as drug delivery to target tissues, and examines the technological angle emerging at the intersection of nanotechnology and artificial intelligence (AI) in

DOI: 10.4018/979-8-3373-5268-8.ch007

response to this problem. This review begins by examining the problems inherent in traditional antibiotic architecture, such as systemic toxicity, low bioavailability, poor stability, and, most importantly, the inability to penetrate protected sites such as bacterial biofilms or intracellular bunkers.

1. THE ANTIMICROBIAL RESISTANCE CRISIS AND THE INADEQUACY OF TRADITIONAL TREATMENTS

1.1. The Rise of Antibiotic Resistance (AMR) as a Global Health Threat

Undoubtedly, the discovery of antibiotics is one of the greatest medical achievements of the last century. Tuberculosis, pneumonia, and syphilis widely affected societies and caused significant human loss. However, with the discovery of penicillin and subsequent development of various antibiotics, these became completely treatable diseases (Aminov, 2010). Antibiotics are now incorporated into treatment regimens for many diseases, either as primary or adjunctive therapy, but their indiscriminate use has become a global health crisis known as Antimicrobial Resistance (AMR). The World Health Organization (WHO) defines this problem as one of the greatest threats to human health, requiring urgent intervention (World Health Organization, 2024a). There are some dangerous bacterial strains that even the most powerful antibiotics we have at our disposal are ineffective against. This group, which the World Health Organization (WHO) has labeled "superbugs," has been identified as the most urgent and highest priority health threats to humanity. The pathogens classified as these high-priority threats are (World Health Organization, 2024b):

- Carbapenem-resistant (a very powerful antibiotic) Acinetobacter baumannii (CRAB)
- Carbapenem-resistant Pseudomonas aeruginosa (CRPA)
- Extended-spectrum β-lactamase (ESBL)-producing (which break down other antibiotics) or carbapenem-resistant Enterobacterales (CRE)

This crisis has two significant outcomes. First, systematic analyses demonstrate that bacterial AMR directly causes millions of deaths (Murray et al., 2022). Second, it places severe economic pressure on global health systems (Jonas et al., 2017). The innovation gap in developing new antibiotics to address this threat is a particular concern. According to WHO 2024 statistics and the AMR Pipeline Tracker, while there were more than 30 candidate systemic antibiotics in the 1990s, today this number is limited to 10-12 in advanced clinical trials (World Health Organization,

2024c). This demonstrates that existing solutions are rapidly depleting and that maintaining or improving drug efficacy is critical.

1.2. Pharmacokinetic and Pharmaceutical Limitations of Existing Antibiotics

Superbugs resulting from bacterial evolution are not the only problem in the antibiotic crisis. Another problem is that existing antibiotics have inherent shortcomings: they either fail to fight bacteria in the body or are impractical for use in the body (Parvin et al., 2025). Some drugs are eliminated from the body too quickly, preventing them from reaching the infected tissue (e.g., the lungs) in sufficient quantities. This creates pharmacokinetic disadvantages (Galm & Shen, 2007). Furthermore, many naturally discovered potential antibiotics are highly toxic to human cells, causing toxicity. Furthermore, because they are insoluble in water, they are very difficult to convert into formulation (pills or injections) for patient administration, presenting pharmaceutical disadvantages. In other words, even if a drug that kills bacteria in a laboratory environment (in vitro) is discovered and found to be successful, turning it into an effective treatment for a living patient (in vivo) is very difficult and points to a different level of problem, which brings the crisis to a different dimension (Brown & Wright, 2016).

1.2.1. Systemic Toxicity and Side Effects

Traditional antibiotics are generally administered systemically, which can be likened to a bombardment. The drug is released throughout the body (brain, liver, kidneys, etc.) with the expectation of healing a 1 cm^2 infected area in the lungs, but unfortunately, healthy organs are subject to some toxicity. This off-target distribution causes side effects called systemic toxicity, such as in the kidneys (nephrotoxicity) or the ear (ototoxicity) (Gomez-Simmonds et al., 2016). This situation becomes more complex when resistant bacteria are encountered. For example, while 1 unit of drug would be sufficient for normal bacteria, 10 units of drug are required for resistant bacteria. When this 10-unit dose is administered to a patient, the drug is far exceeded within its “therapeutic window,” and becomes toxic. The therapeutic window is defined as the safe range between the minimum effective dose at which a drug begins to heal a patient and the minimum toxic dose at which it begins to harm (toxicity). In this example, the doctor would either risk sacrificing the patient's kidneys (nephrotoxicity) to treat the infection or administering a low dose to protect the kidneys and allowing the treatment to fail or, worse, allow resistance to grow even further. Some frequently cited antibiotics include aminoglycosides (Gentamicin, Amikacin, etc.) and Vancomycin, which are effective against multidrug-resistant bacteria but are notorious for causing nephrotoxicity (kidney damage) and ototoxicity (permanent deafness) (Krause et al., 2016). Vancomycin is also another example, which is related to cause nephrotoxicity at the high doses needed to treat resistant infections (Rybak et al., 2020). Therefore, developing "targeted drug delivery systems" that deliver the drug directly to target tissues (e.g., by loading the antibiotic into a nanoparticle) is just as critical as, or even more so than, discovering new antibiotics (Liew et al., 2022).

1.2.2. Low Bioavailability and Poor Tissue Penetration

Many new-generation antibiotic agents, especially natural photosensitizers (PSs), face fundamental problems such as low solubility and low bioavailability (Wang et al., 2025). Even if a potentially effective antibiotic (or next-generation agent) is discovered, the discovery may fail if the drug (1) cannot enter the bloodstream or (2) cannot reach the target tissue (e.g., an abscess or bone) from the bloodstream. Most antibiotic candidates are oily (lipophilic) in nature, but the basis of our bodies and bloodstreams is water (hydrophilic). In this case, if the drug is administered orally, it is difficult for it to dissolve in the stomach or intestines, causing it to be directly excreted without being absorbed. The amount that does indirectly enter the bloodstream may remain minimal, such as 1-2%. Alternatively, when administered systemically (intravenously), the drug must first be dissolved in a liquid. If it is insoluble in water, preparing the necessary injection formulation may be impossible. Furthermore, simply mixing a drug into the bloodstream doesn't mean it penetrates the target tissue. Infections are rarely directly contained within the bloodstream, and it's very difficult for it to reach the deep tissues, lung alveoli, and bone tissue where infections usually occur. This situation is called poor penetration (Patra et al., 2018). In addition to factors related to the drug and human body structure, the biofilms bacteria produce to protect themselves pose a distinct disadvantage. These hard or slimy structures produced by bacteria keep bacteria safe, and even if the antibiotic leaves the bloodstream and reaches the tissue, it cannot penetrate these structures (Liu et al., 2024). Intracellular bunkers, where bacteria like Salmonella or tuberculosis infiltrate our own immune cells (macrophages) to hide from the immune system, further complicate treatment (Abed & Couvreur, 2014). This is because the antibiotic must now cross not only the bacterium but also the membrane of the human cell where it hides. In short, the reason for treatment failure isn't that the drug is weak against the bacterium, but rather that the drug never physically encounters the bacterium. The main reason why new and powerful molecules (such as PSs) discovered cannot be translated into practical treatments is these pharmacological and logistical barriers regarding solubility, bioavailability and penetration to the target tissue (Gupta et al., 2013).

1.2.3. Stability Issues of New Agents Like Antimicrobial Peptides (AMPs)

We have a wonder weapon called antimicrobial peptides (AMPs), which are among the most promising candidates in the fight against antibiotic resistance and are excellent at destroying bacteria in the laboratory. What makes AMPs so special lies in their working principle. They generally target the bacterial cell membrane (outer wall) directly and puncture, dissolve, or rupture it. Changing the structure of the outer membrane is much more difficult for bacteria than mutating a single enzyme, so it is virtually impossible for bacteria to develop resistance to AMPs (Magana et al., 2020). However, AMPs are peptides, the small building blocks of proteins, and our bodies (especially our bloodstream and digestive systems) perceive peptides as food, nutrients, or waste that needs to be broken down. Enzymes called proteases, found in our bodies, exist precisely to break down these peptides, and they attack the AMPs as soon as they are injected into the patient (e.g., intravenously), inactivating this peptide chain within seconds or minutes. In short, they disappear in the bloodstream before they ever reach the site of infection (e.g., the lungs or an abscess). Therefore, their remarkable in vitro effectiveness, as seen in examples like Pexiganan and iseganan, has not been translated into in vivo treatment (Mahlapuu et al., 2016). The most practical way to make AMPs successful is to encapsulate them in a system that will protect them from this enzymatic attack—in other words, to package them within carrier systems (e.g., nanoparticles) (Teixeira et al., 2020).

1.3. Opportunities Offered by Nanomedicine in Targeted Drug Delivery

Considering the above, it is clearly observed that the AMR crisis is not limited to new drug discovery; it also poses a serious challenge in the effective delivery of existing and candidate drugs to target tissues. At this point, one of the most rational approaches is nanotechnology-based drug delivery systems (Liew et al., 2022).

Nanocarriers provide the following benefits by encapsulating antibiotic agents (small molecules or peptides such as AMPs) or binding them to their surfaces:

1. **Protection and Stability:** They protect the drug against enzymes and chemical reactions in the bloodstream (Teixeira et al., 2020).
2. **Improved Pharmacokinetics:** They extend the drug's circulation time, giving it time to reach target tissues and thus increasing its bioavailability.
3. **Targeting and Reduced Toxicity:** They prevent the drug from penetrating into healthy tissues, eliminating the potential for toxicity to healthy tissues while

simultaneously increasing localized efficacy at the site of infection (Abed & Couvreur, 2014).

Given these findings, nanocarriers are becoming an important tool for revitalizing the potential of existing and candidate antibiotics and enabling critical agents like AMPs to reach their clinical potential.

2. NANOCARRIER SYSTEMS FOR TARGETED ANTIBIOTIC DELIVERY

Nanocarriers are structured materials with dimensions ranging from 1 to 1000 nanometers designed to deliver antibiotics and other therapeutic agents to their target (Su & Kang, 2020). The clinical failure of many potent antibiotics stems from poor solubility, systemic toxicity, or rapid in vivo degradation. Nanocarrier systems are designed to overcome precisely these obstacles and are potent agents in the fight against AMR (Dyett et al., 2024). The following sections will detail the basic functions of these nanocarriers, their main types, and how they are used against the most challenging infectious conditions where conventional treatments have failed.

2.1. The Role of Nanocarriers in Antibiotic Delivery

The primary function of nanocarriers can be thought of as a cargo transport system, where the antibiotic molecule is transported as cargo and protected throughout the journey until it reaches the target tissue. Nanocarriers can load antibiotics using three main methods:

- **Physical encapsulation:** the drug is trapped within the core of the carrier,
- **Chemical conjugation:** the drug is bound to the carrier surface by a covalent bond,
- **Adsorption:** the drug adheres or clings to the carrier surface (without a chemical bond) through weaker physical forces (Lee & Yeo, 2015).

Nanoparticles possess many unique properties; most notably their small size, high stability, the capacity to carry both hydrophilic (water-soluble) and hydrophobic (fat-soluble) drugs (Zheng et al., 2007), their ability to provide slow and sustained drug release (Wang et al., 2025), and their suitability for various administration routes (e.g., respiratory tract, vascular access, etc.).

More importantly, nanocarriers have the potential to overcome existing drug resistance mechanisms developed by bacteria (Baptista et al., 2018). Bacteria have two main defenses against antibiotics: (1) not taking in the drug and (2) immediately pumping out the drug once it has entered. At this point, nanocarriers introduce the drug into the cell by hiding it from these defenses or by tricking the system (e.g., disguising itself as food). Once the drug reaches the cell in high concentrations, the bacteria develop resistance. its mechanism breaks down and becomes meaningless and becomes sensitive again (Parvin et al., 2025).

Figure 1. Methods of loading drugs into nanoparticles

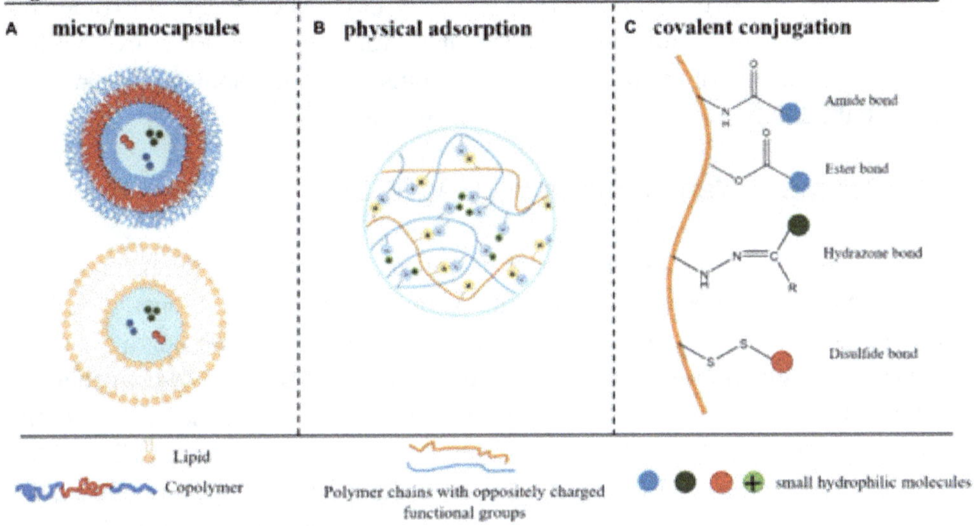

Adapted from (Li et al., 2020).

2.2. Basic Nanocarrier Platforms: Structures, Advantages, and Disadvantages

Various nanocarrier systems have been developed for antibiotic delivery, which will be discussed below.

2.2.1. Liposomes

A liposome is essentially a hollow, artificial sac or capsule, nanometer-sized, composed of one or more phospholipid bilayers. The wall of this capsule is a bilayer membrane composed of fatty molecules called phospholipids, the same basic material as our cell membranes. This allows the liposome to carry cargo (antibiotics) both in the aqueous space within it (water-soluble drugs) and in the fatty portion of its wall (fat-soluble drugs). This property makes it an excellent natural carrier for protecting drugs in the body and delivering them to their target. The greatest advantage of liposomes is their biocompatibility and their tendency not to activate the immune system. The surface structure is chemically well-suited for PEGylation (prolonged circulation time) or for affixing location information to target tissues (Bozzuto & Molinari, 2015). However, it also has some disadvantages. (1) Low Physical Stability: The wall of liposomes is essentially a natural phospholipid and has a fragile structure. (2) Formulation Reproducibility Issues: Making liposomes is a complex process, and it is very difficult to produce liposomes of the exact same size, thickness, and structure every time. (3) Low Drug Encapsulation Efficiency: During production, it is difficult to fully encapsulate the antibiotic used within the vesicle, and some of the drug may remain outside the liposome and be wasted (Andra et al., 2022).

2.2.2. Polymeric Nanoparticles (PLGA, PCL)

These carriers are microscopic beads designed to carry drugs, the primary material of which is a polymer (i.e., long chains of molecules linked together). They can be composed of natural (e.g., chitosan, alginate) or synthetic polymers (e.g., poly(lactide-co-glycolide) (PLGA), polycaprolactone (PCL)). They were generally developed to overcome the disadvantages of liposomes, such as stability and encapsulation efficiency. However, the most important feature of these carriers is that they can precisely regulate when, at what rate, and for how long they release the drug into the body. This is called controlled release (Anwar et al., 2021). In short, instead of delivering the drug all at once, the polymer can be adjusted to release the drug slowly over 10 days. However, there are some drawbacks. (1) Natural Materials (Chitosan/Algae) Inconsistency: The material in one production batch (e.g., algae taken on Monday) is exactly the same as the material in the next batch (algae taken on Wednesday), making it difficult to develop a standard medical product. (2) Material or Production Residue Toxicity: Some polymers (materials) used or chemical solvents used during production (production residues) can be toxic to the body. (3) Risk of Not Being Excreted from the Body: If the material is not biodegradable (can be dissolved in the body), these particles can accumulate in organs such as the kidneys or liver, causing long-term damage (Castro et al., 2022).

2.2.3. Dendrimers and Micelles

Dendrimers and micelles are two specialized nanocarrier systems. Dendrimers have a structure that branches outward from the center in a perfect and symmetrical manner, like a perfect tree branch or snowflake. All dendrimers are created layer by layer in the laboratory through chemical synthesis with great precision and are identical in size (Abbasi et al., 2014). Micelles are structures formed by the self-assembly of amphiphilic (both water- and oil-loving) molecules in an aqueous environment. They have a hydrophobic core and a hydrophilic shell. The process of micelle formation is a completely natural physical-chemical process, but they are engineered in the laboratory with synthetic polymers for stability and control. Both dendrimers and micelles are used to encapsulate normally water-insoluble antibiotics within their oily interiors, making them transportable in the blood (Negut & Bita, 2023).

2.2.4. Inorganic Nanoparticles (Gold, Mesoporous Silica)

Nanocarriers made from metal/ceramic materials are also available; examples include gold (AuNPs), silica (porous glass) (MSNs), and iron oxide (magnetic) (SPIONs). Their high stability is a significant advantage compared to organic carriers such as liposomes or polymers. MSNs, notable for their tunable pore structures, are suitable candidates for controlled release profiles. Thanks to their optical and magnetic properties, gold and iron oxide nanocarriers can precisely identify where a drug is located in the body (diagnostics) and initiate treatment at that point simultaneously. These structures, which combine both properties in one framework, are called Theranostic (Therapy + Diagnostic/Diagnostic) (Ding et al., 2024). However, the lack of biodegradability of these materials raises concerns about their long-term accumulation in organs and potential chronic toxicity (Casals et al., 2019).

Figure 2. Different Types of Nanocarriers

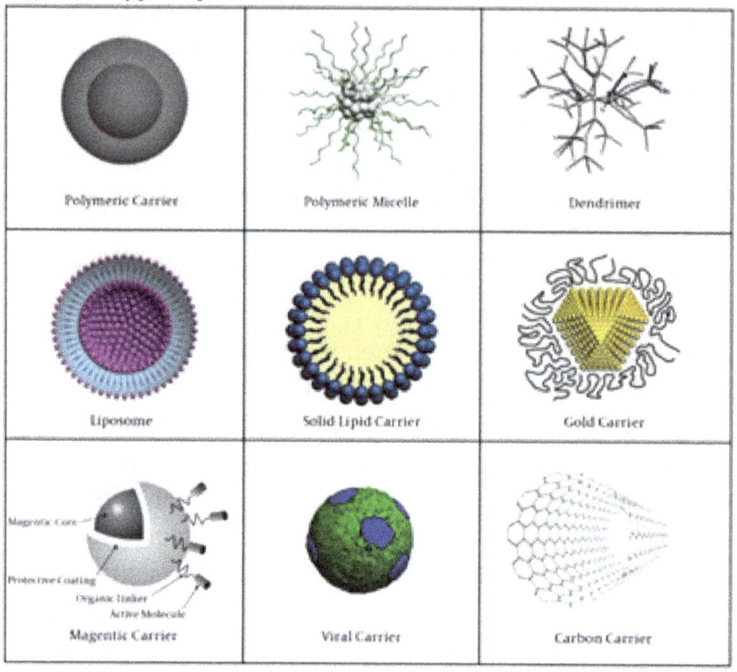

Adapted from (Khodabandehloo et al., 2016).

2.3. The Success of Nanocarriers in Challenging Infections: Biofilm Penetration

Biofilms are one of the biggest obstacles encountered in the treatment of bacterial infections. Bacteria attempt to form an impenetrable barrier by grouping together and secreting a biological concrete or slimy armor called polysaccharide matrix (EPS). Dental plaque is the most well-known example. This matrix protects bacteria like a physical wall against both antibiotics and the host's immune system. Drugs cannot penetrate these dense and sticky biofilms. Furthermore, within this fortress-like environment, bacteria create a low-oxygen, acidic environment, slowing their metabolism and entering a dormant state. Even if drugs penetrate this wall, most antibiotics fail because they can only kill actively dividing bacteria (Roy et al., 2022).

Nanocarriers have been programmed with different properties to overcome biofilms. By exploiting their physical advantages, small-sized nanocarriers can penetrate through pores or weak points in this matrix (Sedighi et al., 2024). Additionally, the outer surfaces of the nanocarrier can be charged with different electrical charges if desired. For example, the surface of a biofilm (EPS) is generally negatively (-)

charged. If the surface of the carrier is positively (+) charged (e.g., with chitosan), the carrier adheres strongly to the wall and is drawn in, rather than hitting it (Cao et al., 2024).

Triggerable Release, on the other hand, is a strategy for using the nanocarrier almost like a smart bomb. As the nanocarrier circulates in the body (normal pH, 7.4), the drug does not leave the carrier and does not harm healthy tissues. When it penetrates the biofilm and detects the low pH (acidic environment) or a specific enzyme, the lock is released (Zeng et al., 2025). The antibiotic is released at the maximum dose right within the biofilm, right in the middle of the bacteria. This is the pinnacle of targeted therapy.

2.4. Combating Intracellular Pathogens

Pathogens like Salmonella (typhoid), Listeria (food poisoning), and most notoriously, Mycobacterium tuberculosis (tuberculosis), don't just circulate in the blood and fight the immune system. They are facultative intracellular parasites. This means they can live both outside and (more importantly) inside cells. The location bacteria choose to hide demonstrates the cleverness of this strategy. Macrophages find foreign invaders that enter our immune system (such as bacteria), engulf them (a process called phagocytosis), and break them down in acid-filled sacs (lysosomes) within them. These clever pathogens have evolved to evade this deadly process: Some (like Salmonella) avoid the formation of this acidic sac (lysosome) and live in a comfortable "pouch" inside the macrophage (Flannagan et al., 2009). Some (like Listeria) escape from that sac and migrate directly into the cell's primary fluid (cytosol), where they multiply freely (Mounier et al., 1990).

The result is recurrent and chronic infections. In this case, the antibiotic prescribed by the doctor (e.g., gentamicin, vancomycin) circulates in the bloodstream, that is, outside the cells, killing all freely circulating bacteria. The bacteria, however, reside within the cell membrane. The medication has no effect on the "dormant" bacteria hiding inside the macrophages. As soon as the patient stops taking the medication, the bacteria within those macrophages reappear, escape the macrophage, and restart the infection. This is a recurrent (recurring) infection. The bacteria remain in that shelter until the immune system weakens, creating a chronic infection (Abed & Couvreur, 2014).

2.4.1. Targeting Macrophages: The "Trojan Horse" Strategy

Combating intracellular pathogens requires not only a drug that kills bacteria, but also a delivery system capable of delivering that drug inside our own immune cells (macrophages). Nanocarriers use a Trojan Horse strategy to solve this problem(Abed

& Couvreur, 2014). The macrophage's job is to detect any foreign or unrecognized particles (bacteria, viruses, dead cell debris, and nanoparticles) circulating in the blood and destroy them through engulfment (phagocytosis) (Krpetic et al., 2010). Researchers have turned this natural process into an advantage. Antibiotic-loaded nanocarriers are deliberately engulfed by macrophages. Once inside the host cell, the carrier releases its antibiotic cargo into the intracellular environment (cytosol or phagosome), combating the bacteria where they normally find safety (intracellular sanctuary) (Chen et al., 2023). In short, it's a strategy that allows the cell to take in the antibiotic itself, rather than trying to force it into the cell.

In a study conducted by Yang et al. (2017), solid lipid nanoparticles (SLNs) loaded with enrofloxacin were used to treat macrophages infected with Salmonella. The results showed that this nanoparticle formulation increased antibiotic accumulation within macrophages by approximately 40-fold compared to the free drug, demonstrating a much higher efficacy in killing intracellular bacteria.

3. SMART TARGETING MECHANISMS: PASSIVE AND ACTIVE STRATEGIES

As discussed above, nanocarrier systems solve pharmaceutical challenges such as increasing drug solubility (water-insoluble antibiotics) or preventing rapid degradation in the blood (e.g., AMPs), transforming them into effective treatments. However, if a targeting mechanism is not incorporated into the carrier, it circulates untargeted in the blood and can accumulate in healthy organs, causing toxicity and undesirable side effects (Anselmo & Mitragotri, 2021). The process of specifically targeting nanocarriers to the site of infection is called targeted drug delivery.

The primary goal of targeting strategies is to reduce the amount of drug delivered to healthy tissues while simultaneously maximizing the amount targeted to the site of infection. This significantly prevents systemic toxicity (e.g., nephrotoxicity), a serious problem in conventional treatments, while radically increasing the effectiveness (therapeutic index) of the treatment. The passive and active strategies discussed in this section are approaches used to achieve this goal (Attia et al., 2019).

3.1. Passive Targeting: The Role of the EPR Effect at Infection Sites

Passive targeting refers to the spontaneous accumulation of a nanocarrier at the site of infection, utilizing the body's natural biological processes. When an infection or inflammation begins in the body, the physiology of that area begins to change significantly compared to its previously healthy state. Natural defense mechanisms are activated, and blood vessels in the area dilate and become highly permeable to allow immune cells such as macrophages to reach the infected area more quickly. Simultaneously, the lymphatic drainage system, responsible for removing fluid and waste from the area, slows down significantly. Nanocarriers, taking advantage of their nanometer size, exploit this natural process, first penetrating permeable tissues and then becoming trapped. This is called the Enhanced Permeability and Retention (EPR) effect (Shi et al., 2020).

Figure 3. Demonstration of passive targeting with the EPR effect

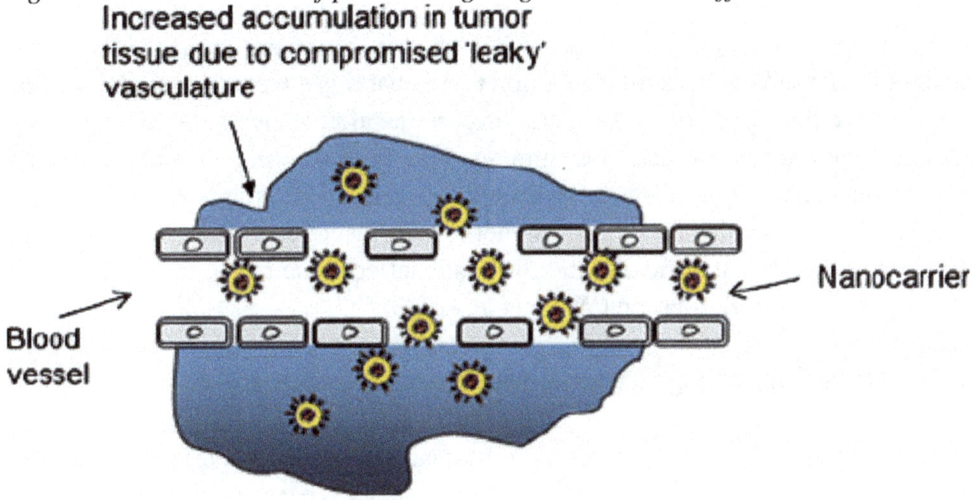

Adapted from (Sawant & Torchilin, 2012).

3.1.1. Pathophysiological Parallels Between Cancer and Inflammation

The EPR effect was first described in cancer biology. Tumors require nutrition to survive and continue growing. Therefore, they build their own supply lines (blood vessels) to obtain the oxygen and nutrients they need, allowing them to continue growing uncontrollably. Because of the rapid and hasty angiogenesis (new blood vessel formation), these new vessels are not organized and robust like normal, healthy vessels; on the contrary, they are disorganized, poor-quality, and leaky. Furthermore, lymphatic drainage in the tumor tissue is impaired. These two factors (high permeability and impaired drainage) cause nanoparticles within a certain size range to leak out of the vessels and passively accumulate (retain) in the tumor tissue (Desai, 2012).

Studies have revealed that this phenomenon is not specific to cancer but also applies to sites of bacterial infection and inflammation. This similarity observed between cancer and infection/inflammation sites has shown that the knowledge gained on the EPR effect in cancer nanomedicine can now be used in the fight against AMR (Azzopardi et al., 2013).

3.1.2. Mechanisms at Infection Sites

When the body is invaded by bacteria, the immune system initiates a powerful defense response known as inflammation to combat this attack. Symptoms such as redness, swelling, warmth, or a general fever in the affected area are inflammatory responses and are a sign that your immune system is working. Antibiotic-loaded nanocarriers passively concentrate at the site of infection without requiring any specific targeting structure. This means higher local drug concentration in the infected area, better therapeutic efficacy, and fewer side effects due to the reduced amount of drug in systemic circulation (Wang et al., 2023).

3.2. Active Targeting: Ligand-Mediated Recognition

Active targeting involves directing a drug carrier (nanoparticle) to a target site by attaching a specific address label to it. This strategy is based on the simple "lock-and-key" principle. The "Lock" on the Target (Receptor): Infected cells in our bodies (e.g., macrophages) or bacteria have specialized receptors on their surfaces that are absent (or very rare) on our healthy cells. The "Key" on the Carrier (Ligand): A special key (ligand) that fits exactly into that lock is attached to the outer surface of the nanocarrier (Bazak et al., 2015). Locking (High Affinity): These keys can be antibodies, peptides, or other molecules. As the carrier travels through the body, it hits millions of healthy cells (because their locks don't fit) but doesn't adhere to them. Once it finds the right target bacterium or infected cell, the key (ligand) perfectly locks onto that lock (receptor). This way, the drug doesn't just travel to the general site of infection; it locks directly onto the correct enemy cell (pathogen) in that area. This prevents the drug from harming healthy cells and focuses the entire dose on the enemy (Mi et al., 2020).

Figure 4. Demonstration of ligand-based antigen targeting

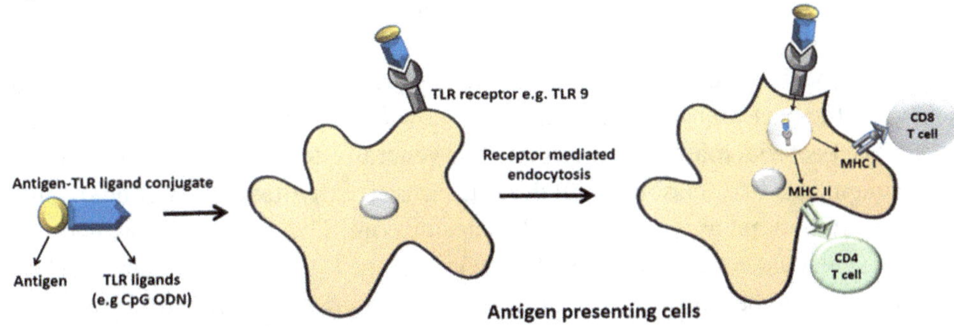

Adapted from (Shrestha et al., 2018).

3.2.1. Targeting Bacterial Surfaces

The goal of this approach is to attach the nanocarrier directly to the surface of the bacteria. This is a highly selective strategy that focuses the drug directly on the undesirable pathogen. Bacterial cell walls possess unique molecular properties that are completely different from human cell membranes, and this is used to their advantage. For example, the peptidoglycan structure of Gram-positive bacteria or the lipopolysaccharide (LPS) layer of Gram-negative bacteria are bacterial-specific structures and are not found in healthy human cells (Koatale et al., 2024). When the surface of the nanocarrier is coated with ligands that recognize these specific structures, the system directly targets and adheres to the target bacteria, bypassing healthy tissues in the body (Sabnis & Edwards, 2023). This direct adhesion increases the local concentration (density) of the antibiotic directly on the pathogen's surface by thousands of times, preventing the drug from contacting healthy tissues (and thus its toxicity). The following example of Vancomycin is one of the most classic applications of this principle.

3.2.1.1. Case Study: Vancomycin-D-Ala-D-Ala Interaction

One of the best examples of active targeting is the use of the antibiotic vancomycin as a ligand (Rashid et al., 2021). Vancomycin is a perfect key that naturally fits into a lock (receptor) called D-Ala-D-Ala, which Gram-positive bacteria (e.g., S. aureus) use to build their cell walls. Essentially, vancomycin molecules were placed as ligands on the outer surface of a nanocapsule filled with ciprofloxacin (a unique and powerful antibiotic). This vancomycin-coated nanocarrier (Rashid et al., 2021) ignores healthy cells when administered into the body. However, upon encountering the target Gram-positive bacterium, the Vancomycin keys irrevocably lock onto the D-Ala-D-Ala locks in the bacterial wall like a magnet.

After this stage, the capsule does not release the drug immediately; it has a "three-stage lock" that is triggered sequentially. Stage 1 (Target Found): First, Vancomycin locks onto the bacterial wall. The capsule is now fixed to the target. Stage 2 (Acid Lock): The bacterial infection environment is generally acidic (low pH). This acidic environment acts like scissors to cut the acid-sensitive (hydrazone) chemical bonds that connect the Vancomycin shell to the capsule core. The first lock is opened, and the outer shell (Vancomycin) is shed. Stage 3 (Enzyme Lock): Once the shell is shed, the actual core of the capsule (PCL polymer) is exposed. This core is designed to be broken down by the lipase enzyme that bacteria secrete to protect themselves. The bacteria's own enzyme unknowingly unlocks the second lock. Finally, as the nucleus ruptures, the actual ciprofloxacin antibiotic stored within is released precisely onto the bacteria's surface at maximum concentration. This "hierarchical" (multi-layered) approach—passively navigating to the site (EPR), then actively latching onto the target (vancomycin), and finally releasing the drug under conditions specific to that target (pH and enzyme)—has been remarkably successful. In an experiment on infected mice, this method dramatically increased survival rates in mice treated with the nanocarrier (vancomycin-coated, pH- and enzyme-sensitive) antibiotic system compared to those treated with normal antibiotics alone (Rashid et al., 2021).

3.2.1.2. Other Ligands

The vancomycin-D-Ala-D-Ala interaction is a classic example demonstrating the effectiveness of this direct binding method. However, vancomycin targets Gram-positive bacteria to a limited extent. To target the broad spectrum of pathogens involved in the AMR crisis (especially Gram-negative bacteria such as P. aeruginosa) or Vancomycin-resistant strains (e.g., VRSA), a diverse set of keys (ligands) that can recognize different receptors, or locks, on different bacterial surfaces is needed. For this purpose, peptides, antibodies, and even other antibiotics, each recognizing a different bacterial signature, are being investigated as ligands.

- **Peptides:** Various peptides that bind specifically to bacterial surfaces. For example, the ubiquicidin (UBI 29–41) peptide is used to target various bacteria such as S. aureus and P. aeruginosa (Jiang et al., 2025).
- **Antibodies:** Monoclonal antibodies or antibody fragments (Fab) developed against bacterial surface antigens or biofilm components can direct nanocarriers to their targets with high specificity (Miranda et al., 2023).
- **Other Antibiotics:** Other antibiotics that bind to specific targets, such as polymyxin B (targets LPS in Gram-negative bacteria) and daptomycin (tar-

gets Gram-positive cell membranes), can also be used as ligands (Jiang et al., 2021).

3.2.2. Targeting Infected Host Cell Receptors

Bacteria such as Salmonella, Tuberculosis, or Francisella hide within macrophages, and to reach the pathogens, the nanocarrier must target the infected cell, not the bacteria. Macrophages produce much more specific receptors on their surfaces, called mannose receptors. This causes the nanocarriers to become coated with mannose (a sugar molecule). Macrophages engulfed these sugar-coated (mannose) liposomes 2.5 times more than normal liposomes. Because the carriers entered the cell, the antibiotic they contained (Streptomycin) was 16 times more effective at killing the hidden Francisella tularensis bacteria than conventional treatment (Andrade et al., 2013). For example, lactoferrin (Lf) is a naturally occurring iron-carrying glycoprotein in our bodies. Many of our cells, including macrophages, require iron. Therefore, they have a special lactoferrin receptor on their surface to allow this iron cargo (lactoferrin) to enter. The researchers took this lactoferrin protein and used it as both the main material of the nanocarrier (capsule) and the key (ligand) itself. When the macrophage sees this lactoferrin carrier, it mistakes it for a needed iron cargo and quickly takes it in. In short, the macrophage has actually invited this drug-covered carrier inside (targeted delivery) (Andima et al., 2024).

4. STIMULUS-RESPONSIVE NANOCARRIERS: "ON-DEMAND" ANTIBIOTIC RELEASE

Traditional controlled-release systems release drugs at a slow but steady rate over time. This is not always ideal in a dynamic and time-consuming process such as infection. Stimulus-responsive systems, also known as smart nanocarriers, have been developed to offer an alternative. These carriers are designed to keep the drug locked inside them stably under normal physiological conditions (e.g., in the bloodstream at pH 7.4). However, when they encounter the differentiated pathophysiological microenvironment of the infection site or a specific externally applied signal, they undergo a conformational change and rapidly release their payload. This "on-demand" release ensures that the drug is released at maximum concentration precisely where and when it is needed, minimizing systemic toxicity and helping to prevent the development of resistance (Mura et al., 2013).

These triggering stimuli fall into two main categories: internal (endogenous) and external (exogenous).

4.1. Systems Sensitive to Internal (Endogenous) Stimuli

These systems autonomously respond to the biochemical and physiological hallmarks of disease or infection, requiring no external hardware or intervention. Just as a smoke detector detects smoke (the hallmark of fire) and autonomously sounds the alarm without you having to press a button, these systems detect unique chemical hallmarks and autonomously initiate treatment. These chemical hallmarks are outlined below.

4.1.1. pH Sensitivity

Infection Microenvironment: Bacterial infection sites, biofilms, and even phagosomes (spaces) within macrophages are significantly acidic compared to normal tissues (pH ~7.4). The pH in these areas typically ranges from 5.0 to 6.5. The infection site becomes acidic due to the accumulation of acidic waste products due to the intense activity of both bacteria and our immune cells (Zeng et al., 2025).

Mechanisms:

1. **Charge Reversal:** Nanocarriers are designed to remain biocompatible (negatively or neutrally charged) in the bloodstream. However, their structures incorporate chemical components that are protonated at low pH (e.g., amine groups). When the carrier reaches the acidic environment of the infection, its surface charge rapidly becomes positive. This positive charge allows it to bind to negatively charged bacterial membranes or the biofilm matrix through a strong electrostatic attraction. This both increases the local concentration of the drug reaching the target and facilitates the uptake of the carrier by bacteria (Deiss-Yehiely et al., 2023).
2. **Breakage of Acid-Sensitive Bonds:** The chemical bonds that bind the drug to the carrier or hold the carrier's polymeric structure together (e.g., hydrazone bonds, acetals) are designed to be unstable at acidic pH and undergo rapid hydrolysis. When the pH drops, these bonds are unlocked, and the drug is rapidly released (Gao et al., 2010).
3. **Shape Change:** Some nanoassembly systems can change their morphology in response to environmental pH. For example, systems that transform into nanofibers in response to the pH of a chronic wound have exhibited enhanced antibacterial activity due to this shape change. For example, systems have been developed that transform into microscopic filaments (nanofibers) when they sense the acidic environment of a non-healing (chronic) wound. Simply by adopting this new filamentous shape, they have been much more effective at killing bacteria (Li et al., 2022).

4.1.2. Enzyme Sensitivity

Infection Microenvironment: Infection sites are rich in specialized enzymes (molecular scissors) not found in healthy tissues, either secreted by the bacteria themselves (e.g., lipase, collagenase, and hyaluronidase secreted by P. aeruginosa) or by activated immune cells (Hu et al., 2014).

Mechanism: The nanocarrier is designed to contain peptide or polymer bonds (substrates) that are recognized, cleaved, and fragmented by these specific enzymes (proteases, lipases, etc.). Because these enzymes are not present in normal tissues, the carrier remains intact as it passes through these tissues. Upon reaching the infection site, at precisely the right time and place, the high concentration of enzymes acts as molecular scissors, cleaving the carrier and releasing a high dose of the drug (Yeh et al., 2020).

4.1.3. Redox/Reactive Oxygen Species (ROS) Sensitive Release

Infection Microenvironment: The site of inflammation and infection resembles a chemical battlefield. In this area, two types of molecules are significantly higher than normal. First, Reactive Oxygen Species (ROS) are oxidizing molecules produced by our immune cells to kill bacteria (e.g., Hydrogen Peroxide) (Liu et al., 2020). Second, Glutathione (GSH) is an antioxidant molecule produced by the body to balance oxidative stress.

Mechanism: The carrier is locked by specific bonds sensitive to these two chemical excesses. The disulfide bonds used to hold the capsule together are immediately broken by the high concentration of GSH molecules. The thiol groups added to the capsule are immediately oxidized by the high concentration of Hydrogen Peroxide. Breaking the lock disrupts the structure of the nanocarrier (capsule), causing it to disintegrate and releasing the drug inside precisely to the desired infection site (Hadipour et al., 2017).

4.2. Systems Controlled by External (Exogenous) Stimuli

Internal stimuli (pH, enzyme-sensitive) autonomously detonate when they detect the chemical signature of the disease, resulting in the release of the drug. However, sometimes the clinician is required to have complete control over the timing and dose of the drug release, as if they were holding a remote control. This is precisely what exogenous stimuli systems are. These carriers (nanoparticles) remain in a dormant state (inert) and do not release the drug, even if they reach the site of infection in the body. However, they are activated by a specific externally applied physical signal (light, magnetism, ultrasound) (Mi, 2020).

4.2.1. Light (NIR) and Photodynamic Triggering

Mechanism: Near-Infrared (NIR) light is generally used in these systems because NIR light can penetrate biological tissues deeper than other wavelengths of light. Components that absorb NIR light (e.g., gold nanoparticles, dyes such as indocyanine green, upconversion nanoparticles (UCNPs)) are incorporated into the nanocarrier. This light, delivered externally to the body, activates the nanocarrier, allowing it to spread to the desired location.

Effects:

- **Photothermal Release:** Absorption of NIR light causes local heating of the carrier (photothermal effect). If the carrier is made of a heat-sensitive polymer (e.g., poly(N-isopropylacrylamide) - PNIPAM), this heating causes the polymer to collapse, disintegrate, and rapidly release the drug to the desired location (Tang & Wang, 2021).
- **Combined Therapy:** This heat (hyperthermia) or light-induced ROS production (photodynamic therapy - PDT) not only releases the drug, but also acts as a key, opening the lid of the waiting nanocarrier (capsule) and triggering the release of the antibiotic stored inside. This allows the bacteria to do both jobs simultaneously. In short, the bacteria are simultaneously hit by both the external heat/light attack and the internally released antibiotic, creating a much more powerful overall effect (Chen et al., 2025).

4.2.2. Magnetic Field

Mechanism: Magnetic components such as superparamagnetic iron oxide nanoparticles are incorporated into the carrier.

Effects:

- **Magnetic Targeting:** The system has two parts: (1) Magnetic particles such as iron oxide (SPION) are loaded into the core of the drug-carrying nanocarrier. (2) The doctor places a strong, fixed (static) magnet from outside the patient's body, directly over the infection site (for example, over an infected knee prosthesis). As the “magnetic” carriers, administered into the bloodstream, pass through that site, they are physically captured and attracted by this external magnet. This forces the drug to activate precisely at the site of the infection, rather than circulating through healthy organs (Kakar et al., 2017).
- **Magnetic Hyperthermia Release:** This strategy is used to unlock the drug. The goal is not to attract the drug, but to heat it. The carrier is sealed with

a heat-sensitive lock (which melts or opens with temperature), and magnetic particles are placed inside. When the doctor wants the medication to be released, they apply an oscillating (pulsating) magnetic field, not a constant one. This specific signal causes the tiny magnets inside the carrier to vibrate rapidly and heat up instantly (magnetic hyperthermia). This localized heat melts or opens the heat-sensitive lock, releasing the medication precisely onto the target (Jauregui et al., 2018).

4.2.3. Ultrasound-Mediated Release (Triggered by the Power of Sound)

This strategy uses sound as a non-invasive (non-surgical) physical disruptor that can penetrate deep into the body to trigger the release of the drug.

Mechanism: Experts use a device called High-Intensity Focused Ultrasound (HIFU). This device focuses sound waves on a specific infection site deep within the body (where light or magnets cannot reach), much like a magnifying glass focuses sunlight on a single point. The drug (antibiotic) is loaded into "sound-sensitive" capsules such as liposomes or micelles and administered into the body. As these capsules circulate in the blood, the intense sound waves create microscopic bubbles in the fluid at the site when they pass through the focused sound wave. These micro-explosions physically disrupt or disrupt and disperse the drug capsules (liposomes) passing through, releasing the drug at the targeted millimeter point (Wang et al., 2025).

Effects: The benefits of ultrasound aren't limited to simply "breaking" the capsule; it also has a direct therapeutic effect. Direct Damage (Hyperthermia): This vibration created by intense sound waves also produces localized heat (hyperthermia) at the focused point. This heat directly weakens the bacteria or biofilm. Wall Weakening (Sonoporation): More importantly, the intense shock (cavitation) created by ultrasound temporarily loosens or punctures the bacteria's own cell wall or biofilm shield. This is a true double-whammy strategy, simultaneously weakening the bacteria's defensive shield (wall) and delivering a high dose of antibiotic directly to the weakenThe impact of ed shield (Liu et al., 2024).

Figure 5. A visual representation of stimuli-responsive nanoparticles to treat different illnesses.

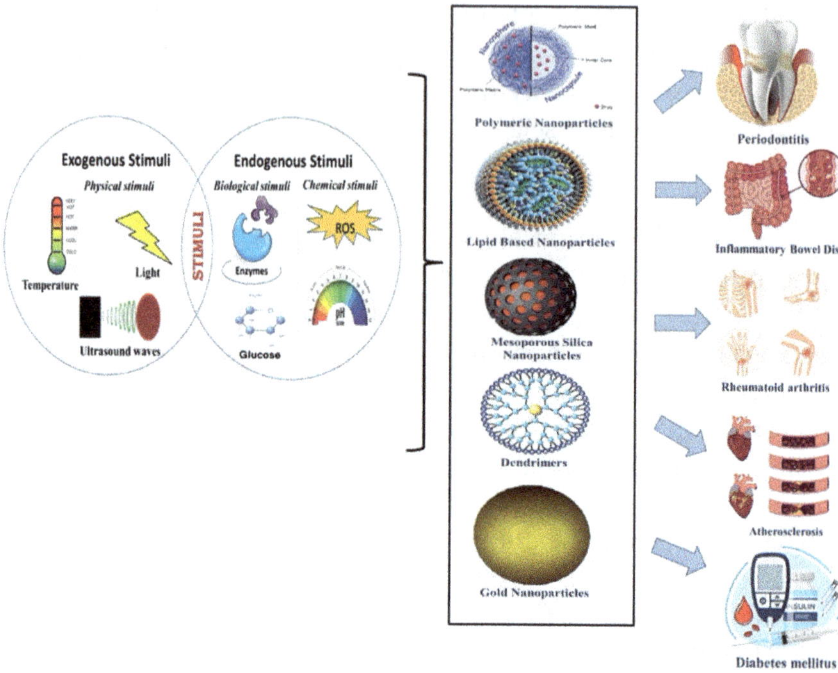

Adapted from (Fatima et al., 2024).

5. THE ROLE OF ARTIFICIAL INTELLIGENCE (AI) IN NANO CARRIER DESIGN AND OPTIMIZATION

The design of active-targeting, multi-stimulus-responsive smart nanocarriers, as described in the previous sections, is a complex optimization problem requiring the combined consideration of a vast array of factors. Traditional pharmaceutical development methods are inadequate to cope with this level of complexity. Artificial intelligence (AI) and machine learning (ML) are creating a paradigm shift in this field, transforming the entire process from rational design to optimization.

5.1. Challenges of Traditional Formulation Development

While nanocarriers have the potential to revolutionize drug delivery, a fundamental problem is that smart carriers are not single components but multifunctional, hierarchical systems. This makes them a complex and multifactorial problem to solve

(Shi et al., 2017). For example, consider a single system that combines the mechanisms discussed in previous sections: a carrier with vancomycin (ligand) attached to its surface for active targeting, a pH-sensitive polymer and enzyme-responsive linkages for internal stimuli response at the infection site, and a gold core sensitive to NIR light for external triggering.

For such a complex system to perform as expected in vivo (i.e., pharmacokinetics and pharmacodynamics - PK/PD), numerous critical interacting parameters must be precisely optimized: nanoparticle size, surface chemistry and surface electrical charge, ligand density on the surface, drug loading capacity, and drug release profile. Each of these parameters influences the others, creating an enormous design space (Anselmo & Mitragotri, 2015). Within this vast range of possible combinations, finding the optimal formulation through traditional trial-and-error or one-factor-at-a-time approaches is an extremely slow, costly, and often inefficient process (Waghule et al., 2021).

5.2. Predictive Modeling with Machine Learning (ML)

Artificial intelligence and machine learning specialize in working with large and complex datasets, providing powerful tools for automating and accelerating this design process by analyzing formulation datasets. Another key driver of success is the ability to eliminate both laboratory and human errors and obstacles associated with trial and error. AI can learn the complex, nonlinear relationships between formulation parameters (inputs) and the resulting nanoparticle properties (outputs), creating repeatable experimental environments where parameters can be analyzed in depth.

5.2.1. Optimization of Nanoparticle Properties: In Silico Prediction

The inadequacy of traditional trial-and-error methods in the face of this vast design space is precisely where Artificial Intelligence (AI) and Machine Learning (ML) come into play. ML models have a high potential to tackle levels of complexity that are nearly impossible for human researchers to manually handle, and to detect hidden patterns and nonlinear relationships within multivariate data (Chou et al., 2025).

To this end, researchers use ML models such as support vector machines (SVM) or random forest algorithms, along with deep learning techniques such as artificial neural networks (ANNs). These models analyze hundreds of previously generated experimental data sets (training sets) to learn and model the complex mathematical relationships between formulation parameters (inputs; e.g., polymer concentration, mixing speed, lipid ratio) and the resulting nanoparticle properties (outputs; e.g., particle size, surface charge, drug release kinetics) (Rezvantalab et al., 2024). Machine learning methods, in particular, stand out for their success in discovering and explaining the dark-box mechanisms and feature sets of AI applications.

Once this predictive model is established, AI bypasses the most costly and slow part of the process: conducting hundreds of new laboratory experiments. AI can rapidly scan millions of possible formulation configurations in silico (in a computer environment) in seconds to find the best starting recipe to achieve desired therapeutic goals (e.g., maximum targeting efficiency, minimum systemic toxicity, a specific half-life). Furthermore, small changes to feature sets (e.g., pH changes, temperature changes) yield repeatedly applicable experimental sets.

This approach not only identifies the most promising candidates for experimental testing, but also reduces the R&D process from months or years of experimentation to weeks of computational time, saving significant resources and time (Kapoor et al., 2024).

5.2.2. In Silico Screening and Pharmacokinetic (PK/PD) Prediction

Indeed, the role of AI goes beyond predicting the best prescription or physical properties. ML models can also be trained to predict the behavior and interactions of nanocarriers in biological environments such as blood plasma, cell membranes, or the immune system.

AI algorithms can be used to predict the pharmacokinetic (PK), pharmacodynamic (PD) profiles, and toxicity of potential drug candidates by analyzing large biological datasets such as genomics or proteomics (Ha et al., 2025).

This capability significantly reduces the need for costly, ethically controversial, and time-consuming animal testing (preclinical studies). For example, in the development of Lipid Nanoparticles (LNPs) carrying mRNA vaccines for COVID-19, ML models have significantly increased screening efficiency by predicting the transfection (cell entry) efficiency of new LNPs with high accuracy (Ding et al., 2023). The same principle applies to the rational design of nanoparticles carrying antibiotics.

5.3. Artificial Intelligence for Optimizing Manufacturing Processes

The impact of AI is not limited to the laboratory design phase. It is also being used to address one of the biggest obstacles to transitioning to the clinic, manufacturing scalability and batch-to-batch consistency, which will be discussed in Section 7. Even the slightest variation in laboratory processes impacts manufacturing precision. ML models can analyze complex manufacturing data to identify critical manufacturing parameters controlling particle size, drug loading, and release kinetics. This helps optimize and standardize manufacturing processes, reducing the need for costly and time-consuming experimental optimization and enabling the production of stable results (Djuriš et al., 2021).

Artificial intelligence is moving from being an "interesting addition" to nanotechnology to an essential tool for rationally designing and manufacturing hierarchical systems responsive to multiple stimuli. By reducing the number of experiments and identifying the most promising candidates through computer-based screening, AI is transforming nanocarrier development from an art to an engineering discipline (Bae et al., 2025).

Figure 6. An illustration of the complete process of AI-based nanosensor design.

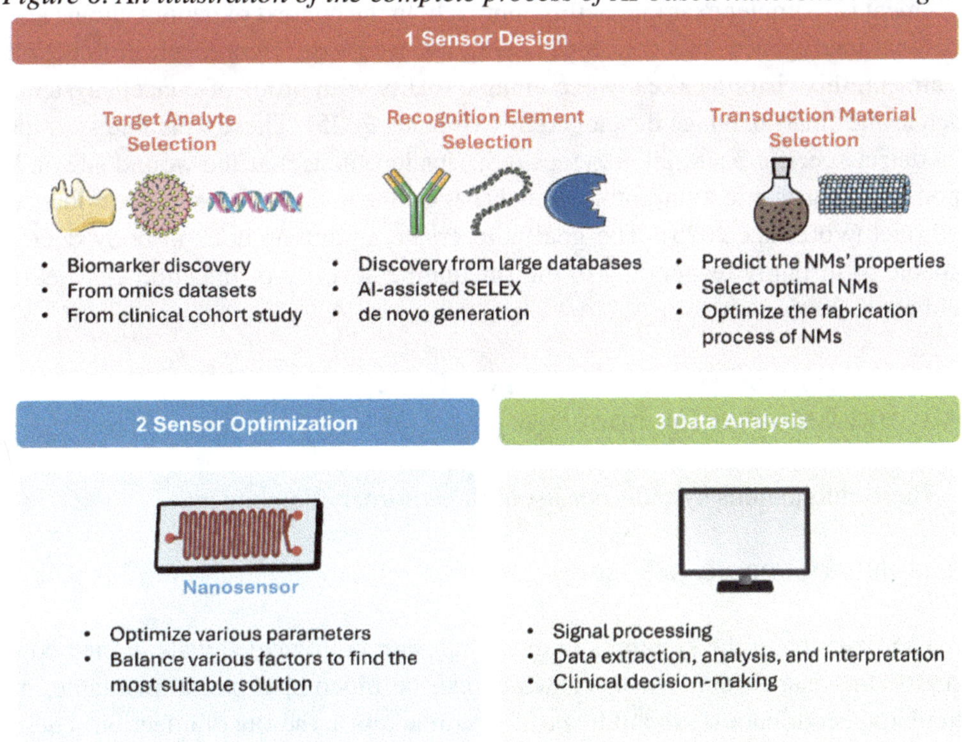

Adapted from (Yin, 2025).

6. NEXT-GENERATION SOLUTION: ARTIFICIAL INTELLIGENCE-GUIDED REAL-TIME DELIVERY SYSTEMS

We discussed how AI is used to design nanocarriers above. This section focuses on how AI is used to operate these carriers. This represents a revolution in antibiotic treatment, representing a shift from a reactive approach to a proactive and autonomous intervention, disrupting traditional methods.

6.1. Concept: Closed-Loop Therapeutic Systems

Closed-loop systems are an engineering paradigm that combines diagnosis and treatment in a single autonomous system (Wu et al., 2025). These systems continuously monitor the patient's physiological state, analyze this data, and dynamically adjust treatment according to patient-specific needs in real time (Wu et al., 2025).

The integration of insulin pumps and continuous glucose monitors in diabetes treatment is a simpler example of this approach. In the context of nanomedicine and artificial intelligence, this concept is being taken to a different level. In infection treatment, this vision is already becoming a reality with proof-of-concept systems such as the smart bandage developed by (Vo et al., 2025). These wearable systems can detect specific bacterial enzymes (e.g., hyaluronidase) at the wound site and, upon detecting an infection, autonomously trigger the release of an antibiotic-loaded hydrogel (Vo et al., 2025). The goal is to create an autonomous in-body doctor capable of instantly responding to the fluctuating severity of infection or disease (Wu et al., 2025).

6.2. System Components: Sensor, Processor (AI), and Actuator (Nanocarrier)

These autonomous systems consist of three primary components:

6.2.1. Nanosensors (Sensing)

The first step in the system is to detect relevant biological signals in the body. Smart nanosensors, or biosensors, which can be biocompatible, implantable, or circulating, continuously monitor specific biomarkers at the site of infection. These markers can be inflammatory markers (e.g., cytokines) (Wu et al., 2025), pathological pH or Reactive Oxygen Species (ROS) (Wu et al., 2025), or specific bacterial molecules. For example, in a smart implant prototype developed by (Liang et al., 2023) for orthopedic implant infections, the sensors can measure both the pH decrease caused by bacterial metabolism and the electrical impedance change caused by biofilm accumulation in real time.

6.2.2. Artificial Intelligence Algorithms (Interpretation and Decision Making)

The raw data stream from nanosensors is typically transmitted wirelessly (e.g., via a wearable device or implant) to an AI model (Wu et al., 2025). The role of AI here is critical:

- **Interpretation:** AI algorithms analyze this noisy biological data, distinguishing meaningful signals from noise (Wu et al., 2025). Noise removal is one of the most important steps in AI, and the underlying principle is that not every piece of data entering the model is valuable enough and can even hinder learning.
- **Pattern Recognition:** Instead of simply looking at a momentary value (e.g., pH is currently 6.8), AI recognizes patterns and trends in the data. This is where AI goes beyond simple if-then statements and breaks away from the rule-based algorithmic structure. Models such as reinforcement learning can be trained to interpret sensor data and make optimal decisions to release or withhold medication (Wu et al., 2025). For example, it is possible to intervene in instantaneous decisions, such as a rapid drop in pH over the past hour or a rise in a specific inflammatory marker.
- **Decision Making:** When the AI interprets these patterns and detects a pathological signal (e.g., worsening or recurring infection), it decides whether a therapeutic intervention (drug release) is necessary and, more importantly, how much drug should be released (dosing) (Laubenbacher et al., 2021).

Figure 7. AI-ML Assisted Design for Point-of-Care Technologies.

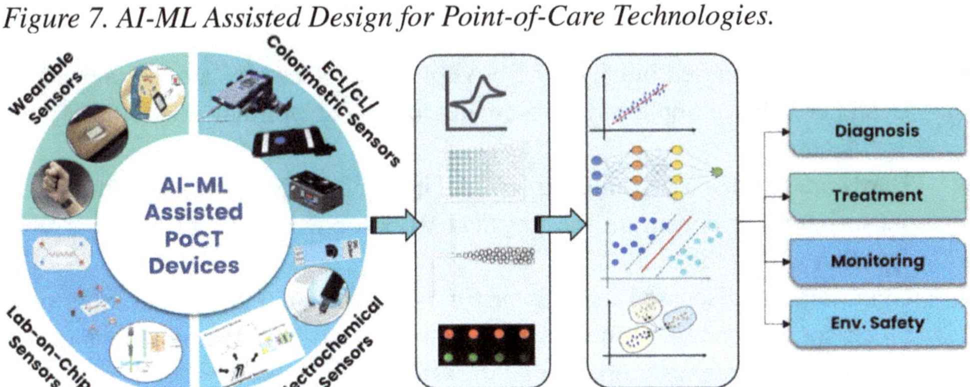

Adapted from (Bhaiyya et al., 2024).

6.2.3. Autonomous Triggering (Action/Actuation)

When the AI makes an intervention decision, it triggers the actuator component of the system. These actuators are nanocarriers responsive to external stimuli, as described in Chapter 4 (Wu et al., 2025). The AI sends a signal to the nanocarriers to release the drug. In the smart implant study mentioned above (Liang et al., 2023), when the AI processor detected a biofilm, it directly sent an electrochemical signal to release the antibiotic. This signal triggered drug release from the surface of the carrier. This external signal could be generating a localized magnetic field through an implanted system, triggering photothermal release by illuminating an LED (light), or sending a focused ultrasound wave. The drug is released, fighting the infection (bacteria), and the nanosensors detect changing conditions (e.g., pH returning to normal) and send this new data back to the AI, closing the loop.

6.3. Paving the Way for Personalized Antibiotic Therapies

These closed-loop systems have the potential to bring an end to the static dosing schedules (e.g., "take a 500 mg pill every 8 hours") that form the basis of modern antibiotic therapy (Wu et al., 2025). Static, one-for-all dosing fails to account for individual patient differences or the dynamic nature of the infection, leading either to underdosing and, further, to the development of resistance, or to overdosing and, undesirably, toxicity.

AI-driven closed-loop systems, on the other hand, completely personalize treatment. Treatment is tailored to each patient's unique and immediate response to infection (Laubenbacher et al., 2021).This is particularly revolutionary when symptoms fluctuate or resistance development is a critical risk (Wu et al., 2025).

This vision is a technological culmination of all the previous chapters:

- Design (as discussed in Chapter 5): AI primarily serves as a computer-aided Designer; This complex system creates the most optimal prescription and architecture.
- Hardware (as discussed in Chapter 2): The nanocarrier (actuator) and nano-sensor (sensor) constitute the physical workhorses of this design that will function within the body.
- Operation (as discussed in Chapter 6): This time, AI takes on the role of a processor, interpreting sensor data in real time and issuing commands to the actuator, autonomously operating the system.

This is not just a matter of hardware; it is also a matter of data modeling. The concept of creating "digital twins of infection" based on continuous physiological data (sensor data) from the patient will allow AI to simulate and personalize optimal antimicrobial treatment regimens in silico for each patient (Laubenbacher et al., 2021). This is the path that combines diagnosis and treatment (Wu et al., 2025), and ushers in the era of autonomous, personalized medicine (Laubenbacher et al., 2021).

Figure 8. The design flow of personalized antibiotics guided by patient data analysis.

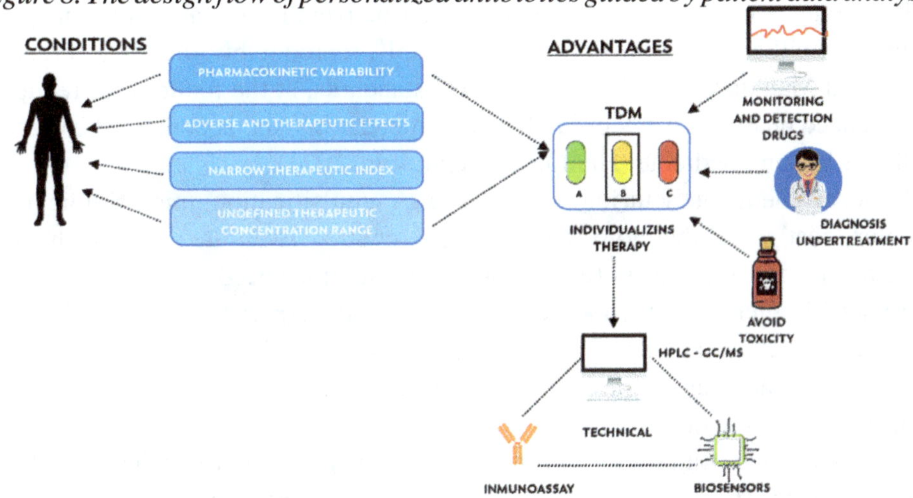

Adapted from (Garzón et al., 2020).

7. CHALLENGES, REGULATIONS, AND FUTURE PERSPECTIVES

Despite the tremendous potential offered by nanotechnology and AI-based strategies, the transition of these systems from the laboratory bench to the patient bed (clinical translation) is not only challenging scientifically but also fraught with practical, economic, and legal hurdles. This section will address these hurdles and practical challenges (Anselmo & Mitragotri, 2021).

7.1. Barriers to the Transfer of Nanomedicine to the Clinic

Despite laboratory and animal model successes and proof of concept, a significant challenge remains in the translational gap in medicine regarding the transfer of nanoplatforms to the patient. The issue is no longer simply whether the drug works, but also the practical, safe, and economic feasibility of these complex drug-device combination products. The primary hurdles typically relate to large-scale production (manufacturability), long-term biosafety (toxicity and immunogenicity), and the complex regulatory frameworks specific to these new technologies (Metselaar & Lammers, 2020).

7.1.1. Scalability and Large-Scale Manufacturing (GMP)

Producing a nanoparticle formulation that delivers excellent results in small quantities (milligrams) in the laboratory, reliably, reproducibly, and in accordance with Good Manufacturing Practices (GMP) standards, at the tonne scale required for advanced phase experiments and the commercial market, is perhaps the biggest challenge in nanomedicine (Hamilton & Kingston, 2024).

Nanoparticle manufacturing processes (e.g., emulsification, microfluidics) are complex, costly, and therefore require precise control. Even the smallest changes in the manufacturing process (e.g., temperature, stirring speed, flow pressure) can unpredictably alter critical properties of the final product (particle size, morphology, drug loading, release profile). This is because the final product is only possible with the precise and accurate alignment of many factors. This problem of batch-to-batch inconsistency is a major hurdle (Gu et al., 2024).

This problem of batch-to-batch inconsistency is precisely the reflection of the enormous design space challenge mentioned in Chapter 5 in the manufacturing phase. Attempting to optimize a GMP process using traditional trial-and-error methods is extremely slow, costly, and, as expected, inefficient when so many variables are involved.

This scalability dilemma is one of the areas where Artificial Intelligence (AI) and Machine Learning (ML) models play a critical role (Chou et al., 2025). Researchers now train a production model using data from hundreds of production experiments using ML algorithms (e.g., artificial neural networks or random forests). These models learn the complex, nonlinear relationships between inputs such as polymer concentration, temperature, and flow rate, and outputs such as particle size and drug loading efficiency, and then adapt them to new features.

The traditional trial-and-error approach is gradually being replaced by a modern approach called Quality by Design (QbD). Using laboratory-scale data, AI can predict and define in a computerized environment the optimal process window required to achieve the same quality product in large-scale (ton) production (Djuriš et al., 2021). This helps remove the biggest obstacle to scalability by reducing the time to develop a repeatable (robust) manufacturing process at GMP standards from months to weeks.

7.1.2. Biocompatibility, Nanotoxicity, and Long-Term Safety

Nanoparticles must be meticulously evaluated not only for the drug they carry but also for the safety of the carrier itself.

- **Biocompatibility:** The carrier must not trigger undesirable immune responses (immunogenicity) or activate the blood clotting system (Zolnik et al., 2010).
- **Nanotoxicity:** Especially for non-biodegradable inorganic nanoparticles (e.g., gold, silica, iron oxide), how these particles will be eliminated from the body and whether they will accumulate in organs over the long term (potential chronic toxicity) is a significant concern (Fadeel et al., 2018).

Unlike traditional drugs, the complex structure of nanomaterials makes it extremely difficult to predict their biological interactions. Traditional animal toxicity testing is costly, slow, and ethically controversial (Hartung, 2018). This is referred to as the toxicity screening bottleneck and has the potential to be one of the areas where Artificial Intelligence (AI) can make one of its most valuable contributions. As mentioned in Section 5.2.2, AI models are now being trained using Quantitative Structure-Activity Relationship (QSAR) and Quantitative Structure-Property Relationship (QSPR) models to learn the complex relationship between a nanoparticle's physical/chemical properties and its biological effects (Ha et al., 2025).Using these 'trained' models, AI can screen thousands of potential nanoparticle designs and predict which ones carry a high risk of toxicity or immunogenicity before they are even synthesized in the laboratory (Sriram et al., 2024). This allows researchers to focus on only the safest candidates and radically speeds up the safety assessment process by significantly reducing the number, cost, and time of rigorous preclinical animal testing required for transition to clinical applications (Ha et al., 2025).

7.1.3. Regulatory Processes (FDA, EMA) and Lack of Standardization

The regulatory pathways for nanomedicines are much more complex than for traditional small molecule drugs and are still evolving (Allan et al., 2021). Agencies such as the US Food and Drug Administration (FDA) and the European Medicines Agency (EMA) require comprehensive data and detailed reports on the safety, efficacy, toxicity, biodistribution, and potential environmental impacts of these innovative nanomaterials.

Further complicating this process is the lack of industry-wide, standardized protocols and quality control measures for the characterization and testing of nanomaterials (Mulvaney et al., 2016). This lack of standardization is one of the largest barriers to regulatory approval and another critical point where Artificial Intelligence (AI)-driven nanoinformatics platforms come into play (Maojo et al., 2012).

As discussed in Sections 7.1.1 (Manufacturing) and 7.1.2 (Toxicity), AI's use of Quality by Design (QbD) approaches to ensure batch-to-batch consistency and computational toxicity prediction forms the basis for the "comprehensive data" required by regulatory agencies (FDA/EMA). Additionally, AI is being used to create large, standardized databases that can analyze and harmonize heterogeneous (inconsistent) characterization and safety data from different laboratories (Mulvaney et al., 2016). Regulatory agencies are increasingly adopting computational models and databases (e.g., AI-QSAR) to assess the risk of a nanomaterial with an evidence-weighted approach (Lin et al., 2022).

7.2. Special Challenges of AI-Driven Systems

While the general challenges of transferring nanomedicine to the clinic, such as production, safety, and regulatory approval, are already complex enough, AI-driven closed-loop systems that autonomously manage treatment take these challenges to a new level. This is because computers are becoming active players in the diagnostic and therapeutic processes, making their own decisions. This device-drug-software combination comes with its own unique technological, ethical, and legal hurdles, in addition to the general nanomedicine challenges discussed earlier.

7.2.1. Data Security and Patient Privacy

AI systems, especially personalized models—those derived primarily from studies conducted on human subjects—rely on enormous amounts of sensitive patient data, including genetic information, medical history, lifestyle factors, or simply demographic details, and, most importantly, real-time physiological data.

- **Privacy:** While ethics committee reviews aim to ensure the confidentiality of personal health data and protect it from cyberattacks, this remains one of the most significant challenges facing these efforts. Achieving 100% success is a difficult area (Price & Cohen, 2019).
- **Patient Consent (Informed Consent):** Although patient confidentiality and anonymization structures are implemented, patients must be informed of how their data will be used by an autonomous decision-making mechanism, and they must consent to this. Consent forms can sometimes be a difficult procedure to implement and even significantly reduce the number of patients reached (Moy et al., 2024).

7.2.2. Algorithmic Transparency and Reliability

AI (especially deep learning models) operate like a black box. This means that the steps involved in reaching a decision cannot be clearly explained and vary slightly each time they are run. This presents a significant trust issue for the medical community and regulatory bodies.

A doctor or patient's ability to understand in detail why an autonomous system decides to administer an antibiotic at that moment and at that dose is crucial for interpretability. Providing transparency in AI decision-making is critical for the clinical acceptance of these systems and the determination of legal liability. This is a frequently cited challenge and obstacle for AI-based systems.

7.2.3. Technical Integration and Long-Term Stability Challenges

The engineering of AI-driven closed-loop systems is extremely complex because it must integrate biologically based structures, electronic hardware (sensor/actuator), and software architectures (AI model) Laubenbacher et al., 2021). The main technical hurdles are:

- **Feedback Loop:** Real-time synchronization of nanosensor readings as they are transferred to the model is critical. The system must know when to detect, for example, a changing pH value and stop intervention. Calibration errors or delays in the system can lead to over- or underdosing.
- **Data Transmission and Latency:** The wireless transmission of data from the implant to an external processor through the tissue, and the interpretation of noisy biological signals (such as MRI) by the AI takes time. This delay can cause the system to react to a past state rather than the body's current state, rendering the treatment ineffective (Khan et al., 2020).
- **Long-Term Functionality:** One of the biggest obstacles is maintaining the device's functionality within the body for extended periods. Biofouling, the physical blockage of the sensor's entry points by proteins and cells over time, can make readings impossible. Similarly, the system's limited drug reservoir and constant power (battery) requirements create significant engineering challenges, such as how to refill or charge these components without surgical intervention (Segurado et al., 2022).

7.3. Future Vision: Nanorobotics and Autonomous Therapeutic Agents

The logical next step in this field is nanorobots, microscopic machines capable of not only drug delivery but also active locomotion (propulsion), real-time disease monitoring, optimized drug delivery, and precision surgical interventions.

Biohybrid micro/nanorobots driven by magnetic fields or using biological motors (e.g., magnetotactic bacteria), and even DNA-based nanomachines designed for specific molecular tasks (DNA origami) are no longer science fiction but active areas of research.

However, these futuristic systems inherit all the regulatory, scalability, and security challenges faced by nanocarriers, and more. Approval of a nanodrug is challenging. Approval of an AI medical device is challenging. A closed-loop nanorobot that is both an autonomous AI device and a nanodrug presents an unprecedented challenge for regulatory agencies. This practical and legal obstacle often poses a greater challenge than the initial scientific discovery.

8. CONCLUSION AND EVALUATION

This chapter examines the profound technological evolution that has occurred in pharmaceutical sciences and medicine in response to the global crisis of antimicrobial resistance (AMR). Fundamental limitations of traditional antibiotics, such as systemic toxicity, poor bioavailability, and, most importantly, their inability to penetrate protected areas such as biofilms and intracellular shelters, have made the AMR crisis a drug delivery challenge as much as a drug discovery challenge.

The evolution of technologies developed in response to this challenge demonstrates a clear progression from simplicity to complexity:

- **First Generation (Encapsulation):** Passive carriers such as liposomes and polymeric nanoparticles addressed fundamental stability and pharmacokinetic issues by protecting the antibiotic and increasing its solubility.
- **Second Generation (Targeting):** Treatment became more specific with intelligent targeting mechanisms. Passive targeting (EPR effect at infection sites) directed the carrier to the correct tissue, while active targeting (ligand-mediated) targeted it to a specific bacterial cell or infected macrophage.
- **Third Generation (Smart Release):** Stimulus-responsive systems have gained the ability not only to deliver drugs to the right location but also to release them on demand at the right time, responding to the unique biochemical triggers of the infection microenvironment (low pH, enzymes).

The main argument of this section is that the fourth and most revolutionary phase of this evolution is the integration of artificial intelligence (AI) into this field. AI is not just an optimization tool but also a paradigm-shifting force. The role of AI is twofold:

- **Rational Design:** AI and machine learning have become essential tools for managing the immense combinatorial complexity in the design of the aforementioned multifunctional (active-targeting, multi-stimulus-responsive) systems. AI enables the rational engineering of these systems through in silico modeling and predictive optimization.
- **Autonomous Operation:** AI not only plays a role in design but also functions as the brain of closed-loop systems that define the future of therapy by interpreting real-time biological data from nanosensors and commanding nanocarriers to autonomously release drugs. This convergence has the potential to fundamentally transform antibiotic therapy. It transforms treatment from a static, reactive, and periodic dosing regimen (e.g., three times daily) to a proactive, dynamic, personalized, and autonomous intervention that suits all patients.

However, this bright future vision must be balanced against the immense translational challenges that stand in the way of the clinic. As discussed in detail, scalability and GMP-compliant manufacturing, long-term nanotoxicity and safety, complex regulatory pathways, and the unique ethical (data privacy) and technical (transparency) challenges presented by AI-driven autonomous systems are the major obstacles to the widespread clinical use of these technologies.

In conclusion, the intersection of nanotechnology and artificial intelligence presents an unprecedented opportunity in the fight against AMR. As scientific progress continues apace, the central challenge of the next decade will be building the manufacturing and regulatory bridges that will bring these revolutionary platforms from the laboratory to the patients waiting for them in a safe, effective, and accessible manner.

REFERENCES

Abbasi, E., Aval, S. F., Akbar, A., & Nojoom, M. (2014). Dendrimers: Synthesis, applications, and properties. *Nanoscale Research Letters*, *9*(1), 247. DOI: 10.1186/1556-276X-9-247 PMID: 24994950

Abed, N., & Couvreur, P. (2014). Nanocarriers for antibiotics: A promising solution to treat intracellular bacterial infections. *International Journal of Antimicrobial Agents*, *43*(6), 485–496. DOI: 10.1016/j.ijantimicag.2014.02.009 PMID: 24721232

Allan, J., Belz, S., Hoeveler, A., Hugas, M., Okuda, H., Patri, A., Rauscher, H., Silva, P., Slikker, W., Sokull-Kluettgen, B., Tong, W., & Anklam, E. (2021, June). Regulatory landscape of nanotechnology and nanoplastics from a global perspective. *Regulatory Toxicology and Pharmacology*, *122*, 104885. DOI: 10.1016/j.yrtph.2021.104885 PMID: 33617940

Aminov, R. I. (2010, December 8). A brief history of the antibiotic era: Lessons learned and challenges for the future. *Frontiers in Microbiology*, *1*, 134. DOI: 10.3389/fmicb.2010.00134 PMID: 21687759

Andima, M., Boese, A., Paul, P., Koch, M., Loretz, B., & Lehr, C. M. (2024). Targeting intracellular bacteria with dual drug-loaded lactoferrin nanoparticles. *ACS Infectious Diseases*, *10*(5), 1696–1710. DOI: 10.1021/acsinfecdis.4c00045 PMID: 38577780

Andra, V. V. S. N. L., Pammi, S. V. N., Bhatraju, L. V. K. P., & Ruddaraju, L. K. (2022). A Comprehensive Review on Novel Liposomal Methodologies, Commercial Formulations, Clinical Trials and Patents. *BioNanoScience*, *12*(1), 274–291. DOI: 10.1007/s12668-022-00941-x PMID: 35096502

Andrade, F., Rafael, D., Videira, M., Ferreira, D., Sosnik, A., & Sarmento, B. (2013, November). Nanotechnology and pulmonary delivery to overcome resistance in infectious diseases. *Advanced Drug Delivery Reviews*, *65*(13-14), 1816–1827. DOI: 10.1016/j.addr.2013.07.020 PMID: 23932923

Anselmo AC, Mitragotri S. (2021). A Review of Clinical Translation of Inorganic Nanoparticles. AAPS J. 2015 Sep;17(5):1041-54. Epub 2015 May 9. PMID: 25956384; PMCID: PMC4540735. DOI: DOI: 10.1208/s12248-015-9780-2

Anselmo, A. C., & Mitragotri, S. (2021). Nanoparticles in the clinic: An update. *Bioengineering & Translational Medicine*, *6*(3), e10246. DOI: 10.1002/btm2.10246 PMID: 34514159

Anwar, M., Muhammad, F., & Akhtar, B. (2021). Biodegradable nanoparticles as drug delivery devices. *Journal of Drug Delivery Science and Technology, 64*, 102638. DOI: 10.1016/j.jddst.2021.102638

Attia MF, Anton N, Wallyn J, Omran Z, Vandamme TF. An overview of active and passive targeting strategies to improve the nanocarriers efficiency to tumour sites. J Pharm Pharmacol. 2019 Aug;71(8):1185-1198. . Epub 2019 May 3. PMID: 31049986. DOI: https://doi.org/.DOI: 10.1111/jphp.13098

Azzopardi EA, Ferguson EL, Thomas DW. The enhanced permeability retention effect: a new paradigm for drug targeting in infection. J Antimicrob Chemother. 2013 Feb;68(2):257-74. . Epub 2012 Oct 10. PMID: 23054997.DOI: https://doi .org/.DOI: 10.1093/jac/dks379

Bae, H., Ji, H., Konstantinov, K., Sluyter, R., Ariga, K., Kim, Y. H., & Kim, J. H. (2025). Artificial intelligence-driven nanoarchitectonics for smart targeted drug delivery. *Advanced Materials, 10239*(42), e10239. Advance online publication. DOI: 10.1002/adma.202510239 PMID: 40772340

Baptista, P. V., McCusker, M. P., Carvalho, A., Ferreira, D. A., Mohan, N. M., Martins, M., & Fernandes, A. R. (2018, July 2). Nano-Strategies to Fight Multidrug Resistant Bacteria-"A Battle of the Titans". *Frontiers in Microbiology, 9*, 1441. DOI: 10.3389/fmicb.2018.01441 PMID: 30013539

Bazak R, Houri M, El Achy S, Kamel S, Refaat T. Cancer active targeting by nanoparticles: a comprehensive review of literature. J Cancer Res Clin Oncol. 2015 May;141(5):769-84. Epub 2014 Jul 9. PMID: 25005786; PMCID: PMC4710367. 1DOI: https://doi.org/DOI: 0.1007/s00432-014-1767-3

Bhaiyya, M., Panigrahi, D., Rewatkar, P., & Haick, H. (2024). Role of machine learning assisted biosensors in point-of-care-testing for clinical decisions. *ACS Sensors, 9*(9), 4495–4519. DOI: 10.1021/acssensors.4c01582 PMID: 39145721

Bozzuto, G., & Molinari, A. (2015). Liposomes as nanomedical devices. *International Journal of Nanomedicine, 10*, 975–999. DOI: 10.2147/IJN.S68861 PMID: 25678787

Brown, E. D., & Wright, G. D. (2016). Antibacterial drug discovery in the resistance era. *Nature, 529*(7586), 336–343. DOI: 10.1038/nature17042 PMID: 26791724

Brown, E. D., & Wright, G. D. (2016). Antibacterial drug discovery in the resistance era. *Nature, 529*(7586), 336–343. DOI: 10.1038/nature17042 PMID: 26791724

Cao, Y. H., Cai, W. J., He, X. W., Song, H. L., Gao, J., Yang, Y. L., & Zhou, J. (2024). A review of advances & potential of applying nanomaterials for biofilm inhibition. npj. *NPJ Clean Water, 7*(1), 131. DOI: 10.1038/s41545-024-00423-5

Casals, E., Casals, G., Puntes, V., & Rosenholm, J. M. (2019). Biodistribution, excretion, and toxicity of inorganic nanoparticles. In *Theranostic bionanomaterials* (pp. 3–26). Elsevier., DOI: 10.1016/B978-0-12-815341-3.00001-8

Castro, K. C. D., Costa, J. M., & Campos, M. G. N. (2022). Drug-loaded polymeric nanoparticles: A review. *International Journal of Polymeric Materials*, *71*(1), 1–13. DOI: 10.1080/00914037.2020.1798436

Chen, X., Lin, Z., Cheng, N., Mo, Y., Lu, L., Hou, J., Li, Z., Nie, X., Gao, S., & Hua, Q. (2025). Recent advances in NIR-II photothermal and photodynamic therapies for drug-resistant wound infections. *Materials Today. Bio*, *101871*, 101871. Advance online publication. DOI: 10.1016/j.mtbio.2025.101871 PMID: 40496729

Chen Y, He X, Chen Q, He Y, Chen F, Yang C, Wang L. Nanomaterials against intracellular bacterial infection: from drug delivery to intrinsic biofunction. Front Bioeng Biotechnol. 2023 Apr 27;11:1197974. . PMID: 37180049; PMCID: PMC10174311. DOI: https://doi.org/DOI: 10.3389/fbioe.2023.1197974

Chou, W. C., Canchola, A., Zhang, F., & Lin, Z. (2025, July-August). Machine Learning and Artificial Intelligence in Nanomedicine. *Wiley Interdisciplinary Reviews. Nanomedicine and Nanobiotechnology*, *17*(4), e70027. DOI: 10.1002/wnan.70027 PMID: 40813104

Chou, W. C., Canchola, A., Zhang, F., & Lin, Z. (2025). Machine learning and artificial intelligence in nanomedicine. *Wiley Interdisciplinary Reviews. Nanomedicine and Nanobiotechnology*, *17*(4), e70027. DOI: 10.1002/wnan.70027 PMID: 40813104

Deiss-Yehiely, E., Cárcamo-Oyarce, G., Berger, A. G., Ribbeck, K., & Hammond, P. T. (2023, August 14). pH-Responsive, Charge-Reversing Layer-by-Layer Nanoparticle Surfaces Enhance Biofilm Penetration and Eradication. *ACS Biomaterials Science & Engineering*, *9*(8), 4794–4804. DOI: 10.1021/acsbiomaterials.3c00481 PMID: 37390118

Desai N. Challenges in development of nanoparticle-based therapeutics. AAPS J. 2012 Jun;14(2):282-95. . Epub 2012 Mar 10. PMID: 22407288; PMCID: PMC3326161. DOI: https://doi.org/DOI: 10.1208/s12248-012-9339-4

Ding, D. Y., Zhang, Y., Jia, Y., & Sun, J. (2023). Machine learning-guided lipid nanoparticle design for mRNA delivery. arXiv preprint arXiv:2308.01402. DOI: https://doi.org//arXiv.2308.01402DOI: 10.48550

Ding, X., Ma, J., Fan, T., Issa, R., Li, Y., Weng, D., Zhang, D., & Chen, Y. (2024). Inorganic nanoparticles-based strategies for the microbial detection in infectious diseases. *Interdisciplinary Medicine*, *2*(2), e20230045. DOI: 10.1002/INMD.20230045

Djuriš, J., Kurćubić, I., & Ibrić, S. (2021). Review of machine learning algorithms application in pharmaceutical technology. Archives of Pharmacy, 71(Notebook 4), 302-317. DOI: DOI: 10.5937/arhfarm71-32499

Dyett, B. P., Sarkar, S., Yu, H., Strachan, J., Drummond, C. J., & Conn, C. E. (2024, May 15). Overcoming Therapeutic Challenges of Antibiotic Delivery with Cubosome Lipid Nanocarriers. *ACS Applied Materials & Interfaces*, *16*(19), 24191–24205. DOI: 10.1021/acsami.4c00921 PMID: 38690584

Fadeel, B., Farcal, L., Hardy, B., Vázquez-Campos, S., Hristozov, D., Marcomini, A., Lynch, I., Valsami-Jones, E., Alenius, H., & Savolainen, K. (2018). Advanced tools for the safety assessment of nanomaterials. *Nature Nanotechnology*, *13*(7), 537–543. DOI: 10.1038/s41565-018-0185-0 PMID: 29980781

Fatima, M., Almalki, W. H., Khan, T., Sahebkar, A., & Kesharwani, P. (2024). Harnessing the Power of Stimuli-Responsive Nanoparticles as an Effective Therapeutic Drug Delivery System. *Advanced Materials*, *36*(24), 2312939. DOI: 10.1002/adma.202312939 PMID: 38447161

Flannagan, R. S., Cosío, G., & Grinstein, S. (2009). Antimicrobial mechanisms of phagocytes and bacterial evasion strategies. *Nature Reviews. Microbiology*, *7*(5), 355–366. DOI: 10.1038/nrmicro2128 PMID: 19369951

Galm, U., & Shen, B. (2007, October). Natural product drug discovery: The times have never been better. *Chemistry & Biology*, *14*(10), 1098–1104. DOI: 10.1016/j.chembiol.2007.10.004 PMID: 17961822

Galm, U., & Shen, B. (2007, October). Natural product drug discovery: The times have never been better. *Chemistry & Biology*, *14*(10), 1098–1104. DOI: 10.1016/j.chembiol.2007.10.004 PMID: 17961822

Gao, W., Chan, J. M., & Farokhzad, O. C. (2010, December 6). pH-Responsive nanoparticles for drug delivery. *Molecular Pharmaceutics*, *7*(6), 1913–1920. DOI: 10.1021/mp100253e PMID: 20836539

Garzón, V., Bustos, R. H., & Pinacho, G., D. (. (2020). Personalized medicine for antibiotics: The role of nanobiosensors in therapeutic drug monitoring. *Journal of Personalized Medicine*, *10*(4), 147. DOI: 10.3390/jpm10040147 PMID: 32993004

Gomez-Simmonds, A., Nelson, B., Eiras, D. P., Loo, A., Jenkins, S. G., Whittier, S., Calfee, D. P., Satlin, M. J., Kubin, C. J., & Furuya, E. Y. (2016, May 23). Combination Regimens for Treatment of Carbapenem-Resistant Klebsiella pneumoniae Bloodstream Infections. *Antimicrobial Agents and Chemotherapy*, *60*(6), 3601–3607. DOI: 10.1128/AAC.03007-15 PMID: 27044555

Gu, K., Liang, Y., Su, J., Sun, P., Peng, J., Miao, N., & Zhang, J. (2024). Deep Learning Models for Colloidal Nanocrystal Synthesis. *ACS Nano*. Advance online publication. DOI: 10.1021/acsnano.5c09134 PMID: 41193409

Gupta, S., Kesarla, R., & Omri, A. (2013, December 26). Formulation strategies to improve the bioavailability of poorly absorbed drugs with special emphasis on self-emulsifying systems. *ISRN Pharmaceutics, 848043*, 1–16. Advance online publication. DOI: 10.1155/2013/848043 PMID: 24459591

Ha, E., Ha, S. M., Gerelkhuu, Z., Kim, H. Y., & Yoon, T. H. (2025). AI-based nanotoxicity data extraction and prediction of nanotoxicity. *Computational and Structural Biotechnology Journal, 29*, 138–148. DOI: 10.1016/j.csbj.2025.03.052 PMID: 40255458

Hadipour Moghaddam, S. P., Saikia, J., Yazdimamaghani, M., & Ghandehari, H. (2017). Redox-responsive polysulfide-based biodegradable organosilica nanoparticles for delivery of bioactive agents. *ACS Applied Materials & Interfaces, 9*(25), 21133–21146. DOI: 10.1021/acsami.7b04351 PMID: 28609092

Hamilton, S., & Kingston, B. R. (2024). Applying artificial intelligence and computational modeling to nanomedicine. *Current Opinion in Biotechnology, 85*, 103043. DOI: 10.1016/j.copbio.2023.103043 PMID: 38091874

Hartung, T. (2018, December 1). Perspectives on In Vitro to In Vivo Extrapolations. *Applied In Vitro Toxicology, 4*(4), 305–316. DOI: 10.1089/aivt.2016.0026 PMID: 31890748

Hu, Q., Katti, P. S., & Gu, Z. (2014, November 7). Enzyme-responsive nanomaterials for controlled drug delivery. *Nanoscale, 6*(21), 12273–12286. DOI: 10.1039/C4NR04249B PMID: 25251024

Jauregui, R., Srinivasan, S., Vojtech, L. N., Gammill, H. S., Chiu, D. T., Hladik, F., Stayton, P. S., & Lai, J. J. (2018, October 10). Stayton PSLai JJ. Temperature-Responsive Magnetic Nanoparticles for Enabling Affinity Separation of Extracellular Vesicles. *ACS Applied Materials & Interfaces, 10*(40), 33847–33856. DOI: 10.1021/acsami.8b09751 PMID: 30152229

Jiang, Q., Yan, X., Jiao, D., Zhang, J., Wu, Y., & Cheng, Y. (2021). Polymyxin B-modified conjugated oligomer nanoparticle for targeted identification and enhanced photodynamic antimicrobial therapy. *Chemical Communications, 57*(85), 11244–11247. DOI: 10.1039/D1CC04389G PMID: 34633009

Jonas, O. & Team, W. (2017). Drug-Resistant Infections: A Threat to Our Economic Future.

Kakar, S., Jain, A., & Singh, R. (2017). Magnetic Microspheres: A Novel Targeting Delivery System. *FABAD Journal of Pharmaceutical Sciences*, *42*(3), 209–223.

Kapoor, D. U., Sharma, J. B., Gandhi, S. M., Prajapati, B. G., Thanawuth, K., Limmatvapirat, S., & Sriamornsak, P. (2024). AI-driven design and optimization of nanoparticle-based drug delivery systems. Science, Engineering and Health Studies, 24010003-24010003. DOI: DOI: 10.69598/sehs.18.24010003

Khan, S. R., Pavuluri, S. K., Cummins, G., & Desmulliez, M. P. Y. (2020, June 19). Wireless Power Transfer Techniques for Implantable Medical Devices: A Review. *Sensors (Basel)*, *20*(12), 3487. DOI: 10.3390/s20123487 PMID: 32575663

Khodabandehloo, H., Zahednasab, H., & Hafez, A. A. (2016). Nanocarriers usage for drug delivery in cancer therapy. *Iranian Journal of Cancer Prevention*, *9*(2), e3966. DOI: 10.17795/ijcp-3966 PMID: 27482328

Koatale, P. C., Welling, M. M., Ndlovu, H., Kgatle, M., Mdanda, S., Mdlophane, A., Okem, A., Takyi-Williams, J., Sathekge, M. M., & Ebenhan, T. (2024). insights into Peptidoglycan-Targeting Radiotracers for Imaging Bacterial infections: Updates, challenges, and future perspectives. *ACS Infectious Diseases*, *10*(2), 270–286. DOI: 10.1021/acsinfecdis.3c00443 PMID: 38290525

Krause, K. M., Serio, A. W., Kane, T. R., & Connolly, L. E. (2016). Aminoglycosides: An Overview. *Cold Spring Harbor Perspectives in Medicine*, *6*(6), a027029. DOI: 10.1101/cshperspect.a027029 PMID: 27252397

Krpetic, Z., Porta, F., Caneva, E., Dal Santo, V., & Scarì, G. (2010). Phagocytosis of biocompatible gold nanoparticles. *Langmuir*, *26*(18), 14799–14805. DOI: 10.1021/la102758f PMID: 20795674

Laubenbacher, R., Sluka, J. P., & Glazier, J. A. (2021, March 12). Using digital twins in viral infection. *Science*, *371*(6534), 1105–1106. DOI: 10.1126/science.abf3370 PMID: 33707255

Lee, J. H., & Yeo, Y. (2015). Controlled drug release from pharmaceutical nanocarriers. *Chemical Engineering Science*, *125*, 75–84. DOI: 10.1016/j.ces.2014.08.046 PMID: 25684779

Li, Q., Li, X., & Zhao, C. (2020). Strategies to obtain encapsulation and controlled release of small hydrophilic molecules. *Frontiers in Bioengineering and Biotechnology*, *8*, 437. DOI: 10.3389/fbioe.2020.00437 PMID: 32478055

Li, Z., Zhu, Y., & Matson, J. B. (2022, May 3). pH-Responsive Self-Assembling Peptide-Based Biomaterials: Designs and Applications. *ACS Applied Bio Materials, 5*(10), 4635–4651. Advance online publication. DOI: 10.1021/acsabm.2c00188 PMID: 35505454

Liang, S., Hu, M., Li, B., Xia, D., Liang, C., Peng, F., & Wang, D. (2023, June 28). Smart Implant with Bacteria Monitoring and Killing Ability for Orthopedic Applications. *ACS Applied Materials & Interfaces, 15*(25), 29889–29901. DOI: 10.1021/acsami.3c03599 PMID: 37318286

Liew, K. B., Janakiraman, A. K., Sundarapandian, R., Khalid, S. H., Razzaq, F. A., Ming, L. C., Khan, A., Kalusalingam, A., & Ng, P. W. (2022, March). A review and revisit of nanoparticles for antimicrobial drug delivery. *Journal of Medicine and Life, 15*(3), 328–335. DOI: 10.25122/jml-2021-0097 PMID: 35449993

Lin, Z., & Chou, W. C. (2022, August 25). Machine Learning and Artificial Intelligence in Toxicological Sciences. *Toxicological Sciences, 189*(1), 7–19. DOI: 10.1093/toxsci/kfac075 PMID: 35861448

Liu, H.Y., Prentice, E.L. & Webber, M.A. Mechanisms of antimicrobial resistance in biofilms. npj Antimicrob Resist 2, 27 (2024). DOI: 10.1038/s44259-024-00046-3

Liu, J., Li, Y., Chen, S., Lin, Y., Lai, H., Chen, B., & Chen, T. (2020, September 18). Biomedical Application of Reactive Oxygen Species-Responsive Nanocarriers in Cancer, Inflammation, and Neurodegenerative Diseases. *Frontiers in Chemistry, 8*, 838. DOI: 10.3389/fchem.2020.00838 PMID: 33062637

Liu, J. D., Van Treeck, K. E., Marston, W. A., Papadopoulou, V., & Rowe, S. E. (2024, October 16). Ultrasound-Mediated Antibiotic Delivery to In Vivo Biofilm Infections: A Review. *ChemBioChem, 25*(20), e202400181. DOI: 10.1002/cbic.202400181 PMID: 38924307

Magana, M., Pushpanathan, M., Santos, A. L., Leanse, L., Fernandez, M., Ioannidis, A., Giulianotti, M. A., Apidianakis, Y., Bradfute, S., Ferguson, A. L., Cherkasov, A., Seleem, M. N., Pinilla, C., de la Fuente-Nunez, C., Lazaridis, T., Dai, T., Houghten, R. A., Hancock, R. E. W., & Tegos, G. P. (2020). The value of antimicrobial peptides in the age of resistance. *The Lancet. Infectious Diseases, 20*(9), e216–e230. DOI: 10.1016/S1473-3099(20)30327-3 PMID: 32653070

Mahlapuu, M., Håkansson, J., Ringstad, L., & Björn, C. (2016). Antimicrobial peptides: An emerging class of therapeutics. *Frontiers in Cellular and Infection Microbiology, 6*, 194. DOI: 10.3389/fcimb.2016.00194 PMID: 28083516

Maojo, V., Fritts, M., de la Iglesia, D., Cachau, R. E., Garcia-Remesal, M., Mitchell, J. A., & Kulikowski, C. (2012). Nanoinformatics: A new area of research in nanomedicine. *International Journal of Nanomedicine*, •••, 3867–3890. DOI: 10.2147/IJN.S24582 PMID: 22866003

Metselaar, J. M., & Lammers, T. (2020, June). Challenges in nanomedicine clinical translation. *Drug Delivery and Translational Research*, *10*(3), 721–725. DOI: 10.1007/s13346-020-00740-5 PMID: 32166632

Mi, P. (2020, March 15). Stimuli-responsive nanocarriers for drug delivery, tumor imaging, therapy and theranostics. *Theranostics*, *10*(10), 4557–4588. DOI: 10.7150/thno.38069 PMID: 32292515

Mi, P., Cabral, H., & Kataoka, K. (2020). Ligand-installed nanocarriers toward precision therapy. *Advanced Materials*, *32*(13), 1902604. DOI: 10.1002/adma.201902604 PMID: 31353770

Miranda Calderon, L. G., Alejo, T., Santos, S., Mendoza, G., Irusta, S., & Arruebo, M. (2023, August 30). Antibody-Functionalized Polymer Nanoparticles for Targeted Antibiotic Delivery in Models of Pathogenic Bacteria Infecting Human Macrophages. *ACS Applied Materials & Interfaces*, *15*(34), 40213–40227. DOI: 10.1021/acsami.3c07367 PMID: 37596966

Mounier, J., Ryter, A., Coquis-Rondon, M., & Sansonetti, P. J. (1990). Intracellular and cell-to-cell spread of Listeria monocytogenes involves interaction with F-actin in the enterocytelike cell line Caco-2. Infection and immunity, 58(4), 1048-1058. DOI: (Pizarro-Cerdá et al., 1990).(Mounier et al., 1990).DOI: 10.1128/iai.58.4.1048-1058.1990

Moy, S., Irannejad, M., Manning, S. J., Farahani, M., Ahmed, Y., Gao, E., Prabhune, R., Lorenz, S., Mirza, R., & Klinger, C. (2024, April 2). Patient Perspectives on the Use of Artificial Intelligence in Health Care: A Scoping Review. *Journal of Patient-Centered Research and Reviews*, *11*(1), 51–62. DOI: 10.17294/2330-0698.2029 PMID: 38596349

Mulvaney, P., Parak, W. J., Caruso, F., & Weiss, P. S. (2016). Standardizing nanomaterials. *ACS Nano*, *10*(11), 9763–9764. DOI: 10.1021/acsnano.6b07629 PMID: 27934085

Mura, S., Nicolas, J., & Couvreur, P. (2013). Stimuli-responsive nanocarriers for drug delivery. *Nature Materials*, *12*(11), 991–1003. DOI: 10.1038/nmat3776 PMID: 24150417

Murray, C. J., Ikuta, K. S., Sharara, F., Swetschinski, L., Aguilar, G. R., Gray, A., & Tasak, N. (2022). Global burden of bacterial antimicrobial resistance in 2019: A systematic analysis. *Lancet*, *399*(10325), 629–655. DOI: 10.1016/S0140-6736(21)02724-0 PMID: 35065702

Negut, I., & Bita, B. (2023, March 17). Polymeric Micellar Systems-A Special Emphasis on "Smart" Drug Delivery. *Pharmaceutics*, *15*(3), 976. DOI: 10.3390/pharmaceutics15030976 PMID: 36986837

Parvin, N., Joo, S. W., & Mandal, T. K. (2025). Nanomaterial-Based Strategies to Combat Antibiotic Resistance: Mechanisms and Applications. *Antibiotics (Basel, Switzerland)*, *14*(2), 207. DOI: 10.3390/antibiotics14020207 PMID: 40001450

Parvin, N., Joo, S. W., & Mandal, T. K. (2025). Nanomaterial-Based Strategies to Combat Antibiotic Resistance: Mechanisms and Applications. *Antibiotics (Basel, Switzerland)*, *14*(2), 207. DOI: 10.3390/antibiotics14020207 PMID: 40001450

Parvin, N., Joo, S. W., & Mandal, T. K. (2025). Nanomaterial-based strategies to combat antibiotic resistance: Mechanisms and applications. *Antibiotics (Basel, Switzerland)*, *14*(2), 207. DOI: 10.3390/antibiotics14020207 PMID: 40001450

Patra, J. K., Das, G., Fraceto, L. F., Campos, E. V. R., Rodriguez-Torres, M. D. P., Acosta-Torres, L. S., Diaz-Torres, L. A., Grillo, R., Swamy, M. K., Sharma, S., Habtemariam, S., & Shin, H. S. (2018, September 19). Nano based drug delivery systems: Recent developments and future prospects. *Journal of Nanobiotechnology*, *16*(1), 71. DOI: 10.1186/s12951-018-0392-8 PMID: 30231877

Price, W. N.II, & Cohen, I. G. (2019). Privacy in the age of medical big data. *Nature Medicine*, *25*(1), 37–43. DOI: 10.1038/s41591-018-0272-7 PMID: 30617331

Rashid, M., Rabbi, M. A., Ara, T., Hossain, M. M., Islam, M. S., Elaissari, A., Ahmad, H., & Rahman, M. M. (2021). Vancomycin conjugated iron oxide nanoparticles for magnetic targeting and efficient capture of Gram-positive and Gram-negative bacteria. *RSC Advances*, *11*(57), 36319–36329. DOI: 10.1039/D1RA04390K PMID: 35492746

Rezvantalab, S., Mihandoost, S., & Rezaiee, M. (2024). Machine learning assisted exploration of the influential parameters on the PLGA nanoparticles. *Scientific Reports*, *14*(1), 1114. DOI: 10.1038/s41598-023-50876-w PMID: 38212322

Roy R, Tiwari M, Donelli G, Tiwari V. (2022) Strategies for combating bacterial biofilms: A focus on anti-biofilm agents and their mechanisms of action. Virulence. 2018 Jan 1;9(1):522-554. DOI: . PMID: 28362216; PMCID: PMC5955472.DOI: 10.1080/21505594.2017.1313372

Rybak, M. J., Le, J., Lodise, T. P., Levine, D. P., Bradley, J. S., Liu, C., & Tamma, P. D. (2020). Therapeutic monitoring of vancomycin for serious methicillin-resistant Staphylococcus aureus infections: A revised consensus guideline and review by the American Society of Health-System Pharmacists, the Infectious Diseases Society of America, the Pediatric Infectious Diseases Society, and the Society of Infectious Diseases Pharmacists. *American Journal of Health-System Pharmacy*, *77*(11), 835–864. DOI: 10.1093/ajhp/zxaa036 PMID: 32191793

Sabnis, A., & Edwards, A. M. (2023). Lipopolysaccharide as an antibiotic target. Biochimica et Biophysica Acta (BBA)-. *Biochimica et Biophysica Acta. Molecular Cell Research*, *1870*(7), 119507. DOI: 10.1016/j.bbamcr.2023.119507 PMID: 37268022

Sawant, R. R., & Torchilin, V. P. (2012). Challenges in development of targeted liposomal therapeutics. *The AAPS Journal*, *14*(2), 303–315. DOI: 10.1208/s12248-012-9330-0 PMID: 22415612

Sedighi, O., Bednarke, B., Sherriff, H., & Doiron, A. L. (2024). Nanoparticle-based strategies for managing biofilm infections in wounds: A comprehensive review. *ACS Omega*, *9*(26), 27853–27871. DOI: 10.1021/acsomega.4c02343 PMID: 38973924

Segurado, O. G., Jiang, R., & Pipe, S. W. (2022, September). Challenges and opportunities when transitioning from in vivo gene replacement to in vivo CRISPR/Cas9 therapies - a spotlight on hemophilia. *Expert Opinion on Biological Therapy*, *22*(9), 1091–1098. DOI: 10.1080/14712598.2022.2090241 PMID: 35708146

Shi, J., Kantoff, P. W., Wooster, R., & Farokhzad, O. C. (2017). Cancer nanomedicine:progress, challenges and opportunities. *Nature Reviews. Cancer*, *17*(1), 20–37. DOI: 10.1038/nrc.2016.108 PMID: 27834398

Shi, Y., van der Meel, R., Chen, X., & Lammers, T. (2020). The EPR effect and beyond: Strategies to improve tumor targeting and cancer nanomedicine treatment efficacy. *Theranostics*, *10*(17), 7921–7924. DOI: 10.7150/thno.49577 PMID: 32685029

Shrestha, A., Sadeyen, J. R., & Iqbal, M. (2018). Enhancing protective efficacy of poultry vaccines through targeted delivery of antigens to antigen-presenting cells. *Vaccines*, *6*(4), 75. DOI: 10.3390/vaccines6040075 PMID: 30445683

Sriram, T., Chakraborty, T., & Prasanna, P. M. (2024). Artificial intelligence powered insights into nanotoxicology. *Int. J. Adv. Life Sci. Res*, *7*(2), 68–80. DOI: 10.31632/ijalsr.2024.v07i02.005

Su, S., & Kang, M, P. (2020). Recent advances in nanocarrier-assisted therapeutics delivery systems. *Pharmaceutics*, *12*(9), 837.

Tang, Y., & Wang, G. (2021). NIR light-responsive nanocarriers for controlled release. *Journal of Photochemistry and Photobiology C, Photochemistry Reviews*, *47*, 100420. DOI: 10.1016/j.jphotochemrev.2021.100420

Teixeira, M. C., Carbone, C., Sousa, M. C., Espina, M., Garcia, M. L., Sanchez-Lopez, E., & Souto, E. B. (2020). Nanomedicines for the Delivery of Antimicrobial Peptides (AMPs). *Nanomaterials (Basel, Switzerland)*, *10*(3), 560. DOI: 10.3390/nano10030560 PMID: 32244858

Vo, D.-K., & Trinh, K. T. L. (2025). Advances in Wearable Biosensors for Wound Healing and Infection Monitoring. *Biosensors (Basel)*, *15*(3), 139. DOI: 10.3390/bios15030139 PMID: 40136936

Waghule, T., Dabholkar, N., Gorantla, S., Rapalli, V. K., Saha, R. N., & Singhvi, G. (2021). Quality by design (QbD) in the formulation and optimization of liquid crystalline nanoparticles (LCNPs): A risk based industrial approach. *Biomedicine and Pharmacotherapy*, *141*, 111940. DOI: 10.1016/j.biopha.2021.111940 PMID: 34328089

Wang, C., Yang, Y., Cao, Y., Liu, K., Shi, H., Guo, X., Liu, W., Hao, R., Song, H., & Zhao, R. (2023). Nanocarriers for the delivery of antibiotics into cells against intracellular bacterial infection. *Biomaterials Science*, *11*(2), 432–444. DOI: 10.1039/D2BM01489K PMID: 36503914

Wang, H., Liu, Y., Li, Y., Zhao, Y., & Li, X. (2025, August 28). Ultrasound-Responsive Drug Delivery System for Cancer Therapy. *Macromolecular Bioscience*, *e00082*(11), e00082. Advance online publication. DOI: 10.1002/mabi.202500082 PMID: 40873253

Wang, Y., Huang, R., Feng, S., & Mo, R. (2025). Advances in nanocarriers for targeted drug delivery and controlled drug release. *Chinese Journal of Natural Medicines*, *23*(5), 513–528. DOI: 10.1016/S1875-5364(25)60861-2 PMID: 40383609

Wang, Y., Li, T., Zhou, H., & Liu, C. (2025, May 29). Advancements in Nanocarrier Delivery Systems for Photodynamic Therapy in Lung Cancer. *International Journal of Nanomedicine*, *20*, 6853–6874. DOI: 10.2147/IJN.S521444 PMID: 40458748

World Health Organization. (2024b, May 17). WHO bacterial priority pathogens list, 2024. URL:https://www.who.int/publications/i/item/9789240093461

World Health Organization. (2024a, February 7). Antimicrobial resistance. WHO Fact Sheets. URL:https://www.who.int/news-room/fact-sheets/detail/antimicrobial-resistance

World Health Organization. (2024c, May 14). 2023 Antibacterial agents in clinical and preclinical development: an overview and analysis. URL:https://www.who.int/publications/i/item/9789240094000

Wu, Y., Wang, N., Xiong, P., Wang, R., Deng, J., & Ouyang, D. (2025). Artificial intelligence for drug delivery: Yesterday, today and tomorrow. *Acta Pharmaceutica Sinica. B*. Advance online publication. DOI: 10.1016/j.apsb.2025.09.022

Yang, H., Wang, W., Wang, Y., Ke, L., Wang, L., & Li, R. (2017). Enhanced intracellular delivery and antibacterial efficacy of enrofloxacin-loaded docosanoic acid solid lipid nanoparticles against intracellular Salmonella. *Scientific Reports*, *7*(1), 41104. DOI: 10.1038/srep41104 PMID: 28112240

Yeh, Y. C., Huang, T. H., Yang, S. C., Chen, C. C., & Fang, J. Y. (2020, April 24). Nano-Based Drug Delivery or Targeting to Eradicate Bacteria for Infection Mitigation: A Review of Recent Advances. *Frontiers in Chemistry*, *8*, 286. DOI: 10.3389/fchem.2020.00286 PMID: 32391321

Yin, S. (2025). Artificial Intelligence-Assisted Nanosensors for Clinical Diagnostics: Current Advances and Future Prospects. *Biosensors (Basel)*, *15*(10), 656. DOI: 10.3390/bios15100656 PMID: 41149307

Zeng, Z., Jiang, J., Sun, Y., Li, W., Zheng, D., Lin, H., & Zhao, C. (2025). Infected microenvironment responsive nanoprodrug for the specific therapeutics of Helicobacter pylori clearance. *Biomaterials*, *123415*, 123415. Advance online publication. DOI: 10.1016/j.biomaterials.2025.123415 PMID: 40411983

Zolnik, B. S., González-Fernández, Á., Sadrieh, N., & Dobrovolskaia, M. A. (2010). Minireview: Nanoparticles and the immune system. *Endocrinology*, *151*(2), 458–465. DOI: 10.1210/en.2009-1082 PMID: 20016026

Chapter 8
Nanocarrier Technologies for Targeted Delivery

Jyotishman Tamuli
http://orcid.org/0009-0000-7367-694X
Assam University, Silchar, India

Nirbanya Devi
http://orcid.org/0009-0003-2832-201X
Assam University, Silchar, India

Jharna Tamuli
Tezpur University, Napam, India

Khemnath Patir
http://orcid.org/0000-0002-6708-4590
Assam University, Silchar, India

Sarbani Giri
http://orcid.org/0000-0001-7792-8374
Assam University, Silchar, India

ABSTRACT

Targeted delivery is the delivery of a drug to a particular location of the body to achieve an effective therapeutic response with negligible side effects. Based on the ability of nanoparticles to sustain in different biological environments, they have emerged as a crucial component of the targeted delivery system. By utilizing different nanocarrier technologies, the pharmacokinetics and pharmacodynamics of the selected medication can be enhanced. This chapter highlights the various nanocarrier technologies for targeted delivery of antibiotics, which have been successfully applied for years in the site-specific delivery of drugs. The fundamentals of various

DOI: 10.4018/979-8-3373-5268-8.ch008

nanocarriers, including liposomes, polymers, dendrimers, and biomimetics, have been discussed in detail. The state of current knowledge for various nanocarrier platforms for targeted drug delivery in antibiotic-resistant infection, together with their clinical status, has also been compiled and discussed in this study. Moreover, the sustainable development goals (SDGs) have been thoughtfully incorporated.

1. INTRODUCTION

Antimicrobial resistance (AMR) is a naturally occurring condition that results from mutations in bacterial genes and the involvement of foreign DNA encoding determinants by the horizontal gene transfer process. However, due to the overuse of antibiotics such as fluoroquinolones, tetracyclines, β-lactams, aminoglycosides, etc. in recent times, AMR has become a threat to public safety that demands immediate attention from the scientific community. According to recent WHO statistics, every year, 700,000 people die due to drug-resistant disorders (Watkins, 2022; Aggarwal et al., 2024). Due to overutilization, misemployment, and poor stewardship of antibiotics, antibiotic resistance also affects the gut microbiome, which interferes with immunotherapies and has the potential to cause secondary infections. In order to lower the risk of antibiotic resistance and ensure successful treatment, the objective of antibiotic stewardship is to enhance and monitor the proper use of antibiotics, thereby protecting the overall public health (Hand, 2013). Precision medicine is a paradigm that could address the challenges of antibiotic stewardship. It involves the study of the altered biology of patients using genetic and molecular profiling and guiding them in therapy accordingly. It is a treatment strategy that links current patient data to a suggested course of action through a set of decision rules (Bulman et al., 2022).

Targeted delivery refers to the delivery of a medicine to a specific site of infection for effective treatment with minimal side effects. So, scientists are now focusing on the safe delivery of the drug molecules to the point of action and their controlled release (Wang et al., 2022a). By utilizing different innovative nanocarriers such as liposomes, polymers, dendrimers, and biomimetics, target-specific drug delivery can be attained, which enhances the administration routes and biodistribution of pharmaceuticals with minimal immunogenicity and side effects (Shah et al., 2021). Over the past two decades, nanocarrier technologies have been investigated for the selective transport of drug molecules to targeted cells with minimal proximity to healthy tissues. Nanocarriers have unique characteristics such as tiny size and high surface area-to-volume ratio, which enable them to transport a variety of drugs with excellent loading capacity (Sarode & Mahajan, 2024). By employing these nanocarrier technologies for target-specific drug delivery, different demerits of conventional therapies can be overcome. Nanocarriers can encapsulate different kinds of drugs, enhance the biocompatibility, and carry them efficiently to the site of infection, and concurrently decreasing toxicity (Majumder and Minko, 2021). Nowadays, artificial intelligence (AI) approaches are applied to streamline the formulation of drug delivery systems and their outcomes, bridging the experimental knowledge gaps (Nuhn, 2023). Therefore, AI models may be employed as a suitable alternative in predicting different optimization conditions of the nanocarrier systems required in drug delivery. In this book chapter, AMR towards antibiotics in drug delivery systems has been highlighted. To overcome these issues, different types of nanocarriers, such as liposomes, polymers, dendrimers, and biomimetics, are incorporated in the medicines for targeted delivery. Further, AI applications in the design and optimization of the nanocarriers in drug delivery systems are also included.

2. ANTIBIOTIC RESISTANCE CHALLENGE

Antibiotics are a class of medicines that are used for treating bacterial infections, which kill the bacteria or prevent them from reproducing (Hutchings et al., 2019). Different types of antibiotics, such as macrolides, fluoroquinolones, tetracyclines, β-lactams, aminoglycosides, etc., are prescribed for the treatment of infections caused by bacteria such as *E. coli, Campylobacter spp., Enterococcus spp., Salmonella spp., Listeria spp.,* and *Staphylococcus spp.* However, every year thousands of deaths are reported due to the antibiotic-resistant issues of bacteria (Chmiel et al., 2022). Nowadays, bacterial resistance has become a serious problem, as the commonly used antibiotics are ineffective, affecting public health worldwide. The presence of multi-resistant strains in the environment and the transmission of drug resistance genes between the bacterial species, includes commensal flora or microbiota and pathogenic ones, like foodborne pathogens, favors the spread of multi-resistance among bacteria in humans as well as in animals. Due to the significant rise in multidrug-resistant bacterial strains, the WHO has acknowledged this occurrence as a major global health threat (Palme et al., 2008; Aggarwal et al., 2024). Researchers have reported the mechanism of bacterial resistance towards antibiotics. Bacteria acquire antibiotic resistance by removing the antibiotics from the cell, modifying the cell components, enzymatic modifications of the antibiotics, and generating alternative pathways of the metabolites (Van Hoek et al., 2011). Therefore, research on prevention and development of methods for the identification, control, and treatment of bacterial resistance is essential in the field of medical science. Nowadays, nanocarriers are found to be beneficial in transporting the antibiotic drug molecules to the bacterial-infected area, leading to effective treatment of the disease (Xu et al.,

2025). The details of different nanocarriers used in drug delivery for the treatment of diseases are discussed below.

3. LIPOSOMES

Liposomes are tiny spherical artificial vesicles having particle sizes ranging from 30 nm to several micrometers. Due to their small size, biocompatibility, and biodegradability properties, liposomes are used as nanocarriers in site-specific antibiotic treatment, gene therapy, and cancer therapy (Akbarzadeh et al., 2013; Allen & Cullis, 2013). For example, a combination of colistin and ciprofloxacin liposomal formulations was found to be effective for treating the *Pseudomonas aeruginosa* bacteria. Colistin (8mg/L) and ciprofloxacin (4 mg/L and 8mg/L) could achieve moderate antibacterial activities against *Pseudomonas aeruginosa* bacteria that results in rapid regrowth of the bacteria. However, the combination solutions of colistin (8mg/L) and ciprofloxacin (8mg/L) destroy *Pseudomonas aeruginosa* bacteria from the infected mouse lung cells (Wang et al., 2018).

3.1 Structure of Liposomes

Liposome-based nanocarriers are generally spherical, self-assembled vesicular nanostructures and consist of one or more phospholipid bilayers surrounding an internal aqueous core. It is often stabilized by cholesterol to improve the membrane rigidity and fluidity (Allen & Cullis, 2013). **Figure 1** illustrates the various components of a liposome-based nanocarrier, including the core structure, integrated cholesterol, and the phospholipid bilayer with hydrophilic heads and hydrophobic fatty tails. Various components of the liposome-based nanocarrier are discussed below:

- ***Core:*** It serves as a reservoir for hydrophilic drugs and protects them from enzymatic degradation. Hydrophobic drugs incorporate into the lipid bilayer, which ensures loading of a wide range of therapeutics in the liposome (Nsairat et al., 2022).
- ***Cholesterol:*** It is integrated into the lipid bilayer to improve the stability of the liposome. It integrates into the phospholipid bilayer and reduces the membrane fluidity and permeability, which results in the prevention of drug leakage (Allen & Cullis, 2013).
- ***Phospholipid bilayer:*** A phospholipid molecule consists of a hydrophilic head and hydrophobic fatty tails. They spontaneously arrange into bilayers when dispersed in an aqueous medium. The tails orient towards the inward direction, while the heads face towards the surrounding water. This unique

arrangement results in a closed structure that is capable of encapsulating both water- and lipid-soluble compounds (Bangham et al., 1965).

Figure 1. Structure of liposome-based nanocarrier with its functional components such as core, phospholipid bilayer, and cholesterol

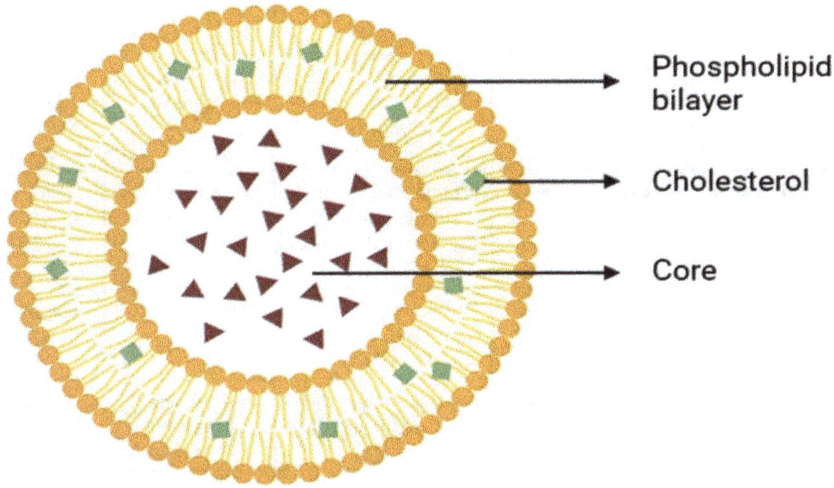

(Created with BioRender.com)

3.2 Classification of Liposome-Based Nanocarrier for Antibiotic Delivery

Liposome-based nanocarriers can be classified based on several parameters, including size and lamellarity, lipid composition, and preparation methods. Each type possesses a distinctive structure and properties, which are suited for specific biomedical applications. **Table 1** comprises the classification of liposome-based nanocarriers based on size, lamellarity, composition, and surface modification. Additionally, their subtypes, characteristic features and drug delivery applications have been included.

Table 1. Classification of Liposome-Based Nanocarriers for Antibiotic Delivery.

Category	Subtypes	Characteristics	Applications	References
Classification based on Size & Lamellarity	Small Unilamellar Vesicles (SUVs)	Diameter 20–100 nm, single bilayer	Diagnostic imaging	Pattni et al., 2015; Akbarzadeh et al., 2013
	Large Unilamellar Vesicles (LUVs)	Diameter >100 nm, single bilayer	Encapsulation of hydrophilic drugs	Pattni et al., 2015; Akbarzadeh et al., 2013;
	Multilamellar Vesicles (MLVs)	500–5000 nm, multiple concentric bilayers	Slow-release vaccines	Pattni et al., 2015; Akbarzadeh et al., 2013
Classification based on Composition & Surface Modification	Conventional Liposomes	Natural phospholipids and cholesterol, no surface modification	Basic drug carriers	Pattni et al., 2015; Torchilin, 2005
	Sterically Stabilized (Stealth)	PEGylated evade RES and have long circulation	Cancer therapy	Pattni et al., 2015; Torchilin, 2005
	Immunoliposomes	Surface-conjugated with antibodies or ligands	Cancer therapy, diagnostic imaging	Pattni et al., 2015; Torchilin, 2005
	Cationic Liposomes	Positively charged, interacts with nucleic acids and membranes	Gene therapy	Pattni et al., 2015; Torchilin, 2005
	pH/Stimuli-Responsive Liposomes	Release triggered by pH, enzymes, temperature, or redox	Tumor-targeted therapy	Pattni et al., 2015; Torchilin, 2005

3.3 Mechanisms of Action of Liposome-Based Antibiotic Delivery

Liposome-based nanocarriers can deliver antibiotics to the infected site through several key mechanisms. The types of mechanisms are discussed below:

I. Passive targeting:

In this type of targeted delivery, nanocarriers are primarily dependent on the enhanced permeability and retention (EPR) effect. It occurs in the infection site, where more permeable blood vessels and decreased lymphatic drainage are present. As a result, nanocarriers circulating in the blood tend to accumulate in the infected site, allowing for site-specific drug delivery (Allen & Cullis, 2013).

II. Active targeting:

In an active targeting mechanism, modification takes place on the liposome surface through particular ligands, including antibodies or peptides, which can identify and attach to the specific markers present on the infected cell. Through the receptor-mediated mechanisms, liposome-based nanocarriers can enter the cell, consequently increasing drug deposition at the targeted site and enhancing the treatment (Sercombe et al., 2015).

III. Stimuli-responsive release:

Stimuli-responsive liposome-based nanocarriers are designed to release their drug content in response to certain conditions found in the infected tissues. Whenever they sense changes in pH, the presence of bacterial enzymes, or other biochemical signals, they immediately release the antibiotic with pinpoint precision. This mechanism assists in overcoming the challenges brought on by bacterial biofilms and drug resistance (Bozzuto & Molinari, 2015).

Despite their clinical potential, liposomal nanocarriers have faced several limitations in targeted drug delivery due to their physicochemical instability, polydispersity, and immunostimulation, which restrict their application and therapeutic efficacy in precision medicine. Additionally, challenges in large-scale manufacturing, reproducibility, and high production costs further limited their translational potential (Liu et al., 2022; Allen & Cullis, 2013).

4. POLYMERS

Polymers are macromolecules widely utilized in various therapeutic areas, such as infection management, immune response modulation, and regenerative medicine, due to their physicochemical flexibility, solubility, biodegradability, and responsiveness to different physiological stimuli. To serve as drug carriers, polymers can transform their architectures and form different structures, including nanoparticles, micelles, and nanogels (Duncan, 2003). Different physicochemical characteristics of the polymer, including molecular weight, hydrophobicity, and degradability, also determine their loading efficiency, biocompatibility, and kinetics during release. Based on the properties of the polymers, drugs can be encapsulated within the matrix, absorbed onto the surface of the core structure, or conjugated to the functional groups (Akbarzadeh et al., 2013). In polymer-based nanocarrier technology, the antibiotics are enclosed by the polymers, or they can bind with polymers. In both types of systems, it helps to facilitate sustained, controlled, and targeted release of the antibiotics at targeted locations in the body (Birk et al., 2021). Several biodegradable polymers, such as PLGA, PLA, PEG-based systems, and chitosan, are being used to carry different antibiotics, including vancomycin, ciprofloxacin, and gentamicin (Danhier et al., 2012). With increasing cases of antimicrobial resistance, polymers with some key unique features become important for the treatment of multidrug-resistant disease (Pelgrift & Friedman, 2013). In some polymer-based nanocarrier systems, the polymers are stimuli-responsive, which can release antibiotics in response to infection-specific triggers (Baptista et al., 2021).

4.1 Structure of Polymers

The polymer-based nanocarrier systems employ a variety of polymers, such as natural, synthetic, or semi-synthetic ones. The core structure of nanocarriers serves as the main compartment for drug encapsulation, while the polymer part can be arranged in different forms, like micelles, as depicted in **Figure 2**. Primarily, their structures depend on the arrangement of the polymeric matrix, which determines their effectiveness in targeted delivery of pharmaceuticals (Kumari et al., 2010). The architecture of the polymer may be arranged in different types, such as linear, branched, or cross-linked, which eventually can regulate how the drugs will load, release, or penetrate into the targeted site (Zielinska et al., 2020). The hydrophilic drugs have the tendency to bind to the outer surface or be entrapped within the polymeric micelles; on the other hand, the polymeric core structure incorporates the hydrophobic drugs (Ahmad et al., 2022).

Figure 2. Structure of polymer-based nanocarrier with functional components like polymeric micelles and core structure

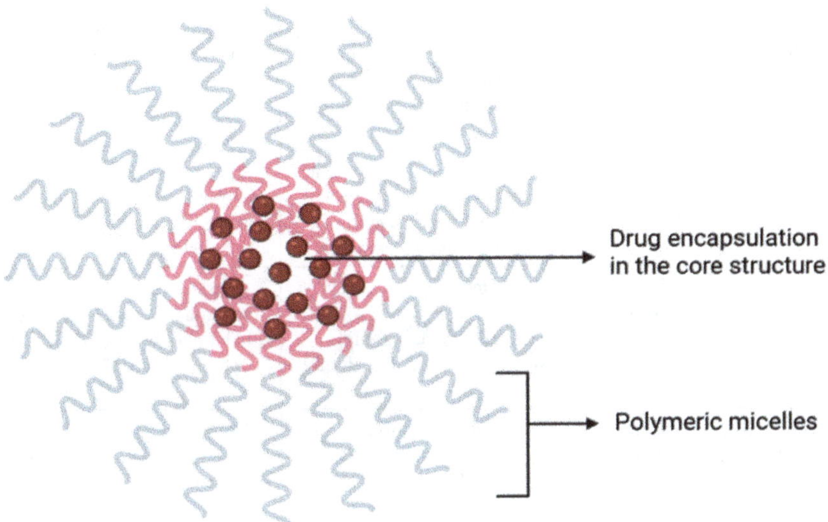

(Created with BioRender.com)

4.2 Classification of Polymer-Based Nanocarriers for Antibiotic Delivery

Based on the structural composition, pattern of organization, and functional properties, polymer-based nanocarriers are categorized into several types. **Table 2** encompasses the classification of nanocarriers based on their structural organization, polymer source, drug-polymer interaction, and stimulus responsiveness. Furthermore, their different subtypes, characteristics, and applications in drug delivery have also been discussed thoroughly.

Table 2. Classification of Polymer-Based Nanocarriers for Antibiotic Delivery.

Category	Subtypes	Characteristics	Applications	References
Based on Structural Organization	Polymeric nanoparticles (PNPs)	Solid colloidal systems (10–1000 nm) can be nanospheres or nanocapsules and are suitable for both hydrophilic and hydrophobic drugs.	Ciprofloxacin- or rifampicin-loaded PLGA nanoparticles for sustained release.	Akbarzadeh et al., 2013
	Nanogels / Hydrogels	Cross-linked 3D polymeric networks; high water content; responsive to pH, temperature, or enzymes.	pH-responsive chitosan nanogels for site-specific antibiotic release.	Thambi et al., 2016
	Polymeric nanofibers	One-dimensional fibers fabricated via electrospinning; high surface area and porosity for sustained drug release.	Tetracycline- or gentamicin-loaded PLGA nanofibers for wound healing.	Sill & Recum, 2008
Based on Polymer Source	Natural polymers	Biocompatible, biodegradable, and mucoadhesive; may have antimicrobial activity.	Chitosan, alginate, and gelatin nanoparticles for localized antibiotic delivery.	George & Abraham, 2006
	Synthetic polymers	Chemically defined, tunable degradation rate and release kinetics; high mechanical stability.	PLGA, PEG, and PCL-based nanocarriers for ciprofloxacin and ampicillin.	George & Abraham, 2006
	Semi-synthetic polymers	Modified natural polymers with enhanced solubility and stability.	PEGylated chitosan or carboxymethyl cellulose nanoparticles for antibiotic encapsulation.	Prabaharan, 2011
Based on Drug-Polymer Interaction	Encapsulation	A drug trapped inside a polymeric matrix or vesicular core protects the drug from degradation.	Rifampicin encapsulated in PLGA nanoparticles; ampicillin in chitosan nanocapsules.	Prabaharan, 2011
	Adsorption / Surface loading	Drugs adsorbed or ionically bound to the nanoparticle surface allow fast release.	Gentamicin adsorbed onto chitosan nanoparticles.	Grenha et al., 2007
	Covalent conjugation	Drug covalently linked to polymer backbone (prodrug concept); offers prolonged release.	Penicillin–PEG conjugates; amoxicillin–dextran conjugates.	Duncan, 2003
Based on Stimulus Responsiveness	pH-responsive	Release triggered in acidic microenvironment (infection sites).	Chitosan-based nanogels for infection-triggered drug release.	Rizwan et al., 2017

continued on following page

Table 2. Continued

Category	Subtypes	Characteristics	Applications	References
	Temperature-responsive	Release occurs at elevated temperature (fever, inflammation).	PNIPAM-based nanogels loaded with antibiotics.	Rizwan et al., 2017
	Enzyme-responsive	Degrades in the presence of bacterial or host enzymes.	Dextranase-sensitive dextran nanocarriers for bacterial infections.	Rizwan et al., 2017
	Redox-responsive	Controlled release due to redox potential differences between healthy and infected tissues.	Thiolated polymer nanoparticles for intracellular antibiotic delivery.	Meng et al., 2009

4.3 Mechanisms of Action of Polymer-Based Antibiotic Delivery

Polymer-based antibiotic delivery is a complex process that involves different mechanisms of action. The detailed mechanisms are explained below:

I. Controlled Release

With the help of polymer-based drug delivery, sustained and regulated release of antibiotics can be obtained. The antibiotics are incorporated into the polymer matrix for prolonged release in the targeted area. In case of stimuli-responsive release of antibiotics, they are covalently linked to polymer chains or attached to the surfaces via hydrophobic or electrostatic interactions (Danhier et al., 2012).

II. Protection and Targeted Delivery

Targeted nanocarrier technology protects the antibiotics from premature degradation and guides them to the site of infection for smooth release. The process of specific targeting can occur by means of an active or passive mechanism of action. In an active mechanism, the ligands and antibodies are mediated through some selective bacterial recognition on the polymer surface. In the passive mechanism, site-specific targeting occurs via enhanced permeability and retention. (Torchilin, 2014).

III. Adhesion and Stimuli Responsiveness

In an adhesion-responsive process, polymers are combined with antibiotics for localization at the specific infection site for a long duration, enhancing the therapeutic activity. In stimuli-responsive processes, polymers release antibiotics in reaction to infection-specific triggers, such as altered pH, temperature shifts, or enzymatic activity (Li et al., 2021).

IV. Combination Therapy

In combination therapy, antibiotics can be co-delivered with polymeric nanocarriers and other therapeutic agents, such as resistance-modulating compounds or anti-biofilm molecules. This results in enhanced therapeutic efficiency and offers an efficient way to combat multidrug-resistant infections (Pelgrift & Friedman, 2013).

Polymeric nanocarriers offer significant advantages for targeted antibiotic delivery. However, their clinical translation is hindered by various challenges such as batch-to-batch variability, complex synthesis, and limited scalability. Issues related to drug loading efficiency and unpredictable in vivo degradation and excretion processes have affected its therapeutic applications in targeted drug delivery (Zielinska et al., 2020; Baptista et al., 2021; Ahmad et al., 2022).

5. DENDRIMERS

Dendrimers are composed of highly branched, globular, three-dimensional nanoscale structures designed through a controlled and sequential process. They are water-soluble, biocompatible, and structurally unique, which allows for both covalent and non-covalent drug encapsulation (Kalomiraki et al., 2015). Due to their well-defined globular structure, they have been widely explored for drug delivery, gene therapy, and cancer therapy since the late 20th century (Sarode & Mahajan, 2024). Dendrimer-based nanocarrier systems utilize a passive targeting mechanism through the enhanced permeability and retention (EPR) to improve the bioavailability and prevent the enclosed therapeutics from degradation. This results in a higher concentration of drugs at the targeted area, improving the therapeutic efficacy and lowering the systemic toxicity or side effects (Wang et al., 2016). Dendrimers show two types of interactions with drug molecules: one is covalent bonding, which is important for their stability and biological activity, and the other is non-covalent interaction, which occurs between charged therapeutic agents like DNA, RNA, or siRNA and the functional groups of dendrimers present on the surface. The non-covalent interaction also helps in site-specific drug delivery (Calabretta et al., 2015).

5.1 Structure of Dendrimers

Dendrimers are a type of synthetic polymer that consists of many branched polymers, forming a tree-shaped structure with a three-dimensional globular shape.

Due to the presence of different functional groups on the surface, they exhibit a polycationic nature, which imparts intrinsic antimicrobial properties. This surface property can be tuned to regulate the half-life, biocompatibility, and biodistribution of dendrimer-based nanocarriers for targeted drug delivery (Svenson & Tomalia, 2012).

Structurally, dendrimers consist of three components, a core, branches, and terminal functional groups, as shown in **Figure 3**. The structure of the dendrimeric nanocarrier with its various components for targeted drug delivery has been thoroughly discussed below:

- ***Core:*** The core of a dendrimer is its central part, which can be a single atom or a multifunctional molecule such as ammonia, ethylenediamine, or polyols. The core part is known as the first-generation dendrimer. Depending upon the core structure, the dendrimers have different structures (Gillies & Fréchet, 2005).
- ***Branches (Generations):*** The formation of branches from the core structure results in the next generation of dendrimers, and the process continues likewise. As the number of generations or branches increases, so does the size and number of reactive sites, **Figure 3**. The internal cavities of the dendrimers also expand with each generation, which acts as a host for drugs or antibiotics (Mintzer et al., 2011).
- ***Terminal Functional Groups (Surface Groups):*** The terminal branches of dendrimers have different reactive functional groups, which participate in various intermolecular interactions, **Figure 3**. These groups also determine the solubility, biocompatibility, stability, and drug encapsulation efficiency of dendrimer-based nanocarriers. The large number of peripheral functional sites provides exceptional versatility for targeted drug delivery (Patri et al., 2002)

Figure 3. Structure of dendrimer-based nanocarrier with functional components like core, branches, and terminal functional groups.

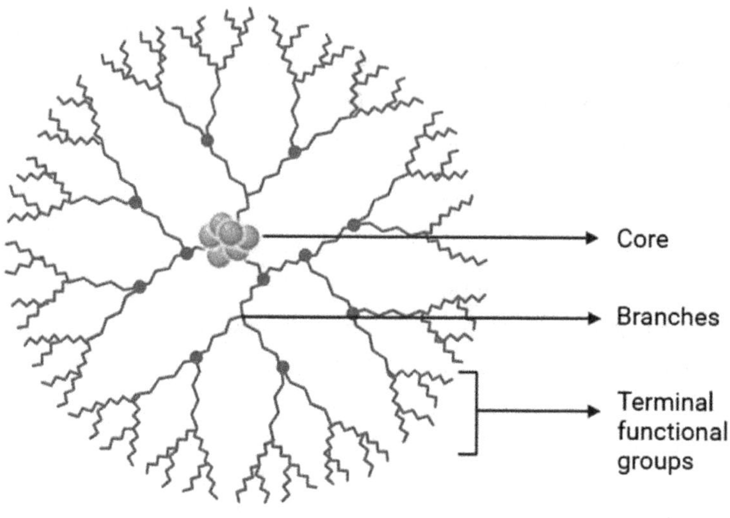

(Created with BioRender.com)

5.2 Classification of Dendrimer-Based Nanocarriers for Antibiotic Delivery

Based on generation, surface modification, drug conjugation, and stimulus response, dendrimer-based nanocarriers can be classified into different categories. **Table 3** comprises different classifications of dendrimer-based nanocarriers, subtypes, characteristics, and applications in antibiotic delivery. Based on differences in generation, surface modification, and stimuli responsiveness, dendrimers can be divided into different categories as presented in **Table 3**.

Table 3. Classification of Dendrimer-Based Nanocarriers for Antibiotic Delivery.

Category	Subtypes	Characteristics	Applications	References
Generation-based Dendrimers	PAMAM, PPI, PLL	Highly branched, monodisperse macromolecules with multiple terminal groups; generation-dependent size and surface charge	Enhanced solubility and stability of antibiotics such as ciprofloxacin and amoxicillin	Kesharwani et al., 2014
Surface-Modified Dendrimers	PEGylated, Acetylated, Carboxylated	Surface modification reduces toxicity and improves biocompatibility, allowing targeted and sustained drug release	PEGylated PAMAM for gentamicin delivery with reduced cytotoxicity	Kesharwani et al., 2014
Stimuli-Responsive Dendrimers	pH-, enzyme-, or redox-responsive	Drug release triggered by environmental changes at infection sites enhances site-specific therapy	pH-sensitive dendrimers for vancomycin delivery in infected tissues	Wang et al., 2016
Hybrid Dendrimer Systems	Dendrimer–liposome or dendrimer–polymer hybrids	Combines properties of both carriers; improved drug loading and permeability	Dendrimer-liposome hybrid delivering azithromycin	Kesharwani et al., 2014

5.3 Mechanisms of Action of Dendrimer-Based Antibiotic Delivery

Dendrimer-based nanocarrier systems work through two different mechanisms. They can encapsulate the therapeutics in their inner cavities, or else they have the potential to bind them to their outer surface with their distinguished features of biofilm penetration or multifunctional adaptability (Chowdhury et al., 2016).

I. ***Encapsulation***

Through non-covalent interactions, such as hydrogen bonding, hydrophobic forces, or electrostatic interactions, dendrimers can entrap the antibiotics within their internal cavities. This encapsulation process enhances the solubility of the drug and protects it from enzymatic degradation (Svenson & Tomalia, 2012).

II. *Conjugation*

By utilizing the covalent and electrostatic interactions, the dendrimer-based nanocarrier framework can directly bind with the drug molecule. This type of conjugation facilitates the stimuli-responsive, target-specific transportation of drugs. The surface-functionalized groups release the conjugated drug in response to alterations in external stimuli like pH, enzymatic activity, or temperature (Abbasi et al., 2014).

III. ***Biofilm Penetration and Membrane Interaction***

With the help of branched and multivalent structures, dendrimers can easily penetrate the bacterial biofilms, which improves the antibiotic therapy. Additionally, the fusogenic properties allow them to disrupt the bacterial membrane. Collectively, all these increase the antibiotic uptake and overall efficiency (Mintzer & Grinstaff, 2011).

IV. ***Multifunctionality***

Dendrimers can be further engineered to transport ligands and peptides along with the therapeutics. This property makes the dendrimer highly versatile for antibiotic resistance (Mintzer & Grinstaff, 2011).

Dendrimeric nanocarriers offer precise structural control for antibiotic delivery. Owing to their high cytotoxicity, low target specificity, and accumulation in non-target tissues, their clinical translation is limited. Additionally, issues related to industrial synthesis and high production costs restricted their application in targeted delivery (Sarode & Mahajan, 2024; Svenson & Tomalia, 2012; Calabretta et al., 2015).

6. BIOMIMETICS

Biomimetic nanocarriers refer to a nanoscale drug delivery system that mimics biological components to improve therapeutic outcomes. Biomimetic nano drug delivery systems (BNDDS) play an important role in bioscience by overcoming the demerits of conventional nanotechnology (Perini et al., 2024). They combine drug-loaded nanostructured core particles with biomimetic features that mimic the properties, structures, and functions of biological entities such as cell membranes, extracellular vesicles, or even entire cells. Biomimetic coatings or carriers enhance biocompatibility and stability by extending drug circulation time, preventing degradation of therapeutic agents, and minimizing immune responses, effectively overcoming key challenges in antibiotic treatment (Sheikhpour, 2017). The biomimetic nanocarriers respond to external stimuli like pH and enzymatic activity and are deposited at the infected site while reducing off-target toxicity. They also penetrate biofilms and help in reaching intracellular bacteria. For example, it increases the half-life of vancomycin, and exosome-based gentamicin drug delivery systems effectively get rid of *Pseudomonas aeruginosa* biofilms by penetrating them (Zhang et al., 2023).

6.1 Structure of Biomimetics

The biomimetic-based nanocarrier system mainly consists of two parts, the core structure, and the biological interface, as illustrated in **Figure 4**. Both the core structure and the biological interface of the biomimetic nanocarrier are explained below:

I. Core Structure

The core structure is designed with nanocarriers, such as exosomes, erythrocytes, liposomes, polymers, and dendrimers, which are effective for circulation and help in depositing the drug molecules at the infected tissues by using different mechanisms, such as the enhanced permeability and retention effect (EPR). The core structure also provides mechanical stability, enhances controlled drug release, and improves the biocompatibility of the drug delivery systems, **Figure 4**. Acting as a versatile platform, the core serves as a carrier for different therapeutics (Sheikhpour, 2017).

II. Biological Interface

The biological interface of biomimetic nanocarriers serves as the outer functional layer that mimics natural cell surfaces, thereby enhancing biocompatibility, circulation time, and immune evasion. This interface is typically composed of cell-derived membranes, such as red blood cells, platelets, stem cells, and cancer cells, or surface-functionalized biomolecules like proteins, peptides, polysaccharides, and antibodies, **Figure 4**. Such coatings retain the structural and functional components of their biological source, including receptors, antigens, and adhesion molecules, which enable specific interactions with target tissues and pathogens. By replicating the surface chemistry and signaling cues of natural cells, the biomimetic shell allows nanocarriers to avoid recognition by the immune system, achieve prolonged systemic circulation, and perform targeted delivery to diseased sites (Ijaz et al., 2024).

Figure 4. Structure of biomimetic-based nanocarrier with functional groups such as core structure and biological interface

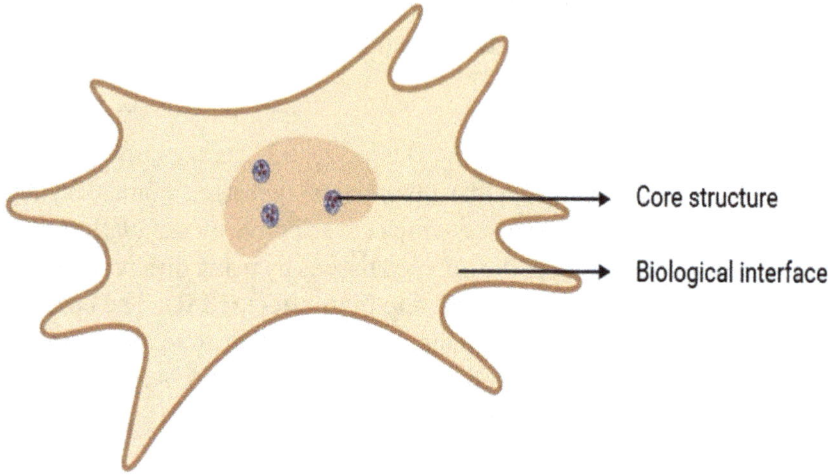

(Created with BioRender.com)

6.2 Classification of Biomimetics-Based Nanocarriers for Antibiotic Delivery

Table 4 comprises various categories of biomimetic-based nanocarriers, based on differences in membrane-coated, extracellular-vesicle, and functionalized systems. Their different subtypes and characteristic features have also been summarized in **Table 4**.

Table 4. Classification of Biomimetics-Based Nanocarriers for Antibiotic Delivery.

Category	Subtypes	Characteristics	Applications	References
Cell Membrane-Coated Nanocarriers	Erythrocyte (RBC) membrane-coated NPs	Prolonged circulation; immune evasion and protection of antibiotics from degradation	Vancomycin-loaded RBC-coated NPs for MRSA infections	Meyer et al., 2015
	Leukocyte membrane-coated NPs	Active homing to inflamed/infected tissues and immune modulation	Macrophage-membrane NPs delivering rifampicin for tuberculosis	Perini et al., 2024
Extracellular Vesicle-Based Systems	Exosomes	High biocompatibility cross biological barriers and deliver antibiotics directly to pathogens	Exosome-mediated ciprofloxacin delivery	Behrouzi et al., 2024
Protein- and Ligand-Functionalized Systems	Aptamer-functionalized NPs	Specific binding to bacterial surface proteins and improved targeting	Aptamer-modified NPs for S. aureus targeting	Chramiec et al., 2019
	Peptide-functionalized NPs	Improve bacterial membrane binding and penetration	Polymyxin-coated NPs against multidrug-resistant bacteria	Goes & Fuhrmann, 2018
Hybrid Biomimetic Systems	Liposome–cell membrane hybrids	Combine the stability of synthetic vesicles with the bio-functionality of membranes	Hybrid RBC-liposome carriers for antibiotic delivery	Perini et al., 2024
	Polymer–biomolecule hybrids	Facilitated receptor-mediated targeting; regulated drug release	Folic acid-conjugated polymeric NPs for localized antibiotic therapy	Boppana et al., 2024

6.3 Mechanisms of Action of Biomimetic-based Antibiotic Delivery

Biomimetic nanocarriers are designed for targeted antibiotic delivery through mimicking the biological systems to improve therapeutic efficiency. The mechanisms can be summarized as follows:

I. Camouflage and Immune Evasion

Biomimetic nanocarriers are coated with natural cell membranes (e.g., red blood cells, platelets, bacteria, or immune cells), which helps them evade detection and clearance by the immune system. This increases circulation time, allowing the nanocarrier to persist long enough in the bloodstream to reach infection sites (Perini et al., 2024).

II. Targeted Delivery

These nanocarriers exploit both passive and active targeting mechanisms:

- Passive targeting leverages the enhanced permeability and retention (EPR) effect to accumulate in infected tissues or biofilms where vascular permeability is higher.
- Active targeting involves surface functionalization with ligands or antibodies that specifically bind to receptors expressed on bacterial cells or infected tissues, ensuring precise localization (Yang et al., 2019).

III. Responsive Drug Release

Biomimetic carriers are often engineered to release antibiotics in response to specific stimuli found in infection microenvironments, such as

- pH variations (acidic conditions at infection sites)
- Redox environment changes
- Presence of bacterial enzymes or toxins, enabling controlled and localized drug release to maximize efficacy and reduce side effects (Yang et al., 2019).

IV. Enhanced Penetration and Uptake

The biological membrane coating enhances penetration through biological barriers and biofilms. The nanocarriers can also fuse with bacterial membranes or host cells harboring intracellular bacteria, enabling intracellular delivery of antibiotics (Wang et al., 2023).

V. Reduction of Resistance and Side Effects

By delivering antibiotics directly to targeted sites and maintaining therapeutic concentrations locally, biomimetic nanocarriers reduce systemic exposure, toxicity, and antibiotic resistance (Sushnitha et al., 2020).

Biomimetic nanocarriers for antibiotic delivery face significant challenges due to their complex and costly fabrication processes, particularly in the isolation and purification of cell membranes or extracellular vesicles, which limit scalability and reproducibility. Additionally, relatively low structural integrity and drug loading capacity further compromise its uses in precision medicine (Sheikhpour, 2017). Zhang et al., 2023).

7. CRITERIA FOR SELECTING OPTIMAL CARRIER PLATFORM

A wide range of physicochemical, biological, and practical factors needs to be considered carefully while choosing a nanocarrier in order to ensure effective and reliable targeted delivery of drugs. Some of the important requirements are as follows:

Biocompatibility and Biodegradability: The nanocarrier has to be non-immunogenic, non-toxic, and biodegradable. In general, liposomes and biomimetic carriers are naturally biodegradable, whereas polymer- and dendrimer-based nano-carrier systems can be designed for biocompatibility by surface modifications or selecting biodegradable monomers (Shah et al., 2021).

Drug Loading Capacity: To be an optimal nanocarrier platform, the system could effectively encapsulate or bind to an adequate portion of the therapeutics. Broadly, liposomes have the potential to effectively encapsulate hydrophobic as well as hydrophilic therapeutics. With the help of an adequate number of functional groups attached, dendrimers and polymers offer significantly higher drug loading potential (Sarode & Mahajan, 2024).

Controlled and Targeted Release: The rate, time frame, and site of the entrapped drug release should be accurately controlled by the nanocarrier system. With the help of stimuli-responsive polymers and dendrimers, ligand-functionalized liposomes, and biomimetic nanocarrier systems, controlled and targeted administration of antibiotics has been obtained (Shah et al., 2021).

Physicochemical Stability: The structural and chemical stability of the therapeutics is highly important during the whole process of targeted delivery, starting from their formulation and storage till their site-specific release. Usually, polymers and dendrimers are chemically stable by nature, whereas liposomes often need lyophilization or stabilizers in order to keep their structure intact (Danaei et al., 2018).

Surface Functionalization: By changing the surface functional groups of the nanocarriers, it is possible to improve the overall targeting efficiency and limit the immune recognition process. With the implementation of biomimetic coating or ligand conjugation, surface functionalization can be accomplished (Shah et al., 2021).

Particle Size and Surface Charge: Generally, the size of the nanocarrier systems should be between 50 and 200 nm for smooth release and cellular uptake. Liposomes and biomimetic nanocarrier systems don't need any further modifications, while polymers and dendrimers require accurate customization for sizes and charges to increase their stability and biodistribution (Danaei et al., 2018).

Pharmacokinetic and Biodistribution Profile: The selected nanocarrier systems should increase the drug circulation time, which ultimately results in a consistent and prolonged effect of the therapeutics at the targeted site. This also lowers the widespread use of the drug. Till now, polymeric micelles and PEGylated liposomal systems have achieved these features (Allen & Cullis, 2013).

Scalability and Reproducibility: For clinical translation, the synthesis procedure should be easily scalable, affordable, and highly reproducible. Typically, polymer- and dendrimer-based nanocarrier systems offer higher reproducibility, while biomimetic systems need some intricate preparatory steps (Shah et al., 2021).

Regulatory and Safety Considerations: The nanocarrier material must adhere to safety regulations and have an established toxicological profile. Liposomal carriers have FDA approval; however, polymer- and dendrimer-based systems are under evaluation. Biomimetic carriers also have an emerging regulatory framework (Allen & Cullis, 2013; Anselmo & Mitragotri, 2016).

8. COMPARATIVE ANALYSIS OF DIFFERENT NANOCARRIER PLATFORMS

A systematic and comprehensive comparative analysis of various nanocarrier platforms is essential to differentiate their efficiency in drug delivery **Table 5**. Liposomes, polymeric nanoparticles, dendrimers, and biomimetic nanocarriers exhibit significant differences in physicochemical properties, drug loading capacity, stability, biocompatibility, drug release behavior, and targeting efficiency (Allen and Cullis, 2013). Understanding these variations is particularly important in antibiotic delivery, where overcoming antimicrobial resistance, enhancing bioavailability, and achieving site-specific targeting remain critical challenges. **Table 5** comprises the comparative analysis of previously discussed nanocarriers. Several advantages, limitations, and their relevance in targeted drug delivery system has been summarized in **Table 5**.

Table 5. Comparative analysis of different nanocarrier systems

Types of nanocarriers	Advantages	Limitations	Relevance in drug delivery	References
Liposomes	Greater solubility Increased half-life Resistance against chemotherapeutics	Lower stability Polydispersity Immunostimulation	Effective against biofilms, enhances drug penetration	Liu et al., 2022; Allen & Cullis, 2013
Polymers	Increases the solubility of drugs Reduces the side effects of drugs Controlled and modifiable properties	Production is expensive Hard to remove from the body Handling and fabrication are difficult.	Enables sustained antibiotic release and improved pharmacokinetics	Zielinska et al., 2020; Baptista et al., 2021; Ahmad et al., 2022
Dendrimers	Higher drug loading capacity Easily excreted out from the body High biocompatibility and biodistribution	Higher toxicity Complex synthesis process Lower target specificity	Useful for membrane targeting and intracellular antibiotic delivery	Sarode & Mahajan, 2024; Svenson & Tomalia, 2012; Calabretta et al., 2015
Biomimetics	High resemblance to biomolecules Mimic natural cellular components Minimal side effects	Lower structural integrity High cost and manufacturing complexity Lower drug loading efficiency	Superior targeting of infection sites and reduced off-target effects	Sheikhpour, 2017; Zhang et al., 2023

9. AI-ASSISTED FORMULATION DESIGN AND OPTIMIZATION

Recent advances in AI have transformed the design and optimization of nanocarriers for targeted drug delivery. Traditional approaches are often time-consuming and resource-intensive. AI-driven modeling addresses these limitations by enabling rapid prediction of physicochemical properties, formulation stability, drug-loading efficiency, and toxicity profiles based on different experimental datasets (Alshawwa et al., 2022; Prabhu & Selvakumar, 2025). AI involves the utilization of machine learning or its subset, such as deep learning and natural language processing. The learning processes are divided into supervised, unsupervised, and reinforcement learning (Rao et al., 2024). In supervised learning (SL), the model is used to predict new data from a labeled data set input, while in unsupervised learning (USL), the input data is unlabeled, and it identifies the hidden patterns within it (Greener et al., 2022). On the other hand, reinforcement learning (RL) is the teaching of an agent to make decisions by communicating with its environment and receiving feedback in the form of rewards or penalties (Rao et al., 2024). SL algorithms analyze biological data, such as omics data, to find the targets and patterns for studying the growth of diseases. USL encompasses clustering and dimensionality reduction of intricate data for enhanced exploration. Clustering identifies the patient's data, gene profiles, and chemical structure (Kant et al., 2025). RL assists in modifying the molecular structure and dose tuning to maximize the medicinal effectiveness and minimize the toxicity (Dalbanjan et al., 2025). Different types of algorithms used for constructing models under SL are linear regression, logistic regression, random forests (RF), support vector machines (SVM), k-nearest neighbours (KNNs), neural networks (NNs), such as artificial neural networks (ANNs), deep neural networks (DNNs), conventional neural networks (CNNs), recursive neural networks (RNNs), gradient boosting machines (GBMs), long short-term memory networks (LSTMs), and Bayesian statistics. On the other hand, principal component analysis (PCA), vibrational autoencoders (VAEs), clustering algorithms, generative adversarial networks (GANs), and hierarchical clustering fall under USL (Dara et al., 2022). For example, Han et al. (2023) explored ML for predicting the liposome properties. Various ML algorithms like LGBM, RF, SVR, and PLS were evaluated to predict the essential characteristics of liposomes, such as size, PDI, zeta potential, and encapsulation capability. The analysis of the results shows that drug molecules with logS [-3, -6], molecular complexity [500, 1000], and XLogP3 have higher affinity for liposome encapsulation. Further, Naproxen (NAP) and Palmatine HCl (PAL) were used as insoluble and water-soluble molecules in the liposome formulations to validate the predictive ability. The experimental and predicted values were found to be consistent, indicating the effectiveness of the ML models. ANNs successfully model the library of polymeric nanoparticles, such as PLGA and PVA, with particle sizes of 70-400 nm. The percentage differences for the particle prediction were 2%, 4%, and 6% for the training, validation, and testing data. The polymer surface activity was found to have the highest effect on the particle size, followed by viscosity and hydrophobicity (Youshia et al., 2017). Wang and co-workers (2022) developed ML to improve the delivery of mRNA vaccines with lipid nanoparticles (LNPs). They have used ML algorithms to predict the different parameters, such as cholesterol, desaturated phosphatidylcholine, structure, fraction of ionizable lipid, antigen protein, and cap as inputs with the binding immunoglobulin G titer as output. 325 data samples of mRNA vaccine LNPs formulations with IgG titer are collected. The light GBM model could mRNA vaccine LNPs formulations with good performance (Wang et al., 2022b). Similarly, genomic nanoparticle trafficking networks identified that gene expression of SLC46A3 inversely predicted the cellular adhesion of LNPs out of 35 different nanoparticle types in vivo and in vitro. This results in highlighting the potential of integrating high-throughput screening and ML algorithms for nano-formulation design in the field of drug transport (Boehnke et al., 2022). Furthermore, the ML algorithms were integrated with meta-analysis to study the cellular uptakes of poly β-amino ester (pBAE) nanoparticles in 4 different cancer cell lines (OVCAR-4, Panc02, 4TI, and human dermal fibroblasts). It successfully predicted that the nanoparticle uptake by macrophages is facilitated by several specific corona proteins, such as apolipoproteins and complement proteins. The gradient boost trees and neural network models are found to be best suited for predicting the uptake capability (Loecher et al., 2023). Lu and co-workers trained a convolutional neural network to identify 14 microbial species. In this case, training sets of 300 bacterial Ramanomes per species were cultured under different growth conditions to account for intra-species variability. The model could predict the highest probability species out of all 14 species. The accuracy was 95% for all the species, of which the highest and lowest accuracy were obtained for *E. coli* (99%) and *P. aeruginosa* (81%), respectively (Lu et al., 2020). The Ran group utilized a combination of encoding methods for whole genome sequencing (WGS) and ML algorithms to predict AMR using bacterial genomic information. WGS data for 1000 *E. coli* strains resistant to ciprofloxacin, cefotaxime, and ceftazidime were aligned to wild *E. coli* to identify single-nucleotide polymorphisms (SNPs). The SNPS resistance profile was encoded using one-hot encoding label encoding (A, G, C, T, N: 1, 2, 3, 4, 0), which converts SNPs data into an image-like matrix. The results were reinforced with the RF model achieving the highest AUROC (Area Under the Receiver Operating Characteristic Curve) (0.95) for *E. coli* strains resistant to ciprofloxacin among the 12 model configurations (Ren et al., 2022). **Table 6** summarizes the representative examples of AI models applied in nanocarrier designing, target system design, and predictive modeling. Thus, the integration of AI, nanobiotechnology, and precision antibiotic therapy marks a significant transition from conventional empirical antimicrobial treatment to a data-driven and adaptive therapeutic strategy. Within this integrated framework, AI serves as the core decision-making engine, processing complex and multidimensional datasets such as pathogen genomic information, antimicrobial resistance patterns, patient-specific clinical variables, and pharmacokinetic-pharmacodynamics models (Heydari et al., 2024; Arnold et al., 2025). Based on this comprehensive analysis, AI can screen the drug molecules, optimize the formulation, and predict the most suitable antibiotic candidates, enabling rational nanocarrier development for delivery systems (Han et al., 2025; Aghajanpour et al., 2025). **Figure 5** illustrates the schematic representation of integration between AI, nanotechnology, and precision antibiotic therapy.

Figure 5. Schematic representation of integration between AI, nanotechnology, and precision antibiotic therapy

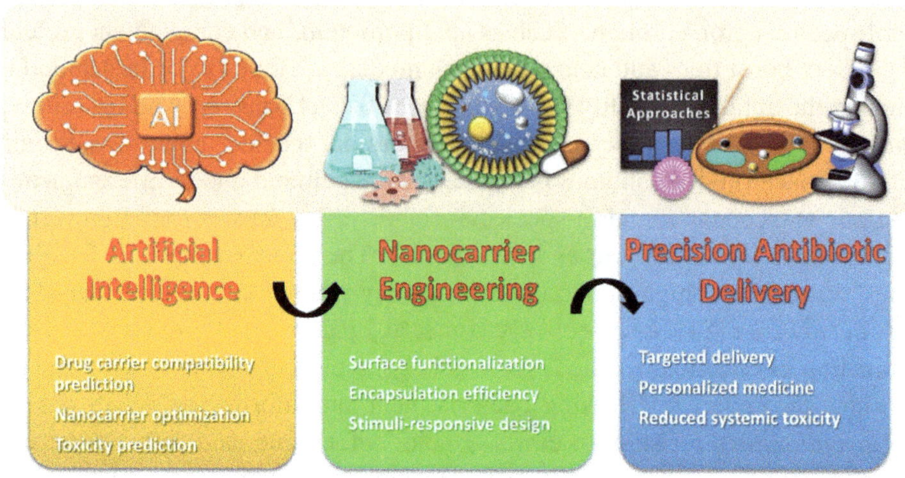

Table 6. Representative examples of AI models applied in nanocarrier designing, target system design, and predictive modeling

AI technique	Applications	References
DNNs	Predict nanocarrier drug compatibility Optimize nanoparticle synthesis.	Chen & Lu, 2022 Berrada et al., 2021
RFs	Predict clinically relevant physicochemical properties of the nanoparticles.	Henser-Brown hill et al., 2024; Perez et al., 2024
AutoML platforms	Rapid screening of material toxicity	Xiao et al., 2024
GANs, VAEs	Predict crystal structures or chemical compositions, Nanocarrier design	Watson et al., 2023
Bayesian statistics	Optimize nanoparticle synthesis parameters	Iwano et al., 2022
CNNs	Predict ligand-binding motifs and screen aptamer sequences	Raiann et al., 2024
SVMs	Binding affinity, classification	Rodriguez-Perez et al., 2022

Nanocarrier-based targeted delivery systems represent a transformative approach toward personalized antimicrobial therapy, particularly when integrated with advanced computational modeling and artificial intelligence-driven design (Ahmad et al., 2022; Gao et al., 2024). Different AI models now enable rapid prediction of critical parameters such as stability, biodistribution, and controlled drug release, significantly accelerating optimization while reducing experimental burden and cost (Chen & Lu, 2022; Berrada et al., 2021). This data-driven framework facilitates the customization of nanocarriers, especially liposomal systems, based on pathogen profile, infection duration, and disease progression (Alshawwa et al., 2022). Concurrently, hybrid nanoplatforms incorporating polymers, metallic nanoparticles, and bioactive peptides, along with co-delivery strategies combining antibiotics and resistance-modulating agents, offer promising solutions to combat multidrug-resistant infections (Wang et al., 2020). The convergence of nanotechnology with immune modulation, antimicrobial peptides, and gene-editing tools such as CRISPR-Cas9 is poised to redefine precision antimicrobial therapy (Bikard & Barrangou, 2017). Ultimately, the successful clinical translation of these systems will depend on the seamless integration of data-driven design, biocompatibility, and scalable manufacturing within a unified precision medicine framework.

AI-assisted formulation and design of nanocarriers for targeted delivery, although increasingly influential, face several critical limitations. The performance of AI models is highly dependent on the availability of high-quality, standardized, and diverse datasets, which are often limited in nanomedicine research. Additionally, many AI models lack interpretability, which makes it difficult to establish mechanistic insights and regulatory clearance for clinical uses. Various challenges related to model generalizability, data bias, and integration with experimental validation further limit their credibility into precision medicine framework.

10. TRANSLATIONAL AND REGULATORY CHALLENGES

Translational and regulatory challenges delay the clinical success of nanomedicines regardless of the promising experimental results. Due to the intrinsic physicochemical complexity, polydispersity, and dynamic properties of different nanoparticles, it becomes complicated for the large-scale manufacturing process (Fadeel et al., 2018). Additionally, the lack of standardized evaluation methods for assessing different critical parameters, such as pharmacokinetics, pharmacodynamics, biodistribution, immune response, and long-term toxicity studies, limits the regulatory process. Therefore, effective clinical translation of nanomedicines requires more harmonized regulatory guidelines, improved and predictive testing models, and adaptive scientific evaluation strategies (Patra et al., 2018).

I. Safety evaluation and toxicity assessment

The safety evaluation of nanomedicine is a crucial step of clinical translation, mainly due to the unique properties and characteristic interaction of nanomaterials with biological systems. The variations in biodistribution, cellular uptake, clearance, and potential toxicity of different nanomedicines necessitate the comprehensive assessment through in vivo, in vitro, and in silico models to evaluate cytotoxicity, genotoxicity, immunotoxicity, and long-term effects (Fadeel et al., 2018). The problems of nanoparticle toxicity cannot be ignored, considering the wide use of nanomedicine in healthcare. For example, liposomes or polymers have toxicity dependent on their size, charge, or composition (Lamberti et al., 2014). The positively charged liposomes interact with lipoproteins, serum proteins, or extracellular matrix, causing aggregation or release of the agents that are loaded before reaching the target cells. If the dose is high, liposomes may damage the reticuloendothelial system, granulomas, hepatomegaly, or splenomegaly (De et al., 2008). The addition of chitosan as a stabilizer to polylactide-co-glycolic acid (PLGA) medicine results in enhanced local toxicity due to high concentrations or aggregations (Nicolete et al., 2011). The nanoparticle surface charge on polymeric particles also has effects on their toxicity. Comparison of the lung toxicity of positively and negatively charged polyethylene glycol-polylactic acid (PEG-PLA) in BALB/c mice displays that damage was widespread and localized (Harush et al., 2010). Based on the above issues, nowadays the nanoparticles are combined with natural products to reduce the toxicity (David et al., 2015).

II. Manufacturing scalability and quality control

Manufacturing scalability and quality control remain a major hurdle in translational nanomedicine. Different techniques, such as microfluidic technologies, continuous flow reactors, and microreactors, are used for synthesizing nanoparticle-based drug delivery systems (Khalid-Salako et al., 2025). For example, the Anderluzzi group applied a microfluidizer-based mixing process to synthesize the desired size of nanoparticles and polydispersity (Hou et al., 2021). As nanomedicines are very sensitive, small changes in the production conditions can significantly affect the therapeutic performance and safety. Therefore, maintaining batch-to-batch consistency and reproducibility of nanomedicine at the industrial level of manufacturing is a challenging task (Operti et al., 2021). The nature of materials used, nanoparticle synthesis process, solvent used, cost-effectiveness, toxicity evaluation of the nanoparticle, and biodegradability of the nanocarriers are some important components associated with the scaling up of nanomedicine from the laboratory to the industrial level (Paliwal et al., 2014). As nanomedicines are highly specialized nano-units that deliver medications extremely precisely to the target site, an effective quality control and manufacturing assessment method is of utmost importance (Hock et al., 2011).

III. Regulatory frameworks for Nanomedicines

The regulatory framework for nanomedicine is formulated by the U.S. Food and Drug Administration (FDA) and the European Medicines Agency (EMA) after critical considerations of the quality, safety, and efficiency of the drug molecules, such as pharmacology, biodegradation, environmental toxicity, and biocompatibility (Csoka et al., 2021). Unlike the new products, the nanocarrier formulations proceed to the clinical trials; their manufacturing process and characterization techniques need to be submitted by the manufacturers to enable proper screening to be useful in vivo activities and finally be open to global markets (Dri et al., 2023). The transdisciplinary nature of nanomedicine creates difficulties and suggests a flexible regulatory framework (Bawa, 2011). Good Manufacturing Practice (GMP) is one of the most important components of the regulatory framework for nanomedicine. It ensures the overall quality, safety, and consistency during large-scale production. However, the implementation of GMP for nanomedicines is quite challenging due to the complex physicochemical properties, higher sensitivity to minor production variations, and difficulties in maintaining batch-to-batch reproducibility (Csoka et al., 2021).

Translational and regulatory advancement of nanocarrier-based antibiotic delivery systems remains challenging due to the complex interplay between their physicochemical properties, biological interactions, and clinical performance. The lack of universally accepted standards for characterization, safety evaluation, and efficacy assessment creates uncertainty in regulatory approval pathways. Furthermore, ensuring reproducible large-scale manufacturing under GMP conditions remains technically demanding and resource-intensive. Addressing these challenges requires standard regulatory frameworks, robust characterization strategies, and improved alignment between preclinical findings and clinical outcomes.

11. CLINICAL APPLICATIONS OF DIFFERENT NANOCARRIER SYSTEMS

Nanocarrier-based targeted drug delivery systems have advanced from experimental research into clinically approved therapies, offering significant benefits over conventional formulations. Several of them are in clinical use, while others are in advanced stages of development or clinical trials. **Table 7** summarizes key examples of different types of clinically used nanocarrier, their formulations, encapsulated drugs, clinical applications, and their therapeutic advantages.

Table 7. Clinical applications of different nanocarrier-based targeted delivery systems.

Nanocarrier Type	Formulations	Drug Encapsulated	Clinical Applications	Key Advantages	References
Liposomal Carrier	AmBisome	Amphotericin B	Systemic fungal infections	Reduced nephrotoxicity, improved tolerability vs. conventional amphotericin B	Adler-Moore & Proffitt, 2002
	Arikayce (Amikacin Liposome Inhalation Suspension)	Amikacin	*Mycobacterium avium complex* (MAC) lung infections; cystic fibrosis	Targeted lung delivery, reduced systemic side effects, improved local concentration	Griffith et al., 2018
Polymeric Carrier	PLGA nanoparticles	Vancomycin	*S. aureus* biofilm-associated infections	Sustained release, enhanced biofilm penetration	Nouruzi et al., 2023
	PEG–PLA micelles	Levofloxacin	Respiratory infections	Enhanced solubility, improved bioavailability	Derbali et al., 2019
Dendrimer-Based Systems	VivaGel (SPL-7013 Gel)	Polylysine dendrimer	Treatment of bacterial vaginosis	Effective antimicrobial gel, well-tolerated in clinical use	Tyssen et al., 2010
	PAMAM dendrimers (unmodified/modified)	PAMAM	General antibiotic and drug delivery	Controlled release, improved drug solubility	Svenson & Tomalia, 2012
Biomimetic Systems	RBC membrane–coated liposomes	Vancomycin	*Staphylococcus aureus* (MRSA)	Prolonged circulation, immune evasion, enhanced MRSA killing in vitro and in vivo	Hu et al., 2015
	Enzyme-responsive hybrid biomimetics	Ciprofloxacin	Bacterial infection/ wound contexts	Triggered release at infection sites, responsive delivery	Bourgat Y. et al., 2021

12. CONCLUSION AND FUTURE PROSPECTS

This chapter provides a comprehensive analysis of the current understanding of AMR to antibiotics while critically evaluating how various nanocarrier systems, such as liposomes, polymers, dendrimers, and biomimetics, offer a complementary structural and functional approach for targeted drug delivery. Comparative evaluation of all these nanocarrier platforms indicates that their effectiveness is highly therapeutic application-specific, based on physicochemical properties and biological attributes. Additionally, various AI-assisted models, including RF, SVM, K-NNs, ANNs, DNNs, CNNs, GBMs, Bayesian statistics, PCA, VAEs, and GANs, are critically analyzed for the designing and optimization of a nanocarrier for targeted delivery. Although these models are suitable for predicting and formulating nanocarrier systems, due to their low-quality data and standardization challenge, their clinical uses in precision medicine are limited. Furthermore, translational and regulatory challenges in terms of large-scale production, reproducibility, and toxicity assessment are evaluated. Finally, several examples of clinically approved nanocarrier-based drugs for targeted delivery are included, which highlights the experimental advancement to commercialization.

AI-based predictive frameworks are increasingly employed in antibiotic discovery to expedite virtual compound screening, structural refinement, and antimicrobial peptide development. Nevertheless, their translational effectiveness remains constrained by several methodological challenges. Model accuracy is strongly influenced by the quality, scale, and heterogeneity of training datasets, which are often biased toward extensively studied pathogens and conventional chemical scaffolds. Such dataset limitations reduce the model's capacity to generalize to structurally novel molecules or emerging microbial threats. Furthermore, the limited interpretability of deep neural network architectures and inadequate modeling of host-pathogen interactions diminish predictive robustness. Despite promising in silico outcomes, many candidates fail during preclinical and clinical validation due to biological complexity, manufacturing scalability barriers, regulatory uncertainties, and elevated production costs.

CONFLICT OF INTEREST

The authors declare no conflict of interest.

ACKNOWLEDGEMENT

The authors confirm that any use of AI-assisted tools in the preparation of tables, figures, or text has been carefully reviewed and validated. All data and results presented in this chapter are original, accurate, and supported by the cited references.

REFERENCES

Abbasi, E., Akbarzadeh, A., & Milani, M. (2014). Dendrimers for biofilm-associated infections. *Journal of Microbiological Methods, 103*, 1–8. DOI: 10.1016/j.mimet.2014.05.009

Adler-Moore, J., & Proffitt, R. (2002). AmBisome: Liposomal formulation, structure, mechanism of action and pre-clinical experience. *The Journal of Antimicrobial Chemotherapy, 49*(1, suppl_1), 21–30. DOI: 10.1093/jac/49.suppl_1.21 PMID: 11801577

Aggarwal, R., Mahanjan, P., Pandiya, S., Bajaj, A., Verma, S. K., Yadav, P., Kharat, A. S., Khan, A. U., Dua, M., & Johri, A. K. (2024). Antibiotic resistance: A global crisis, problems and solutions. *Critical Reviews in Microbiology, 50*(5), 896–921. DOI: 10.1080/1040841X.2024.2313024 PMID: 38381581

Aghajanpour, S., Ebrahimnejad, P., Amiri, H., Jeelani, H., Esfandyari-Manesh, M., Henschel, A., Singh, H., Dinarvand, R., & Hassan, S. (2025). Utilizing machine learning for predicting drug release from polymeric drug delivery systems. *Computers in Biology and Medicine, 188*, 109756. DOI: 10.1016/j.compbiomed.2025.109756 PMID: 39978092

Ahmad, A., Imran, M., & Sharma, N. (2022). Precision Nanotoxicology in Drug Development: Current Trends and Challenges in Safety and Toxicity Implications of Customized Multifunctional Nanocarriers for Drug-Delivery Applications. *Pharmaceutics, 14*(11), 2463. DOI: 10.3390/pharmaceutics14112463 PMID: 36432653

Akbarzadeh, A., Rezaei-Sadabady, R., Davaran, S., Joo, S. W., Zarghami, N., Hanifehpour, Y., Samiei, M., Kouhi, M., & Nejati-Koshki, K. (2013). Liposome: Classification, preparation, and applications. *Nanoscale Research Letters, 8*(1), 102. DOI: 10.1186/1556-276X-8-102 PMID: 23432972

Allen, T. M., & Cullis, P. R. (2013). Liposomal drug delivery systems: From concept to clinical applications. *Advanced Drug Delivery Reviews, 65*(1), 36–48. DOI: 10.1016/j.addr.2012.09.037 PMID: 23036225

Alshawwa, S. Z., Kassem, A. A., Farid, R. M., Labib, G. S., & Mostafa, S. K. (2022). Nanocarrier drug delivery systems: Characterization, limitations, future perspectives, and implementation of artificial intelligence. *Pharmaceutics*, *14*(4), 883. DOI: 10.3390/pharmaceutics14040883 PMID: 35456717

Anselmo, A. C., & Mitragotri, S. (2016). Nanoparticles in the clinic: An update. *Bioengineering & Translational Medicine*, *1*(1), 10–29. DOI: 10.1002/btm2.10003 PMID: 29313004

Arnold, A., Mclellan, S., Stokes, J. M. (2025). How AI can help us beat AMR, *npj Antimicrobials and Resistance, 3*(1), 18. https://doi.org/.DOI: 10.1038/s44259-025-00085-4

Bangham, A. D., Standish, M. M., & Watkins, J. C. (1965). Diffusion of univalent ions across the lamellae of swollen phospholipids. *Journal of Molecular Biology*, *13*(1), 238–252. DOI: 10.1016/S0022-2836(65)80093-6 PMID: 5859039

Baptista, M., Fernandes, S. N., & Godinho, M. H. (2021). Stimuli-responsive polymeric nanocarriers for controlled drug delivery. *Advanced Functional Materials*, *31*(30), 2101170. DOI: 10.1002/adfm.202101170

Bawa, R. (2011). Regulating nanomedicine-can the FDA handle it? *Current Drug Delivery*, *8*(3), 227–234. DOI: 10.2174/156720111795256156 PMID: 21291376

Behrouzi, E., Abesi, F., Ghorbani, H., & Gholinia, H. (2024). Association between gubernacular canals characteristics and teeth eruption status: A cone-beam computed tomography study. *Journal of Clinical and Experimental Dentistry*, *16*(2), e145–e150. DOI: 10.4317/jced.61169 PMID: 38496806

Bikard, D., & Barrangou, R. (2017). Using CRISPR-Cas systems as antimicrobials. *Current Opinion in Microbiology*, *37*, 155–160. DOI: 10.1016/j.mib.2017.08.005 PMID: 28888103

Birk, S. E., Boisen, A., & Nielsen, L. H. (2021). Polymeric nano and microparticulate drug delivery systems for treatment of biofilms. *Advanced Drug Delivery Reviews*, *174*, 30–52. DOI: 10.1016/j.addr.2021.04.005 PMID: 33845040

Boehnke, N., Straehla, J. P., Safford, H. C., Kocak, M., Rees, M. G., Ronan, M., Rosenberg, D., Adelmann, C. H., Chivukula, R. R., Nabar, N., Berger, A. G., Lamson, N. G., Cheah, J. H., Li, H., Roth, J. A., Koehler, A. N., & Hammond, P. T. (2022). Massively parallel pooled screening reveals genomic determinants of nanoparticle delivery. *Science*, *377*(6604), eabm5551. Advance online publication. DOI: 10.1126/science.abm5551 PMID: 35862544

Boppana, S. H., Kutikuppala, L. V. S., Sharma, S., Madhavrao, C., Rangari, G., Misra, A. K., Kandi, V., Mishra, S., Singh, P. K., Rabaan, A. A., Mohapatra, R. K., & Kudrat-E-Zahan, M. (2024). Current approaches in smart nano-inspired drug delivery: A narrative review. *Health Science Reports*, *7*(4), e2065. DOI: 10.1002/hsr2.2065 PMID: 38660006

Bourgat, Y., Mikolai, C., Stiesch, M., Klahn, P., & Menzel, H. (2021). Enzyme-responsive nanoparticles and coatings made from alginate/peptide ciprofloxacin conjugates as drug release system. *Antibiotics (Basel, Switzerland)*, *10*(6), 653. DOI: 10.3390/antibiotics10060653 PMID: 34072352

Bulman, Z. P., Patel, S., Vyas, V., & Bhavnani, S. M. (2022). Research priorities towards precision antibiotic therapy to improve patient care. *The Journal of Antimicrobial Chemotherapy*, *77*(8), 1999–2008. DOI: 10.1093/jac/dkac137 PMID: 35777386

Calabretta, M. K., Kumar, A., McDermott, A. M., & Cai, C. (2015). Antimicrobial properties of dendrimers and dendrimer-biocide conjugates. *Therapeutic Delivery*, *6*(4), 395–414. DOI: 10.4155/tde.15.1

Chen, X., & Lu, H. (2022). Intelligent control of nanoparticle synthesis on microfluidic chips with machine learning. *NPG Asia Materials*, *14*(1), 69. DOI: 10.1038/s41427-022-00416-1

Chmiel, R. U., Marek, A., Pysniak, D. S., Wieczorek, K., Dec, M., Nowaczek, A., & Osek, J. (2022). Antibiotic Resistance in Bacteria-A Review. *Antibiotics (Basel, Switzerland)*, *11*(8), 1079. DOI: 10.3390/antibiotics11081079 PMID: 36009947

Chramiec, A., & Vunjak-Novakovic, G. (2019). Tissue-engineered models of healthy and malignant human bone marrow. *Advanced Drug Delivery Reviews*, *140*, 78–92. DOI: 10.1016/j.addr.2019.04.003 PMID: 31002835

Csóka, I., Ismail, R., Jójárt-Laczkovich, O., & Pallagi, E. (2021). Regulatory considerations, challenges and risk-based approach in nanomedicine development. *Current Medicinal Chemistry*, *28*(36), 7461–7476. DOI: 10.2174/0929867328666210406115529 PMID: 33823761

Dalbanjan, N. P., Korgaonkar, K., Eelager, M. P., Gonal, B. N., Kadapure, A. J., Arakera, S. B., & Praveen Kumar, S. K. (2025). In-silico strategies in nano-drug design: Bridging nanomaterials and pharmacological applications. *Nano TransMed.*, *4*, 100091. DOI: 10.1016/j.ntm.2025.100091

Danaei, M., Dehghankhold, M., Ataei, S., Hasanzadeh Davarani, F., Javanmard, R., Dokhani, A., Khorasani, S., & Mozafari, M. R. (2018). Impact of particle size and polydispersity index on the clinical applications of lipidic nanocarrier systems. *Pharmaceutics*, *10*(2), 57. DOI: 10.3390/pharmaceutics10020057 PMID: 29783687

Danhier, F., Ansorena, E., Silva, J. M., Coco, R., Le Breton, A., & Préat, V. (2012). PLGA-based nanoparticles: An overview of biomedical applications. *Journal of Controlled Release*, *161*(2), 505–522. DOI: 10.1016/j.jconrel.2012.01.043 PMID: 22353619

Dara, S., Dhamerchela, S., Jadav, S. S., Babu, C. M., & Ashan, M. J. (2022). Machine Learning in Drug Discovery: A Review. *Artificial Intelligence Review*, *55*(3), 1947–1999. DOI: 10.1007/s10462-021-10058-4 PMID: 34393317

David, B., Wolfender, J. L., & Dias, D. A. (2015). The pharmaceutical industry and natural products: Historical status and new trends. *Phytochemistry Reviews*, *14*(2), 299–315. DOI: 10.1007/s11101-014-9367-z

Derbali, R. M., Aoun, V., Moussa, G., Frei, G., Tehrani, S. F., Campos Del'Orto, J., Hildgen, P., Roullin, V. G., & Leblond, J. (2019). Tailored nanocarriers for the pulmonary delivery of levofloxacin against Pseudomonas aeruginosa: A comparative study. *Molecular Pharmaceutics*, *16*(5), 1906–1916. DOI: 10.1021/acs.molpharmaceut.8b01256 PMID: 30900903

Dri, D. A., Rinaldi, F., Carafa, M., & Marianecci, C. (2023). Nanomedicines and nanocarriers in clinical trials: Surfing through regulatory requirements and physico-chemical critical quality attributes. *Drug Delivery and Translational Research*, *13*(3), 757–769. DOI: 10.1007/s13346-022-01262-y PMID: 36450964

Duncan, R. (2003). The dawning era of polymer therapeutics. *Nature Reviews. Drug Discovery*, *2*(5), 347–360. DOI: 10.1038/nrd1088 PMID: 12750738

Fadeel, B., Farcal, L., Hardy, B., Vázquez-Campos, S., Hristozov, D., Marcomini, A., Lynch, I., Valsami-Jones, E., Alenius, H., & Savolainen, K. (2018). Advanced tools for the safety assessment of nanomaterials. *Nature Nanotechnology*, *13*(7), 537–543. DOI: 10.1038/s41565-018-0185-0 PMID: 29980781

Gao, X. J., Ciura, K., Ma, Y., Mikolajczyk, A., Jagiello, K., Wan, X., Gao, Y., Zhang, J., Zhang, S., & Puzyn, T. (2024). Toward the integration of machine learning and molecular modeling for designing drug delivery nanocarriers. *Advanced Materials*, *36*(45), 2407793. DOI: 10.1002/adma.202407793 PMID: 39252670

George, M., & Abraham, T. E. (2006). Polyionic hydrocolloids for the intestinal delivery of protein drugs: Alginate and chitosan-A review. *Journal of Controlled Release*, *114*(1), 1–14. DOI: 10.1016/j.jconrel.2006.04.017 PMID: 16828914

Gillies, E. R., & Fréchet, J. M. (2005). Dendrimers and dendritic polymers in drug delivery. *Drug Discovery Today*, *10*(1), 35–43. DOI: 10.1016/S1359-6446(04)03276-3 PMID: 15676297

Goes, A., & Fuhrmann, G. (2018). Biogenic and biomimetic carriers as versatile transporters to treat infections. *ACS Infectious Diseases*, *4*(6), 881–892. DOI: 10.1021/acsinfecdis.8b00030 PMID: 29553240

Greener, J. G., Kandathil, S. M., Moffat, L., & Jones, D. T. (2022). A guide to machine learning for biologists. *Nature Reviews. Molecular Cell Biology*, *23*(1), 40–55. DOI: 10.1038/s41580-021-00407-0 PMID: 34518686

Grenha, A., Grainger, C. I., Dailey, L. A., Seijo, B., Martin, G. P., Remunan-Lopez, C., & Forbes, B. (2007). Chitosan nanoparticles are compatible with respiratory epithelial cells in vitro. *European Journal of Pharmaceutical Sciences*, *31*(2), 73–84. DOI: 10.1016/j.ejps.2007.02.008 PMID: 17408932

Griffith, D. E., Eagle, G., Thomson, R., Aksamit, T., Hasegawa, N., Morimoto, K., Addrizzo-Harris, J., O'Donnell, A. E., Marras, T. K., Flume, P. A., Loebinger, M. R., Morgan, L., Codecasa, L. R., Hill, A. T., Ruoss, S. J., Yim, J.-J., Ringshausen, F. C., Field, S. K., Philley, J. V., & Winthrop, K. L. (2018). Amikacin Liposome Inhalation Suspension for Treatment-Refractory Lung Disease Caused by Mycobacterium avium Complex (CONVERT). A Prospective, Open-Label, Randomized Study. *American Journal of Respiratory and Critical Care Medicine*, *198*(12), 1559–1569. DOI: 10.1164/rccm.201807-1318OC PMID: 30216086

Han, R., Ye, Z., Zhang, Y., Cheng, Y., Zheng, Y., & Ouyang, D. (2023). Predicting liposome formulations by the integrated machine learning and molecular modeling approaches. *Asian Journal of Pharmaceutical Sciences*, *18*(3), 100811. DOI: 10.1016/j.ajps.2023.100811 PMID: 37274923

Han, Y., Kim, D. H., & Pack, S. P. (2025). Nanomaterials in drug delivery: Leveraging artificial intelligence and big data for predictive design. *International Journal of Molecular Sciences*, *26*(22), 11121. DOI: 10.3390/ijms262211121 PMID: 41303604

Hand, K. (2013). Antibiotic stewardship. *Clinical Medicine (London, England)*, *13*(5), 499–503. DOI: 10.7861/clinmedicine.13-5-499 PMID: 24115710

Harush-Frenkel, O., Bivas-Benita, M., Nassar, T., Springer, C., Sherman, Y., Avital, A., Altschuler, Y., Borlak, J., & Benita, S. (2010). A safety and tolerability study of differently charged nanoparticles for local pulmonary drug delivery. *Toxicology and Applied Pharmacology*, *246*(1), 83–90. DOI: 10.1016/j.taap.2010.04.011 PMID: 20417650

Henser-Brownhill, T., Martin, L., Samangouei, P., Ladak, A., Apostolidou, M., Nagel, B., & Kwok, A. (2024). In silico screening accelerates nanocarrier design for efficient mRNA delivery. *Advancement of Science*, *11*(30), 2401935. DOI: 10.1002/advs.202401935 PMID: 38837626

Heydari, S., Masoumi, N., Esmaeeli, E., Ayyoubzadeh, S. M., Ghorbani-Bidkorpeh, F., & Ahmadi, M. (2024). Artificial intelligence in nanotechnology for treatment of diseases. *Journal of Drug Targeting*, *32*(10), 1247–1266. DOI: 10.1080/1061186X.2024.2393417 PMID: 39155708

Hock, S. C., Ying, Y. M., & Wah, C. L. (2011). A review of the current scientific and regulatory status of nanomedicines and the challenges ahead. *PDA Journal of Pharmaceutical Science and Technology*, *65*(2), 177–195. DOI: 10.5731/pda-jpst.2011.00754 PMID: 21502077

Hou, X., Zaks, T., Langer, R., & Dong, Y. (2021). Lipid nanoparticles for mRNA delivery. *Nature Reviews. Materials*, *6*(12), 1078–1094. DOI: 10.1038/s41578-021-00358-0 PMID: 34394960

Hu, C. M. J., Zhang, L., Aryal, S., Cheung, C., Fang, R. H., & Zhang, L. (2015). Erythrocyte membrane-camouflaged nanoparticles for antibiotic delivery. *Proceedings of the National Academy of Sciences of the United States of America*, *108*(27), 10980–10985. DOI: 10.1073/pnas.1106634108 PMID: 21690347

Hutchings, M. I., Truman, A. W., & Wilkinson, B. (2019). Antibiotics: Past, present and future. *Current Opinion in Microbiology*, *51*, 72–80. DOI: 10.1016/j.mib.2019.10.008 PMID: 31733401

Ijaz, S., Li, J., & Gong, C. (2024). Cell-derived biomimetic nanocarriers for drug delivery: Classification, mechanisms, and applications. *Journal of Nanobiotechnology*, *22*, 85. DOI: 10.1186/s12951-024-02144-8

Iwano, N., Adachi, T., Aoki, K., Nakamura, Y., & Hamada, M. (2022). Generative aptamer discovery using RaptGen. *Nature Computational Science*, *2*(6), 378–386. DOI: 10.1038/s43588-022-00249-6 PMID: 38177576

Kalomiraki, M., Thermos, K., & Chaniotakis, N. A. (2015). Dendrimers as tunable vectors of drug delivery systems and biomedical and ocular applications. *International Journal of Nanomedicine*, *10*, 2403–2418. DOI: 10.2147/IJN.S78820 PMID: 26730187

Kant, S., Deepika, , & Roy, S. (2025). Artificial intelligence in drug discovery and development: Transforming challenges into opportunities. *Discover Pharmaceutical Sciences*, *1*(1), 7. DOI: 10.1007/s44395-025-00007-3

Kesharwani, P., Jain, K., & Jain, N. K. (2014). Dendrimer toxicity and safety: A critical review. *Progress in Polymer Science*, *39*(2), 268–307. DOI: 10.1016/j.progpolymsci.2013.07.005

Khalid-Salako, F., Salimi Khaligh, S., Fathi, F., Demirci, O. C., Öncer, N., Kurt, H., & Yuce, M. (2025). The nanocarrier landscape-evaluating key drug delivery vehicles and their capabilities: A translational perspective. *ACS Applied Materials & Interfaces*, *17*(26), 37383–37403. DOI: 10.1021/acsami.5c07366 PMID: 40526827

Kumari, A., Yadav, S. K., & Yadav, S. C. (2010). Biodegradable polymeric nanoparticles-based drug delivery systems. *Colloids and Surfaces. B, Biointerfaces*, *75*(1), 1–18. DOI: 10.1016/j.colsurfb.2009.09.001 PMID: 19782542

Lamberti, M., Zappavigna, S., Sannolo, N., Porto, S., & Caraglia, M. (2014). Advantages and risks of nanotechnologies in cancer patients and occupationally exposed workers. *Expert Opinion on Drug Delivery*, *11*(7), 1087–1101. DOI: 10.1517/17425247.2014.913568 PMID: 24773227

Li, D., Gao, C., Kuang, M., Xu, M., Wang, B., Luo, Y., Teng, L., & Xie, J. (2021). Nanoparticles as drug delivery systems of RNAi in cancer therapy. *Molecules (Basel, Switzerland)*, *26*(8), 2380. DOI: 10.3390/molecules26082380 PMID: 33921892

Liu, P., Chen, G., & Zhang, J. (2022). A review of liposomes as a drug delivery system: Current status of approved products, regulatory environments, and future perspectives. *Molecules (Basel, Switzerland)*, *27*(4), 1372. DOI: 10.3390/molecules27041372 PMID: 35209162

Loecher, A., Bruyns-Haylett, M., Ballester, P. J., Borros, S., & Oliva, N. (2023). A machine learning approach to predict cellular uptake of pBAE polyplexes. *Biomaterials Science*, *11*(17), 5797–5808. DOI: 10.1039/D3BM00741C PMID: 37401742

Lu, W., Chen, X., Wang, L., Li, H., & Fu, Y. V. (2020). Combination of an artificial intelligence approach and laser tweezers Raman spectroscopy for microbial identification. *Analytical Chemistry*, *92*(9), 6288–6296. DOI: 10.1021/acs.analchem.9b04946 PMID: 32281780

Majumder, J., & Minko, T. (2021). Multifunctional and stimuli-responsive nanocarriers for targeted therapeutic delivery. *Expert Opinion on Drug Delivery*, *18*(2), 205–227. DOI: 10.1080/17425247.2021.1828339 PMID: 32969740

Mekki-Berrada, F., Ren, Z., Huang, T., Wong, W. K., Zhang, F., Xie, J., Tian, I. P. S., Jayavelu, S., Mahfoud, Z., Bash, D., Hippalgaonker, K., Khan, S., Buonassisi, T., Li, Q., & Wang, X. (2021). Two-step machine learning enable optimized nanoparticle synthesis. *npj Computational Materials, 7*, 55. https://doi.org/.DOI: 10.1038/s4152-021-00520-w

Meng, F., Hennink, W. E., & Zhong, Z. (2009). Reduction-sensitive polymers and bioconjugates for biomedical applications. *Biomaterials*, *30*(12), 2180–2198. DOI: 10.1016/j.biomaterials.2009.01.026 PMID: 19200596

Meyer, R. A., Sunshine, J. C., & Green, J. J. (2015). Biomimetic particles as therapeutics. *Trends in Biotechnology*, *33*(9), 514–524. DOI: 10.1016/j.tibtech.2015.07.001 PMID: 26277289

Mintzer, M. A., & Grinstaff, M. W. (2011). Biomedical applications of dendrimers: A tutorial. *Chemical Society Reviews*, *40*(1), 173–190. DOI: 10.1039/B901839P PMID: 20877875

Nicolete, R., dos Santos, D. F., & Faccioli, L. H. (2011). The Uptake of PLGA Micro or Nanoparticles by Macrophages Provokes a Distinct in Vitro Inflammatory Response. *International Immunopharmacology*, *11*(10), 1557–1563. DOI: 10.1016/j.intimp.2011.05.014 PMID: 21621649

Nouruzi, E., Hosseini, S. M., Asghari, B., Mahjoub, R., Zare, E. N., Shahbazi, M. A., Kalhori, F., & Arabestani, M. R. (2023). Effect of poly (lactic-co-glycolic acid) polymer nanoparticles loaded with vancomycin against Staphylococcus aureus biofilm. *BMC Biotechnology*, *23*(1), 39. DOI: 10.1186/s12896-023-00811-8 PMID: 37723466

Nsairat, H., Khater, D., Sayed, U., Odeh, F., Al Bawab, A., & Alshaer, W. (2022). Liposomes: Structure, composition, types, and clinical applications. *Heliyon*, *8*(5), e09394. DOI: 10.1016/j.heliyon.2022.e09394 PMID: 35600452

Nuhn, L. (2023). Artificial Intelligence Assists Nanoparticles to Enter Solid Tumors. *Nature Nanotechnology*, *18*(6), 550–551. DOI: 10.1038/s41565-023-01382-7 PMID: 37081083

Operti, M. C., Bernhardt, A., Grimm, S., Engel, A., Figdor, C. G., & Tagit, O. (2021). PLGA-based nanomedicines: Challenges and opportunities in industrial scale-up. *International Journal of Pharmaceutics*, *605*, 120807. DOI: 10.1016/j.ijpharm.2021.120807 PMID: 34144133

Ortiz-Perez, A., Tilborg, D. V., Meel, R. V. D., Grisoni, F., & Albertazzi, L. (2024). Machine learning-guided high throughput nanoparticle design. *Digital Discovery*, *3*(7), 1280–1291. DOI: 10.1039/D4DD00104D

Paliwal, R., Babu, R. J., & Palakurthi, S. (2014). Nanomedicine scale-up technologies: Feasibilities and challenges. *AAPS PharmSciTech*, *15*(6), 1527–1534. DOI: 10.1208/s12249-014-0177-9 PMID: 25047256

Palme, J. B., Kristiansson, E., & Larsson, D. G. J. (2008). Environment factors influencing the development and spread of antibiotic resistance. *FEMS Microbiology Reviews*, *42*(1), 68–80. DOI: 10.1093/femsre/fux053

Patra, J. K., Das, G., Fraceto, L. F., Campos, E. V. R., Rodriguez-Torres, M. P., Acosta-Torres, L. S., Diaz-Torres, L. A., Grillo, R., Swamy, M. K., Sharma, S., Habtermarium, S., & Shin, H. S. (2018). Nano-based drug delivery systems: Recent developments and future prospects. *Journal of Nanobiotechnology*, *16*(1), 1–33. DOI: 10.1186/s12951-018-0392-8 PMID: 30231877

Patri, A. K., Kukowska-Latallo, J. F., & Baker, J. R. (2002). Targeted drug delivery with dendrimers: Comparison of the release kinetics of covalently conjugated drug and non-covalent drug inclusion complex. *Advanced Drug Delivery Reviews*, *54*(2), 299–314. DOI: 10.1016/j.addr.2005.09.014 PMID: 16290254

Pattni, B. S., Chupin, V. V., & Torchilin, V. P. (2015). New developments in liposomal drug delivery. *Chemical Society Reviews*, *115*(19), 10938–10966. DOI: 10.1021/acs.chemrev.5b00046 PMID: 26010257

Pelgrift, R. Y., & Friedman, A. J. (2013). Nanotechnology as a therapeutic tool to combat microbial resistance. *Advanced Drug Delivery Reviews*, *65*(13–14), 1803–1815. DOI: 10.1016/j.addr.2013.07.011 PMID: 23892192

Perini, H. F., Matos, B. S., de Oliveira, C. J. F., & da Silva, M. V. (2024). Biomimetic nanocarriers: Integrating natural functions for advanced therapeutic applications. *Beilstein Journal of Nanotechnology*, *15*, 1619–1626. DOI: 10.3762/bjnano.15.127 PMID: 39717696

Prabaharan, M. (2011). Chitosan-based nanoparticles for tumor-targeted drug delivery. *International Journal of Biological Macromolecules*, *48*(2), 177–183. DOI: 10.1016/j.ijbiomac.2014.10.052 PMID: 25450550

Prabhu, A. E., & Selvakumar, V. S. (2025). A Review of AI-Driven Methods for Cytotoxicity Analysis and Predictive Modeling in the Treatment of Cancer Using Nanomaterials. In *2025 International Conference on Electronics, Computing, Communication and Control Technology (ICECCC)* (pp. 1-6). IEEE. https://doi.org/DOI: 10.1109/ICECCC65144.2025.11064249

Rao, L., Yuan, Y., Shen, X., Yu, G., & Chen, X. (2024). Designing nanotheranostics with machine learning. *Nature Nanotechnology*, *19*(12), 1769–1781. DOI: 10.1038/s41565-024-01753-8 PMID: 39362960

Ren, X., Chakraborty, T., Doijad, S., Falgenhacer, L., Falgenhauer, J., Woesmann, A., Houschild, A. C., Schwengers, O., & Heider, D. (2022). Prediction of antimicrobial resistance based on whole-genome sequencing and machine learning. *Bioinformatics (Oxford, England)*, *38*(2), 325–334. DOI: 10.1093/bioinformatics/btab681 PMID: 34613360

Rizwan, M., Yahya, R., Hassan, A., Yar, M., Azzahari, A. D., Selvanathan, V., Sonsudin, F., & Abouloula, C. N. (2017). pH-sensitive hydrogels in drug delivery: Brief history, properties, swelling, and release mechanism, material selection, and applications. *Polymers*, *9*(4), 137. DOI: 10.3390/polym9040137 PMID: 30970818

Rodríguez-Pérez, R., & Bajorath, J. (2022). Evolution of support vector machine and regression modeling in chemoinformatics and drug discovery. *Journal of Computer-Aided Molecular Design*, *36*(5), 355–362. DOI: 10.1007/s10822-022-00442-9 PMID: 35304657

Sarode, R. J., & Mahajan, H. S. (2024). Dendrimers for drug delivery: An overview of its classes, synthesis, and applications. *Journal of Drug Delivery Science and Technology*, *98*, 105896. DOI: 10.1016/j.jddst.2024.105896

Sercombe, L., Veerati, T., Moheimani, F., Wu, S. Y., Sood, A. K., & Hua, S. (2015). Advances and challenges of liposome-assisted drug delivery. *Frontiers in Pharmacology*, *6*, 286. DOI: 10.3389/fphar.2015.00286 PMID: 26648870

Shah, A., Aftab, S., Nisar, J., Ashiq, M. N., & Iftikhar, F. J. (2021). Nanocarriers for targeted drug delivery. *Journal of Drug Delivery Science and Technology*, *62*, 102426. DOI: 10.1016/j.jddst.2021.102426

Shalaby, K. S., Soliman, M. E., Casettari, L., Bonacucina, M., Cespi, G., Palmieri, G. F., Sammour, O. A., & El Shamy, A. A. (2014). Determination of factors controlling the particle size and entrapment efficiency of noscapine in PEG/PLA nanoparticles using artificial neural networks. *International Journal of Nanomedicine*, *9*, 4953–4964. DOI: 10.2147/IJN.S68737 PMID: 25364252

Sheikhpour, M., Barani, L., & Kasaeian, A. (2017). Biomimetics in drug delivery systems: A critical review. *Journal of Controlled Release*, *253*, 97–109. DOI: 10.1016/j.jconrel.2017.03.026 PMID: 28322976

Sill, T. J., & Recum, H. A. (2008). Electrospinning: Applications in drug delivery and tissue engineering. *Biomaterials*, *29*(13), 1989–2006. DOI: 10.1016/j.biomaterials.2008.01.011 PMID: 18281090

Stokes, J. M., Yang, K., Swanson, K., Jin, W., Cubillos-Ruiz, A., Donghia, N. M., MacNair, C. R., French, S., Carfrae, L. A., Bloom-Ackermann, Z., Tran, V. M., Chiappino-Pepe, A., Badran, A. H., Andrews, I. W., Chory, E. J., Church, G. M., Brown, E. D., Jaakkola, T. S., Barzilay, R., & Collins, J. J. (2020). A deep learning approach to antibiotic discovery. *Cell*, *180*(4), 688–702.e13. DOI: 10.1016/j.cell.2020.01.021 PMID: 32084340

Sushnitha, M., Evangelopoulos, M., Tasciotti, E., & Taraballi, F. (2020). Cell membrane–based biomimetic nanoparticles and the immune system: Immunomodulatory interactions to therapeutic applications. *Frontiers in Bioengineering and Biotechnology*, *8*, 627. DOI: 10.3389/fbioe.2020.00627 PMID: 32626700

Svenson, S., & Tomalia, D. A. (2012). Dendrimers in biomedical applications—Reflections on the field. *Advanced Drug Delivery Reviews*, *64*, 102–115. DOI: 10.1016/j.addr.2012.09.030 PMID: 16305813

Thambi, T., Park, J. H., & Lee, D. S. (2016). Stimuli-responsive polymersomes for cancer therapy. *Biomaterials Science*, *4*(1), 55–69. DOI: 10.1039/C5BM00268K PMID: 26456625

Torchilin, V. P. (2005). Recent advances with liposomes as pharmaceutical carriers. *Nature Reviews. Drug Discovery*, *4*(2), 145–160. DOI: 10.1038/nrd1632 PMID: 15688077

Torchilin, V. P. (2014). Multifunctional, stimuli-sensitive nanoparticulate systems for drug delivery. *Nature Reviews. Drug Discovery*, *13*(11), 813–827. DOI: 10.1038/nrd4333 PMID: 25287120

Tyssen, D., Henderson, S. A., Johnson, A., Sterjovski, J., Moore, K., La, J., Zanin, M., Sonza, S., Karellas, P., Giannis, M. P., Krippner, G., Wesselingh, S., McCarthy, T., Gorry, P. R., Ramsland, P. A., Cone, R., Paull, J. R. A., Lewis, G. R., & Tachedjian, G. (2010). Structure Activity Relationship of Dendrimer Microbicides with Dual Action Antiviral Activity. *PLoS One*, *5*(9), e12309. DOI: 10.1371/journal.pone.0012309 PMID: 20808791

Van Hoek, A. H., Mevius, D., Guerra, B., Mullany, P., Roberts, A. P., & Aarts, H. J. (2011). Acquired antibiotic resistance genes: An overview. *Frontiers in Microbiology, 2*, 203. DOI: 10.3389/fmicb.2011.00203 PMID: 22046172

Wang, C., Yang, Y., Cao, Y., Liu, K., Shi, H., Guo, X., Liu, W., Hao, R., Song, H., & Zhao, R. (2023). Nanocarriers for the delivery of antibiotics into cells against intracellular bacterial infection. *Biomaterials Science, 11*(2), 432–444. DOI: 10.1039/D2BM01489K PMID: 36503914

Wang, H., Huang, Q., Chang, H., Xiao, J., & Cheng, Y. (2016). Stimuli-responsive dendrimers in drug delivery. *Biomaterials Science, 4*(3), 375–390. DOI: 10.1039/C5BM00532A PMID: 26806314

Wang, L., Hu, C., & Shao, L. (2017). The antimicrobial activity of nanoparticles: Present situation and prospects for the future. *International Journal of Nanomedicine, 12*, 1227–1249. DOI: 10.2147/IJN.S121956 PMID: 28243086

Wang, S., Yu, S., Lin, Y., Zou, P., Chai, G., Yu, H. H., Wickremasinghe, H., Shetty, N., Ling, J., Li, J., & Velkov, T. (2018). Co-delivery of ciprofloxacin and colistin in liposomal formulations with enhanced in vitro antimicrobial activities against multidrug-resistant *Pseudomonas aeruginosa. Pharmaceutical Research, 35*(10), 187. DOI: 10.1007/s11095-018-2464-8 PMID: 30094660

Wang, W., Feng, S., Ye, Z., Gao, H., Lin, J., & Ouyang, D. (2022a). Prediction of lipid nanoparticles for mRNA vaccines by the machine learning algorithm. *Acta Pharmaceutica Sinica. B, 12*(6), 2950–2962. DOI: 10.1016/j.apsb.2021.11.021 PMID: 35755271

Wang, X., Li, C., Wang, Y., Chen, H., Zhang, X., Luo, C., Zhou, W., Li, L., Teng, L., Yu, H., & Wang, J. (2022b). Smart drug delivery systems for precise cancer therapy. *Acta Pharmaceutica Sinica. B, 12*(11), 4098–4121. DOI: 10.1016/j.apsb.2022.08.013 PMID: 36386470

Watkins, R. R. (2022). Antibiotic stewardship in the era of precision medicine. *JAC-Antimicrobial Resistance, 4*(3), dlac066. Advance online publication. DOI: 10.1093/jacamr/dlac066 PMID: 35733911

Watson, J. L., Juergens, D., Bennett, N. R., Trippe, B. L., Yim, J., Eisenach, H. E., & Baker, D. (2023). De novo design of protein structure and function with RFdiffusion. *Nature, 620*(7976), 1089–1100. DOI: 10.1038/s41586-023-06415-8 PMID: 37433327

Xiao, X., Trinh, T. X., Gerelkhuu, Z., Ha, E., & Yoon, T. H. (2024). Automated machine learning in nanotoxicity assessment: A comparative study of predictive model performance. *Computational and Structural Biotechnology Journal, 25*, 9–19. DOI: 10.1016/j.csbj.2024.02.003 PMID: 38414794

Xu, Z., Xie, Y., Chen, W., & Deng, W. (2025). Nanocarrier-Based Systems for Targeted Delivery: Current Challenges and Future Directions. *MedComm, 6*(9), e70337. DOI: 10.1002/mco2.70337 PMID: 40859956

Yang, G., Chen, S., & Zhang, J. (2019). Bioinspired and biomimetic nanotherapies for the treatment of infectious diseases. *Frontiers in Pharmacology, 10*, 751. DOI: 10.3389/fphar.2019.00751 PMID: 31333467

Youshia, J., Ali, M. E., & Lamprecht, A. (2017). Artificial neural network-based particle size prediction of polymeric nanoparticles. *European Journal of Pharmaceutics and Biopharmaceutics, 119*, 333–342. DOI: 10.1016/j.ejpb.2017.06.030 PMID: 28694160

Zhang, J., Tang, W., Zhang, X., Song, Z., & Tong, T. (2023). An overview of stimuli-responsive intelligent antibacterial nanomaterials. *Pharmaceutics, 15*(8), 2113. DOI: 10.3390/pharmaceutics15082113 PMID: 37631327

Zielinska, A., Carreiró, F., Oliveira, A. M., Neves, A., Pires, B., Venkatesh, D. N., & Souto, E. B. (2020). Polymeric nanoparticles: Production, characterization, toxicology and ecotoxicology. *Molecules (Basel, Switzerland), 25*(16), 3731. DOI: 10.3390/molecules25163731 PMID: 32824172

Chapter 9

AI, Nanobiotechnology, and the Future of Precision Antibiotics: Tailored Therapies for MDR Pathogens

Dini Agustina
http://orcid.org/0000-0002-1861-0056
Faculty of Medicine, University of Jember, Indonesia

Bagus Hermansyah
http://orcid.org/0000-0002-0946-2936
Faculty of Medicine, University of Jember, Indonesia

ABSTRACT

This chapter presents innovative, precision-driven strategies addressing infections caused by multidrug-resistant pathogens, focusing on bacteria and protozoa. It highlights integration of artificial intelligence, genomics, nanobiotechnology, and the One Health framework to improve antimicrobial therapies. Artificial intelligence advances resistance prediction and treatment optimization, while genomics deciphers resistance mechanisms and host-pathogen interactions. Nanobiotechnology facilitates targeted drug delivery to difficult-to-reach infection sites. The chapter emphasizes implementation challenges, especially in low- and middle-income countries, such as infrastructure, expertise, regulatory barriers, and social acceptance. Real-world examples illustrate successful multidisciplinary collaborations adapting these technologies to diverse clinical and resource settings. The chapter outlines

DOI: 10.4018/979-8-3373-5268-8.ch009

a comprehensive framework for sustainable antimicrobial resistance control, uniting molecular science, technological innovation, and systemic health strategies to enhance global health outcomes.

1. INTRODUCTION

The capacity of bacteria and other microorganisms to adapt and develop resistance to antimicrobial agents poses a serious threat to a cornerstone of contemporary healthcare: the reliability of antimicrobial treatment. Although resistance arises through natural selection, its acceleration is largely anthropogenic. The excessive and inappropriate use of antibiotics in community settings, veterinary practice, and agriculture amplifies selective pressure, while inadequate sanitation, limited access to clean water, and the circulation of substandard or falsified medicines further intensify resistance dynamics (Ahmed et al., 2024; Nadimpalli et al., 2018). AMR should therefore not be understood solely as a microbiological phenomenon, but as the outcome of interacting ecological pressures, behavioural patterns, and structural weaknesses in health systems. The central challenge lies in disrupting this multi-layered selection process before resistance genes become entrenched at population and ecosystem levels.

Behavioural and prescribing practices play a decisive role in shaping resistance trajectories. Incomplete antibiotic courses, inappropriate prescribing, and self-medication expose microorganisms to subtherapeutic concentrations that favour survival of partially resistant strains. Such exposure does not merely fail to eradicate infection; it actively selects for adaptive traits. The widespread availability of low-quality antimicrobial products further compounds this problem by maintaining persistent selective pressure without achieving bactericidal thresholds (Getahun et al., 2022). These patterns highlight a systemic failure in antimicrobial stewardship, rather than isolated clinical misjudgements.

Microorganisms employ multiple strategies to survive antibiotic exposure, including producing enzymes that degrade or modify antimicrobial drugs, restricting drug entry, altering metabolic pathways or therapeutic targets, increasing efflux pump activity, forming protective biofilms, and exchanging resistance genes via mobile genetic elements (Ahmed et al., 2024). Bacteria rely on genetic adaptations to maintain resistance, on the other hand, protozoa employ a different survival strategy called persistence. The interplay between persistence and resistance represents a major unresolved obstacle in parasitic disease control, particularly in endemic regions (Capela et al., 2019). Distinguishing true genetic resistance from phenotypic tolerance is therefore essential for designing effective interventions.

This situation underscores the necessity of a dual strategic orientation: preserving the clinical utility of existing antimicrobials while actively disrupting pathogen dormancy, persistence, and adaptive resistance. While conventional regimens remain indispensable, their long-term sustainability is increasingly undermined by biofilm formation, intracellular sequestration, and metabolic quiescence. Multi-omics platforms, including genomics, transcriptomics, proteomics, and metabolic profiling, now allow for systematic identification of persistence-associated pathways and regulatory networks rather than isolated resistance genes alone. However, translating these molecular insights into therapeutic interventions remains uneven, as many identified targets are still at exploratory or preclinical stages (Abu-El-Ruz et al., 2025).

To more effectively limit the rapid growth of global resistance, a multidisciplinary strategy is required that brings molecular biology, bioinformatics, and health policy together under a One Health framework, supported by close collaboration involving researchers, the pharmaceutical industry, and public health institutions (Nadimpalli et al., 2018). Recent technological advances are transforming AMR management through four key pillars. Artificial intelligence (AI) supports predictive modelling of resistance patterns and optimisation of personalised therapies using multi-domain clinical and ecological data. Genomics identifies resistance biomarkers and clarifies host–pathogen interactions through multi-omics integration. Nanotechnology enhances the specificity and efficacy of antimicrobial drug delivery (Solanki et al., 2024). The One Health framework situates these technologies within ecological and socio-political systems for sustainable AMR control (Arnold et al., 2025; Cesaro et al., 2025; Nadimpalli et al., 2018; Solanki et al., 2024; Tousif & Ahmad, 2015). The primary objective of this chapter is to examine novel, precision-focused strategies for managing infections caused by multidrug-resistant (MDR) pathogens and the ongoing challenges in both microbiology and parasitology. We explore how these four major pillars can be implemented together to improve targeted antimicrobial therapies for bacterial and protozoan infections.

By interlinking immune and metabolic profiling with AI-assisted predictive analytics, we propose a translational continuum that bridges conventional antimicrobial stewardship with data-driven precision interventions. Within this continuum, nanocarrier platforms are positioned not as standalone innovations, but as delivery accelerators contingent upon genomic stratification and algorithmic risk assessment. Using real-life examples, especially from low- and middle-income countries (LMICs), this chapter also reflects on practical, regulatory, and infrastructure hurdles involved in integrating advanced technologies of various healthcare settings. Ultimately, this chapter advances a systems-oriented framework that integrates molecular insight, computational modeling, targeted delivery technologies, and governance alignment. Such integration is essential to shift AMR management from reactive containment toward anticipatory, precision-based, and globally sustainable antimicrobial strategies.

Within the broader scope of this volume, which examines how artificial intelligence, nanobiotechnology, and related genomic and microbiome-based approaches are reshaping antimicrobial therapy, this chapter positions multidrug-resistant bacterial and protozoan infections as a critical proving ground for precision antibiotics. It brings together core themes highlighted across the book—antimicrobial therapy, predictive modelling, nanocarrier technologies, genomics, microbiomes, and sustainable development—by asking how these advances can be operationalized into tailored treatment strategies that are both scientifically robust and implementable in resource-constrained health systems. In doing so, the chapter complements contributions that focus more specifically on AI architectures, microbiome-informed drug design, and nanomedicine platforms, and translates them into an integrated clinical and policy framework for real-world MDR management.

2. TECHNOLOGICAL ADVANCES PUSHING PRECISION THERAPY

The healthcare system is changing rapidly as new technologies reshape how multidrug-resistant (MDR) pathogens are diagnosed and treated. This chapter discusses how artificial intelligence (AI), genomics, and nanobiotechnology are influencing clinical practice. Rather than simply presenting these tools as innovations, it evaluates how useful and realistic they are in everyday healthcare settings.

Traditional treatment of MDR infections often relies on broad-spectrum antibiotics and delayed laboratory confirmation. This approach is becoming less effective as resistance increases. AI-based systems can analyse large clinical datasets to help predict resistance patterns and guide antibiotic selection earlier. However, their accuracy depends on data quality, proper validation, and regulatory oversight. Without these safeguards, AI tools may produce unreliable results.

Genomic technologies allow rapid identification of resistance genes and transmission patterns. This improves targeted therapy and outbreak control. Still, high costs, technical infrastructure needs, and limited laboratory capacity restrict their widespread use, especially in low-resource settings where MDR infections are common. Nanobiotechnology offers new therapeutic strategies, such as targeted drug delivery and antimicrobial nanomaterials. While early studies are promising, many applications remain experimental. Questions about long-term safety, large-scale production, and regulatory approval must be addressed before routine clinical use.

Overall, AI, genomics, and nanobiotechnology should not be viewed as separate solutions. Their real potential lies in integration, combining predictive analytics, molecular diagnostics, and targeted therapies within antimicrobial stewardship frameworks. The success of these technologies depends not only on innovation but also on validation, cost-effectiveness, regulatory feasibility, and equitable implementation. Precision treatment for MDR pathogens therefore requires a balanced approach: scientific advancement supported by strong clinical evidence, clear regulation, and sustainable healthcare integration (Ayesiga et al., 2024; Babu & Nutman, 2016; Canales et al., 2025).

These developments align with the broader themes explored throughout this volume, where AI-based predictive modelling, genomics, microbiome science, and nanomedicine are examined as complementary foundations of future precision antibiotics, rather than as isolated innovations.

Artificial Intelligence: From Data to Decision

In acute care settings, clinicians managing suspected sepsis have long faced a structural limitation: culture-based diagnostics require time, often several days, before pathogen identity and resistance profiles are confirmed. This delay sustains diagnostic uncertainty and can compromise early antimicrobial targeting. AI-driven systems attempt to shift this timeline. By integrating vital signs, laboratory parameters, prior microbiology records, and increasingly genomic information, predictive models can estimate likely pathogens and resistance risks within minutes (Arnold et al., 2025; Rabaan et al., 2022).

The real value isn't just about working faster, but also it comes from the AI's ability to "learn" from millions of cases globally. Machine learning systems repeatedly identify subtle patterns between hospitals, communities, and even across countries by learning from large-scale clinical and microbiological datasets. Models trained on extensive electronic medical records and bacterial genomes can now predict which antibiotics an infection is likely to resist, helping save both lives and resources. As these systems continue to improve, they are reshaping antibiotic stewardship and paving the way for entirely new medication classes ones built not only to fight bacteria, but to anticipate resistance itself (Arnold et al., 2025; Olatunji et al., 2024).

The deployment of artificial intelligence in antimicrobial discovery is changing the way new therapies are identified and optimized. Antimicrobial peptides, which are conserved across many living organisms and form a natural component of the host's defense, have emerged as a promising candidate. By leveraging AI and deep-learning tools, researchers are now able to identify and construct these peptides quickly, capitalizing on their low propensity to induce resistance and their ability to target pathogens through multiple mechanisms. For instance, recent models trained on large peptide-activity datasets have been aimed at generating novel candidate sequences with improved predicted stability and antimicrobial potency compared to empirically derived peptides. This flexibility stems not only from the peptides' inherent biological roles but also from AI's capability to fine-tune the relationship of their structure to specific microbes or resistance patterns. However, most AI-designed antimicrobial peptides remain in the preclinical or early translational stages, and their true clinical value depends on rigorous evaluation of safety, efficacy, manufacturability, and cost in real-life settings.

Taken together, artificial intelligence represents a transformative tool in infectious disease diagnostics and antimicrobial innovation, but its maturity varies across applications. Predictive clinical decision-support systems are closer to implementation, albeit with concerns regarding external validity and bias. In contrast, AI-designed therapeutics remain largely experimental, constrained by toxicity, manufacturing complexity, and regulatory ambiguity. The decisive challenge is not whether AI can generate promising candidates, but whether these innovations can demonstrate reproducible clinical benefit, regulatory feasibility, and economic sustainability at scale.

Advancing Antimicrobial Stewardship Through AI-Driven Diagnostics and Predictive Analytics

1. Bacterial Identification and Resistance Profiles

Quick identification of infectious microorganisms and their resistance patterns is essential for determining the right treatment. Unfortunately, conventional culture-based methods can require up to 72 hours to establish resistance, which frequently postpones timely intervention. Whole Genome Sequencing (WGS) is now widely employed to detect antibiotic-resistance genes directly from clinical samples. When WGS-derived genomic data are combined with artificial intelligence (AI), studies such as the one by Ren et al. have shown impressive predictive results. Their research demonstrated that machine learning models, especially Random Forest (RF), can precisely forecast resistance in *E. coli* isolates to multiple antibiotics, reaching AUROC values as high as 0.95 (Ren et al., 2022).

Apart from genetic data, a phenotype-based method was also created to rapidly differentiate sensitive from resistant bacteria through a convolutional neural network (CNN) applied to specially stained microscopic images (Figure 1.b). This system can provide results in approximately 30 minutes, although it still needs an initial culture step to obtain the cell samples (Zagajewski et al., 2023). A growing approach in clinical practice involves using MALDI-TOF MS, which reads bacterial mass spectra and employs thousands of them to train a deep-learning model capable of predicting antibiotic resistance (Figure 1.c). In a retrospective analysis, this model accurately guided changes to antibiotic therapy in 89% of cases (Osthoff et al., 2017).

Regulatory and implementation barriers further constrain translational maturity. Few AI-based antimicrobial susceptibility tools have undergone prospective multicentre validation or regulatory approval. Clinical integration demands explainability, traceability of model decisions, cybersecurity safeguards, and alignment with antimicrobial stewardship frameworks. From a cost-effectiveness perspective, benefits must outweigh investments in sequencing infrastructure, computational resources, data storage, and personnel training, particularly in low- and middle-income settings where resistance burden is often highest.

Overall, the results highlight the considerable potential of AI and ML to enhance and refine antibiotic resistance forecasting. Nonetheless, real-world applications still demand thorough validation to guarantee safety and effectiveness, alongside ongoing efforts to address data bias and adapt to local variations.

2. Sepsis Prediction

The implementation of AI in sepsis forecasting holds critical significance since detecting sepsis at an early stage facilitates quicker antibiotic intervention, which substantially increases patient survival rates. Early identification proves vital because each hour of delay in administering antibiotics after sepsis onset elevates mortality risk by approximately 9%. Since bacterial sepsis frequently mirrors non-infectious systemic inflammatory conditions, medical professionals may encounter difficulties determining whether a patient genuinely has sepsis or a comparable disorder. Within this framework, artificial intelligence (AI) presents a highly prospective instrument to assist clinicians in reaching faster and more precise sepsis diagnoses (Liu et al., 2017; Rudd et al., 2020). When appropriately implemented, AI-driven tools may support antimicrobial stewardship by helping clinicians distinguish high-risk patients who require urgent escalation from those in whom early de-escalation or diagnostic reassessment is justified.

The integration of AI into Electronic Health Records (EHRs) has enabled large-scale predictive modelling using routinely collected clinical data, including laboratory parameters, vital signs, medication records, and free-text clinical documentation. By training on high-dimensional datasets, machine learning algorithms can identify non-linear relationships and temporal patterns that are often imperceptible to clinicians. Importantly, the added value of AI lies not only in early prediction but in probabilistic reasoning. Instead of providing binary outputs, well-designed systems generate calibrated risk estimates, thereby allowing clinicians to incorporate algorithmic outputs into broader clinical assessment (Goh et al., 2021; Schinkel et al., 2019; Xiao et al., 2018).

Nevertheless, leveraging EHR data does not always proceed smoothly. Several obstacles persist: (Yu et al., 2019; Zhang et al., 2021)

- Unstructured and Inconsistent Data: EHR data comprises structured information (lab results) and unstructured documentation (clinical notes), which are frequently measured irregularly and lack standardization, rendering them challenging to analyze automatically through AI.

- Deficient Technological Infrastructure: Numerous health-care facilities, especially in resource-limited regions, suffer from inadequate hardware, unreliable internet connectivity, and servers that aren't powerful enough to handle complex AI operations.
- Data Privacy and Security: Medical records hold sensitive information that needs to be safeguarded with advanced cybersecurity measures to avoid data breaches or misuse.
- Resistance and Lack of Human Resource Steadiness: Some healthcare professionals still lack skills with AI technology and tend to be reluctant to adopt new systems because of inadequate training and trust issues.
- Complex Regulations: Varying health data management regulations across countries or regions can pose challenges for widespread standardized AI implementation.
- Potential Bias and Limited Model Generalization: AI models that are trained on data from a specific population or hospital may not perform well when applied to other populations, requiring methods to reduce bias and improve reliability.

These challenges can be addressed through: (Boussina et al., 2024; Yu et al., 2019; Zhang et al., 2021)

- Adaptive Model Development and Extensive Validation: The bidirectional long short-term memory (BiLSTM) approach leverages temporal sequencing by incorporating both prior and subsequent clinical events, allowing dynamic risk prediction. When trained on multi-hospital datasets in the US, such models achieved high discriminative performance, with area under the curve (AUC) values approaching 0.94, indicating strong separation between septic and non-septic patients. The Sepsis Early Risk Assessment (SERA) algorithm integrates structured data with natural language processing of clinical narratives. By extracting weighted clinical themes, it can estimate sepsis risk hours before overt clinical deterioration. This hybrid approach highlights the potential of combining quantitative biomarkers with qualitative clinical insight. The COMPOSER model attempts to mitigate generalization failures by incorporating mechanisms that evaluate whether incoming data resemble the training distribution. Predictions are withheld when model confidence is low, reducing inappropriate alerts. When implemented at UC San Diego Hospital, COMPOSER showed impressive real-world results by reducing sepsis mortality by 17% and increasing adherence to sepsis protocols by 10% (see Figure 1.a).

- Human Resources Training and Mentoring: Structured education enhances algorithm literacy and fosters collaborative human-AI decision-making rather than passive alert dependence.
- Data and Work Procedure Standardization: Harmonized data formats facilitate interoperability and reduce model fragility.
- Technology Infrastructure Escalation: Strengthening computational capacity is essential to ensure equitable access.
- Multiple-Layered Data Security: Encryption, authentication systems, and transparent governance mechanisms are necessary to maintain public trust.
- Regulatory and Policy Togetherness: Alignment between national and international AI governance frameworks would streamline safe and scalable deployment.

Figure 1. Contemporary Artificial Intelligence Model Framework in Diagnosing Bacterial Infection

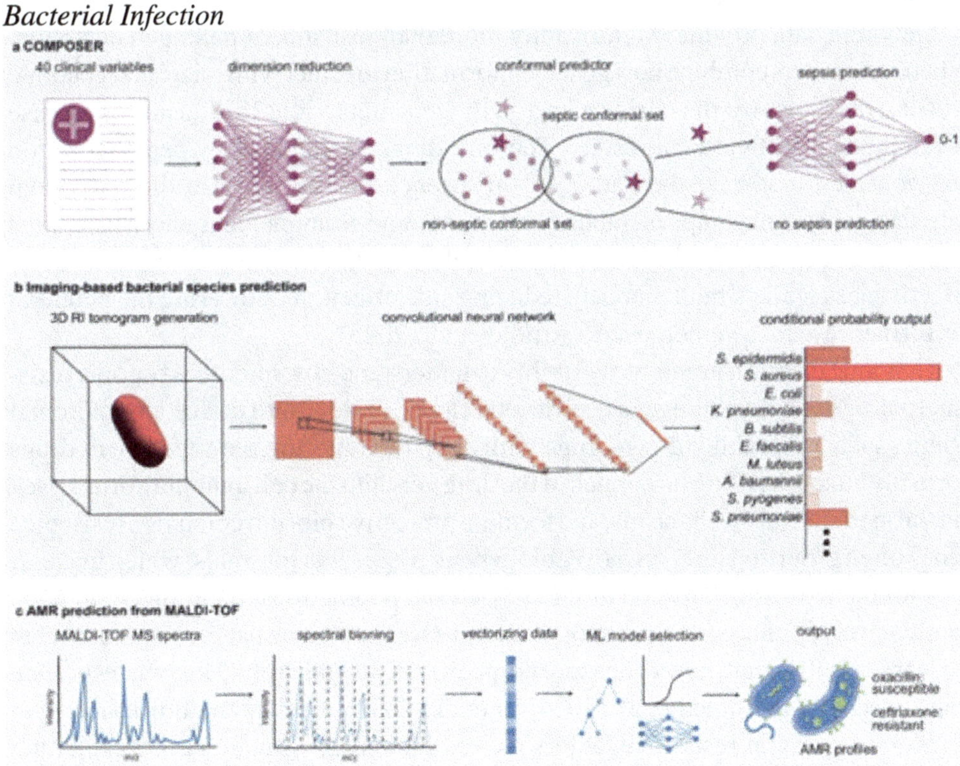

Source: (Arnold et al., 2025)

Genomics: The Language of Resistance

Genomic technologies have quickly evolved from specialized research tools into practical platforms that can now be implemented in regional hospitals and, more frequently, even in remote areas, enabling the rapid identification of pathogens and the resistance-related mutations within their genetic material (Auganova et al., 2025; Mallik et al., 2025). Genomics has essentially transformed our comprehension of antimicrobial resistance (AMR). Through the complete sequencing of infectious microorganisms' entire genomes, researchers can pinpoint with accuracy the specific genes and mutations that drive resistance, virulence, and survival under pharmaceutical pressure. The application of whole-genome sequencing (WGS) to *Mycobacterium tuberculosis* strains across Southeast Asia proved crucial in mapping outbreak clusters down to particular transmission chains, thereby enabling more precisely targeted public health interventions (Daniyarov et al., 2021). Similarly, distinctive genomic indicators like modifications in the pfkelch13 gene serve as dependable markers of resistance to primary antimalarial drugs, offering clinicians immediate actionable information to adjust therapeutic approaches before treatment complications arise (Thawong et al., 2025).

Genomic data provides significantly more than just simple pathogen detection. When researchers combine host genetic information together with microbial genomic profiles, they uncover the intricate host-pathogen interactions that influence disease development and treatment results. These findings enable therapy personalization that considers not only the pathogen's resistence profile but also the individual patient's genetic makeup, metabolic functions, and immune characteristics. As a result, medical treatments become progressively more tailored, improving therapeutic effectiveness while simultaneously reducing side effects and lowering the potential for further resistence emergence. (Arnold et al., 2025).

Genomic research grants us the ability to understand how microbes become resistant to drugs. These pathogens use various clever strategies to dodge antimicrobial agents, such as breaking down drugs with enzymes, altering the sites where drugs normally attach, limiting how much of the drug gets into the cell, pumping drugs back out through special mechanisms, and forming protective biofilm colonies. Resistence genes often group together on moveable genetic pieces like plasmids, which makes it easy for them to jump between different bacterial populations. By doing systematic genomic watchfulness, we can spot these tell-tale molecular patterns and use what we learn to help create new drugs meant specifically to get around known resistence tricks (see fig 2) (Ahmed et al., 2024; Li et al., 2023). Lots of multidrug-resistant germs carry several resistence factors packed into single genetic packages, letting them survive against a wide range of modern medicines. Notable examples include *methicillin-resistant Staph aureus* (MRSA), *carbapenem-resistant Klebsiella pneu-*

moniae and *E. coli*, multi-drug-resistant TB, and *Acinetobacter baumannii*. Similar resistence patterns show up in non-bacterial germs too, like Candida species that don't respond to fluconazole and HIV strains with antiviral resistence mutations.

Figure 2. Antimicrobial resistance mechanisms in bacteria

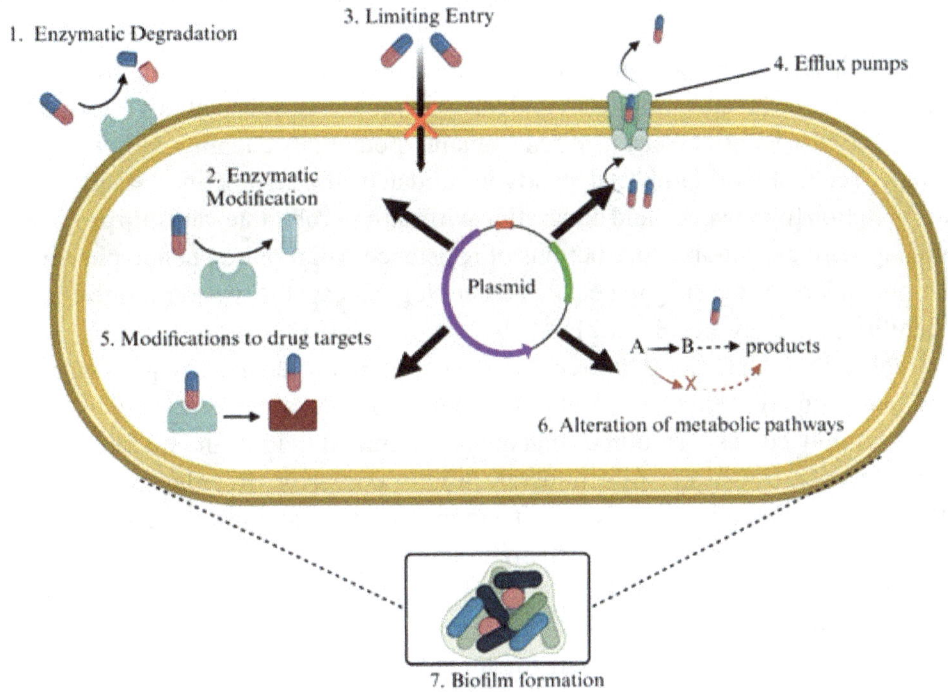

Source: (Ahmed et al., 2024).

Protozoan parasites like Plasmodium sp., Leishmania sp., and Trypanosoma sp. present major hurdles in clinical management. These pathogens can acquire drug-resistance mutations, yet they also employ a survival tactic called persistency, wherein they enter dormant, low-metabolism stages that help them survive drug exposure. Genomic analyses have helped uncover the molecular mechanisms governing this dormancy and resistance, pointing toward novel therapeutic approaches for these elusive life stages (Canales et al., 2025). Persistency cells in protozoa display remarkable flexibility by modulating their internal environment within the host. They typically accomplish this by lowering their metabolic rate, reducing DNA replication and protein synthesis, and furthermore, they respond to host cues that trigger a dormant state to ensure long-term survival. This adaptability enables protozoa such as *Balamuthia mandrillaris* and *Acanthamoeba spp.* to form double-walled cysts able to endure extreme temperatures, UV radiation, and even antimicrobial agents

(Rodríguez-Almonacid et al., 2023). From a clinical perspective, this dormant state can exacerbate disease progression. The persistence mechanisms seen in protozoa share biological parallels with those in bacteria, wherein organisms minimize their activity to withstand treatment pressure. This widespread phenomenon across species opens avenues to develop more targeted therapies aimed at shared dormancy pathways and their molecular regulators (Capela et al., 2019; Rodríguez-Almonacid et al., 2023).

In clinical practice, genomics increasingly informs diagnostic and therapeutic decisions. Direct sequencing from patient specimens can substantially shorten the time required to identify resistance determinants, potentially enabling earlier targeted therapy. Yet, real-world impact depends on validated clinical pipelines, standardized interpretation frameworks, and integration with antimicrobial stewardship programs. In many settings, genomic predictions of resistance still require phenotypic confirmation, underscoring the gap between technological capability and routine clinical reliability (Daniyarov et al., 2021).

Implementation remains uneven. Large-scale genomic analysis requires specialized bioinformatics expertise, stable computational infrastructure, and sustained reagent supply chains—resources that are often limited in low- and middle-income countries, where resistant infections are highly prevalent. Regulatory oversight, data governance standards, and cost-effectiveness evaluations are also insufficiently harmonized across jurisdictions, constraining scalability. Without structured policy frameworks and sustainable financing models, genomic medicine risks widening existing health inequities (Arnold et al., 2025).

Looking forward, portable sequencing platforms integrated with AI-driven analytics may enhance point-of-care decision-making, particularly in remote or resource-constrained environments. When combined with advances in nanobiotechnology for targeted drug delivery, these approaches could support more precise host–pathogen–guided therapies. However, their clinical adoption will depend on rigorous validation, regulatory alignment, affordability, and demonstrable outcome improvement rather than technological promise alone (Li et al., 2023).

Nanobiotechnology: A New Scale of Precision

Addressing multidrug-resistant (MDR) infections requires strategies that overcome the pharmacokinetic and biological limitations of conventional antibiotics, including poor tissue penetration, intracellular persistence, and biofilm protection. Nanobiotechnology offers a targeted delivery platform at the 1-100 nm scale that enhances drug stability, bioavailability, and site-specific accumulation (Solanki et al., 2024).

Nanocarriers such as liposomes, polymeric nanoparticles, dendrimers, and metal-based particles encapsulate antimicrobials to reduce premature degradation and limit systemic toxicity. Their principal advantage lies in improved penetration into biofilms and intracellular compartments, key reservoirs of persistent infection. However, while these mechanisms are well demonstrated in preclinical models, comparative clinical superiority over optimized conventional regimens remains limited (Li et al., 2023).

Metal nanoparticles, particularly silver and gold, exhibit intrinsic antimicrobial activity through membrane disruption, oxidative stress induction, and interference with microbial replication pathways. Surface functionalization further enhances selective binding to bacterial structures. Despite strong in vitro efficacy, concerns regarding long-term toxicity, biodistribution, and environmental accumulation continue to restrict widespread therapeutic endorsement (Li et al., 2023; Solanki et al., 2024).

Nanocarriers also enable combination strategies, co-delivering antibiotics with immunomodulatory agents or incorporating stimuli-responsive release systems triggered by pH or enzymatic signals at infection sites. These approaches aim to synchronize pathogen eradication with immune modulation. Yet, most remain at experimental or early translational stages, and robust clinical outcome data are sparse. Emerging regional applications illustrate both promise and limitation. Nanoparticle-stabilized antimalarials in sub-Saharan Africa demonstrate improved thermal stability and pharmacokinetics, suggesting relevance for resource-limited climates. Early trials of liposomal antibiotics for neonatal MDR *Klebsiella pneumoniae* in Southeast Asia indicate improved tolerability and bacterial clearance. Nevertheless, these examples reflect early-phase implementation rather than standardized global practice (Solanki et al., 2024).

Translational barriers remain substantial. Manufacturing complexity, scalability constraints, high production costs, and evolving regulatory frameworks slow clinical integration. In low- and middle-income countries, infrastructure limitations further reduce feasibility. Regulatory harmonization and cost-effectiveness evaluation remain unresolved challenges. Integration with artificial intelligence and genomics may enhance design precision and personalization. AI-assisted modelling can optimize nanoparticle size and surface properties, while genomic profiling enables pathogen- and host-specific therapeutic targeting. However, the central challenge lies not in technological capability but in demonstrating reproducible clinical benefit, safety, and affordability at scale (Li et al., 2023).

The Mosaic of Integration

Fighting infections caused by multidrug-resistant (MDR) pathogens clearly requires more than just stand-alone tech fixes. Real progress comes from knitting together various innovations into a single, synergistic framework. By melding artificial intelligence (AI), genomics, nano-biotechnology, and the One Heath approach, this chapter proposes an integrated strategy to tackle microbial threats that cover molecular mechanisms, clinical decision processes, and system-level determinants.

At the analytical core lies data convergence. AI platforms can synthesize genomic data, electronic health records, environmental indicators, and transmission dynamics to model resistance trajectories and identify outbreak signals earlier than conventional surveillance (Arnold et al., 2025). However, predictive accuracy depends on data quality and contextual calibration. Genomics provides mechanistic resolution, identifying resistance mutations, mobile genetic elements, and regulatory adaptations, but its value is realized only when linked to predictive analytics and therapeutic decision support. Nanobiotechnology, in turn, operationalizes these insights by improving targeted drug delivery and biofilm penetration while attempting to minimize systemic toxicity (Li et al., 2023; Solanki et al., 2024). Importantly, most nano-based antimicrobials remain at varying stages of translational maturity, and few have achieved broad clinical validation.

Rather than functioning as isolated pillars, these domains should operate within a feedback-driven cycle: genomic surveillance informs AI prediction; AI refines therapeutic targeting and nanomedicine design; clinical outcomes recalibrate both algorithms and molecular targets. The strength of this model lies in adaptive learning. Its central challenge lies in regulatory harmonization, validation standards, and real-world scalability. The One Health framework expands this integration beyond the clinic. Because resistance circulates across human, animal, and environmental reservoirs, genomic surveillance must extend to hospitals, livestock systems, wildlife interfaces, and water ecosystems (Nadimpalli et al., 2018; Thawong et al., 2025). Without cross-sector coordination, technological advances risk addressing downstream manifestations rather than upstream transmission drivers.

Implementation disparities are most visible in low- and middle-income countries (LMICs), where MDR burden is high but infrastructural capacity is uneven. AI systems require locally trained datasets to avoid algorithmic bias. Genomic surveillance demands laboratory infrastructure and workforce stability. Nanomedicine development must align with manufacturing feasibility and cost constraints (Anaya et al., 2025). Ethical governance presents an additional unresolved issue: large-scale genomic and clinical data integration raises concerns regarding consent, privacy, and equitable benefit-sharing (Meagher et al., 2022). Technological sophistication without public trust undermines sustainability.

In Southeast Asia, portable genomic sequencing combined with AI-assisted outbreak analytics improved detection and containment of multidrug-resistant *Klebsiella pneumoniae*, though long-term sustainability depends on financing and regulatory support (Anaya et al., 2025). In sub-Saharan Africa, nanoparticle-based antimalarial development is increasingly coupled with genomic surveillance and AI-guided treatment algorithms, yet scale-up remains contingent on cost-effectiveness and supply chain resilience (Canales et al., 2025). In Latin America, participatory governance models have strengthened acceptance of AI-assisted applications for tuberculosis and Chagas disease by embedding ethical oversight into data-sharing structures (Anaya et al., 2025). These cases demonstrate feasibility but also highlight dependency on political commitment and sustained funding.

Realizing this integrated paradigm requires more than technological investment. It demands multidisciplinary training, interoperable data standards, cross-border regulatory alignment, and financing models that support affordable production and equitable access (Anaya et al., 2025). The long-term objective is not technological accumulation but adaptive capacity: a health system able to anticipate resistance trends, intervene early, recalibrate therapies, and respond proportionally to emerging MDR threats. The viability of this framework ultimately depends on rigorous validation, economic realism, and ethical governance rather than on transformative rhetoric alone. Table 1 synthesises the five major technological domains examined in this chapter — artificial intelligence, genomics, nanobiotechnology, One Health frameworks, and precision-oriented antimicrobial stewardship — across six critical dimensions: principal applications, key limitations, readiness levels, regulatory considerations, and feasibility in low- and middle-income country contexts.

Table 1. Structured overview of precision technologies for MDR pathogens

Domain	Main applications in MDR/AMR	Key limitations	Readiness level	Regulatory / ethical issues	Feasibility in LMICs
Artificial intelligence (AI)	• Rapid resistance prediction from WGS, MALDI-TOF, microscopy • Sepsis early warning from EHR • Outbreak detection and risk stratification.	• Poor generalisation across sites/ populations • Risk of misclassification and inappropriate antibiotic changes • Need for high-quality digital data and IT infrastructure.	• Some tools in real-world pilots (e.g. sepsis models); many still proof-of-concept or retrospective only.	• Limited AI-specific regulation • Data protection, consent, and algorithmic bias concerns.	• Constrained by partial digitisation, connectivity, and expertise • Most suitable for phased pilots in referral centres linked to HIS strengthening.
Genomics / multi-omics	• Detection of resistance genes/mutations • Transmission mapping (e.g. MDR TB, pfkelch13 malaria) • One Health genomic surveillance.	• High capital and running costs • Need for bioinformatics skills and stable lab/data systems • Sensitive data with long-term storage implications.	• Routine in many reference labs; limited use at frontline facilities, especially in LMICs.	• Unclear data ownership and cross-border sharing rules • Need for robust consent and benefit-sharing frameworks.	• Feasible via regional/national reference centres with strong partnerships • Impact depends on integration into guidelines and surveillance, not sequencing alone.
Nanobiotechnology	• Targeted delivery into biofilms and intracellular niches • Metal nanoparticles with intrinsic antimicrobial effects • Nano-formulated antimalarials adapted to tropical settings.	• Complex and costly manufacturing • Limited long-term safety data • Sparse local production capacity and QC systems.	• Selected products in trials or early use; many concepts remain preclinical.	• Evolving nanomedicine guidance • Challenges in evaluating complex formulations and invisible risks.	• High potential for priority indications; realistic primarily where local or regional manufacturing and clear use-cases exist.
One Health / integrated frameworks	• Cross-sector AMR surveillance (human, animal, environment) • Combined AI–genomics–nano responses to outbreaks.	• Demands strong cross-sector governance and stable funding • Risk of fragmented pilots if not embedded in national AMR plans.	• Conceptually mature; practical implementation uneven and mostly project-based.	• Need for harmonised standards and data-sharing agreements across sectors and borders.	• Feasible where political will, donor alignment, and local leadership are present; case studies show success in selected regions.
Antimicrobial stewardship programmes (ASP) in a precision context	• Use of rapid diagnostics/AI/genomics to refine empirical therapy, shorten broad-spectrum use, and guide local guidelines/ formularies.	• Dependent on data systems, clinical buy-in, and leadership • Risk of "data-rich, action-poor" implementation.	• Widely established concept; integration of precision tools still emerging and mostly in pilot projects.	• Need for clear mandates, accountability, and transparent use of data for audit and feedback.	• Relatively low-cost, high-yield lever when adapted to resource constraints; precision tools can be layered in as infrastructure improves.

3. IMPLEMENTATION CHALLENGES AND LMIC CASE STUDIES

Infrastructure, Data, and Expertise Gaps

Implementing advanced precision therapies in low- and middle-income countries (LMICs) requires more than technology transfer; it demands systemic readiness. In many settings, foundational infrastructure remains insufficient, not only in tertiary hospitals but also in peripheral clinics and mobile units. While basic microbiology laboratories may exist, next-generation sequencing capacity, reliable cold storage, stable electricity, and secure digital connectivity are often limited. Even where equipment is available, frequent power disruptions, fragile supply chains, and high maintenance costs compromise sustainability rather than merely delaying implementation (Anaya et al., 2025).

Beyond infrastructure, data ecosystem fragmentation represents a more structural constraint. Precision approaches depend on integrated clinical, microbiological, and genomic data streams. However, incomplete digitization and reliance on paper-based records restrict the availability of high-quality, locally representative datasets for AI training and validation. This not only weakens algorithmic reliability but also risks reinforcing bias through the use of externally derived models that do not reflect local epidemiology (Anaya et al., 2025; Bilal et al., 2025). The central challenge therefore lies not simply in adopting AI tools, but in establishing interoperable data systems capable of supporting them. Workforce capacity constitutes a parallel limitation. Advanced diagnostics require interdisciplinary expertise spanning laboratory sciences, bioinformatics, quality assurance, and data analytics. Yet many LMICs face persistent brain drain and limited curricular integration of omics and digital health competencies. The mismatch between rapid technological advancement and slower educational reform constrains translational maturity and long-term scalability (Anaya et al., 2025; Horgan et al., 2025).

Connectivity and logistics further influence feasibility. Cloud-based AI platforms rely on stable, high-speed internet, while nanobiotechnology applications depend on robust cold-chain systems. In environments characterized by network instability and climate-sensitive storage challenges, diagnostic turnaround times and therapeutic reliability are directly affected (Anaya et al., 2025). These constraints highlight that technological readiness alone does not determine implementation success; contextual adaptation is equally critical. Crucially, implementation strategies must be pathogen-specific and system-responsive. Sequencing priorities, partnership models, and regulatory pathways will differ between malaria-endemic regions in West Africa and tuberculosis programs in Peru. Scalable innovation in LMICs will therefore depend on locally anchored solutions, such as portable sequencing platforms or hybrid remote training models, supported by sustained institutional investment.

Without structural alignment between infrastructure, data governance, workforce development, and policy frameworks, precision technologies risk remaining pilot initiatives rather than becoming integrated components of routine care.

Economic and Policy Constraints

Policymakers in many LMICs face sustained pressure to justify that emerging health technologies are both cost-effective and clinically meaningful. When positioned against foundational public health priorities, such as water and sanitation, immunization, and nutrition, precision-based interventions are often framed as optional rather than essential. A central challenge lies in the limited availability of context-specific cost–benefit analyses. Much of the economic and clinical evidence supporting precision medicine, AI-driven diagnostics, and nanomedicine is derived from high-income settings, reducing its policy relevance and weakening arguments for domestic investment and scale-up (Anaya et al., 2025; Horgan et al., 2025).

Regulatory fragility further constrains implementation. Only a small number of LMICs have developed clear frameworks governing clinical AI deployment, genomic data governance, or approval pathways for nanobiotechnology. In the absence of coherent national standards, approval processes become inconsistent and protracted, oversight mechanisms remain weak, and substandard or insufficiently validated products are more likely to enter the market. This regulatory uncertainty not only compromises patient safety but also undermines professional and public trust (Mani et al., 2025).

The problem is compounded when international guidance, such as that issued by the World Health Organization or the U.S. Food and Drug Administration, is adopted without contextual adaptation. Direct policy transplantation often creates operational misalignment, delays implementation, and obscures accountability structures. Without locally tailored regulatory architecture and structured post-market surveillance, the translational pathway for AI diagnostics and nanomedicine remains unstable, limiting long-term sustainability and equitable integration into national health systems.

Cultural, Ethical, and Social Barriers

Socio-cultural dynamics critically determine whether precision medicine is implemented responsibly or rejected despite technical feasibility. The integration of AI, genomics, and nanotechnology requires the collection and cross-border exchange of sensitive genetic and health data, raising unresolved questions about data ownership, informed consent, and equitable benefit sharing. These concerns are particularly acute in communities with histories of exploitation or weak legal protections, where fears of genetic surveillance, discrimination, or external data control can undermine participation and trust (Patil et al., 2025). The central challenge lies not only in protecting privacy, but in demonstrating tangible community benefit.

Clinical translation presents additional barriers. AI-derived recommendations and genomics-based resistance profiling may be scientifically robust yet difficult to operationalize in resource-limited or culturally diverse settings. Complex probabilistic outputs do not automatically translate into clinical confidence, especially where literacy levels vary or multiple local languages are spoken. Similarly, novel nanomedicine delivery systems, such as patches, inhalers, or long-acting formulations, may provoke suspicion when perceived as "invisible" or poorly explained interventions. In such contexts, technological sophistication without culturally grounded communication risks reinforcing mistrust rather than improving care (Meagher et al., 2022).

These limitations highlight that implementation readiness is as much social as technical. Community-engaged governance models, where patients and local stakeholders participate in protocol design, data oversight, and benefit-sharing frameworks, have shown greater potential to enhance legitimacy and long-term adoption (Anaya et al., 2025; Meagher et al., 2022). Critically, participatory approaches reposition precision medicine from a top-down innovation model toward a co-developed public health strategy. Without deliberate trust-building mechanisms, even highly advanced technologies may fail to achieve sustainable integration.

Southeast Asia: Data-Driven Outbreak Response - A Case Study

A compelling case study emerges from Southeast Asia's recent fight against MDR Klebsiella. Hospitals in both urban and rural settings worked together with universities and government agencies to rapidly implement their local genomic tools. When neonatal deaths suddenly spiked, teams set up small "pop-up" sequencing stations sometimes just a generator-powered room with a portable sequencer and Wi-Fi. With backing from local & international experts, field teams uploaded bacterial genomes to global platforms and quickly traced the outbreak's origin back to a single hospital supplier. Instead of enforcing a broad antibiotic restriction, health authorities were able to isolate specific product batches for quarantine, and clinicians could adjust their prescribing practices in real time (Anaya et al., 2025; Horgan et al., 2025).

This approach demonstrates what becomes possible when local health systems are given flexible tools, working partnerships, and the credibility to use molecular data for instant decision-making. The ripple effects include better training for nurses and technicians, data-sharing agreements between hospitals, and constant growth in surveillance capabilities in the future.

Sub-Saharan Africa: Nanobiotechnology and Malaria Therapy - A Paradigm Friction

In sub-Saharan Africa, malaria remains a major health concern, especially since conventional treatments are becoming less effective against drug-resistant strains. Through collaboration between Nigerian, Kenyan, and international researchers, antimalarial medications using nanoparticles have been developed locally. These nanoparticles are manufactured using plant extracts and remain stable across different temperatures, requiring minimal imported ingredients and eliminating the need for cold-chain logistics. Clinical testing at various healthcare facilities demonstrated improved treatment outcomes, reduced side effects, and higher patient adherence to treatment regimens (Canales et al., 2025; Patil et al., 2025).

Implementation beyond just new technology: local stakeholders adapted clinical guidelines to fit community practices (ex. using traditional remedies to illustrate how the novel treatment works). The scientific and social framework was just as notable as the technical aspect. By forming proprietary craftsmanship in nanotechnologies, these collabs are also priming local industries for manufacturing advancements across infectious disease indications.

Latin America: Community-Based Genomics for Infectious Disease

Latin America, particularly the Andean and Amazon regions, has served as a hotbed for innovative "bottom-up" strategies when it comes to rolling out new infectious disease tools. In nations such as Peru & Columbia, local health clinics and research institutions have been forced to collaborate closely to build co-governance frameworks for biobanking patient specimens and conducting genomic analyses on various pathogens. Community councils at the local level have been granted a meaningful seat at the table when it comes to establishing research agendas, drafting data-sharing protocols, and making certain that any resulting advantages trickle down to the community level (Anaya et al., 2025).

By weaving social values, context-specific ethics, and accountability mechanisms into every facet of the work, these initiatives were able to break down traditional barriers to participation. The outcomes have been remarkable: improved patient recruitment, quicker diagnoses, and better detection of MDR tuberculosis infections and Chagas disease. This success also prompted local policymakers to proactively embrace new technologies. Most notably, genomic intervention came to be viewed as something "of the people, by the people," solidifying its role in both scientific and community settings.

4. CONCLUSION AND FUTURE DIRECTIONS

The accelerating threat of antimicrobial resistance (AMR) represents not only a clinical emergency but also a systemic test of global health capacity. Advances in artificial intelligence (AI), genomics, and nanobiotechnology are shifting antimicrobial strategies from empirical, population-based approaches toward precision-guided interventions across molecular, clinical, and epidemiological levels. Rapid diagnostics, targeted therapeutics, and advanced delivery platforms demonstrate substantial promise. However, their real-world impact depends less on technological sophistication than on regulatory coherence, infrastructure readiness, workforce competence, and stewardship frameworks. Without these enabling conditions, innovation risks remaining aspirational rather than operational (Arnold et al., 2025; Mallik et al., 2025; Okeke et al., 2024). In the author's assessment, the gap between innovation and implementation therefore constitutes the most consequential frontier in contemporary AMR research.

Experience from low- and middle-income countries (LMICs) underscores a central tension: scientific advancement does not automatically translate into equitable implementation. Structural barriers, fragmented data systems, limited laboratory capacity, financing constraints, and uneven regulatory oversight, often outweigh technical limitations. At the same time, locally driven adaptations such as mobile diagnostics, community-based genomic surveillance, and context-specific nanotherapeutics illustrate that innovation can emerge from constrained settings. The critical issue, therefore, is not whether precision technologies exist, but whether they can be embedded into sustainable delivery systems that reflect local epidemiology and economic realities (Anaya et al., 2025; Horgan et al., 2025).

Translational maturity remains uneven across domains. While AI-assisted surveillance and CRISPR-based diagnostics show increasing feasibility, many novel antimicrobials and biologics remain financially inaccessible or poorly aligned with region-specific pathogen burdens. Moreover, enthusiasm for emerging modalities, such as phage therapy or monoclonal antibodies, often outpaces regulatory preparedness and cost-effectiveness evaluation. A major unresolved challenge lies in bridging discovery pipelines with procurement mechanisms that ensure affordability, stewardship, and supply-chain reliability in high-burden settings (Fayyad-Kazan, 2025; Okeke et al., 2024; Wheeler et al., 2023). Among the technologies examined, artificial intelligence emerges as uniquely positioned to bridge these translational gaps — not as a standalone solution, but as an integrative layer connecting genomic, clinical, and epidemiological intelligence.

AI is likely to play a pivotal integrative role, linking genomic, clinical, and epidemiological data to enable earlier detection of resistance trends and more precise therapeutic decisions. Yet algorithmic performance depends on representative datasets, transparent validation, and governance safeguards. Without deliberate inclusion of LMIC-derived data and clear accountability structures, predictive systems may reinforce rather than reduce inequities (Arnold et al., 2025; Ayesiga et al., 2024).

Ultimately, technological convergence alone will not resolve AMR. Sustainable progress requires coordinated governance, regional regulatory harmonization, ethical data-sharing frameworks, workforce development in genomics and data science, and community engagement that builds trust in precision approaches. The decisive shift must be from isolated innovation to system-level integration.

As a summary for this chapter, AMR remains a dynamic and adaptive threat, but the available toolkit, spanning AI, genomics, and nanobiotechnology, has become increasingly sophisticated and interconnected. The central challenge is no longer purely scientific; it is translational and structural. Precision antimicrobial strategies will achieve durable impact only if embedded within resilient, equitable health systems capable of scaling innovation responsibly across diverse global contexts. In this sense, this chapter contributes to the collective goal of the volume AI, Nanobiotechnology, and the Future of Precision Antibiotics by articulating how converging advances in biomedicine, biotechnology, and data science can be deliberately harnessed not merely to achieve technological sophistication, but to develop precision antimicrobial strategies that are globally accessible, contextually grounded, and resilient enough to meet the evolving demands of a world facing persistent microbial threats.

REFERENCES

Abu-El-Ruz, R., AbuHaweeleh, M. N., Hamdan, A., Rajha, H. E., Sarah, J. M., Barakat, K., & Zughaier, S. M. (2025). Artificial Intelligence in Bacterial Infections Control: A Scoping Review. *Antibiotics (Basel, Switzerland)*, *14*(3), 256. DOI: 10.3390/antibiotics14030256 PMID: 40149067

Ahmed, S. K., Hussein, S., Qurbani, K., Ibrahim, R. H., Fareeq, A., Mahmood, K. A., & Mohamed, M. G. (2024). Antimicrobial resistance: Impacts, challenges, and future prospects. *Journal of Medicine, Surgery, and Public Health*, *2*, 100081. DOI: 10.1016/j.glmedi.2024.100081

Anaya, J. M., Herrán, M., & Pino, L. E. (2025). Challenges and opportunities for precision medicine in developing nations. *Expert Review of Precision Medicine and Drug Development*, *10*(1), 1–15. DOI: 10.1080/23808993.2025.2505796

Arnold, A., McLellan, S., & Stokes, J. M. (2025). How AI can help us beat AMR. *Npj Antimicrobials and Resistance 2025 3:1*, *3*(1), 1–15. DOI: 10.1038/s44259-025-00085-4

Auganova, D., Atavliyeva, S., Gharbi, N., Zholdybayeva, E., Skiba, Y., Akisheva, A., Tsepke, A., Alenova, A., Guyeux, C., Wirth, T., Sola, C., Tarlykov, P., & Tarlykov, P. *. (2025). Genomic characterization and epidemiology of Mycobacterium tuberculosis lineage 2 isolates from Kazakhstan. *Scientific Reports 2025 15:1*, *15*(1), 1–10. DOI: 10.1038/s41598-025-22485-2

Ayesiga, I., Yeboah, M. O., Okoro, L. N., Edet, E. N., Gmanyami, J. M., Ovye, A., Atimango, L., Gadzama, B. N., Kembabazi, E., & Atwau, P. (2024). Artificial intelligence-enhanced biosurveillance for antimicrobial resistance in sub-Saharan Africa. *International Health*, *17*(5), 795–803. DOI: 10.1093/inthealth/ihae081 PMID: 39545538

Babu, S., & Nutman, T. B. (2016). Helminth-Tuberculosis Co-infection: An Immunologic Perspective. *Trends in Immunology*, *37*(9), 597–607. DOI: 10.1016/j.it.2016.07.005 PMID: 27501916

Bilal, H., Khan, M. N., Khan, S., Shafiq, M., Fang, W., Khan, R. U., Rahman, M. U., Li, X., Lv, Q. L., & Xu, B. (2025). The role of artificial intelligence and machine learning in predicting and combating antimicrobial resistance. *Computational and Structural Biotechnology Journal*, *27*, 423–439. DOI: 10.1016/j.csbj.2025.01.006 PMID: 39906157

Boussina, A., Shashikumar, S. P., Malhotra, A., Owens, R. L., El-Kareh, R., Longhurst, C. A., Quintero, K., Donahue, A., Chan, T. C., Nemati, S., & Wardi, G. (2024). Impact of a deep learning sepsis prediction model on quality of care and survival. *Npj Digital Medicine 2024 7:1, 7*(1), 1–9. DOI: 10.1038/s41746-023-00986-6

Canales, C. C. S., Marquez Cazorla, J. I., Marquez Cazorla, R. M., Sábio, R. M., Santos, H. A., & Pavan, F. R. (2025). Combating Gram-negative infections: The role of antimicrobial peptides and nanotechnology in overcoming antibiotic resistance. *Materials Today. Bio, 35*, 102381. DOI: 10.1016/j.mtbio.2025.102381 PMID: 41142413

Capela, R., Moreira, R., & Lopes, F. (2019). An Overview of Drug Resistance in Protozoal Diseases. *International Journal of Molecular Sciences, 20*(22), 5748. DOI: 10.3390/ijms20225748 PMID: 31731801

Cesaro, A., Hoffman, S. C., Das, P., & de la Fuente-Nunez, C. (2025). Challenges and applications of artificial intelligence in infectious diseases and antimicrobial resistance. *Npj Antimicrobials and Resistance 2025 3:1, 3*(1), 1–10. DOI: 10.1038/s44259-024-00068-x

Daniyarov, A., Molkenov, A., Rakhimova, S., Akhmetova, A., Yerezhepov, D., Chingissova, L., Bismilda, V., Toksanbayeva, B., Rakisheva, A., Akilzhanova, A., Kozhamkulov, U., & Kairov, U. (2021). Genomic Analysis of Multidrug-Resistant Mycobacterium tuberculosis Strains From Patients in Kazakhstan. *Frontiers in Genetics, 12*, 683515. DOI: 10.3389/fgene.2021.683515 PMID: 34858467

Fayyad-Kazan, M. (2025). *CRISPR-Cas Systems: Transformative Precision Tools to Combat Antimicrobial Resistance in Multidrug-Resistant Gram-Negative Pathogens.* https://www.genesispub.org/crispr-cas-systems-transformative-precision-tools-to-combat-antimicrobial-resistance-in-multidrug-resistant-gram-negative-pathogens

Getahun, Y. A., Ali, D. A., Taye, B. W., & Alemayehu, Y. A. (2022). Multidrug-Resistant Microbial Therapy Using Antimicrobial Peptides and the CRISPR/Cas9 System. *Veterinary Medicine (Auckland), 13*, 173–190. DOI: 10.2147/VMRR.S366533 PMID: 35983086

Goh, K. H., Wang, L., Yeow, A. Y. K., Poh, H., Li, K., Yeow, J. J. L., & Tan, G. Y. H. (2021). Artificial intelligence in sepsis early prediction and diagnosis using unstructured data in healthcare. *Nature Communications, 12*(1), 711. Advance online publication. DOI: 10.1038/s41467-021-20910-4 PMID: 33514699

Horgan, D., Tanner, M., Aggarwal, C., Thomas, D., Grover, S., Basel-Salmon, L., Dienstmann, R., Tan, T. J. Y., Park, W.-Y., Abu Rasheed, H. M., Siu, L. L., Ma, B., Ortiz-López, R., Van den Bulcke, M., Taucher, S. C., Ferris, A., Starling, N., Malapelle, U., Longshore, J., & Subbiah, V. (2025). Precision Oncology: A Global Perspective on Implementation and Policy Development. *JCO Global Oncology, 11*(11), e2400416. Advance online publication. DOI: 10.1200/GO-24-00416 PMID: 39847746

Li, M., Liu, Y., Gong, Y., Yan, X., Wang, L., Zheng, W., Ai, H., & Zhao, Y. (2023). Recent advances in nanoantibiotics against multidrug-resistant bacteria. *Nanoscale Advances, 5*(23), 6278–6317. DOI: 10.1039/D3NA00530E PMID: 38024316

Liu, V. X., Fielding-Singh, V., Greene, J. D., Baker, J. M., Iwashyna, T. J., Bhattacharya, J., & Escobar, G. J. (2017). The Timing of Early Antibiotics and Hospital Mortality in Sepsis. *American Journal of Respiratory and Critical Care Medicine, 196*(7), 856–863. DOI: 10.1164/rccm.201609-1848OC PMID: 28345952

Mallik, S., Dash, S., Mishra, S., Mohanty, R. B., Sahoo, S., Patra, A., Sahoo, S., Mishra, R., & Mohanty, J. N. (2025). Threat of multidrug resistant bacteria: The role of genomic solutions and a call to action. *Gene Reports, 41*, 102298. DOI: 10.1016/j.genrep.2025.102298

Mani, S., Lalani, S. R., & Pammi, M. (2025). Genomics and multiomics in the age of precision medicine. *Pediatric Research 2025 97:4, 97*(4), 1399–1410. DOI: 10.1038/s41390-025-04021-0

Meagher, K. M., Watson, S., Suh, G. A., & Virk, A. (2022). The New Precision Stewards? *Journal of Personalized Medicine, 12*(8), 1308. DOI: 10.3390/jpm12081308 PMID: 36013256

Nadimpalli, M., Delarocque-Astagneau, E., Love, D. C., Price, L. B., Huynh, B. T., Collard, J. M., Lay, K. S., Borand, L., Ndir, A., Walsh, T. R., Guillemot, D., De Lauzanne, A., Kerleguer, A., Tarantola, A., Piola, P., Chon, T., Lach, S., Ngo, V., Touch, S., & Abdou, A. Y. (2018). Combating Global Antibiotic Resistance: Emerging One Health Concerns in Lower- and Middle-Income Countries. *Clinical Infectious Diseases, 66*(6), 963–969. DOI: 10.1093/cid/cix879 PMID: 29346620

Okeke, I. N., de Kraker, M. E. A., Van Boeckel, T. P., Kumar, C. K., Schmitt, H., Gales, A. C., Bertagnolio, S., Sharland, M., & Laxminarayan, R. (2024). The scope of the antimicrobial resistance challenge. *Lancet, 403*(10442), 2426–2438. DOI: 10.1016/S0140-6736(24)00876-6 PMID: 38797176

Olatunji, I., Bardaji, D. K. R., Miranda, R. R., Savka, M. A., & Hudson, A. O. (2024). Artificial intelligence tools for the identification of antibiotic resistance genes. *Frontiers in Microbiology*, *15*, 1437602. DOI: 10.3389/fmicb.2024.1437602 PMID: 39070267

Osthoff, M., Gürtler, N., Bassetti, S., Balestra, G., Marsch, S., Pargger, H., Weisser, M., & Egli, A. (2017). Impact of MALDI-TOF-MS-based identification directly from positive blood cultures on patient management: A controlled clinical trial. *Clinical Microbiology and Infection*, *23*(2), 78–85. DOI: 10.1016/j.cmi.2016.08.009 PMID: 27569710

Patil, S., Singh, I., Verma, I. K., Kumar, A., Sharma, J., Ratn, A., Dhakad, M. S., & Sharma, D. (2025). Vaccines as Potential Frontliners Against Antimicrobial Resistance (AMR): A Focused Review. *Infection and Drug Resistance*, *18*, 5023–5041. DOI: 10.2147/IDR.S544665 PMID: 41020145

Rabaan, A. A., Alhumaid, S., Al Mutair, A., Garout, M., Abulhamayel, Y., Halwani, M. A., Alestad, J. H., Al Bshabshe, A., Sulaiman, T., Alfonaisan, M. K., Almusawi, T., Albayat, H., Alsaeed, M., Alfaresi, M., Alotaibi, S., Alhashem, Y. N., Temsah, M. H., Ali, U., & Ahmed, N. (2022). Application of Artificial Intelligence in Combating High Antimicrobial Resistance Rates. *Antibiotics (Basel, Switzerland)*, *11*(6), 784. DOI: 10.3390/antibiotics11060784 PMID: 35740190

Ren, Y., Chakraborty, T., Doijad, S., Falgenhauer, L., Falgenhauer, J., Goesmann, A., Hauschild, A. C., Schwengers, O., & Heider, D. (2022). Prediction of antimicrobial resistance based on whole-genome sequencing and machine learning. *Bioinformatics (Oxford, England)*, *38*(2), 325–334. DOI: 10.1093/bioinformatics/btab681 PMID: 34613360

Rodríguez-Almonacid, C. , Salazar-Bravo, J. ;, Karamysheva, Z. N., Soares, H., Nolasco, S., Tarannum, A., Camilo Rodríguez-Almonacid, C., Salazar-Bravo, J., & Karamysheva, Z. N. (2023). Molecular Mechanisms of Persistence in Protozoan Parasites. *Microorganisms 2023, Vol. 11, Page 2248*, *11*(9), 2248. DOI: 10.3390/microorganisms11092248

Rudd, K. E., Johnson, S. C., Agesa, K. M., Shackelford, K. A., Tsoi, D., Kievlan, D. R., Colombara, D. V., Ikuta, K. S., Kissoon, N., Finfer, S., Fleischmann-Struzek, C., Machado, F. R., Reinhart, K. K., Rowan, K., Seymour, C. W., Watson, R. S., West, T. E., Marinho, F., Hay, S. I., & Naghavi, M. (2020). Global, regional, and national sepsis incidence and mortality, 1990–2017: Analysis for the Global Burden of Disease Study. *Lancet*, *395*(10219), 200–211. DOI: 10.1016/S0140-6736(19)32989-7 PMID: 31954465

Schinkel, M., Paranjape, K., Nannan Panday, R. S., Skyttberg, N., & Nanayakkara, P. W. B. (2019). Clinical applications of artificial intelligence in sepsis: A narrative review. *Computers in Biology and Medicine*, *115*, 103488. Advance online publication. DOI: 10.1016/j.compbiomed.2019.103488 PMID: 31634699

Solanki, R., Makwana, N., Kumar, R., Joshi, M., Patel, A., Bhatia, D., & Sahoo, D. K. (2024). Nanomedicines as a cutting-edge solution to combat antimicrobial resistance. *RSC Advances*, *14*(45), 33568–33586. DOI: 10.1039/D4RA06117A PMID: 39439838

Thawong, N., Srilohasin, P., Phelan, J. E., Phornsiricharoenphant, W., Tongsima, S., Suriyaphol, P., Prammananan, T., Faksri, K., Sawaengdee, W., Wang, L., Hinthong, W., Hibberd, M. L., Campino, S., Wattanapokayakit, S., Mahasirimongkol, S., Chaiprasert, A., & Clark, T. G. (2025). Genomic decoding of drug-resistant tuberculosis transmission in Thailand over three decades. *Scientific Reports 2025 15:1, 15*(1), 1–11. DOI: 10.1038/s41598-025-15093-7

Tousif, S., & Ahmad, S. (2015). Challenges of Tuberculosis Treatment with DOTS: An Immune Impairment Perspective. *Journal of Cell Science & Therapy*, *06*(05). Advance online publication. DOI: 10.4172/2157-7013.1000223

Wheeler, N. E., Price, V., Cunningham-Oakes, E., Tsang, K. K., Nunn, J. G., Midega, J. T., Anjum, M. F., Wade, M. J., Feasey, N. A., Peacock, S. J., Jauneikaite, E., & Baker, K. S. (2023). Innovations in genomic antimicrobial resistance surveillance. *The Lancet Microbe*, *4*(12), e1063–e1070. DOI: 10.1016/S2666-5247(23)00285-9 PMID: 37977163

Xiao, C., Choi, E., & Sun, J. (2018). Opportunities and challenges in developing deep learning models using electronic health records data: A systematic review. *Journal of the American Medical Informatics Association : JAMIA*, *25*(10), 1419–1428. DOI: 10.1093/jamia/ocy068 PMID: 29893864

Yu, Y., Si, X., Hu, C., & Zhang, J. (2019). A Review of Recurrent Neural Networks: LSTM Cells and Network Architectures. *Neural Computation*, *31*(7), 1235–1270. DOI: 10.1162/neco_a_01199 PMID: 31113301

Zagajewski, A., Turner, P., Feehily, C., El Sayyed, H., Andersson, M., Barrett, L., Oakley, S., Stracy, M., Crook, D., Nellåker, C., Stoesser, N., & Kapanidis, A. N. (2023). Deep learning and single-cell phenotyping for rapid antimicrobial susceptibility detection in Escherichia coli. *Communications Biology*, *6*(1), 1164. Advance online publication. DOI: 10.1038/s42003-023-05524-4 PMID: 37964031

Zhang, D., Yin, C., Hunold, K. M., Jiang, X., Caterino, J. M., & Zhang, P. (2021). An interpretable deep-learning model for early prediction of sepsis in the emergency department. *Patterns (New York, N.Y.)*, *2*(2), 100196. Advance online publication. DOI: 10.1016/j.patter.2020.100196 PMID: 33659912

Conclusion

The collective contributions within this volume suggest that antimicrobial resistance (AMR) is no longer a challenge that can be contained within the traditional silos of microbiology or pharmacology. As we conclude this exploration, it becomes increasingly evident that the persistence of resistant pathogens is not merely a biological phenomenon but a reflection of a systemic fragmentation in how we design, deliver, and evaluate therapeutic interventions. The evidence presented across these chapters points toward a fundamental shift: the future of infection control may depend less on the discovery of isolated molecules than on the convergence of predictive intelligence, precision engineering, and ecological awareness.

THE SYNTHESIS OF CONVERGENCE: BEYOND ISOLATED INNOVATIONS

The conceptual journey of this book has been anchored by the Reyed Precision Convergence Framework for Antimicrobial Resistance (RPCF-AMR). This framework was introduced not as a static solution, but as a dynamic interpretive model to align disparate domains that have historically operated in isolation. By synthesizing the predictive capabilities of Artificial Intelligence (AI) with the site-specific delivery potential of nanobiotechnology, we move toward a "coordinated therapeutic logic."

One of the most significant interpretations emerging from this synthesis is the re-evaluation of antimicrobial efficacy. The data discussed suggests that efficacy is not an intrinsic, fixed property of a chemical compound. Instead, it appears to be an emergent outcome—a result of how effectively a therapeutic system can navigate the host's biological barriers, respond to the local microenvironment, and interact with complex microbial communities. This "context-aware" approach acknowledges that a compound's success is dictated by its environment as much as its molecular structure.

AI AND NANOTECHNOLOGY: FROM SPECULATION TO FUNCTIONAL ALIGNMENT

Throughout the chapters, Artificial Intelligence has been showcased as more than a computational tool for data processing. It serves as a "navigational compass" within the vast complexity of genomic and proteomic landscapes. However, as several contributors have noted, the power of AI-driven prediction is only as impactful as the physical systems capable of executing those predictions. Here, nanobiotechnology bridges the gap.

The transition from conventional drug delivery to "smart" nanocarrier systems—whether liposomal, polymeric, or biomimetic architectures—offers a degree of selectivity that was previously unattainable. Yet, a nuanced reading of current research indicates that no single platform is universally optimal. The performance of these systems remains sensitive to physiological constraints and translational hurdles. This realization reinforces the book's central theme: progress will likely depend on matching specific technological tools to precise clinical scenarios, rather than seeking a "one-size-fits-all" silver bullet.

The Polybiome and the Ecological Mandate

Perhaps the most critical shift articulated in this volume is the transition from a "pathogen-centric" view to an "ecological" perspective. The introduction of the polybiome concept underscores that antimicrobial interventions do not occur in a vacuum; they ripple through complex microbial ecosystems.

Recognizing that the host's microbiome, environmental reservoirs, and gene exchange networks are active participants in the resistance narrative is essential for long-term sustainability. The findings suggest that future strategies must prioritize "microbiome-sparing" interventions—therapies that clear infection while preserving the integrity of beneficial microbial communities. This alignment with One Health principles is not merely an ethical choice but a functional necessity for preventing the cyclical recurrence of resistance across human, animal, and environmental domains.

Translational Realities and the Path Ahead

While the theoretical and experimental successes of AI and nanomedicine are compelling, the path to clinical implementation remains a frontier requiring rigorous validation. We must distinguish between the "computational potential" of a model and its "clinical reliability." The challenges of scalability, regulatory standardizations, and the inherent variability of human biology remain formidable.

Future research must therefore move beyond proof-of-concept studies toward integrative development pipelines. These pipelines should ideally combine real-time experimental feedback with iterative computational modeling. The goal is to create a "closed-loop" system where clinical data informs AI models, which in turn optimize the next generation of nanodelivery systems.

FINAL REFLECTION: INTEGRATION AS THE NEW FRONTIER

In summary, the crisis of AMR demands a disciplined convergence of technologies, biological insights, and translational practices. The RPCF-AMR framework, as detailed in this work, provides a structured pathway for this integration, yet it remains open to refinement as new data emerges.

The future of antimicrobial therapy will likely be defined by alignment—the seamless coordination between what is discovered in the lab, how it is delivered to the site of infection, and how it respects the broader biological context. We hope this volume serves as a generative reference, prompting researchers and clinicians to look past disciplinary boundaries and work toward a future where precision antibiotics are not just a possibility, but a sustainable reality.

Reyed M. Reyed

Genetic Engineering and Biotechnology Research Institute (GEBRI), Egypt &
City of Scientific Research and Technological Applications (SRTA-City), Egypt

Compilation of References

Abbas, I. M., Mahdi, S. A., & Jabir, M. S. (2025). Insights into the antibacterial and anticancer activity of novel silver nanoparticles mediated bacterial colicin: Sustainable nanotechnology, in silico and dynamic study. *Plasmonics*, *20*, 10985–10999. DOI: 10.1007/s11468-025-03153-7

Abbasi, E., Akbarzadeh, A., & Milani, M. (2014). Dendrimers for biofilm-associated infections. *Journal of Microbiological Methods*, *103*, 1–8. DOI: 10.1016/j.mimet.2014.05.009

Abbasi, E., Aval, S. F., Akbar, A., & Nojoom, M. (2014). Dendrimers: Synthesis, applications, and properties. *Nanoscale Research Letters*, *9*(1), 247. DOI: 10.1186/1556-276X-9-247 PMID: 24994950

Abdelsattar, A. S., Dawoud, A., & Helal, M. A. (2021). Interaction of nanoparticles with biological macromolecules: A review of molecular docking studies. *Nanotoxicology*, *15*(1), 66–95. DOI: 10.1080/17435390.2020.1842537

Abed, N., & Couvreur, P. (2014). Nanocarriers for antibiotics: A promising solution to treat intracellular bacterial infections. *International Journal of Antimicrobial Agents*, *43*(6), 485–496. DOI: 10.1016/j.ijantimicag.2014.02.009 PMID: 24721232

Abegunde, S. M., Afolayan, B. O., & Ilesanmi, T. M. (2024). Ensuring sustainable plant-assisted nanoparticles synthesis through process standardization and reproducibility: Challenges and future directions – A review. *Sustainable Chemistry One World*, *3*, 100014. DOI: 10.1016/j.scowo.2024.100014

Abu-El-Ruz, R., AbuHaweeleh, M. N., Hamdan, A., Rajha, H. E., Sarah, J. M., Barakat, K., & Zughaier, S. M. (2025). Artificial Intelligence in Bacterial Infections Control: A Scoping Review. *Antibiotics (Basel, Switzerland)*, *14*(3), 256. DOI: 10.3390/antibiotics14030256 PMID: 40149067

Abughalia, A., Flynn, M., Clarke, P. F. A., Fayne, D., & Gobbo, O. L. (2025). The Use of Computational Approaches to Design Nano delivery Systems. *Nanomaterials (Basel, Switzerland)*, *15*(17), 1354. DOI: 10.3390/nano15171354 PMID: 40938032

Adler-Moore, J., & Proffitt, R. (2002). AmBisome: Liposomal formulation, structure, mechanism of action and pre-clinical experience. *The Journal of Antimicrobial Chemotherapy*, *49*(1, suppl_1), 21–30. DOI: 10.1093/jac/49.suppl_1.21 PMID: 11801577

Afkhami, F., Chen, Y., Walsh, L. J., Peters, O. A., & Xu, C. (2024). Application of nanomaterials in endodontics. *BME Frontiers, 5*, 0043. https://doi.org/DOI: 10.34133/bmef.0043

Afroze, C. A., Ahmed, M. N., Jahan, R., & Rahmatullah, M. (2026). Marine-derived prospective antibiotic armamentarium: From anti-quorum sensing to anti-biofilm and anti-biofouling biometabolites. *Heliyon*, *12*(2), e44524. DOI: 10.1016/j.heliyon.2026.e44524

Afzal, O., Altamimi, A. S. A., Nadeem, M. S., Alzarea, S. I., Almalki, W. H., Tariq, A., Mubeen, B., Murtaza, B. N., Iftikhar, S., Riaz, N., & Kazmi, I. (2022). Nanoparticles in drug delivery: From history to therapeutic applications. *Nanomaterials (Basel, Switzerland)*, *12*(24), 4494. DOI: 10.3390/nano12244494

Agboola, O. E., Agboola, O., Ayinla, Z. A., Agboola, S. S., Olaiya, O. E., Oyinloye, O. M., Fajana, O. M., Idowu, O. T., Omotuyi, O. I., Ilesanmi, O. S., & Oyinloye, B. E. (2025). Integrated virtual screening for anti-caries compounds from neem: Dual-target inhibition of biofilm formation and bacterial DNA replication. *Biomedicines*, *13*(9), 2202. DOI: 10.3390/biomedicines13092202

Aggarwal, R., Mahanjan, P., Pandiya, S., Bajaj, A., Verma, S. K., Yadav, P., Kharat, A. S., Khan, A. U., Dua, M., & Johri, A. K. (2024). Antibiotic resistance: A global crisis, problems and solutions. *Critical Reviews in Microbiology*, *50*(5), 896–921. DOI: 10.1080/1040841X.2024.2313024 PMID: 38381581

Aghajanpour, S., Ebrahimnejad, P., Amiri, H., Jeelani, H., Esfandyari-Manesh, M., Henschel, A., Singh, H., Dinarvand, R., & Hassan, S. (2025). Utilizing machine learning for predicting drug release from polymeric drug delivery systems. *Computers in Biology and Medicine*, *188*, 109756. DOI: 10.1016/j.compbiomed.2025.109756 PMID: 39978092

Agüero-Chapin, G., Antunes, A., & Marrero-Ponce, Y. (2026). A 2026 update on computational approaches to the discovery and design of antimicrobial peptides. *Antibiotics (Basel, Switzerland)*, *15*(2), 203. DOI: 10.3390/antibiotics15020203

Agu, P. C., Afiukwa, C. A., Orji, O. U., Ezeh, E. M., Ofoke, I. H., Ogbu, C. O., Ugwuja, E. I., & Aja, P. M. (2023). Molecular docking as a tool for the discovery of molecular targets of nutraceuticals in diseases management. *Scientific Reports*, *13*(1), 13398. DOI: 10.1038/s41598-023-40160-2

Ahmad, A., Imran, M., & Sharma, N. (2022). Precision Nanotoxicology in Drug Development: Current Trends and Challenges in Safety and Toxicity Implications of Customized Multifunctional Nanocarriers for Drug-Delivery Applications. *Pharmaceutics*, *14*(11), 2463. DOI: 10.3390/pharmaceutics14112463 PMID: 36432653

Ahmed, S. K., Hussein, S., Qurbani, K., Ibrahim, R. H., Fareeq, A., Mahmood, K. A., & Mohamed, M. G. (2024). Antimicrobial resistance: Impacts, challenges, and future prospects. *Journal of Medicine, Surgery, and Public Health*, *2*, 100081. DOI: 10.1016/j.glmedi.2024.100081

Ahsan, A., Thomas, N., Barnes, T. J., Subramaniam, S., Loh, T. C., Joyce, P., & Prestidge, C. A. (2024). Lipid nanocarriers-enabled delivery of antibiotics and antimicrobial adjuvants to overcome bacterial biofilms. *Pharmaceutics*, *16*(3), 396. DOI: 10.3390/pharmaceutics16030396 PMID: 38543290

Akbarzadeh, A., Rezaei-Sadabady, R., Davaran, S., Joo, S. W., Zarghami, N., Hanifehpour, Y., Samiei, M., Kouhi, M., & Nejati-Koshki, K. (2013). Liposome: Classification, preparation, and applications. *Nanoscale Research Letters*, *8*(1), 102. DOI: 10.1186/1556-276X-8-102 PMID: 23432972

Al Mazrouei, N., Ahmed Elnour, A., Badi, S., Abdalla, N. M., Mohamed, M. G., Ali, A. A., Sheikh, A., & Ashmawi, S. (2026). The impact of artificial intelligence on the prescribing, selection, resistance, and stewardship of antimicrobials: A scoping review. *BMC Infectious Diseases*, *26*, 222. DOI: 10.1186/s12879-025-12320-4

Al-Asmi, A., Al-Jabri, F. S., Al-Amrani, S. A., Al Farsi, Y., Al Sabahi, F., Al-Futaisi, A., & Al-Adawi, S. (2024). Medical tourism and neurological diseases: Omani patients' experience seeking treatment abroad. *Oman Medical Journal*, *39*(3), e628. DOI: 10.5001/omj.2024.69

Aldea, A. C., Digu ă, F. C., Presacan, O., Voaide , C., Toma, R. C., & Matei, F. (2025). Detecting antibiotic resistance: Classical, molecular, advanced bioengineering, and AI-enhanced approaches. *Frontiers in Microbiology*, *16*, 1673343. Advance online publication. DOI: 10.3389/fmicb.2025.1673343 PMID: 41113632

AlGain, S., Marra, A. R., Kobayashi, T., Marra, P. S., Celeghini, P. D., Hsieh, M. K., Shatari, M. A., Althagafi, S., Alayed, M., Ranavaya, J. I., Boodhoo, N. A., Meade, N. O., Fu, D., Sampson, M. M., Rodriguez-Nava, G., Zimmet, A. N., Ha, D., Alsuhaibani, M., Huddleston, B. S., & Salinas, J. L. (2025). Can we rely on artificial intelligence to guide antimicrobial therapy? A systematic literature review. *Antimicrobial Stewardship & Healthcare Epidemiology : ASHE*, *5*(1), e90. DOI: 10.1017/ash.2025.47

Alharthi, S. (2025). AI-powered in silico twins: Redefining precision medicine through simulation, personalization, and predictive healthcare. *Saudi Pharmaceutical Journal : SPJ : The Official Publication of the Saudi Pharmaceutical Society*, *34*(1), 1. DOI: 10.1007/s44446-025-00055-x

Alif, M. N., & Zhang, W. (2025). SynOmics: Integrating multiomics data through feature interaction networks. *Briefings in Bioinformatics*, *26*(6), bbaf595. Advance online publication. DOI: 10.1093/bib/bbaf595

Ali, T., Ahmed, S., & Aslam, M. (2023). Artificial intelligence for antimicrobial resistance prediction: Challenges and opportunities toward practical implementation. *Antibiotics (Basel, Switzerland)*, *12*(3), 523. DOI: 10.3390/antibiotics12030523

Al-Khalaifah, H., Rahman, M. H., Al-Surrayai, T., Al-Dhumair, A., & Al-Hasan, M. (2025). A One-Health perspective of antimicrobial resistance (AMR): Human, animals and environmental health. *Life*, *15*(10), 1598. DOI: 10.3390/life15101598

Allan, J., Belz, S., Hoeveler, A., Hugas, M., Okuda, H., Patri, A., Rauscher, H., Silva, P., Slikker, W., Sokull-Kluettgen, B., Tong, W., & Anklam, E. (2021, June). Regulatory landscape of nanotechnology and nanoplastics from a global perspective. *Regulatory Toxicology and Pharmacology*, *122*, 104885. DOI: 10.1016/j.yrtph.2021.104885 PMID: 33617940

Allen, T. M., & Cullis, P. R. (2013). Liposomal drug delivery systems: From concept to clinical applications. *Advanced Drug Delivery Reviews*, *65*(1), 36–48. DOI: 10.1016/j.addr.2012.09.037 PMID: 23036225

Almajidi, Y. Q., Faisal, R. R., Yaseen, A., & Al-hakeem, M. A. L. (2025). GastroPlus® as simulating/conducting software for modeling physiologically based pharmacokinetics & physiologically based biopharmaceutics. *Maaen Journal for Medical Sciences*, *4*(3), 1078. DOI: 10.55810/2789-9136.1078

Alprol, A. E., Hamad, T. M., Sharaf, H. E. R., El-Sayed, H. S., Barakat, K. M., Abouelwafa, A., & Khairy, H. M. (2026). Toxicological impacts of biogenic zinc oxide nanoparticles on blue parrotfish using multibiomarker assessment. *Scientific Reports*, *16*(1), 6546. DOI: 10.1038/s41598-026-36870-y

AlQurashi, D. M., AlQurashi, T. F., Alam, R. I., Shaikh, S., & Tarkistani, M. A. M. (2025). Advanced nanoparticles in combating antibiotic resistance: Current innovations and future directions. *Journal of Nanotheranostics*, *6*(2), 9. DOI: 10.3390/jnt6020009

Alshawwa, S. Z., Kassem, A. A., Farid, R. M., Labib, G. S., & Mostafa, S. K. (2022). Nanocarrier drug delivery systems: Characterization, limitations, future perspectives, and implementation of artificial intelligence. *Pharmaceutics*, *14*(4), 883. DOI: 10.3390/pharmaceutics14040883 PMID: 35456717

Altammar, K. A. (2023). A review on nanoparticles: Characteristics, synthesis, applications, and challenges. *Frontiers in Microbiology*, *14*, 1155622. DOI: 10.3389/fmicb.2023.1155622

Altevogt, B. M., Taylor, P., Akwar, H. T., Graham, D. W., Ogilvie, L. A., Duffy, E., & Essack, S. Y. (2025). A One Health framework for global and local stewardship across the antimicrobial lifecycle. *Communications Medicine*, *5*(1), 414. DOI: 10.1038/s43856-025-01090-4

Aminov, R. (2021). Acquisition and Spread of Antimicrobial Resistance: A *tet*(X) Case Study. *International Journal of Molecular Sciences*, *22*(8), 3905. DOI: 10.3390/ijms22083905

Aminov, R. I. (2010, December 8). A brief history of the antibiotic era: Lessons learned and challenges for the future. *Frontiers in Microbiology*, *1*, 134. DOI: 10.3389/fmicb.2010.00134 PMID: 21687759

An, A. Y., Choi, K. G., Baghela, A. S., & Hancock, R. E. W. (2021). An Overview of Biological and Computational Methods for Designing Mechanism-Informed Anti-biofilm Agents. *Frontiers in Microbiology*, *12*, 640787. DOI: 10.3389/fmicb.2021.640787

Anaya, J. M., Herrán, M., & Pino, L. E. (2025). Challenges and opportunities for precision medicine in developing nations. *Expert Review of Precision Medicine and Drug Development*, *10*(1), 1–15. DOI: 10.1080/23808993.2025.2505796

Andima, M., Boese, A., Paul, P., Koch, M., Loretz, B., & Lehr, C. M. (2024). Targeting intracellular bacteria with dual drug-loaded lactoferrin nanoparticles. *ACS Infectious Diseases*, *10*(5), 1696–1710. DOI: 10.1021/acsinfecdis.4c00045 PMID: 38577780

Andrade, F., Rafael, D., Videira, M., Ferreira, D., Sosnik, A., & Sarmento, B. (2013, November). Nanotechnology and pulmonary delivery to overcome resistance in infectious diseases. *Advanced Drug Delivery Reviews*, *65*(13-14), 1816–1827. DOI: 10.1016/j.addr.2013.07.020 PMID: 23932923

Andrade, S., Ramalho, M. J., Santos, S. B., Melo, L. D. R., Santos, R. S., Guimarães, N., Azevedo, N. F., Loureiro, J. A., & Pereira, M. C. (2023). Fighting methicillin-resistant *Staphylococcus aureus* with targeted nanoparticles. *International Journal of Molecular Sciences*, *24*(10), 9030. DOI: 10.3390/ijms24109030

Andra, V. V. S. N. L., Pammi, S. V. N., Bhatraju, L. V. K. P., & Ruddaraju, L. K. (2022). A Comprehensive Review on Novel Liposomal Methodologies, Commercial Formulations, Clinical Trials and Patents. *BioNanoScience*, *12*(1), 274–291. DOI: 10.1007/s12668-022-00941-x PMID: 35096502

Ansari, M., Shahlaei, M., Hosseinzadeh, S., & Moradi, S. (2024). Recent advances in nanostructured delivery systems for vancomycin. *Nanomedicine (London)*, *19*(23), 1931–1951. DOI: 10.1080/17435889.2024.2377063 PMID: 39143926

Anselmo AC, Mitragotri S. (2021). A Review of Clinical Translation of Inorganic Nanoparticles. AAPS J. 2015 Sep;17(5):1041-54. Epub 2015 May 9. PMID: 25956384; PMCID: PMC4540735. DOI: DOI: 10.1208/s12248-015-9780-2

Anselmo, A. C., & Mitragotri, S. (2021). Nanoparticles in the clinic: An update. *Bioengineering & Translational Medicine*, *6*(3), e10246. DOI: 10.1002/btm2.10246 PMID: 34514159

Anwar, M., Muhammad, F., & Akhtar, B. (2021). Biodegradable nanoparticles as drug delivery devices. *Journal of Drug Delivery Science and Technology*, *64*, 102638. DOI: 10.1016/j.jddst.2021.102638

Arnold, A., McLellan, S., & Stokes, J. M. (2025). How AI can help us beat AMR. *npj antimicrobials and resistance*, *3*(1), 18. https://doi.org/DOI: 10.1038/s44259-025-00085-4

Ashmore, D., Chaudhari, A., Barlow, B., Barlow, B., Harper, T., Vig, K., Miller, M., Singh, S., Nelson, E., & Pillai, S. (2018). Evaluation of *E. coli* inhibition by plain and polymer-coated silver nanoparticles. *Revista do Instituto de Medicina Tropical de São Paulo*, *60*, e18. DOI: 10.1590/s1678-9946201860018

Attia MF, Anton N, Wallyn J, Omran Z, Vandamme TF. An overview of active and passive targeting strategies to improve the nanocarriers efficiency to tumour sites. J Pharm Pharmacol. 2019 Aug;71(8):1185-1198. . Epub 2019 May 3. PMID: 31049986. DOI: https://doi.org/.DOI: 10.1111/jphp.13098

Auganova, D., Atavliyeva, S., Gharbi, N., Zholdybayeva, E., Skiba, Y., Akisheva, A., Tsepke, A., Alenova, A., Guyeux, C., Wirth, T., Sola, C., Tarlykov, P., & Tarlykov, P. *. (2025). Genomic characterization and epidemiology of Mycobacterium tuberculosis lineage 2 isolates from Kazakhstan. *Scientific Reports 2025 15:1, 15*(1), 1–10. DOI: 10.1038/s41598-025-22485-2

Aulifa, D. L., Al Shofwan, A. A., Megantara, S., Fakih, T. M., & Budiman, A. (2024). Elucidation of Molecular Interactions Between Drug-Polymer in Amorphous Solid Dispersion by a Computational Approach Using Molecular Dynamics Simulations. *Advances and Applications in Bioinformatics and Chemistry : AABC, 17*, 1–19. DOI: 10.2147/AABC.S441628

Avanzo, M., Stancanello, J., Pirrone, G., Drigo, A., & Retico, A. (2024). The evolution of artificial intelligence in medical imaging: From computer science to machine and deep learning. *Cancers (Basel), 16*(21), 3702. DOI: 10.3390/cancers16213702

Ayesiga, I., Yeboah, M. O., Okoro, L. N., Edet, E. N., Gmanyami, J. M., Ovye, A., Atimango, L., Gadzama, B. N., Kembabazi, E., & Atwau, P. (2024). Artificial intelligence-enhanced biosurveillance for antimicrobial resistance in sub-Saharan Africa. *International Health, 17*(5), 795–803. DOI: 10.1093/inthealth/ihae081 PMID: 39545538

Azzopardi EA, Ferguson EL, Thomas DW. The enhanced permeability retention effect: a new paradigm for drug targeting in infection. J Antimicrob Chemother. 2013 Feb;68(2):257-74. . Epub 2012 Oct 10. PMID: 23054997.DOI: https://doi .org/.DOI: 10.1093/jac/dks379

Babu, M., & Snyder, M. (2023). Multi-Omics Profiling for Health. *Molecular & Cellular Proteomics, 22*(6), 100561. DOI: 10.1016/j.mcpro.2023.100561 PMID: 37119971

Babu, S., & Nutman, T. B. (2016). Helminth-Tuberculosis Co-infection: An Immunologic Perspective. *Trends in Immunology, 37*(9), 597–607. DOI: 10.1016/j.it.2016.07.005 PMID: 27501916

Bacha, K., Chemotti, C., Mbakidi, J.-P., Deleu, M., & Bouquillon, S. (2023). Dendrimers: Synthesis, Encapsulation Applications and Specific Interaction with the *Stratum Corneum*-A Review. *Macromol, 3*(2), 343–370. DOI: 10.3390/macromol3020022

Bae, H., Ji, H., Konstantinov, K., Sluyter, R., Ariga, K., Kim, Y. H., & Kim, J. H. (2025). Artificial Intelligence-Driven Nanoarchitectonics for Smart Targeted Drug Delivery. *Advanced Materials, 37*(42), e10239. DOI: 10.1002/adma.202510239 PMID: 40772340

Bagewadi, Z. K., Yunus Khan, T. M., Gangadharappa, B., Kamalapurkar, A., Mohamed Shamsudeen, S., & Yaraguppi, D. A. (2023). Molecular dynamics and simulation analysis against superoxide dismutase (SOD) target of *Micrococcus luteus* with secondary metabolites from *Bacillus licheniformis* recognized by genome mining approach. *Saudi Journal of Biological Sciences*, *30*(9), 103753. DOI: 10.1016/j.sjbs.2023.103753

Bah, S. Y., Morang'a, C. M., Kengne-Ouafo, J. A., Amenga–Etego, L., & Awandare, G. A. (2018). Highlights on the Application of Genomics and Bioinformatics in the Fight Against Infectious Diseases: Challenges and Opportunities in Africa. *Frontiers in Genetics*, *9*, 575. Advance online publication. DOI: 10.3389/fgene.2018.00575 PMID: 30538723

Baig, M. S., Ahmad, A., Pathan, R. R., & Mishra, R. K. (2024). Precision Nanomedicine with Bio-Inspired Nanosystems: Recent Trends and Challenges in Mesenchymal Stem Cells Membrane-Coated Bioengineered Nanocarriers in Targeted Nanotherapeutics. *Journal of Xenobiotics*, *14*(3), 827–872. DOI: 10.3390/jox14030047

Baig, N., Kammakakam, I., & Falath, W. (2021). Nanomaterials: A review of synthesis methods, properties, recent progress, and challenges. *Materials Advances*, *2*(6), 1821–1871. DOI: 10.1039/D0MA00807A

Bamford, N. C., MacPhee, C. E., & Stanley-Wall, N. R. (2023). Microbial Primer: An introduction to biofilms - what they are, why they form and their impact on built and natural environments. *Microbiology (Reading, England)*, *169*(8), 001338. DOI: 10.1099/mic.0.001338 PMID: 37526065

Bandodkar, A. J., & Wang, J. (2014). Non-invasive wearable electrochemical sensors: A review. *Trends in Analytical Chemistry*, *76*, 173–181. DOI: 10.1016/j.trac.2015.07.005

Bangham, A. D., Standish, M. M., & Watkins, J. C. (1965). Diffusion of univalent ions across the lamellae of swollen phospholipids. *Journal of Molecular Biology*, *13*(1), 238–252. DOI: 10.1016/S0022-2836(65)80093-6 PMID: 5859039

Bannigan, P., Bao, Z., Hickman, R. J., Aldeghi, M., Häse, F., Aspuru-Guzik, A., & Allen, C. (2023). Machine learning models to accelerate the design of polymeric long-acting injectables. *Nature Communications*, *14*(1), 35. DOI: 10.1038/s41467-022-35343-w PMID: 36627280

Bao, Y., He, J., Song, K., Guo, J., Zhou, X., & Liu, S. (2021). Plant-Extract-Mediated Synthesis of Metal Nanoparticles. *Journal of Chemistry*, *2021*, 1–14. DOI: 10.1155/2021/6562687

Baptista, M., Fernandes, S. N., & Godinho, M. H. (2021). Stimuli-responsive polymeric nanocarriers for controlled drug delivery. *Advanced Functional Materials, 31*(30), 2101170. DOI: 10.1002/adfm.202101170

Baptista, P. V., McCusker, M. P., Carvalho, A., Ferreira, D. A., Mohan, N. M., Martins, M., & Fernandes, A. R. (2018, July 2). Nano-Strategies to Fight Multidrug Resistant Bacteria-"A Battle of the Titans". *Frontiers in Microbiology, 9*, 1441. DOI: 10.3389/fmicb.2018.01441 PMID: 30013539

Barabási, A. L., Gulbahce, N., & Loscalzo, J. (2011). Network medicine: A network-based approach to human disease. *Nature Reviews. Genetics, 12*(1), 56–68. DOI: 10.1038/nrg2918

Barnard, A. S. (2015). Challenges in modeling nanoparticles for drug delivery. *Journal of Physics. Condensed Matter : An Institute of Physics Journal, 28*(2), 023002. DOI: 10.1088/0953-8984/28/2/023002

Barquero, A., Marini, S., Boucher, C., Ruiz, J., & Prosperi, M. (2022). KARGA-Mobile: Android app for portable, real-time, easily interpretable analysis of antibiotic resistance genes *via* nanopore sequencing. *Frontiers in Bioengineering and Biotechnology, 10*, 1016408. DOI: 10.3389/fbioe.2022.1016408 PMID: 36324897

Barrila, J., Crabbé, A., Yang, J., Franco, K., Nydam, S. D., Forsyth, R. J., Davis, R. R., Gangaraju, S., Ott, C. M., Coyne, C. B., Bissell, M. J., & Nickerson, C. A. (2018). Modeling host–pathogen interactions in the context of the microenvironment: Three-dimensional cell culture comes of age. *Infection and Immunity, 86*(9), e00282–e18. DOI: 10.1128/IAI.00282-18

Bass, D., Stentiford, G. D., Wang, H. C., Koskella, B., & Tyler, C. R. (2019). The Pathobiome in Animal and Plant Diseases. *Trends in Ecology & Evolution, 34*(11), 996–1008. DOI: 10.1016/j.tree.2019.07.012

Bawa, R. (2011). Regulating nanomedicine-can the FDA handle it? *Current Drug Delivery, 8*(3), 227–234. DOI: 10.2174/156720111795256156 PMID: 21291376

Bazak R, Houri M, El Achy S, Kamel S, Refaat T. Cancer active targeting by nanoparticles: a comprehensive review of literature. J Cancer Res Clin Oncol. 2015 May;141(5):769-84. Epub 2014 Jul 9. PMID: 25005786; PMCID: PMC4710367. 1DOI: https://doi.org/DOI: 0.1007/s00432-014-1767-3

Behrouzi, E., Abesi, F., Ghorbani, H., & Gholinia, H. (2024). Association between gubernacular canals characteristics and teeth eruption status: A cone-beam computed tomography study. *Journal of Clinical and Experimental Dentistry, 16*(2), e145–e150. DOI: 10.4317/jced.61169 PMID: 38496806

Belay, W. Y., Getachew, M., Tegegne, B. A., Teffera, Z. H., Dagne, A., Zeleke, T. K., Abebe, R. B., Gedif, A. A., Fenta, A., Yirdaw, G., Tilahun, A., & Aschale, Y. (2024). Mechanism of antibacterial resistance, strategies and next-generation antimicrobials to contain antimicrobial resistance: A review. *Frontiers in Pharmacology*, *15*, 1444781. DOI: 10.3389/fphar.2024.1444781 PMID: 39221153

Berros, N., Filaly, Y., El Mendili, F., & El Bouzekri El Idrissi, Y. (2025). Scalable predictive modeling for hospitalization prioritization: A hybrid batch–streaming approach. *Big Data and Cognitive Computing*, *9*(11), 271. DOI: 10.3390/bdcc9110271

Berry, S. B., Gower, M. S., Su, X., Seshadri, C., & Theberge, A. B. (2020). A modular microscale granuloma model for immune–microenvironment signaling studies in vitro. *Frontiers in Bioengineering and Biotechnology*, *8*, 551913. DOI: 10.3389/fbioe.2020.00931

Bess, A., Berglind, F., Mukhopadhyay, S., Brylinski, M., Griggs, N., Cho, T., Galliano, C., & Wasan, K. M. (2022). Artificial intelligence for the discovery of novel antimicrobial agents for emerging infectious diseases. *Drug Discovery Today*, *27*(4), 1099–1107. DOI: 10.1016/j.drudis.2021.10.022 PMID: 34748992

Bhaiyya, M., Panigrahi, D., Rewatkar, P., & Haick, H. (2024). Role of machine learning assisted biosensors in point-of-care-testing for clinical decisions. *ACS Sensors*, *9*(9), 4495–4519. DOI: 10.1021/acssensors.4c01582 PMID: 39145721

Bhati, D., Shukla, P., Shukla, P., Devi, B. N., Sarakula, P., Chopade, P. D., Kundu, H., & Bhatt, P. (2024). Fused and substituted pyrimidine derivatives as potent anticancer agents. *Biochemical and Cellular Archives*, *24*(1). Advance online publication. DOI: 10.51470/bca.2024.24.1.749

Bhattacharyya, S., Walker, D. M., & Harshey, R. M. (2020). Dead cells release a 'necrosignal' that activates antibiotic survival pathways in bacterial swarms. *Nature Communications*, *11*(1), 4157. DOI: 10.1038/s41467-020-17709-0

Bian, Y., & Xie, X. Q. (2021). Generative chemistry: Drug discovery with deep learning generative models. *Journal of Molecular Modeling*, *27*(3), 71. DOI: 10.1007/s00894-021-04674-8 PMID: 33543405

Bikard, D., & Barrangou, R. (2017). Using CRISPR-Cas systems as antimicrobials. *Current Opinion in Microbiology*, *37*, 155–160. DOI: 10.1016/j.mib.2017.08.005 PMID: 28888103

Bilal, H., Khan, M. N., Khan, S., Shafiq, M., Fang, W., Khan, R. U., Rahman, M. U., Li, X., Lv, Q. L., & Xu, B. (2025). The role of artificial intelligence and machine learning in predicting and combating antimicrobial resistance. *Computational and Structural Biotechnology Journal*, *27*, 423–439. DOI: 10.1016/j.csbj.2025.01.006

Birk, S. E., Boisen, A., & Nielsen, L. H. (2021). Polymeric nano- and microparticulate drug delivery systems for treatment of biofilms. *Advanced Drug Delivery Reviews*, *174*, 30–52. DOI: 10.1016/j.addr.2021.04.005

Blanquart, F. (2019). Evolutionary epidemiology models to predict the dynamics of antibiotic resistance. *Evolutionary Applications*, *12*(3), 365–383. DOI: 10.1111/eva.12753

Blondeau, J. M. (2009). What have we learned about antimicrobial use and the risks for *Clostridium difficile*-associated diarrhoea? *The Journal of Antimicrobial Chemotherapy*, *63*(2), 238–242. DOI: 10.1093/jac/dkn477

Boehnke, N., Straehla, J. P., Safford, H. C., Kocak, M., Rees, M. G., Ronan, M., Rosenberg, D., Adelmann, C. H., Chivukula, R. R., Nabar, N., Berger, A. G., Lamson, N. G., Cheah, J. H., Li, H., Roth, J. A., Koehler, A. N., & Hammond, P. T. (2022). Massively parallel pooled screening reveals genomic determinants of nanoparticle delivery. *Science*, *377*(6604), eabm5551. Advance online publication. DOI: 10.1126/science.abm5551 PMID: 35862544

Boiko, D. A., MacKnight, R., Kline, B., & Gomes, G. (2023). Autonomous chemical research with large language models. *Nature*, *624*(7992), 570–578. DOI: 10.1038/s41586-023-06792-0

Bokhari, F. A. S., Mariuzzo, F., & Yan, W. (2024). Antibacterial resistance and the cost of affecting demand: The case of UK antibiotics. *International Journal of Industrial Organization*, *95*, 103082. DOI: 10.1016/j.ijindorg.2024.103082

Bonazzetti, C., Rocchi, E., Toschi, A., Derus, N. R., Sala, C., Pascale, R., Rinaldi, M., Campoli, C., Pasquini, Z. A. I., Tazza, B., Amicucci, A., Gatti, M., Ambretti, S., Viale, P., Castellani, G., & Giannella, M. (2025). Artificial Intelligence model to predict resistances in Gram-negative bloodstream infections. *NPJ Digital Medicine*, *8*(1), 319. Advance online publication. DOI: 10.1038/s41746-025-01696-x PMID: 40442363

Bonifácio, B. V., da Silva, P. B., dos Santos Ramos, M. A., Negri, K. M. S., Bauab, T. M., & Chorilli, M. (2014). Nanotechnology-based drug delivery systems and herbal medicines: A review. *International Journal of Nanomedicine*, *9*, 1–15. DOI: 10.2147/IJN.S57357

Boppana, S. H., Kutikuppala, L. V. S., Sharma, S., Madhavrao, C., Rangari, G., Misra, A. K., Kandi, V., Mishra, S., Singh, P. K., Rabaan, A. A., Mohapatra, R. K., & Kudrat-E-Zahan, M. (2024). Current approaches in smart nano-inspired drug delivery: A narrative review. *Health Science Reports*, *7*(4), e2065. DOI: 10.1002/hsr2.2065 PMID: 38660006

Bose, A., Roy Burman, D., Sikdar, B., & Patra, P. (2021). Nanomicelles: Types, properties and applications in drug delivery. *IET Nanobiotechnology / IET*, *15*(1), 19–27. DOI: 10.1049/nbt2.12018

Bouhedjar, K., Boukelia, A., Khorief Nacereddine, A., Boucheham, A., Belaidi, A., & Djerourou, A. (2020). A natural language processing approach based on embedding deep learning from heterogeneous compounds for quantitative structure–activity relationship modeling. *Chemical Biology & Drug Design*, *96*(3), 961–972. DOI: 10.1111/cbdd.13742

Bourgat, Y., Mikolai, C., Stiesch, M., Klahn, P., & Menzel, H. (2021). Enzyme-responsive nanoparticles and coatings made from alginate/peptide ciprofloxacin conjugates as drug release system. *Antibiotics (Basel, Switzerland)*, *10*(6), 653. DOI: 10.3390/antibiotics10060653

Boussina, A., Shashikumar, S. P., Malhotra, A., Owens, R. L., El-Kareh, R., Longhurst, C. A., Quintero, K., Donahue, A., Chan, T. C., Nemati, S., & Wardi, G. (2024). Impact of a deep learning sepsis prediction model on quality of care and survival. *Npj Digital Medicine 2024 7:1*, *7*(1), 1–9. DOI: 10.1038/s41746-023-00986-6

Boyd, N. K., Teng, C., & Frei, C. R. (2021). Brief Overview of Approaches and Challenges in New Antibiotic Development: A Focus On Drug Repurposing. *Frontiers in Cellular and Infection Microbiology*, *11*, 684515. DOI: 10.3389/fcimb.2021.684515

Bozzuto, G., & Molinari, A. (2015). Liposomes as nanomedical devices. *International Journal of Nanomedicine*, *10*, 975–999. DOI: 10.2147/IJN.S68861 PMID: 25678787

Braga, D. M., & Rawal, B. (2025). Harnessing AI and Quantum Computing for Revolutionizing Drug Discovery and Approval Processes: Case Example for Collagen Toxicity. *JMIR Bioinformatics and Biotechnology*, *6*, e69800. DOI: 10.2196/69800 PMID: 41342151

Branda, F., & Scarpa, F. (2024). Implications of Artificial Intelligence in Addressing Antimicrobial Resistance: Innovations, Global Challenges, and Healthcare's Future. *Antibiotics (Basel, Switzerland)*, *13*(6), 502. DOI: 10.3390/antibiotics13060502 PMID: 38927169

Bratkovič, T., Zahirović, A., Bizjak, M., Rupnik, M., Štrukelj, B., & Berlec, A. (2024). New treatment approaches for *Clostridioides difficile* infections: Alternatives to antibiotics and fecal microbiota transplantation. *Gut Microbes*, *16*(1), 2337312. DOI: 10.1080/19490976.2024.2337312 PMID: 38591915

Brauner, A., Fridman, O., Gefen, O., & Balaban, N. Q. (2016). Distinguishing between resistance, tolerance and persistence to antibiotic treatment. *Nature Reviews. Microbiology*, *14*(5), 320–330. DOI: 10.1038/nrmicro.2016.34

Brooks, C., & Thompson, C. (2017). Predictive modeling in teaching and learning. In *Handbook of Learning Analytics* (pp. 61–68). Society for Learning Analytics Research (SoLAR). https://doi.org/DOI: 10.18608/hla17.005

Brown, D. G., Wobst, H. J., Kapoor, A., Kenna, L. A., & Southall, N. (2022). Clinical development times for innovative drugs. *Nature Reviews. Drug Discovery*, *21*(11), 793–794. DOI: 10.1038/d41573-021-00190-9 PMID: 34759309

Brown, E. D., & Wright, G. D. (2016). Antibacterial drug discovery in the resistance era. *Nature*, *529*(7586), 336–343. DOI: 10.1038/nature17042 PMID: 26791724

Bulman, Z. P., Patel, S., Vyas, V., & Bhavnani, S. M. (2022). Research priorities towards precision antibiotic therapy to improve patient care. *The Journal of Antimicrobial Chemotherapy*, *77*(8), 1999–2008. DOI: 10.1093/jac/dkac137 PMID: 35777386

Bulman, Z. P., Wicha, S. G., Nielsen, E. I., Lenhard, J. R., Nation, R. L., Theuretzbacher, U., Derendorf, H., Tängdén, T., Zeitlinger, M., Landersdorfer, C. B., Bulitta, J. B., Friberg, L. E., Li, J., & Tsuji, B. T. (2022). Research priorities towards precision antibiotic therapy to improve patient care. *The Lancet Microbe*, *3*(10), e795–e802. DOI: 10.1016/S2666-5247(22)00121-5

Cabrera-Andrade, A., López-Cortés, A., Munteanu, C. R., Pazos, A., Pérez-Castillo, Y., Tejera, E., Arrasate, S., & González-Díaz, H. (2020). Perturbation-Theory Machine Learning (PTML) multilabel model of the ChEMBL dataset of preclinical assays for antisarcoma compounds. *ACS Omega*, *5*(42), 27211–27220. DOI: 10.1021/acsomega.0c03356

Cai, D., Gao, W., Li, Z., Zhang, Y., Xiao, L., & Xiao, Y. (2022). Current Development of Nano-Drug Delivery to Target Macrophages. *Biomedicines*, *10*(5), 1203. DOI: 10.3390/biomedicines10051203 PMID: 35625939

Calabretta, M. K., Kumar, A., McDermott, A. M., & Cai, C. (2015). Antimicrobial properties of dendrimers and dendrimer-biocide conjugates. *Therapeutic Delivery*, *6*(4), 395–414. DOI: 10.4155/tde.15.1

Canales, C. C. S., Marquez Cazorla, J. I., Marquez Cazorla, R. M., Sábio, R. M., Santos, H. A., & Pavan, F. R. (2025). Combating Gram-negative infections: The role of antimicrobial peptides and nanotechnology in overcoming antibiotic resistance. *Materials Today. Bio*, *35*, 102381. DOI: 10.1016/j.mtbio.2025.102381 PMID: 41142413

Cao, F., Zhang, L., Wang, H., You, Y., Wang, Y., Gao, N., Ren, J., & Qu, X. (2019). Defect-Rich Adhesive Nanozymes as Efficient Antibiotics for Enhanced Bacterial Inhibition. *Angewandte Chemie International Edition*, *58*(45), 16236–16242. DOI: 10.1002/anie.201908289

Cao, S., Liu, J., & Li, Y. (2025). Harnessing multi-omics and machine learning for predicting immune checkpoint blockade responses: Advances, challenges, and future directions. *Fundamental Research (Beijing)*. Advance online publication. DOI: 10.1016/j.fmre.2025.08.009 PMID: 41647536

Cao, Y. H., Cai, W. J., He, X. W., Song, H. L., Gao, J., Yang, Y. L., & Zhou, J. (2024). A review of advances & potential of applying nanomaterials for biofilm inhibition. npj. *NPJ Clean Water*, *7*(1), 131. DOI: 10.1038/s41545-024-00423-5

Cao, Y., Lancaster, E., Lee, J., & Wu, J. (2026). Integrating explainable AI and One Health: A new frontier in combating infectious diseases. *EBioMedicine*, *126*, 106207. DOI: 10.1016/j.ebiom.2026.106207

Capela, R., Moreira, R., & Lopes, F. (2019). An Overview of Drug Resistance in Protozoal Diseases. *International Journal of Molecular Sciences*, *20*(22), 5748. DOI: 10.3390/ijms20225748 PMID: 31731801

Casals, E., Casals, G., Puntes, V., & Rosenholm, J. M. (2019). Biodistribution, excretion, and toxicity of inorganic nanoparticles. In *Theranostic bionanomaterials* (pp. 3–26). Elsevier., DOI: 10.1016/B978-0-12-815341-3.00001-8

Cascella, M., Montomoli, J., Bellini, V., & Bignami, E. (2023). Evaluating the feasibility of ChatGPT in healthcare: An analysis of multiple clinical and research scenarios. *Journal of Medical Systems*, *47*(1), 33. DOI: 10.1007/s10916-023-01925-4

Castañeda-Barba, S., Top, E. M., & Stalder, T. (2024). Plasmids, a molecular cornerstone of antimicrobial resistance in the One Health era. *Nature Reviews. Microbiology*, *22*(1), 18–32. DOI: 10.1038/s41579-023-00926-x

Castro, K. C. D., Costa, J. M., & Campos, M. G. N. (2022). Drug-loaded polymeric nanoparticles: A review. *International Journal of Polymeric Materials*, *71*(1), 1–13. DOI: 10.1080/00914037.2020.1798436

Cesaro, A., Hoffman, S. C., Das, P., & de la Fuente-Nunez, C. (2025). Challenges and applications of artificial intelligence in infectious diseases and antimicrobial resistance. *Npj Antimicrobials and Resistance 2025 3:1, 3*(1), 1–10. DOI: 10.1038/s44259-024-00068-x

Chakraborty, N., Jha, D., Roy, I., Kumar, P., Gaurav, S. S., Marimuthu, K., Ng, O. T., Lakshminarayanan, R., Verma, N. K., & Gautam, H. K. (2022). Nanobiotics against antimicrobial resistance: Harnessing the power of nanoscale materials and technologies. *Journal of Nanobiotechnology*, *20*(1), 375. DOI: 10.1186/s12951-022-01573-9

Chamundeeswari, M., Jeslin, J., & Verma, M. L. (2019). Nanocarriers for drug delivery applications. *Environmental Chemistry Letters*, *17*(2), 849–865. DOI: 10.1007/s10311-018-00841-1

Changruenngam, S., Modchang, C., & Bicout, D. J. (2022). Modeling of the transmission dynamics of carbapenem-resistant *Klebsiella pneumoniae* in hospitals and design of control strategies. *Scientific Reports*, *12*(1), 3805. DOI: 10.1038/s41598-022-07728-w

Charani, E., McKee, M., Balasegaram, M., Mendelson, M., Singh, S., & Holmes, A. H. (2022). Global burden of antimicrobial resistance: Essential pieces of a global puzzle. *Lancet*, *399*(10344), 2346–2347. DOI: 10.1016/S0140-6736(22)00935-7

Chatterjee, A. K., Chakraborty, R., & Basu, T. (2014). Mechanism of antibacterial activity of copper nanoparticles. *Nanotechnology*, *25*(13), 135101. DOI: 10.1088/0957-4484/25/13/135101

Chaudhary, A., Krishnarth, N., Tripathi, P. N., Chaudhary, A. K., Rejeeth, C., & Sharma, A. (2025). Smart Materials Design for Antibacterial Application. *Pharmaceutical Nanotechnology, 13*(3), 411-431. https://doi.org/DOI: 10.2174/01221 17385274576240209054652

Chen Y, He X, Chen Q, He Y, Chen F, Yang C, Wang L. Nanomaterials against intracellular bacterial infection: from drug delivery to intrinsic biofunction. Front Bioeng Biotechnol. 2023 Apr 27;11:1197974.. PMID: 37180049; PMCID: PMC10174311. DOI: https://doi.org/DOI: 10.3389/fbioe.2023.1197974

Chen, D., Li, D., Su, L., Wang, D. Y., Ren, Y., van der Mei, H. C., Shi, L., Chen, T., & Busscher, H. J. (2026). Stimuli-responsive, antimicrobial-loaded nanocarriers for oral biofilm control and microbiome restoration. *International Journal of Oral Science*, *18*(1), 17. DOI: 10.1038/s41368-025-00422-3

Chen, H., Engkvist, O., Wang, Y., Olivecrona, M., & Blaschke, T. (2018). The rise of deep learning in drug discovery. *Drug Discovery Today, 23*(6), 1241–1250. DOI: 10.1016/j.drudis.2018.01.039 PMID: 29366762

Chen, P., Sun, Z., Wang, J., Liu, X., Bai, Y., Chen, J., Liu, A., Qiao, F., Chen, Y., Yuan, C., Sha, J., Zhang, J., Xu, L.-Q., & Li, J. (2023). Portable nanopore-sequencing technology: Trends in development and applications. *Frontiers in Microbiology, 14*, 1043967. Advance online publication. DOI: 10.3389/fmicb.2023.1043967 PMID: 36819021

Chen, X., Lin, Z., Cheng, N., Mo, Y., Lu, L., Hou, J., Li, Z., Nie, X., Gao, S., & Hua, Q. (2025). Recent advances in NIR-II photothermal and photodynamic therapies for drug-resistant wound infections. *Materials Today. Bio, 101871*, 101871. Advance online publication. DOI: 10.1016/j.mtbio.2025.101871 PMID: 40496729

Chen, X., & Lu, H. (2022). Intelligent control of nanoparticle synthesis on microfluidic chips with machine learning. *NPG Asia Materials, 14*(1), 69. DOI: 10.1038/s41427-022-00416-1

Cheung, H. Y. F., Tantiwong, C., Kale, D., Gibbins, J. M., Watson, S. P., Heemskerk, J. W. M., Sickmann, A., Ahrends, R., & Dunster, J. L. (2025). A computational framework for the investigation of phosphoinositide regulation. *PLoS Computational Biology, 21*(9), e1013477. DOI: 10.1371/journal.pcbi.1013477

Chhabra, S., Taksande, A. B., & Munjewar, P. (2024). The penicillin pioneer: Alexander Fleming's journey to a medical breakthrough. *Cureus, 16*(7). Advance online publication. DOI: 10.7759/cureus.65179 PMID: 39176366

Chmiel, R. U., Marek, A., Pysniak, D. S., Wieczorek, K., Dec, M., Nowaczek, A., & Osek, J. (2022). Antibiotic Resistance in Bacteria-A Review. *Antibiotics (Basel, Switzerland), 11*(8), 1079. DOI: 10.3390/antibiotics11081079 PMID: 36009947

Choudhury, A., Kirti, A., Lenka, S. S., Naser, S. S., Sinha, A., Kumari, S., Kaushik, N. K., Ghosh, A., & Verma, S. K. (2025). Strategic advances in liposomes technology: Translational paradigm in transdermal delivery for skin dermatosis. *Journal of Nanobiotechnology, 23*(1), 576. Advance online publication. DOI: 10.1186/s12951-025-03660-z PMID: 40835930

Chou, W. C., Canchola, A., Zhang, F., & Lin, Z. (2025). Machine learning and artificial intelligence in nanomedicine. *Wiley Interdisciplinary Reviews. Nanomedicine and Nanobiotechnology, 17*(4), e70027. DOI: 10.1002/wnan.70027

Chow, J. C. L. (2024). Quantum Computing in Medicine. *Medical Science, 12*(4), 67. DOI: 10.3390/medsci12040067 PMID: 39584917

Chramiec, A., & Vunjak-Novakovic, G. (2019). Tissue-engineered models of healthy and malignant human bone marrow. *Advanced Drug Delivery Reviews*, *140*, 78–92. DOI: 10.1016/j.addr.2019.04.003 PMID: 31002835

Cicha, I., Priefer, R., Severino, P., Souto, E. B., & Jain, S. (2022). Biosensor-integrated drug delivery systems as new materials for biomedical applications. *Biomolecules*, *12*(9), 1198. DOI: 10.3390/biom12091198

Cloudin, S., Priya, R. S., Prabhu, G. A., Ruba, M., Joel, R., & Vanitha, R. (2026). AI-Enhanced Predictive Analytics and Reliability Management for Wind Turbine Maintenance. In AI-Powered Analysis, Modeling, and Monitoring of Wind Energy Systems (pp. 191-218). IGI Global Scientific Publishing.

Coffel, J., & Nuxoll, E. (2018). BioMEMS for biosensors and closed-loop drug delivery. *International Journal of Pharmaceutics*, *544*(2), 335–349. DOI: 10.1016/j.ijpharm.2018.01.030

Condorelli, C., Nicitra, E., Musso, N., Bongiorno, D., Stefani, S., Gambuzza, L. V., Carchiolo, V., & Frasca, M. (2024). Prediction of antimicrobial resistance of *Klebsiella pneumoniae* from genomic data through machine learning. *PLoS One*, *19*(9), e0309333. DOI: 10.1371/journal.pone.0309333

Conteh, J. S., Nucci, G. E., Fernandez Cabada, T., Mai, B. T., Soni, N., De Donato, F., Pasquale, L., Catalano, F., Prato, M., Manna, L., & Pellegrino, T. (2023). $CuFeS_2$ nanoparticles functionalized with a thermoresponsive polymer for photothermia and externally controlled drug delivery. *ACS Applied Materials & Interfaces*, *15*(19), 22999–23011. DOI: 10.1021/acsami.3c03902

Cook, M. A., & Wright, G. D. (2022). The past, present, and future of antibiotics. *Science Translational Medicine*, *14*(657), eabo7793. Advance online publication. DOI: 10.1126/scitranslmed.abo7793 PMID: 35947678

Corlis, J., Zhu, J., Macul, H., Tiberi, O., Boothe, M. A. S., & Resch, S. C. (2023). Framework for determining the optimal course of action when efficiency and affordability measures differ by perspective in cost-effectiveness analysis-with an illustrative case of HIV treatment in Mozambique. *Cost Effectiveness and Resource Allocation : C/E*, *21*(1), 62. Advance online publication. DOI: 10.1186/s12962-023-00474-4 PMID: 37705101

Corso, G., Stark, H., Jegelka, S., Jaakkola, T., & Barzilay, R. (2024). Graph neural networks. *Nature Reviews. Methods Primers*, *4*(1), 1–23. DOI: 10.1038/s43586-024-00294-7

Craig, M., Jernigan, D., Laserson, K., McBride, S., Fairbanks, J., Sievert, D., Armstrong, P. A., Ewing Ogle, H., & Zucker, H. (2024). Antimicrobial resistance at a crossroads: The cost of inaction. *Lancet*, *404*(10458), 1083–1085. DOI: 10.1016/S0140-6736(24)01705-7 PMID: 39265589

Crintea, A., Dutu, A. G., Sovrea, A., Constantin, A.-M., Samasca, G., Masalar, A. L., Ifju, B., Linga, E., Neamti, L., Tranca, R. A., Fekete, Z., Silaghi, C. N., & Craciun, A. M. (2022). Nanocarriers for drug delivery: An overview with emphasis on vitamin D and K transportation. *Nanomaterials (Basel, Switzerland)*, *12*(8), 1376. DOI: 10.3390/nano12081376

Cross, J. L., Choma, M. A., & Onofrey, J. A. (2024). Bias in medical AI: Implications for clinical decision-making. *PLOS Digital Health*, *3*(11), e0000651. DOI: 10.1371/journal.pdig.0000651 PMID: 39509461

Csóka, I., Ismail, R., Jójárt-Laczkovich, O., & Pallagi, E. (2021). Regulatory considerations, challenges and risk-based approach in nanomedicine development. *Current Medicinal Chemistry*, *28*(36), 7461–7476. DOI: 10.2174/0929867328666210406115529 PMID: 33823761

Cui, Y., Zhao, Y., Tian, Y., Zhang, W., Lü, X., & Jiang, X. (2012). The molecular mechanism of action of bactericidal gold nanoparticles on *Escherichia coli*. *Biomaterials*, *33*(7), 2327–2333. DOI: 10.1016/j.biomaterials.2011.11.057

D'Hondt, S., Oramas, J., & De Winter, H. (2025). A beginner's approach to deep learning applied to VS and MD techniques. *Journal of Cheminformatics*, *17*, 47. DOI: 10.1186/s13321-025-00985-7

Dai, J., Ouyang, Y., Gupte, R., Liu, X. J. A., Li, Y., Yang, F., Chen, S., Provin, T., Van Schaik, E., Samuel, J. E., Jayaraman, A., Zhou, A., de Figueiredo, P., Zhou, J., & Han, A. (2025). Microfluidic droplets with amended culture media cultivate a greater diversity of soil microorganisms. *Applied and Environmental Microbiology*, *91*(3), e0179424. DOI: 10.1128/aem.01794-24

Dalbanjan, N. P., Korgaonkar, K., Eelager, M. P., Gonal, B. N., Kadapure, A. J., Arakera, S. B., & Praveen Kumar, S. K. (2025). In-silico strategies in nano-drug design: Bridging nanomaterials and pharmacological applications. *Nano TransMed.*, *4*, 100091. DOI: 10.1016/j.ntm.2025.100091

Danaei, M., Dehghankhold, M., Ataei, S., Hasanzadeh Davarani, F., Javanmard, R., Dokhani, A., Khorasani, S., & Mozafari, M. R. (2018). Impact of particle size and polydispersity index on the clinical applications of lipidic nanocarrier systems. *Pharmaceutics*, *10*(2), 57. DOI: 10.3390/pharmaceutics10020057 PMID: 29783687

Danhier, F., Ansorena, E., Silva, J. M., Coco, R., Le Breton, A., & Préat, V. (2012). PLGA-based nanoparticles: An overview of biomedical applications. *Journal of Controlled Release*, *161*(2), 505–522. DOI: 10.1016/j.jconrel.2012.01.043 PMID: 22353619

Daniyarov, A., Molkenov, A., Rakhimova, S., Akhmetova, A., Yerezhepov, D., Chingissova, L., Bismilda, V., Toksanbayeva, B., Rakisheva, A., Akilzhanova, A., Kozhamkulov, U., & Kairov, U. (2021). Genomic Analysis of Multidrug-Resistant Mycobacterium tuberculosis Strains From Patients in Kazakhstan. *Frontiers in Genetics*, *12*, 683515. DOI: 10.3389/fgene.2021.683515 PMID: 34858467

Dara, S., Dhamerchela, S., Jadav, S. S., Babu, C. M., & Ashan, M. J. (2022). Machine Learning in Drug Discovery: A Review. *Artificial Intelligence Review*, *55*(3), 1947–1999. DOI: 10.1007/s10462-021-10058-4 PMID: 34393317

Darby, E. M., Trampari, E., Siasat, P., Gaya, M. S., Alav, I., Webber, M. A., & Blair, J. M. A. (2023). Molecular mechanisms of antibiotic resistance revisited. *Nature Reviews. Microbiology*, *21*(5), 280–295. DOI: 10.1038/s41579-022-00820-y PMID: 36411397

Das, K. P., & J, C. (2023). Nanoparticles and convergence of artificial intelligence for targeted drug delivery for cancer therapy: Current progress and challenges. *Frontiers in Medical Technology*, *4*, 1067144. DOI: 10.3389/fmedt.2022.1067144 PMID: 36688144

Das, U. (2025). Generative AI for drug discovery and protein design: The next frontier in AI-driven molecular science. *Medicine in Drug Discovery*, *100213*, 100213. Advance online publication. DOI: 10.1016/j.medidd.2025.100213

David, B., Wolfender, J. L., & Dias, D. A. (2015). The pharmaceutical industry and natural products: Historical status and new trends. *Phytochemistry Reviews*, *14*(2), 299–315. DOI: 10.1007/s11101-014-9367-z

Davis, F. P., Barkan, D. T., Eswar, N., McKerrow, J. H., & Sali, A. (2007). Host–pathogen protein interactions predicted by comparative modeling. *Protein Science : a Publication of the Protein Society*, *16*(12), 2597–2608. DOI: 10.1110/ps.073228407

de la Lastra, J. M. P., Wardell, S. J. T., Pal, T., de la Fuente-Nunez, C., & Pletzer, D. (2024). From Data to Decisions: Leveraging Artificial Intelligence and Machine Learning in Combating Antimicrobial Resistance - a Comprehensive Review. *Journal of Medical Systems*, *48*(1), 71. DOI: 10.1007/s10916-024-02089-5 PMID: 39088151

De Plano, L. M., Morganti, D., Nicotra, G., Calorenni, P., Sciuto, E. L., Oddo, S., & Conoci, S. (2025). Engineered phage-silver nanoparticle complexes as a new tool for targeted therapies. *Scientific Reports*, *15*(1), 36135. Advance online publication. DOI: 10.1038/s41598-025-05069-y PMID: 41102167

Deiss-Yehiely, E., Cárcamo-Oyarce, G., Berger, A. G., Ribbeck, K., & Hammond, P. T. (2023, August 14). pH-Responsive, Charge-Reversing Layer-by-Layer Nanoparticle Surfaces Enhance Biofilm Penetration and Eradication. *ACS Biomaterials Science & Engineering*, *9*(8), 4794–4804. DOI: 10.1021/acsbiomaterials.3c00481 PMID: 37390118

Denton, N., Molloy, M., Charleston, S., Lipset, C., Hirsch, J., Mulberg, A. E., Howard, P., & Marsh, E. D. (2021). Data silos are undermining drug development and failing rare disease patients. *Orphanet Journal of Rare Diseases*, *16*(1), 161. DOI: 10.1186/s13023-021-01806-4

Derbali, R. M., Aoun, V., Moussa, G., Frei, G., Tehrani, S. F., Campos Del'Orto, J., Hildgen, P., Roullin, V. G., & Leblond, J. (2019). Tailored nanocarriers for the pulmonary delivery of levofloxacin against Pseudomonas aeruginosa: A comparative study. *Molecular Pharmaceutics*, *16*(5), 1906–1916. DOI: 10.1021/acs.molpharmaceut.8b01256 PMID: 30900903

Desai N. Challenges in development of nanoparticle-based therapeutics. AAPS J. 2012 Jun;14(2):282-95. . Epub 2012 Mar 10. PMID: 22407288; PMCID: PMC3326161. DOI: https://doi.org/DOI: 10.1208/s12248-012-9339-4

Desai, N., Rana, D., Patel, M., Bajwa, N., Prasad, R., & Vora, L. K. (2025). Nanoparticle Therapeutics in Clinical Perspective: Classification, Marketed Products, and Regulatory Landscape. *Small (Weinheim an der Bergstrasse, Germany)*, *21*(29), e2502315. https://doi.org/DOI: 10.1002/smll.202502315

Dhaarani, R., & Reddy, M. K. (2025). Progressing microbial genomics: Artificial intelligence and deep learning driven advances in genome analysis and therapeutics. *Intelligence-Based Medicine*, *11*, 100251. DOI: 10.1016/j.ibmed.2025.100251

Dhanushya, J., Ruba, M., Ponkarthika, M., Soundarya, B., Sakthi Priya, V., Dhanushri, S., & Harikrishnan, S. (2025, April). AI-Powered Multipurpose Agrobot for Precision Farming: Crop Health Management, Fertilizer Recommendation and Yield Prediction. In *2025 International Conference on Recent Advances in Electrical, Electronics, Ubiquitous Communication, and Computational Intelligence (RAEEUCCI)* (pp. 1-6). IEEE.

Di Salvatore, V., Cernuto, F., Russo, G., & Pappalardo, F. (2025). A computational framework for optimizing mRNA vaccine delivery via AI-guided nanoparticle design and in silico gene expression profiling. *Frontiers in Immunology, 16*, 1628583. DOI: 10.3389/fimmu.2025.1628583

Diamantis, S., Retur, N., Bertrand, B., Lieutier-Colas, F., Carenco, P., & Mondain, V.Promise Professional Community Network On Antimicrobial Resistance. (2022). The Production of Antibiotics Must Be Reoriented: Repositioning Old Narrow-Spectrum Antibiotics, Developing New Microbiome-Sparing Antibiotics. *Antibiotics (Basel, Switzerland), 11*(7), 924. DOI: 10.3390/antibiotics11070924 PMID: 35884178

Diaz, R., Ramalheira, E., Afreixo, V., & Gago, B. (2016). Methicillin-resistant *Staphylococcus aureus* carrying the new mecC gene—a meta-analysis. *Diagnostic Microbiology and Infectious Disease, 84*(2), 135–140. DOI: 10.1016/j.diagmicrobio.2015.10.014

Din, F. U., Aman, W., Ullah, I., Qureshi, O. S., Mustapha, O., Shafique, S., & Zeb, A. (2017). Effective use of nanocarriers as drug delivery systems for the treatment of selected tumors. *International Journal of Nanomedicine, 12*, 7291–7309. DOI: 10.2147/IJN.S146315 PMID: 29042776

Ding, D. Y., Zhang, Y., Jia, Y., & Sun, J. (2023). Machine learning-guided lipid nanoparticle design for mRNA delivery. arXiv preprint arXiv:2308.01402. DOI: https://doi.org//arXiv.2308.01402DOI: 10.48550

Ding, M., Zhao, W., Song, L. J., & Luan, S. F. (2022). Stimuli-responsive nanocarriers for bacterial biofilm treatment. *Rare Metals, 41*(2), 482–498. DOI: 10.1007/s12598-021-01802-4

Ding, X., Ma, J., Fan, T., Issa, R., Li, Y., Weng, D., Zhang, D., & Chen, Y. (2024). Inorganic nanoparticles-based strategies for the microbial detection in infectious diseases. *Interdisciplinary Medicine, 2*(2), e20230045. DOI: 10.1002/INMD.20230045

Dini, I. (2026). Mechanisms of Bacterial Resistance and Innovative Strategies to Overcome Antimicrobial Resistance. *Antibiotics (Basel, Switzerland), 15*(3), 319. DOI: 10.3390/antibiotics15030319

Dinya, M., Dinya, E., & Mórotz, G. M. (2026). From Pathology to Formulation: Designing Biodegradable Polymers for Personalized Drug Delivery. *Pharmaceutics, 18*(3), 330. https://doi.org/DOI: 10.3390/pharmaceutics18030330

Djuriš, J., Kurćubić, I., & Ibrić, S. (2021). Review of machine learning algorithms application in pharmaceutical technology. Archives of Pharmacy, 71(Notebook 4), 302-317. DOI: DOI: 10.5937/arhfarm71-32499

Doga, H., Bose, A., Sahin, M. E., Bettencourt-Silva, J., Pham, A., Kim, E., Andress, A., Saxena, S., Parida, L., Robertus, J. L., Kawaguchi, H., Soliman, R., & Blankenberg, D. (2024). How can quantum computing be applied in clinical trial design and optimization? *Trends in Pharmacological Sciences*, *45*(10), 880–891. DOI: 10.1016/j.tips.2024.08.005 PMID: 39317621

Dongre, D. S., Saha, U. B., & Saroj, S. D. (2025). Exploring the role of gut microbiota in antibiotic resistance and prevention. *Annals of Medicine*, *57*(1), 2478317. DOI: 10.1080/07853890.2025.2478317 PMID: 40096354

Do, T., Vaculciakova, S., Kluska, K., Peris-Díaz, M. D., Priborsky, J., Guran, R., Krężel, A., Adam, V., & Zitka, O. (2024). Antioxidant-related enzymes and peptides as biomarkers of metallic nanoparticles (eco)toxicity in the aquatic environment. *Chemosphere*, *364*, 142988. DOI: 10.1016/j.chemosphere.2024.142988

Dri, D. A., Rinaldi, F., Carafa, M., & Marianecci, C. (2023). Nanomedicines and nanocarriers in clinical trials: Surfing through regulatory requirements and physico-chemical critical quality attributes. *Drug Delivery and Translational Research*, *13*(3), 757–769. DOI: 10.1007/s13346-022-01262-y PMID: 36450964

Duncan, K. R., Crüsemann, M., Lechner, A., Sarkar, A., Li, J., Ziemert, N., Wang, M., Bandeira, N., Moore, B. S., Dorrestein, P. C., & Jensen, P. R. (2015). Molecular networking and pattern-based genome mining improves discovery of biosynthesis-related gene clusters and their products from *Salinispora* species. *Chemistry & Biology*, *22*(4), 460–471. DOI: 10.1016/j.chembiol.2015.03.010

Duncan, R. (2003). The dawning era of polymer therapeutics. *Nature Reviews. Drug Discovery*, *2*(5), 347–360. DOI: 10.1038/nrd1088 PMID: 12750738

Durrant, J. D., & Amaro, R. E. (2015). Machine-learning techniques applied to antibacterial drug discovery. *Chemical Biology & Drug Design*, *85*(1), 14–21. DOI: 10.1111/cbdd.12423

Dyett, B. P., Sarkar, S., Yu, H., Strachan, J., Drummond, C. J., & Conn, C. E. (2024, May 15). Overcoming Therapeutic Challenges of Antibiotic Delivery with Cubosome Lipid Nanocarriers. *ACS Applied Materials & Interfaces*, *16*(19), 24191–24205. DOI: 10.1021/acsami.4c00921 PMID: 38690584

Dzoyem, J. P., Tsemeugne, J., Pone Kamdem, B., Foyou Meupiap, R., Kuate, B. A., Mkounga, P., Fekam Boyom, F., & McGaw, L. J. (2025). Antibacterial, antibiofilm and anti-quorum sensing activities of 1,2,3,5-tetrazine derivatives linked to a benzothiazole moiety. *PLoS One*, *20*(6), e0318135. DOI: 10.1371/journal.pone.0318135

Echavarria Galindo, M., & Lai, Y. (2025). CRISPR-based genetic tools for the study of host-microbe interactions. *Infection and Immunity*, *93*(9), e0051024. DOI: 10.1128/iai.00510-24 PMID: 40757822

Eicher, T., Kinnebrew, G., Patt, A., Spencer, K., Ying, K., Ma, Q., Machiraju, R., & Mathé, A. E. A. (2020). Metabolomics and multiomics integration: A survey of computational methods and resources. *Metabolites*, *10*(5), 202. DOI: 10.3390/metabo10050202

El Arab, R. A., & Al Moosa, O. A. (2025). Systematic review of cost effectiveness and budget impact of artificial intelligence in healthcare. *NPJ Digital Medicine*, *8*(1), 548. Advance online publication. DOI: 10.1038/s41746-025-01722-y PMID: 40858882

Elabiad, M., & Salaheldin, A. (2025). Artificial intelligence in nanotechnology: Transforming innovation at the atomic scale. SSRN *Electronic Journal*. DOI: 10.2139/ssrn.5108426

Elalouf, A., Elalouf, H., & Rosenfeld, A. (2025). Artificial intelligence in drug resistance management. *3 Biotech*, *15*(4), 126. https://doi.org/DOI: 10.1007/s13205-025-04282-w

Elbaiomy, R. G., El-Sappah, A. H., Guo, R., Luo, X., Deng, S., Du, M., Jian, X., Bakeer, M., Li, Z., & Zhang, Z. (2025). Antibiotic resistance: A genetic and physiological perspective. *MedComm*, *6*(11), e70447. DOI: 10.1002/mco2.70447

Elgin, C. (2025). Anticipatory macroeconomic governance: Exploring future-oriented strategies for economic resilience and sustainability. *European Journal of Futures Research*, *13*(4). Advance online publication. DOI: 10.1186/s40309-025-00248-6

Eltaib, L. (2025). Polymeric Nanoparticles in Targeted Drug Delivery: Unveiling the Impact of Polymer Characterization and Fabrication. *Polymers*, *17*(7), 833. DOI: 10.3390/polym17070833 PMID: 40219222

Endale, H., Mathewos, M., & Abdeta, D. (2023). Potential causes of spread of antimicrobial resistance and preventive measures in one health perspective—A review. *Infection and Drug Resistance*, *16*, 7515–7545. DOI: 10.2147/IDR.S428837 PMID: 38089962

Esawy, M. A., & Reyed, R. M. (2026). Biointelligent reframing of colorectal cancer: From somatic mutation to systemic coherence. In Reyed, R., Prabhakar, P., & Haghi, A. (Eds.), *Advancing Personalized Medicine with AI-Driven Microbiome and Nutrigenomics: Innovations and Future Perspectives* (pp. 173–218). IGI Global., DOI: 10.4018/979-8-3373-3121-8.ch003

Esmaeilpour, D., Hamblin, M. R., Cheng, J., Khosravi, A., Liu, J., Zarepour, A., Zarrabi, A., Sillanpää, M., Nazarzadeh Zare, E., Shen, J., & Karimi-Maleh, H. (2026). Artificial intelligence driven protein design and sustainable nanomedicine for advanced theranostics. *Bioactive Materials, 60*, 425–455. DOI: 10.1016/j.bioactmat.2026.01.036 PMID: 41674557

Estapé Senti, M., Ceccaldi, A., Luciani, M., Saber, N., Schurmann, P. J. L., Geerlings, M. W., Holig, P., de Beer, J., Hannus, M., Campbell, F., Winter, M., van Hoeve, W., Westerik, M., Dorst-Mooiman, K. D., Dekker, J., Fuchs, S., Wolbers, F., Segers, T., IJzerman, M., & Schiffelers, R. M. (2025). NANOSPRESSO: Toward personalized, locally produced nucleic acid nanomedicines. *Frontiers in Science, 3*, 1458636. DOI: 10.3389/fsci.2025.1458636

Ezike, T. C., Okpala, U. S., Onoja, U. L., Nwike, C. P., Ezeako, E. C., Okpara, O. J., Okoroafor, C. C., Eze, S. C., Kalu, O. L., Odoh, E. C., Nwadike, U. G., Ogbodo, J. O., Umeh, B. U., Ossai, E. C., & Nwanguma, B. C. (2023). Advances in drug delivery systems, challenges and future directions. *Heliyon, 9*(6), e17488. DOI: 10.1016/j.heliyon.2023.e17488

Fadeel, B., Farcal, L., Hardy, B., Vázquez-Campos, S., Hristozov, D., Marcomini, A., Lynch, I., Valsami-Jones, E., Alenius, H., & Savolainen, K. (2018). Advanced tools for the safety assessment of nanomaterials. *Nature Nanotechnology, 13*(7), 537–543. DOI: 10.1038/s41565-018-0185-0 PMID: 29980781

Fagbemi, O. A., Essuah, M., Ugwu, D. C., Ajibola, A. B., Julius, S. O., Onyeyili, I. N., & Ojo-omoniyi, D. S. (2025). Nanobiotechnology-driven innovations for tackling antimicrobial resistance. *World J. Biol. Pharm. Health Sci, 21*(02), 300-328. https://doi.org/DOI: 10.30574/wjbphs.2025.21.2.0093

Fairburn, S. C., Jehi, L., Bicknell, B. T., Wilkes, B. G., & Panuganti, B. (2025). Applications of quantum computing in clinical care. *Frontiers in Medicine, 12*, 1573016. Advance online publication. DOI: 10.3389/fmed.2025.1573016 PMID: 40337272

Fam, S. Y., Chee, C. F., Yong, C. Y., Ho, K. L., Mariatulqabtiah, A. R., & Tan, W. S. (2020). Stealth Coating of Nanoparticles in Drug-Delivery Systems. *Nanomaterials (Basel, Switzerland), 10*(4), 787. DOI: 10.3390/nano10040787 PMID: 32325941

Fan, L., Chen, C., Zhang, H., Zeng, Y., Li, T., Gao, R., Li, J., Ren, Y., Wu, Z., Bi, F., Chu, Y., Li, H., Hu, J., Xu, J., & Xu, Y. (2025). Atmospheric detection, prevalence, transmission, health and ecological consequences of antibiotic resistance genes and resistant bacteria: A comprehensive review. *Emerging Contaminants, 11*(3), 100514. DOI: 10.1016/j.emcon.2025.100514

Farah, H., Kadhim-Abosaoda, M., Mohaisen-Mousa, H., Renuka Jyothi, S., Priyadarshini-Nayak, P., Bethanney Janney, J., Singh, G., Singh-Chauhan, A., & Kumar-Mishra, M. (2026). Nanomedicine Strategies Against Biofilm-Associated Infections: Advances, Challenges, and Translational Barriers. *MicrobiologyOpen*, *15*(1), e70210. DOI: 10.1002/mbo3.70210

Fasseeh, A. N., Korra, N., Elezbawy, B., Sedrak, A. S., Gamal, M., Eldessouki, R., Eldebeiky, M., George, M., Seyam, A., Abourawash, A., Khalifa, A. Y., Shaheen, M., Abaza, S., & Kaló, Z. (2024). Framework for developing cost-effectiveness analysis threshold: The case of Egypt. *The Journal of the Egyptian Public Health Association*, *99*(1), 12. Advance online publication. DOI: 10.1186/s42506-024-00159-7 PMID: 38825614

Fatima, M., Almalki, W. H., Khan, T., Sahebkar, A., & Kesharwani, P. (2024). Harnessing the Power of Stimuli-Responsive Nanoparticles as an Effective Therapeutic Drug Delivery System. *Advanced Materials*, *36*(24), 2312939. DOI: 10.1002/adma.202312939 PMID: 38447161

Fayyad-Kazan, M. (2025). *CRISPR-Cas Systems: Transformative Precision Tools to Combat Antimicrobial Resistance in Multidrug-Resistant Gram-Negative Pathogens*. https://www.genesispub.org/crispr-cas-systems-transformative-precision-tools-to-combat-antimicrobial-resistance-in-multidrug-resistant-gram-negative-pathogens

Fernandes, F. O. L., Crima, S., & Ferrinho, P. (2025). Is it feasible to develop a one health surveillance system in a fragile state? The case of Guinea-Bissau. *One health (Amsterdam, Netherlands)*, *21*, 101269. DOI: 10.1016/j.onehlt.2025.101269

Ferreira, L. G., Dos Santos, R. N., Oliva, G., & Andricopulo, A. D. (2015). Molecular docking and structure-based drug design strategies. *Molecules (Basel, Switzerland)*, *20*(7), 13384–13421. DOI: 10.3390/molecules200713384

Flannagan, R. S., Cosío, G., & Grinstein, S. (2009). Antimicrobial mechanisms of phagocytes and bacterial evasion strategies. *Nature Reviews. Microbiology*, *7*(5), 355–366. DOI: 10.1038/nrmicro2128 PMID: 19369951

Floss, H. G., & Yu, T. W. (2005). Rifamycin-mode of action, resistance, and biosynthesis. *Chemical Reviews*, *105*(2), 621–632. DOI: 10.1021/cr030112j

Forli, S., Huey, R., Pique, M. E., Sanner, M. F., Goodsell, D. S., & Olson, A. J. (2016). Computational protein–ligand docking and virtual drug screening with the AutoDock suite. *Nature Protocols*, *11*(5), 905–919. DOI: 10.1038/nprot.2016.051

Franklin, A. M., Weller, D. L., Durso, L. M., Bagley, M., Davis, B. C., Frye, J. G., Grim, C. J., Ibekwe, A. M., Jahne, M. A., Keely, S. P., Kraft, A. L., McConn, B. R., Mitchell, R. M., Ottesen, R. R., Sharma, M., Strain, E. A., Tadesse, D. A., Tate, H., Wells, J. E., & Garland, J. L. (2024). A One Health approach for monitoring antimicrobial resistance: Developing a national freshwater pilot effort. *Frontiers in Water*, *6*, 1359109. DOI: 10.3389/frwa.2024.1359109

Friedman, A., Phan, J., Schairer, D., Champer, J., Qin, M., Pirouz, A., Blecher-Paz, K., Oren, A., Liu, P., Modlin, R., & Kim, J. (2012). Antimicrobial and anti-inflammatory activity of chitosan–alginate nanoparticles: A targeted therapy for cutaneous pathogens. *The Journal of Investigative Dermatology*, *132*(4), 1234–1245. DOI: 10.1038/jid.2012.399

Fuoco, D. (2025). The stealth effect from a medicinal chemist perspective: Definition and updates. *Frontiers in Drug Delivery*, *5*, 1564120. DOI: 10.3389/fddev.2025.1564120 PMID: 40837707

Galm, U., & Shen, B. (2007, October). Natural product drug discovery: The times have never been better. *Chemistry & Biology*, *14*(10), 1098–1104. DOI: 10.1016/j.chembiol.2007.10.004 PMID: 17961822

Gamża, A. M., Hagenaars, T. J., Koene, M. G. J., & de Jong, M. C. M. (2023). Combining a parsimonious mathematical model with infection data from tailor-made experiments to understand environmental transmission. *Scientific Reports*, *13*(1), 38817. DOI: 10.1038/s41598-023-38817-z

Gao, F., Shao, T., Yu, Y., Xiong, Y., & Yang, L. (2021). Surface-bound reactive oxygen species generating nanozymes for selective antibacterial action. *Nature Communications*, *12*(1), 745. DOI: 10.1038/s41467-021-20965-3

Gao, W., Chan, J. M., & Farokhzad, O. C. (2010, December 6). pH-Responsive nanoparticles for drug delivery. *Molecular Pharmaceutics*, *7*(6), 1913–1920. DOI: 10.1021/mp100253e PMID: 20836539

Gao, X. J., Ciura, K., Ma, Y., Mikolajczyk, A., Jagiello, K., Wan, Y., Gao, Y., Zheng, J., Zhong, S., Puzyn, T., & Gao, X. (2024). Toward the integration of machine learning and molecular modeling for designing drug delivery nanocarriers. *Advanced Materials*, *36*(45), 2407793. DOI: 10.1002/adma.202407793

Garzón, V., Bustos, R. H., & Pinacho, G., D. (. (2020). Personalized medicine for antibiotics: The role of nanobiosensors in therapeutic drug monitoring. *Journal of Personalized Medicine*, *10*(4), 147. DOI: 10.3390/jpm10040147 PMID: 32993004

Gatherer, D. (2010). So what do we truly mean when we say that systems biology is holistic? *BMC Systems Biology*, *4*, 22. DOI: 10.1186/1752-0509-4-22

Gaurav, A., Bakht, P., Saini, M., Pandey, S., & Pathania, R. (2023). Role of bacterial efflux pumps in antibiotic resistance, virulence, and strategies to discover novel efflux pump inhibitors. *Microbiology (Reading, England)*, *169*(5), 001333. DOI: 10.1099/mic.0.001333 PMID: 37224055

GBD 2021 Antimicrobial Resistance Collaborators (2024). Global burden of bacterial antimicrobial resistance 1990-2021: a systematic analysis with forecasts to 2050. *Lancet (London, England)*, *404*(10459), 1199–1226. DOI: 10.1016/S0140-6736(24)01867-1

George, M., & Abraham, T. E. (2006). Polyionic hydrocolloids for the intestinal delivery of protein drugs: Alginate and chitosan-A review. *Journal of Controlled Release*, *114*(1), 1–14. DOI: 10.1016/j.jconrel.2006.04.017 PMID: 16828914

Gerilovych, A., Shevchenko, N., Pishchanskyi, O., Aliekseieva, H., Rosada, M., Gerilovych, I., & Okaievych, O. (2026). Operationalizing the One Health approach in a conflict-affected setting: A scientometric review of policy foundations, systemic gaps, and future pathways in Ukraine. *Veterinary World*, *19*(1), 389–408. DOI: 10.14202/vetworld.2026.389-408

Getahun, Y. A., Ali, D. A., Taye, B. W., & Alemayehu, Y. A. (2022). Multidrug-Resistant Microbial Therapy Using Antimicrobial Peptides and the CRISPR/Cas9 System. *Veterinary Medicine (Auckland)*, *13*, 173–190. DOI: 10.2147/VMRR.S366533 PMID: 35983086

Ghaderzadeh, M., Shalchian, A., Irajian, G., Sadeghsalehi, H., Zahedi Bialvaei, A., & Sabet, B. (2024). Artificial intelligence in drug discovery and development against antimicrobial resistance: A narrative review. *Majallah-i Mikrub/Shinasi-i Pizishki-i Iran*, *18*(3), 135–147. DOI: 10.30699/ijmm.18.3.135

Gheorghita, F. I., Bocanet, V. I., & Iantovics, L. B. (2025). Machine learning-based drug–drug interaction prediction: A critical review of models, limitations, and data challenges. *Frontiers in Pharmacology*, *16*, 1632775. DOI: 10.3389/fphar.2025.1632775

Ghosh, A., Vang, C. K., Brenner, E. P., & Ravi, J. (2025). Unlocking antimicrobial resistance with multiomics and machine learning. *Trends in Microbiology*, *33*(10), 1048–1051. DOI: 10.1016/j.tim.2025.04.017 PMID: 40425396

Giacobbe, D. R., Grossi, A. A., Bassetti, M., & de la Fuente-Nunez, C. (2026). The future of antibiotics and artificial intelligence: Some thoughts from discovery to bedside. *Infectious Diseases and Therapy*, *15*(2), 461–475. DOI: 10.1007/s40121-025-01288-y

Gillies, E. R., & Fréchet, J. M. (2005). Dendrimers and dendritic polymers in drug delivery. *Drug Discovery Today*, *10*(1), 35–43. DOI: 10.1016/S1359-6446(04)03276-3 PMID: 15676297

Ginnan, N., Crandall, S. G., Imchen, M., Dini-Andreote, F., Miyashiro, T. I., Singh, V., Ganda, E., & Bordenstein, S. R. (2025). Ecologically expanding the One Health framework to unify the microbiome sciences. *mBio*, *16*(6), e03147–e24. DOI: 10.1128/mbio.03147-24

Goes, A., & Fuhrmann, G. (2018). Biogenic and biomimetic carriers as versatile transporters to treat infections. *ACS Infectious Diseases*, *4*(6), 881–892. DOI: 10.1021/acsinfecdis.8b00030 PMID: 29553240

Goharshadi, E. K. (2025). A review on the radial distribution function: Insights into molecular structure, intermolecular interactions, and thermodynamic properties. *Journal of Molecular Liquids*, *433*, 127900. DOI: 10.1016/j.molliq.2025.127900

Goh, K. H., Wang, L., Yeow, A. Y. K., Poh, H., Li, K., Yeow, J. J. L., & Tan, G. Y. H. (2021). Artificial intelligence in sepsis early prediction and diagnosis using unstructured data in healthcare. *Nature Communications*, *12*(1), 711. Advance online publication. DOI: 10.1038/s41467-021-20910-4 PMID: 33514699

Golriz Khatami, S., Mubeen, S., Bharadhwaj, V. S., Kodamullil, A. T., Hofmann-Apitius, M., & Domingo-Fernández, D. (2021). Using predictive machine learning models for drug response simulation by calibrating patient-specific pathway signatures. *NPJ Systems Biology and Applications*, *7*(1), 40. DOI: 10.1038/s41540-021-00199-1

Gomez-Simmonds, A., Nelson, B., Eiras, D. P., Loo, A., Jenkins, S. G., Whittier, S., Calfee, D. P., Satlin, M. J., Kubin, C. J., & Furuya, E. Y. (2016, May 23). Combination Regimens for Treatment of Carbapenem-Resistant Klebsiella pneumoniae Bloodstream Infections. *Antimicrobial Agents and Chemotherapy*, *60*(6), 3601–3607. DOI: 10.1128/AAC.03007-15 PMID: 27044555

Gou, F., Liu, J., Xiao, C., & Wu, J. (2024). Research on artificial-intelligence-assisted medicine: A survey on medical artificial intelligence. *Diagnostics (Basel)*, *14*(14), 1472. DOI: 10.3390/diagnostics14141472

Gowthaman, N., Ruba, M., Aswath, S., Hemavarshni, S. N., Bragadeshwaran, R., & Nithish, S. J. (2025, November). Iot Integrated Solution For Smart Vertical Crop Cultivation. In 2025 IEEE 7th International Conference on Computing, Communication and Automation (ICCCA) (pp. 1-6). IEEE

Gray, M., Meehan, J., Ward, C., Langdon, S. P., Kunkler, I. H., Murray, A., & Argyle, D. (2018). Implantable biosensors and their contribution to the future of precision medicine. *Veterinary Journal (London, England)*, *239*, 21–29. DOI: 10.1016/j.tvjl.2018.07.011

Greathouse, K. L., & Choudhury, A. (2025). Precision nutrition and the gut microbiome: Harnessing AI to revolutionize cancer prevention and therapy. *Cell Host & Microbe*, *33*(6), 766–776. DOI: 10.1016/j.chom.2025.05.011

Greener, J. G., Kandathil, S. M., Moffat, L., & Jones, D. T. (2022). A guide to machine learning for biologists. *Nature Reviews. Molecular Cell Biology*, *23*(1), 40–55. DOI: 10.1038/s41580-021-00407-0 PMID: 34518686

Grenha, A., Grainger, C. I., Dailey, L. A., Seijo, B., Martin, G. P., Remunan-Lopez, C., & Forbes, B. (2007). Chitosan nanoparticles are compatible with respiratory epithelial cells in vitro. *European Journal of Pharmaceutical Sciences*, *31*(2), 73–84. DOI: 10.1016/j.ejps.2007.02.008 PMID: 17408932

Griffith, D. E., Eagle, G., Thomson, R., Aksamit, T., Hasegawa, N., Morimoto, K., Addrizzo-Harris, J., O'Donnell, A. E., Marras, T. K., Flume, P. A., Loebinger, M. R., Morgan, L., Codecasa, L. R., Hill, A. T., Ruoss, S. J., Yim, J.-J., Ringshausen, F. C., Field, S. K., Philley, J. V., & Winthrop, K. L. (2018). Amikacin Liposome Inhalation Suspension for Treatment-Refractory Lung Disease Caused by Mycobacterium avium Complex (CONVERT). A Prospective, Open-Label, Randomized Study. *American Journal of Respiratory and Critical Care Medicine*, *198*(12), 1559–1569. DOI: 10.1164/rccm.201807-1318OC PMID: 30216086

Guégan, J.-F., Suzán, G., Kati-Coulibaly, S., Bonpamgue, D. N., & Moatti, J.-P. (2018). Sustainable Development Goal #3, "health and well-being," and the need for more integrative thinking. *Veterinaria México OA*, *5*(2), 443. Advance online publication. DOI: 10.21753/vmoa.5.2.443

Guerra, M. E. S., Destro, G., Vieira, B., Lima, A. S., Ferraz, L. F. C., Hakansson, A. P., Darrieux, M., & Converso, T. R. (2022). *Klebsiella pneumoniae* biofilms and their role in disease pathogenesis. *Frontiers in Cellular and Infection Microbiology*, *12*, 877995. DOI: 10.3389/fcimb.2022.877995

Gu, K., Liang, Y., Su, J., Sun, P., Peng, J., Miao, N., & Zhang, J. (2024). Deep Learning Models for Colloidal Nanocrystal Synthesis. *ACS Nano*. Advance online publication. DOI: 10.1021/acsnano.5c09134 PMID: 41193409

Gupta, S., Kesarla, R., & Omri, A. (2013, December 26). Formulation strategies to improve the bioavailability of poorly absorbed drugs with special emphasis on self-emulsifying systems. *ISRN Pharmaceutics, 848043*, 1–16. Advance online publication. DOI: 10.1155/2013/848043 PMID: 24459591

Gupta, V. P., Reyed, M. R., & Haghi, A. K. (Eds.). (2026). *The blue economy and environmental sustainability: Advancing global governance, innovation, and finance for a resilient future*. Springer Cham., DOI: 10.1007/978-3-032-13638-1

Hadipour Moghaddam, S. P., Saikia, J., Yazdimamaghani, M., & Ghandehari, H. (2017). Redox-responsive polysulfide-based biodegradable organosilica nanoparticles for delivery of bioactive agents. *ACS Applied Materials & Interfaces*, *9*(25), 21133–21146. DOI: 10.1021/acsami.7b04351 PMID: 28609092

Ha, E., Ha, S. M., Gerelkhuu, Z., Kim, H. Y., & Yoon, T. H. (2025). AI-based nanotoxicity data extraction and prediction of nanotoxicity. *Computational and Structural Biotechnology Journal*, *29*, 138–148. DOI: 10.1016/j.csbj.2025.03.052 PMID: 40255458

Hamilton, S., & Kingston, B. R. (2024). Applying artificial intelligence and computational modeling to nanomedicine. *Current Opinion in Biotechnology*, *85*, 103043. DOI: 10.1016/j.copbio.2023.103043 PMID: 38091874

Hand, K. (2013). Antibiotic stewardship. *Clinical Medicine (London, England)*, *13*(5), 499–503. DOI: 10.7861/clinmedicine.13-5-499 PMID: 24115710

Han, R., Ye, Z., Zhang, Y., Cheng, Y., Zheng, Y., & Ouyang, D. (2023). Predicting liposome formulations by the integrated machine learning and molecular modeling approaches. *Asian Journal of Pharmaceutical Sciences*, *18*(3), 100811. DOI: 10.1016/j.ajps.2023.100811 PMID: 37274923

Han, Y., Kim, D. H., & Pack, S. P. (2025). Nanomaterials in drug delivery: Leveraging artificial intelligence and big data for predictive design. *International Journal of Molecular Sciences*, *26*(22), 11121. DOI: 10.3390/ijms262211121

Haredasht, F. N., Maddali, M. V., Ma, S. P., Chang, A., Kim, G. Y. E., Banaei, N., Deresinski, S., Goldstein, M. K., Asch, S. M., & Chen, J. H. (2025). Enhancing antibiotic stewardship: A machine learning approach to predicting antibiotic resistance in inpatient care. *AMIA Annual Symposium Proceedings*, 857–864.

Hartung, T. (2018, December 1). Perspectives on In Vitro to In Vivo Extrapolations. *Applied In Vitro Toxicology*, *4*(4), 305–316. DOI: 10.1089/aivt.2016.0026 PMID: 31890748

Harush-Frenkel, O., Bivas-Benita, M., Nassar, T., Springer, C., Sherman, Y., Avital, A., Altschuler, Y., Borlak, J., & Benita, S. (2010). A safety and tolerability study of differently charged nanoparticles for local pulmonary drug delivery. *Toxicology and Applied Pharmacology*, *246*(1), 83–90. DOI: 10.1016/j.taap.2010.04.011 PMID: 20417650

Hasin, Y., Seldin, M., & Lusis, A. (2017). Multiomics approaches to disease. *Genome Biology*, *18*(1), 83. DOI: 10.1186/s13059-017-1215-1

Hassan, W., Noreen, H., Rehman, S., Kamal, M. A., & da Rocha, J. B. T. (2022). Association of Oxidative Stress with Neurological Disorders. *Current Neuropharmacology*, *20*(6), 1046–1072. DOI: 10.2174/1570159X19666211111141246 PMID: 34781871

Helliwell, R., Ewin, I., Williams, A. D., Levine, D. T., Singer, A. C., Raman, S., Morris, C., & Stekel, D. J. (2025). Rethinking the words hotspot reservoir and pristine in the environmental dimensions of antimicrobial resistance. *npj antimicrobials and resistance, 3*(1), 11. https://doi.org/DOI: 10.1038/s44259-025-00080-9

Hendriksen, R. S., Munk, P., Njage, P., van Bunnik, B., McNally, L., Lukjancenko, O., Röder, T., Nieuwenhuijse, D., Pedersen, S. K., Kjeldgaard, J., Kaas, R. S., Clausen, P. T. L. C., Vogt, J. K., Leekitcharoenphon, P., van de Schans, M. G. M., Zuidema, T., de Roda Husman, A. M., Rasmussen, S., Petersen, B., & Aarestrup, F. M. (2019). Global monitoring of antimicrobial resistance based on metagenomics analyses of urban sewage. *Nature Communications*, *10*(1), 1124. DOI: 10.1038/s41467-019-08853-3

Henser-Brownhill, T., Martin, L., Samangouei, P., Ladak, A., Apostolidou, M., Nagel, B., & Kwok, A. (2024). In silico screening accelerates nanocarrier design for efficient mRNA delivery. *Advancement of Science*, *11*(30), 2401935. DOI: 10.1002/advs.202401935 PMID: 38837626

Hernando-Amado, S., Coque, T. M., Baquero, F., & Martínez, J. L. (2020). Antibiotic Resistance: Moving From Individual Health Norms to Social Norms in One Health and Global Health. *Frontiers in Microbiology*, *11*, 1914. DOI: 10.3389/fmicb.2020.01914

He, S., Nader, K., Abarrategi, J. S., Bediaga, H., Nocedo-Mena, D., Ascencio, E., Casanola-Martin, G. M., Castellanos-Rubio, I., Insausti, M., Rasulev, B., Arrasate, S., & González-Díaz, H. (2024). NANO.PTML model for read-across prediction of nanosystems in neurosciences: Computational model and experimental case of study. *Journal of Nanobiotechnology*, *22*(1), 435. DOI: 10.1186/s12951-024-02660-9

Hetta, H. F., Ramadan, Y. N., Al-Harbi, A. I., Ahmed, A., Battah, E., Abd, B., Ellah, N. H., Zanetti, S., & Donadu, M. G. (2023). Nanotechnology as a promising approach to combat multidrug resistant bacteria: A comprehensive review and future perspectives. *Biomedicines*, *11*(2), 413. DOI: 10.3390/biomedicines11020413

He, W., Yu, Z., Wu, Z., Olesen, A. K., Madsen, J. S., Dechesne, A., Smets, B. F., Nesme, J., & Sørensen, S. J. (2026). Beyond borders: Plasmids drive a shared antibiotic resistome in European urban water systems. *Microbiome*, *14*(1), 39. DOI: 10.1186/s40168-025-02253-0

Heydari, S., Masoumi, N., Esmaeeli, E., Ayyoubzadeh, S. M., Ghorbani-Bidkorpeh, F., & Ahmadi, M. (2024). Artificial intelligence in nanotechnology for treatment of diseases. *Journal of Drug Targeting*, *32*(10), 1247–1266. DOI: 10.1080/1061186X.2024.2393417

Hirani, R., Noruzi, K., Khuram, H., Hussaini, A. S., Aifuwa, E. I., Ely, K. E., Lewis, J. M., Gabr, A. E., Smiley, A., Tiwari, R. K., & Etienne, M. (2024). Artificial intelligence and healthcare: A journey through history, present innovations, and future possibilities. *Life (Basel, Switzerland)*, *14*(5), 557. DOI: 10.3390/life14050557

Hock, S. C., Kian, S. M., & Wah, C. L. (2020). Global challenges in the manufacture, regulation and international harmonization of GMP and quality standards for biopharmaceuticals. *Generics and Biosimilars Initiative Journal*, *9*(2), 52-64. https://doi.org/DOI: 10.5639/gabij.2020.0902.010

Hock, S. C., Ying, Y. M., & Wah, C. L. (2011). A review of the current scientific and regulatory status of nanomedicines and the challenges ahead. *PDA Journal of Pharmaceutical Science and Technology*, *65*(2), 177–195. DOI: 10.5731/pda-jpst.2011.00754 PMID: 21502077

Ho, D., Quake, S. R., McCabe, E. R., Chng, W. J., Chow, E. K., Ding, X., Gelb, B. D., Ginsburg, G. S., Hassenstab, J., Ho, C.-M., Mobley, W. C., Nolan, G. P., Rosen, S. T., Tan, P., Yen, Y., & Zarrinpar, A. (2020). Enabling technologies for personalized and precision medicine. *Trends in Biotechnology*, *38*(5), 497–518. DOI: 10.1016/j.tibtech.2019.12.021 PMID: 31980301

Horgan, D., Tanner, M., Aggarwal, C., Thomas, D., Grover, S., Basel-Salmon, L., Dienstmann, R., Tan, T. J. Y., Park, W.-Y., Abu Rasheed, H. M., Siu, L. L., Ma, B., Ortiz-López, R., Van den Bulcke, M., Taucher, S. C., Ferris, A., Starling, N., Malapelle, U., Longshore, J., & Subbiah, V. (2025). Precision Oncology: A Global Perspective on Implementation and Policy Development. *JCO Global Oncology*, *11*(11), e2400416. Advance online publication. DOI: 10.1200/GO-24-00416 PMID: 39847746

Hornung, R., & Wright, M. N. (2019). Block Forests: Random forests for blocks of clinical and omics covariate data. *BMC Bioinformatics*, *20*(1), 358. DOI: 10.1186/s12859-019-2942-y PMID: 31248362

Hortelao, A. C., Simó, C., Guix, M., Garrido, S. G., Julián, E., Vilela, D., Rejc, L., Ramos-Cabrer, P., Cossío, U., Gómez-Vallejo, V., & Patiño, T. (2021). Swarming behavior and in vivo monitoring of enzymatic nanomotors within the bladder. *Science Robotics*, *6*(52), eabd2823. Advance online publication. DOI: 10.1126/scirobotics.abd2823

Hosni, Z., Achour, S., Saadi, F., Chen, Y., & Al Qaraghuli, M. (2025). Machine learning-driven nanoparticle toxicity. *Ecotoxicology and Environmental Safety*, *299*, 118340. DOI: 10.1016/j.ecoenv.2025.118340 PMID: 40393320

Hosny, S., Alakib, M., Bendary, S. H., Abu-Dief, A. M., & Abbass, L. M. (2026). Eco-friendly synthesis of novel nanocomplexes: Integrated in vitro evaluation, advanced molecular dynamics simulations, network pharmacology, and in silico ADME/toxicity assessment. *Inorganic Chemistry Communications*, *188*(Part 1), 116479. DOI: 10.1016/j.inoche.2026.116479

Hossam Abdelmonem, B., Abdelaal, N. M., Anwer, E. K. E., Rashwan, A. A., Hussein, M. A., Ahmed, Y. F., Khashana, R., Hanna, M. M., & Abdelnaser, A. (2024). Decoding the Role of CYP450 Enzymes in Metabolism and Disease: A Comprehensive Review. *Biomedicines*, *12*(7), 1467. DOI: 10.3390/biomedicines12071467 PMID: 39062040

Hou, X., Zaks, T., Langer, R., & Dong, Y. (2021). Lipid nanoparticles for mRNA delivery. *Nature Reviews. Materials*, *6*(12), 1078–1094. DOI: 10.1038/s41578-021-00358-0 PMID: 34394960

Hrncir, T. (2022). Gut microbiota dysbiosis: Triggers, consequences, diagnostic and therapeutic options. *Microorganisms*, *10*(3), 578. DOI: 10.3390/microorganisms10030578

Huang, K., Fu, T., Glass, L. M., Zitnik, M., Xiao, C., & Sun, J. (2021a). DeepPurpose: A deep learning library for drug–target interaction prediction. *Bioinformatics (Oxford, England)*, *36*(22–23), 5545–5547. DOI: 10.1093/bioinformatics/btaa1005

Huang, Y., Guo, X., Wu, Y., Chen, X., Feng, L., Xie, N., & Shen, G. (2024). Nanotechnology's frontier in combatting infectious and inflammatory diseases: Prevention and treatment. *Signal Transduction and Targeted Therapy*, *9*(1), 34. Advance online publication. DOI: 10.1038/s41392-024-01745-z PMID: 38378653

Huang, Y., Rana, A. P., Wenzler, E., Ozer, E. A., Krapp, F., Bulitta, J. B., Hauser, A. R., & Bulman, Z. P. (2022). Aminoglycoside-resistance gene signatures are predictive of aminoglycoside MICs for carbapenem-resistant *Klebsiella pneumoniae*. *The Journal of Antimicrobial Chemotherapy*, *77*(2), 356–363. DOI: 10.1093/jac/dkab381

Huang, Y., Sokolowski, K., Rana, A., Singh, N., Wang, J., Chen, K., Lang, Y., Zhou, J., Kadiyala, N., Krapp, F., Ozer, E. A., Hauser, A. R., Li, J., Bulitta, J. B., & Bulman, Z. P. (2021b). Generating genotype-specific aminoglycoside combinations with ceftazidime/avibactam for KPC-producing *Klebsiella pneumoniae*. *Antimicrobial Agents and Chemotherapy*, *65*(8), e00692–e21. DOI: 10.1128/AAC.00692-21

Hua, S., de Matos, M. B. C., Metselaar, J. M., & Storm, G. (2018). Current trends and challenges in the clinical translation of nanoparticulate nanomedicines: Pathways for translational development and commercialization. *Frontiers in Pharmacology*, *9*, 790. DOI: 10.3389/fphar.2018.00790

Hu, C. M. J., Zhang, L., Aryal, S., Cheung, C., Fang, R. H., & Zhang, L. (2015). Erythrocyte membrane-camouflaged nanoparticles for antibiotic delivery. *Proceedings of the National Academy of Sciences of the United States of America*, *108*(27), 10980–10985. DOI: 10.1073/pnas.1106634108 PMID: 21690347

Huh, A. J., & Kwon, Y. J. (2011). "Nanoantibiotics": A new paradigm for treating infectious diseases using nanomaterials in the antibiotics resistant era. *Journal of Controlled Release*, *156*(2), 128–145. DOI: 10.1016/j.jconrel.2011.07.002 PMID: 21763369

Huovinen, P., Sundström, L., Swedberg, G., & Sköld, O. (1995). Trimethoprim and sulfonamide resistance. *Antimicrobial Agents and Chemotherapy*, *39*(2), 279–289. DOI: 10.1128/AAC.39.2.279

Hu, Q., Katti, P. S., & Gu, Z. (2014, November 7). Enzyme-responsive nanomaterials for controlled drug delivery. *Nanoscale*, *6*(21), 12273–12286. DOI: 10.1039/C4NR04249B PMID: 25251024

Hutchings, M. I., Truman, A. W., & Wilkinson, B. (2019). Antibiotics: Past, present and future. *Current Opinion in Microbiology*, *51*, 72–80. DOI: 10.1016/j.mib.2019.10.008

Ijaz, S., Li, J., & Gong, C. (2024). Cell-derived biomimetic nanocarriers for drug delivery: Classification, mechanisms, and applications. *Journal of Nanobiotechnology*, *22*, 85. DOI: 10.1186/s12951-024-02144-8

Ingber, D. E. (2022). Human organs-on-chips for disease modelling, drug development and personalized medicine. *Nature Reviews. Genetics*, *23*(8), 467–491. DOI: 10.1038/s41576-022-00466-9 PMID: 35338360

Islam, S., Ahmed, M. M. S., Islam, M. A., Hossain, N., & Chowdhury, M. A. (2025). Advances in nanoparticles in targeted drug delivery–A review. *Results in Surfaces and Interfaces*, *19*, 100529. DOI: 10.1016/j.rsurfi.2025.100529

Issa, K. D., & Muhsin, M. D. (2023). Beta–Lactam drug resistance pattern in Staphylococcus aureus isolates: A review. *Eurasian Journal of Science and Engineering*, *9*(3), 141–153. DOI: 10.23918/eajse.v9i3p13

Iwano, N., Adachi, T., Aoki, K., Nakamura, Y., & Hamada, M. (2022). Generative aptamer discovery using RaptGen. *Nature Computational Science*, *2*(6), 378–386. DOI: 10.1038/s43588-022-00249-6 PMID: 38177576

Jacob, M. C. M., de Carvalho, A. M., Batista, Â. G., Santos Júnior, A. F., Carioca, A. A. F., de Morais, C. M. M., Cazarin, C. B. B., Tregidgo, D., de Oliveira, D. V. B., Marchioni, D. M. L., Giuntini, E. B., de Menezes-Neto, E. J., Pereira, F. O., Moura, G. F., El Bizri, H. R., Bezerra, I. W. L., Aquino, J. S., Silva, J. V. M., Steluti, J., & Albuquerque, U. P. (2026). Prioritizing neglected food species in nutritional studies using expert-knowledge and explainable AI. *Scientific Reports*. Advance online publication. DOI: 10.1038/s41598-026-39484-6

Jain, E., Shin, J. H., Wells, K. M., Pannone, A. F., Hornsby, P. P., & Warren, C. A. (2020). Association between Clostridioides difficile infection and multidrug-resistant organism colonization or infection among hospitalized adults: A case-control study. *American Journal of Infection Control*, *48*(10), 1276–1278. DOI: 10.1016/j.ajic.2020.01.008 PMID: 32145992

Jamkhande, P. G., Ghante, M. H., & Ajgunde, B. R. (2017). Software based approaches for drug designing and development: A systematic review on commonly used software and its applications. *Bulletin of Faculty of Pharmacy, Cairo University*, *55*(2), 203–210. DOI: 10.1016/j.bfopcu.2017.10.001

Jang, Y., Babu, A., Chahal, S., Vasukutty, A., Moon, J. J., Park, I.-K., & Park, H. (2025). AI-guided design of a CXCR4-targeted core–shell nanocarrier for co-delivery of berberine/paclitaxel in cancer therapy. *Journal of Nanobiotechnology*, *23*, 350. DOI: 10.1186/s12951-025-03850-9

Jauregui, R., Srinivasan, S., Vojtech, L. N., Gammill, H. S., Chiu, D. T., Hladik, F., Stayton, P. S., & Lai, J. J. (2018, October 10). Stayton PSLai JJ. Temperature-Responsive Magnetic Nanoparticles for Enabling Affinity Separation of Extracellular Vesicles. *ACS Applied Materials & Interfaces*, *10*(40), 33847–33856. DOI: 10.1021/acsami.8b09751 PMID: 30152229

Jena, G. K., Patra, C. N., Jammula, S., Rana, R., & Chand, S. (2024). Artificial intelligence and machine learning implemented drug delivery systems: A paradigm shift in the pharmaceutical industry. *Journal of Bio-X Research, 7*, 0016. https://doi.org/DOI: 10.34133/jbioxresearch.0016

Jeong, H., Lee, J. Y., Lee, K., Kang, Y. J., Kim, J. T., Avila, R., Tzavelis, A., Kim, J., Ryu, H., Kwak, S. S., Kim, J. U., Banks, A., Jang, H., Chang, J.-K., Li, S., Mummidisetty, C. K., Park, Y., Nappi, S., Chun, K. S., & Rogers, J. A. (2021). Differential cardiopulmonary monitoring system for artifact-canceled physiological tracking of athletes, workers, and COVID-19 patients. *Science Advances*, *7*(23), eabg3092. Advance online publication. DOI: 10.1126/sciadv.abg3092

Jia, H., Hou, W., Wei, L., Xu, B., & Liu, X. (2008). The structures and antibacterial properties of nano-SiO2 supported silver/zinc-silver materials. *Dental Materials : Official Publication of the Academy of Dental Materials*, *24*(2), 244–249. DOI: 10.1016/j.dental.2007.04.015

Jiang, L., & Loo, S. C. J. (2021). Intelligent Nanoparticle-Based Dressings for Bacterial Wound Infections. *ACS Applied Bio Materials*, *4*(5), 3849–3862. DOI: 10.1021/acsabm.0c01168 PMID: 34056562

Jiang, Q., Yan, X., Jiao, D., Zhang, J., Wu, Y., & Cheng, Y. (2021). Polymyxin B-modified conjugated oligomer nanoparticle for targeted identification and enhanced photodynamic antimicrobial therapy. *Chemical Communications*, *57*(85), 11244–11247. DOI: 10.1039/D1CC04389G PMID: 34633009

Jia, Y., Naceur, H., Saadlaoui, Y., Dubar, L., & Bergheau, J. M. (2024). A comprehensive comparison of modeling strategies and simulation techniques applied in powder-based metallic additive manufacturing processes. *Journal of Manufacturing Processes*, *110*, 1–29. DOI: 10.1016/j.jmapro.2023.12.048

Jinan, A. F., Fazeeha, M., & Meera, K. A. (2025). Smart cancer patch: AI-integrated biosensor with adaptive drug release for universal cancer detection and treatment. *International Journal for Research in Applied Science and Engineering Technology, 13*(6), 438–442. Advance online publication. DOI: 10.22214/ijraset.2025.73593

Jin, S., Wang, Y., Deng, W., & Li, J. (2020). Wearable sensors for human health monitoring: From materials to devices. *Biosensors & Bioelectronics, 165*, 112413. DOI: 10.1016/j.bios.2020.112413

Jo, E., Epstein, D. A., Jung, H., & Kim, Y. H. (2023). Understanding the benefits and challenges of deploying conversational AI leveraging large language models for public health intervention. In *Proceedings of the CHI Conference on Human Factors in Computing Systems* (Article 16). DOI: 10.1145/3544548.3581503

Jo, H. E., Kwon, M. S., Whon, T. W., Kim, D. W., Yun, M., Lee, J., Shin, M. Y., Kim, S. H., & Choi, H. J. (2021). Alteration of Gut Microbiota After Antibiotic Exposure in Finishing Swine. *Frontiers in Microbiology, 12*, 596002. DOI: 10.3389/fmicb.2021.596002 PMID: 33643231

Johnson, N. C., & Marín, C. (2025). Functional team selection as a framework for local adaptation in plants and their belowground microbiomes. *The ISME Journal, 19*(1), wraf137. Advance online publication. DOI: 10.1093/ismejo/wraf137

Johnston, C. W., & Badran, A. H. (2022). Natural and engineered precision antibiotics in the context of resistance. *Current Opinion in Chemical Biology, 69*, 102160. DOI: 10.1016/j.cbpa.2022.102160 PMID: 35660248

Jonas, O. & Team, W. (2017). Drug-Resistant Infections: A Threat to Our Economic Future.

Jung, W. K., Koo, H. C., Kim, K. W., Shin, S., Kim, S. H., & Park, Y. H. (2008). Antibacterial activity and mechanism of action of the silver ion in *Staphylococcus aureus* and *Escherichia coli. Applied and Environmental Microbiology, 74*(7), 2171–2178. DOI: 10.1128/AEM.02001-07

Kakar, S., Jain, A., & Singh, R. (2017). Magnetic Microspheres: A Novel Targeting Delivery System. *FABAD Journal of Pharmaceutical Sciences, 42*(3), 209–223.

Kalomiraki, M., Thermos, K., & Chaniotakis, N. A. (2015). Dendrimers as tunable vectors of drug delivery systems and biomedical and ocular applications. *International Journal of Nanomedicine, 10*, 2403–2418. DOI: 10.2147/IJN.S78820 PMID: 26730187

Kamaly, N., Yameen, B., Wu, J., & Farokhzad, O. C. (2016). Degradable controlled-release polymers and polymeric nanoparticles: Mechanisms of controlling drug release. *Chemical Reviews*, *116*(4), 2602–2663. DOI: 10.1021/acs.chemrev.5b00346 PMID: 26854975

Kameswaran, A., Nanthakumar, K., & Ruba, M. (2017). High performance pre-computation based self-controlled precharge-free content-addressable memory. *Journal of Electronics and Communication Systems*, *2*, 1–9.

Kang, S., Antoniewicz, M. R., & Hay, N. (2024). Metabolic and transcriptomic reprogramming during contact inhibition-induced quiescence is mediated by YAP-dependent and YAP-independent mechanisms. *Nature Communications*, *15*(1), 6777. DOI: 10.1038/s41467-024-51117-y

Kantesaria, R., & Panda, H. S. (2026, March 09). (2026). A Review on AI-Based Data-Driven Models for Optimization of Nanocarriers as Drug Delivery Systems. *ACS Biomaterials Science & Engineering*, *12*(3), 1397–1418. Advance online publication. DOI: 10.1021/acsbiomaterials.5c01998 PMID: 41668335

Kant, S., Deepika, , & Roy, S. (2025). Artificial intelligence in drug discovery and development: Transforming challenges into opportunities. *Discover Pharmaceutical Sciences*, *1*(1), 7. DOI: 10.1007/s44395-025-00007-3

Kan, Z., Wen, J., Bonato, V., Webster, J., Yang, W., Ivanov, V., Kim, K. H., Roh, W., Liu, C., Mu, X. J., Lapira-Miller, J., Oyer, J., VanArsdale, T., Rejto, P. A., & Bienkowska, J. (2025). Real-world clinical multi-omics analyses reveal bifurcation of ER-independent and ER-dependent drug resistance to CDK4/6 inhibitors. *Nature Communications*, *16*(1), 932. Advance online publication. DOI: 10.1038/s41467-025-55914-x PMID: 39843429

Kapoor, D. U., Sharma, J. B., Gandhi, S. M., Prajapati, B. G., Thanawuth, K., Limmatvapirat, S., & Sriamornsak, P. (2024). AI-driven design and optimization of nanoparticle-based drug delivery systems. *Science. Engineering and Health Studies*, *18*, 24010003. Advance online publication. DOI: 10.69598/sehs.18.24010003

Kariuki, S. (2024). Global burden of antimicrobial resistance and forecasts to 2050. *Lancet*, *404*(10459), 1172–1173. DOI: 10.1016/S0140-6736(24)01885-3 PMID: 39299259

Kasse, G. E., Cosh, S. M., Humphries, J., & Islam, M. S. (2025). Leveraging artificial intelligence for One Health: Opportunities and challenges in tackling antimicrobial resistance - scoping review. *One Health Outlook*, *7*(1), 51. DOI: 10.1186/s42522-025-00170-8

Kataria, S. (2025). AI-guided design of personalized nanomedicine: A review of data-driven approaches in nanoparticle formulation. *International Journal for Multidisciplinary Research*, *7*(4), 53283. Advance online publication. DOI: 10.36948/ijfmr.2025.v07i04.53283

Katz, L., & Baltz, R. H. (2016). Natural product discovery: Past, present, and future. *Journal of Industrial Microbiology & Biotechnology*, *43*(2–3), 155–176. DOI: 10.1007/s10295-015-1723-5

Kaur, T., & Singh, S. (2021). Controlled release of bi-layered malvidin tablets using 3D printing techniques. *Journal of Pharmaceutical Research International*, *32*(40), 70–78. DOI: 10.9734/jpri/2020/v32i4031034

Kehrein, J., Bunker, A., & Luxenhofer, R. (2024). POxload: Machine learning estimates drug loadings of polymeric micelles. *Molecular Pharmaceutics*, *21*(7), 3356–3374. DOI: 10.1021/acs.molpharmaceut.4c00086

Kelly, C. J., Karthikesalingam, A., Suleyman, M., Corrado, G., & King, D. (2019). Key challenges for delivering clinical impact with artificial intelligence. *BMC Medicine*, *17*(1), 195. Advance online publication. DOI: 10.1186/s12916-019-1426-2

Kesharwani, P., Jain, K., & Jain, N. K. (2014). Dendrimer toxicity and safety: A critical review. *Progress in Polymer Science*, *39*(2), 268–307. DOI: 10.1016/j.progpolymsci.2013.07.005

Khalid-Salako, F., Salimi Khaligh, S., Fathi, F., Demirci, O. C., Öncer, N., Kurt, H., & Yuce, M. (2025). The nanocarrier landscape-evaluating key drug delivery vehicles and their capabilities: A translational perspective. *ACS Applied Materials & Interfaces*, *17*(26), 37383–37403. DOI: 10.1021/acsami.5c07366 PMID: 40526827

Khalifa, M. M. A. H. (2024). Safety of nanoparticles in medicine. In *Nanocarriers in neurodegenerative disorders* (pp. 94–100). CRC Press. DOI: 10.1201/9781003383376-7

Khan, F., Shariq, M., Asif, M., Siddiqui, M. A., Malan, P., & Ahmad, F. (2022). Green Nanotechnology: Plant-Mediated Nanoparticle Synthesis and Application. *Nanomaterials (Basel, Switzerland)*, *12*(4), 673. DOI: 10.3390/nano12040673 PMID: 35215000

Khan, S. R., Pavuluri, S. K., Cummins, G., & Desmulliez, M. P. Y. (2020, June 19). Wireless Power Transfer Techniques for Implantable Medical Devices: A Review. *Sensors (Basel)*, *20*(12), 3487. DOI: 10.3390/s20123487 PMID: 32575663

Khodabandehloo, H., Zahednasab, H., & Hafez, A. A. (2016). Nanocarriers usage for drug delivery in cancer therapy. *Iranian Journal of Cancer Prevention*, *9*(2), e3966. DOI: 10.17795/ijcp-3966 PMID: 27482328

Khokhlov, I., Legashev, L., Bolodurina, I., Shukhman, A., Shoshin, D., & Kolesnik, S. (2024). Prediction of Dynamic Toxicity of Nanoparticles Using Machine Learning. *Toxics*, *12*(10), 750. DOI: 10.3390/toxics12100750 PMID: 39453170

Kim, J., Campbell, A. S., de Ávila, B. E.-F., & Wang, J. (2019). Wearable biosensors for healthcare monitoring. *Nature Biotechnology*, *37*(4), 389–406. DOI: 10.1038/s41587-019-0045-y

Kim, M., Gim, D., Kim, S., Park, S., Eom, T. P., Seol, J., Yeo, J., Jo, C., Seo, G., Ku, H., & Kim, J. H. (2025). From Dysbiosis to Prediction: AI-Powered Microbiome Insights into IBD and CRC. *Gastroenterology Insights*, *16*(3), 34. DOI: 10.3390/gastroent16030034

Kluzek, M., Oppenheimer-Shaanan, Y., Dadosh, T., Morandi, M. I., Avinoam, O., Raanan, C., Goldsmith, M., Goldberg, R., & Klein, J. (2022). Designer liposomic nanocarriers are effective biofilm eradicators. *ACS Nano*, *16*(8), 12451–12468. DOI: 10.1021/acsnano.2c04232

Koatale, P. C., Welling, M. M., Ndlovu, H., Kgatle, M., Mdanda, S., Mdlophane, A., Okem, A., Takyi-Williams, J., Sathekge, M. M., & Ebenhan, T. (2024). insights into Peptidoglycan-Targeting Radiotracers for Imaging Bacterial infections: Updates, challenges, and future perspectives. *ACS Infectious Diseases*, *10*(2), 270–286. DOI: 10.1021/acsinfecdis.3c00443 PMID: 38290525

Koide, R. T. (2023). On holobionts, holospecies, and holoniches: The role of microbial symbioses in ecology and evolution. *Microbial Ecology*, *85*(4), 1143–1149. DOI: 10.1007/s00248-022-02005-9

Koirala, M., Yan, L., Mohamed, Z., & DiPaola, M. (2025). AI-Integrated QSAR Modeling for Enhanced Drug Discovery: From Classical Approaches to Deep Learning and Structural Insight. *International Journal of Molecular Sciences*, *26*(19), 9384. DOI: 10.3390/ijms26199384 PMID: 41096653

Konagurthu, S., Ranathunga, D. T. S., Buchanan, S., Mehta, N. M., & Reynolds, T. (2026). The predictive edge: Modeling and simulation in drug product development. *Advanced Drug Delivery Reviews*, *230*, 115784. DOI: 10.1016/j.addr.2026.115784

Konwar, B., & Kim, K. S. (2025). Recent advances in bacterial hyaluronidase-responsive nano-drug delivery systems for targeting antibiotic-resistant Staphylococcus aureus. Chemical Engineering Journal, 165827. https://doi.org/DOI: 10.1016/j.cej.2025.165827

Krause, K. M., Serio, A. W., Kane, T. R., & Connolly, L. E. (2016). Aminoglycosides: An Overview. *Cold Spring Harbor Perspectives in Medicine*, *6*(6), a027029. DOI: 10.1101/cshperspect.a027029 PMID: 27252397

Krebs, E., & Nosyk, B. (2021). Cost-Effectiveness Analysis in Implementation Science: A Research Agenda and Call for Wider Application. *Current HIV/AIDS Reports*, *18*(3), 176–185. DOI: 10.1007/s11904-021-00550-5 PMID: 33743138

Krithika, S., Jayachandran, T., Raghapriya, N. R., Ruba, M., Janani, C., & Sathiyavathi, S. (2025, April). An Effective Denial of Service Texture Features from Malware Grayscale Image. In 2025 3rd International Conference on Artificial Intelligence and Machine Learning Applications Theme: Healthcare and Internet of Things (AIMLA) (pp. 1-7). IEEE. DOI: 10.1109/AIMLA63829.2025.11041501

Krpetic, Z., Porta, F., Caneva, E., Dal Santo, V., & Scarì, G. (2010). Phagocytosis of biocompatible gold nanoparticles. *Langmuir*, *26*(18), 14799–14805. DOI: 10.1021/la102758f PMID: 20795674

Kumar, R., Kumar, A., Bhardwaj, S., Sikarwar, M., Sriwastaw, S., Sharma, G., & Gupta, M. (2025). Nanotoxicity unveiled: Evaluating exposure risks and assessing the impact of nanoparticles on human health. *Journal of Trace Elements and Minerals*, *13*, 100252. https://doi.org/.DOI: 10.1016/j.jtemin.2025.100252

Kumari, A., Yadav, S. K., & Yadav, S. C. (2010). Biodegradable polymeric nanoparticles-based drug delivery systems. *Colloids and Surfaces. B, Biointerfaces*, *75*(1), 1–18. DOI: 10.1016/j.colsurfb.2009.09.001 PMID: 19782542

Kumavath, R., Gupta, P., Tatta, E. R., Mohan, M. S., Salim, S. A., & Busi, S. (2025). Unraveling the role of mobile genetic elements in antibiotic resistance transmission and defense strategies in bacteria. *Frontiers in Systems Biology*, *5*, 1557413. DOI: 10.3389/fsysb.2025.1557413

Kuperkar, K., Atanase, L. I., Bahadur, A., Crivei, I. C., & Bahadur, P. (2024). Degradable Polymeric Bio(nano)materials and Their Biomedical Applications: A Comprehensive Overview and Recent Updates. *Polymers*, *16*(2), 206. DOI: 10.3390/polym16020206 PMID: 38257005

Lamberti, M., Zappavigna, S., Sannolo, N., Porto, S., & Caraglia, M. (2014). Advantages and risks of nanotechnologies in cancer patients and occupationally exposed workers. *Expert Opinion on Drug Delivery, 11*(7), 1087–1101. DOI: 10.1517/17425247.2014.913568 PMID: 24773227

Laubenbacher, R., Sluka, J. P., & Glazier, J. A. (2021, March 12). Using digital twins in viral infection. *Science, 371*(6534), 1105–1106. DOI: 10.1126/science.abf3370 PMID: 33707255

Laxminarayan, R., Duse, A., Wattal, C., Zaidi, A. K., Wertheim, H. F., Sumpradit, N., Vlieghe, E., Hara, G. L., Gould, I. M., Goossens, H., Greko, C., So, A. D., Bigdeli, M., Tomson, G., Woodhouse, W., Ombaka, E., Peralta, A. Q., Qamar, F. N., Mir, F., & Cars, O. (2013). Antibiotic resistance-the need for global solutions. *The Lancet. Infectious Diseases, 13*(12), 1057–1098. DOI: 10.1016/S1473-3099(13)70318-9 PMID: 24252483

Le Piane, F., Vozza, M., Baldoni, M., & Mercuri, F. (2024). Integrating high-performance computing, machine learning, data management workflows, and infrastructures for multiscale simulations and nanomaterials technologies. *Beilstein Journal of Nanotechnology, 15*, 1498–1521. DOI: 10.3762/bjnano.15.119

Lee, J. H., Choi, B., Jeong, K., Suh, S. W., Rhee, H., Kim, J. H., & Son, D. S. (2026). Diverging regulatory DNA in adaptive medical AI: US agility and EU accountability in lifecycle governance. *Frontiers in Medicine, 13*, 1758708. DOI: 10.3389/fmed.2026.1758708

Lee, J. H., & Yeo, Y. (2015). Controlled drug release from pharmaceutical nanocarriers. *Chemical Engineering Science, 125*, 75–84. DOI: 10.1016/j.ces.2014.08.046 PMID: 25684779

Lemire, J. A., Harrison, J. J., & Turner, R. J. (2013). Antimicrobial activity of metals: Mechanisms, molecular targets and applications. *Nature Reviews. Microbiology, 11*(6), 371–384. DOI: 10.1038/nrmicro3028

Leung, Y. H., Ng, A. M., Xu, X., Shen, Z., Gethings, L. A., Wong, M. T., Chan, C. M., Guo, M. Y., Ng, Y. H., Djurišić, A. B., Lee, P. K., Chan, W. K., Yu, L. H., Phillips, D. L., Ma, A. P., & Leung, F. C. (2014). Mechanisms of antibacterial activity of MgO: Non-ROS mediated toxicity of MgO nanoparticles toward *Escherichia coli. Small, 10*(6), 1171–1183. DOI: 10.1002/smll.201302434

Li, M., Zhao, G., Su, W. K., & Shuai, Q. (2020). Enzyme-responsive nanoparticles for anti-tumor drug delivery. *Frontiers in chemistry, 8*, 647. https://doi.org/DOI: 10.3389/fchem.2020.00647

Liang, S., Hu, M., Li, B., Xia, D., Liang, C., Peng, F., & Wang, D. (2023, June 28). Smart Implant with Bacteria Monitoring and Killing Ability for Orthopedic Applications. *ACS Applied Materials & Interfaces*, *15*(25), 29889–29901. DOI: 10.1021/acsami.3c03599 PMID: 37318286

Li, D., Gao, C., Kuang, M., Xu, M., Wang, B., Luo, Y., Teng, L., & Xie, J. (2021). Nanoparticles as drug delivery systems of RNAi in cancer therapy. *Molecules (Basel, Switzerland)*, *26*(8), 2380. DOI: 10.3390/molecules26082380 PMID: 33921892

Liew, K. B., Janakiraman, A. K., Sundarapandian, R., Khalid, S. H., Razzaq, F. A., Ming, L. C., Khan, A., Kalusalingam, A., & Ng, P. W. (2022, March). A review and revisit of nanoparticles for antimicrobial drug delivery. *Journal of Medicine and Life*, *15*(3), 328–335. DOI: 10.25122/jml-2021-0097 PMID: 35449993

Li, J., Wang, Y., Tang, M., Zhang, C., Fei, Y., Li, M., Li, M., Gui, S., & Guo, J. (2024). New insights into nanotherapeutics for periodontitis: A triple concerto of antimicrobial activity, immunomodulation and periodontium regeneration. *Journal of Nanobiotechnology*, *22*(1), 19. DOI: 10.1186/s12951-023-02261-y

Li, M., Du, C., Guo, N., Teng, Y., Meng, X., Sun, H., Li, S., Yu, P., & Galons, H. (2019). Composition design and medical application of liposomes. *European Journal of Medicinal Chemistry*, *164*, 640–653. DOI: 10.1016/j.ejmech.2019.01.007

Li, M., Liu, Y., Gong, Y., Yan, X., Wang, L., Zheng, W., Ai, H., & Zhao, Y. (2023). Recent advances in nanoantibiotics against multidrug-resistant bacteria. *Nanoscale Advances*, *5*(23), 6278–6317. DOI: 10.1039/D3NA00530E PMID: 38024316

Lim, Y. H., Tiemann, K. M., Hunstad, D. A., Elsabahy, M., & Wooley, K. L. (2016). Polymeric nanoparticles in development for treatment of pulmonary infectious diseases. *Wiley Interdisciplinary Reviews. Nanomedicine and Nanobiotechnology*, *8*(6), 842–871. DOI: 10.1002/wnan.1401

Lin, C.-Y., Chang, Y.-T., Chung, Y.-K., Alalaiwe, A., Yu, H.-P., & Fang, J.-Y. (2025). Dual-membrane bioinspired nanocarriers for targeted therapy of MRSA-induced acute lung injury and bacteremia. *Journal of Materials Chemistry. B*, *13*(19), 5686–5704. DOI: 10.1039/D4TB02742F

Lin, Z., & Chou, W. C. (2022, August 25). Machine Learning and Artificial Intelligence in Toxicological Sciences. *Toxicological Sciences*, *189*(1), 7–19. DOI: 10.1093/toxsci/kfac075 PMID: 35861448

Li, P., Luo, H., Ji, B., & Nielsen, J. (2022). Machine learning for data integration in human gut microbiome. *Microbial Cell Factories*, *21*(1), 241. Advance online publication. DOI: 10.1186/s12934-022-01973-4 PMID: 36419034

Li, Q., Li, X., & Zhao, C. (2020). Strategies to obtain encapsulation and controlled release of small hydrophilic molecules. *Frontiers in Bioengineering and Biotechnology*, *8*, 437. DOI: 10.3389/fbioe.2020.00437 PMID: 32478055

Liu, H.Y., Prentice, E.L. & Webber, M.A. Mechanisms of antimicrobial resistance in biofilms. npj Antimicrob Resist 2, 27 (2024). DOI: 10.1038/s44259-024-00046-3

Liu, Y., Zhang, L., Jiang, Z., Tian, X., Li, P., Wu, P., Du, W., Yuan, B., Xie, C., Bu, G., Zhong, L., Yang, Y., Li, T., Zeng, M., & Sun, C. (2025). Applications of Artificial Intelligence in Biotech Drug Discovery and Product Development. *MedComm*, 6(8). Portico. DOI: 10.1002/mco2.70317

Liu, G.-Y., Yu, D., Fan, M.-M., Zhang, X., Jin, Z.-Y., Tang, C., & Liu, X.-F. (2024). Antimicrobial resistance crisis: Could artificial intelligence be the solution? *Military Medical Research*, *11*(1), 7. Advance online publication. DOI: 10.1186/s40779-024-00510-1 PMID: 38254241

Liu, J. D., Van Treeck, K. E., Marston, W. A., Papadopoulou, V., & Rowe, S. E. (2024, October 16). Ultrasound-Mediated Antibiotic Delivery to In Vivo Biofilm Infections: A Review. *ChemBioChem*, *25*(20), e202400181. DOI: 10.1002/cbic.202400181 PMID: 38924307

Liu, J., Gefen, O., Ronin, I., Bar-Meir, M., & Balaban, N. Q. (2020). Effect of tolerance on the evolution of antibiotic resistance under drug combinations. *Science*, *367*(6474), 200–204. DOI: 10.1126/science.aay3041

Liu, J., Li, Y., Chen, S., Lin, Y., Lai, H., Chen, B., & Chen, T. (2020, September 18). Biomedical Application of Reactive Oxygen Species-Responsive Nanocarriers in Cancer, Inflammation, and Neurodegenerative Diseases. *Frontiers in Chemistry*, *8*, 838. DOI: 10.3389/fchem.2020.00838 PMID: 33062637

Liu, N. N., Jiao, N., Tan, J. C., Wang, Z., Wu, D., Wang, A. J., Chen, J., Tao, L., Zhou, C., Fang, W., Cheong, I. H., Pan, W., Liao, W., Kozlakidis, Z., Heeschen, C., Moore, G. G., Zhu, L., Chen, X., Zhang, G., & Wang, H. (2022). Multikingdom microbiota analyses identify bacterial-fungal interactions and biomarkers of colorectal cancer across cohorts. *Nature Microbiology*, *7*(2), 238–250. DOI: 10.1038/s41564-021-01030-7

Liu, P., Chen, G., & Zhang, J. (2022). A review of liposomes as a drug delivery system: Current status of approved products, regulatory environments, and future perspectives. *Molecules (Basel, Switzerland)*, *27*(4), 1372. DOI: 10.3390/molecules27041372 PMID: 35209162

Liu, S., Bao, J., Lao, X., & Zheng, H. (2018). Novel 3D structure-based model for activity prediction and design of antimicrobial peptides. *Scientific Reports*, *8*(1), 11189. Advance online publication. DOI: 10.1038/s41598-018-29566-5

Liu, V. X., Fielding-Singh, V., Greene, J. D., Baker, J. M., Iwashyna, T. J., Bhattacharya, J., & Escobar, G. J. (2017). The Timing of Early Antibiotics and Hospital Mortality in Sepsis. *American Journal of Respiratory and Critical Care Medicine*, *196*(7), 856–863. DOI: 10.1164/rccm.201609-1848OC PMID: 28345952

Liu, X., Jiang, M., Zhang, J., Li, H., Liu, Y., Zhang, J., Chen, X., Bu, J., Xie, S., Zhang, M., Dong, N., Cao, Q., & Zhou, Z. (2025). Machine learning-based evaluation of risk factors for carbapenem-resistant *Klebsiella pneumoniae* dissemination in neonatal units. *mSystems*, *10*(10), e00909–e00925. DOI: 10.1128/msystems.00909-25

Li, X., Zhang, Y., Zhang, Q., Liu, J., Zhu, Z., Feng, X., Han, L., & Zhang, X. (2024). Strategy and mechanism of One Health governance: Case study of China. *Science in One Health*, *4*, 100098. DOI: 10.1016/j.soh.2024.100098

Li, Z., Zhu, Y., & Matson, J. B. (2022, May 3). pH-Responsive Self-Assembling Peptide-Based Biomaterials: Designs and Applications. *ACS Applied Bio Materials*, *5*(10), 4635–4651. Advance online publication. DOI: 10.1021/acsabm.2c00188 PMID: 35505454

Lloyd-Price, J., Abu-Ali, G., & Huttenhower, C. (2016). The healthy human microbiome. *Genome Medicine*, *8*, 51. DOI: 10.1186/s13073-016-0307-y

Loecher, A., Bruyns-Haylett, M., Ballester, P. J., Borros, S., & Oliva, N. (2023). A machine learning approach to predict cellular uptake of pBAE polyplexes. *Biomaterials Science*, *11*(17), 5797–5808. DOI: 10.1039/D3BM00741C PMID: 37401742

Lopatkin, A. J., Bening, S. C., Manson, A. L., Stokes, J. M., Kohanski, M. A., Badran, A. H., Earl, A. M., Cheney, N. J., Yang, J. H., & Collins, J. J. (2021). Clinically relevant mutations in core metabolic genes confer antibiotic resistance. *Science*, *371*(6531), eaba0862. Advance online publication. DOI: 10.1126/science.aba0862

Lu, J., Choi, K., Eremeev, M., Gobburu, J., Goswami, S., Liu, Q., Mo, G., Musante, C. J., & Shahin, M. H. (2025). Large language models and their applications in drug discovery and development: A primer. *Clinical and Translational Science*, *18*(4), e70205. Advance online publication. DOI: 10.1111/cts.70205

Lu, W., Chen, X., Wang, L., Li, H., & Fu, Y. V. (2020). Combination of an artificial intelligence approach and laser tweezers Raman spectroscopy for microbial identification. *Analytical Chemistry*, *92*(9), 6288–6296. DOI: 10.1021/acs.analchem.9b04946 PMID: 32281780

Lu, Y., Kim, S., & Park, K. (2011). In vitro-in vivo correlation: Perspectives on model development. *International Journal of Pharmaceutics*, *418*(1), 142–148. DOI: 10.1016/j.ijpharm.2011.01.010

Lu, Y., Zhang, B., Wang, Z., He, Y., Ge, H., Ma, H., & Cui, P. (2026). From host-derived pressures to the environmental anti-antimicrobial peptides resistome: Mechanisms, reservoirs and implications for therapeutic peptide design. *Marine Drugs*, *24*(2), 76. DOI: 10.3390/md24020076

Macedo, B., Ribeiro Vaz, I., & Taveira Gomes, T. (2024). MedGAN: Optimized generative adversarial network with graph convolutional networks for novel molecule design. *Scientific Reports*, *14*(1), 1212. DOI: 10.1038/s41598-023-50834-6 PMID: 38216614

Magana, M., Pushpanathan, M., Santos, A. L., Leanse, L., Fernandez, M., Ioannidis, A., Giulianotti, M. A., Apidianakis, Y., Bradfute, S., Ferguson, A. L., Cherkasov, A., Seleem, M. N., Pinilla, C., de la Fuente-Nunez, C., Lazaridis, T., Dai, T., Houghten, R. A., Hancock, R. E. W., & Tegos, G. P. (2020). The value of antimicrobial peptides in the age of resistance. *The Lancet. Infectious Diseases*, *20*(9), e216–e230. DOI: 10.1016/S1473-3099(20)30327-3 PMID: 32653070

Maharjan, R., Kim, N. A., Kim, K. H., & Jeong, S. H. (2025). Transformative roles of digital twins from drug discovery to continuous manufacturing: Pharmaceutical and biopharmaceutical perspectives. *International Journal of Pharmaceutics: X*, *10*, 100409. DOI: 10.1016/j.ijpx.2025.100409

Mahlapuu, M., Björn, C., & Ekblom, J. (2020). Antimicrobial peptides as therapeutic agents: Opportunities and challenges. *Critical Reviews in Biotechnology*, *40*(7), 978–992. DOI: 10.1080/07388551.2020.1796576

Mahlapuu, M., Håkansson, J., Ringstad, L., & Björn, C. (2016). Antimicrobial peptides: An emerging class of therapeutics. *Frontiers in Cellular and Infection Microbiology*, *6*, 194. DOI: 10.3389/fcimb.2016.00194 PMID: 28083516

Maiorov, V. N., & Crippen, G. M. (1994). Significance of root-mean-square deviation in comparing three-dimensional structures of globular proteins. *Journal of Molecular Biology*, *235*(2), 625–634. DOI: 10.1006/jmbi.1994.1017

Maitreya, A., Pal, S., Qureshi, A., Reyed, M. R., & Purohit, H. J. (2022). Nitric oxide–secreting probiotics as sustainable biocleaners for reverse osmosis membrane systems. *Environmental Science and Pollution Research International*, *29*(4), 4911–4929. DOI: 10.1007/s11356-021-17289-6

Majumder, J., & Minko, T. (2021). Multifunctional and stimuli-responsive nanocarriers for targeted therapeutic delivery. *Expert Opinion on Drug Delivery*, *18*(2), 205–227. DOI: 10.1080/17425247.2021.1828339

Makadia, H. K., & Siegel, S. J. (2011). Poly Lactic-co-Glycolic Acid (PLGA) as biodegradable controlled drug delivery carrier. *Polymers*, *3*(3), 1377–1397. DOI: 10.3390/polym3031377

Malik, A. K., Kushwah, S. V., Aslam, M. S., Reyed, M. R., & Haghi, A. K. (Eds.). (2026). *Aligning ESG goals with the UN's sustainable development goals planning: Strategies for equitable financing and world sustainability*. Springer Cham., DOI: 10.1007/978-3-032-01715-4

Mallik, S., Dash, S., Mishra, S., Mohanty, R. B., Sahoo, S., Patra, A., Sahoo, S., Mishra, R., & Mohanty, J. N. (2025). Threat of multidrug resistant bacteria: The role of genomic solutions and a call to action. *Gene Reports*, *41*, 102298. DOI: 10.1016/j.genrep.2025.102298

Manandhar, I., Alimadadi, A., Aryal, S., Munroe, P. B., Joe, B., & Cheng, X. (2021). Gut microbiome-based supervised machine learning for clinical diagnosis of inflammatory bowel diseases. *American Journal of Physiology. Gastrointestinal and Liver Physiology*, *320*(3), G328–G337. DOI: 10.1152/ajpgi.00360.2020 PMID: 33439104

Mani, S., Lalani, S. R., & Pammi, M. (2025). Genomics and multiomics in the age of precision medicine. *Pediatric Research 2025 97:4, 97*(4), 1399–1410. DOI: 10.1038/s41390-025-04021-0

Maojo, V., Fritts, M., de la Iglesia, D., Cachau, R. E., Garcia-Remesal, M., Mitchell, J. A., & Kulikowski, C. (2012). Nanoinformatics: A new area of research in nanomedicine. *International Journal of Nanomedicine*, •••, 3867–3890. DOI: 10.2147/IJN.S24582 PMID: 22866003

Marcos-Zambrano, L. J., Karaduzovic-Hadziabdic, K., Loncar Turukalo, T., Przymus, P., Trajkovik, V., Aasmets, O., Berland, M., Gruca, A., Hasic, J., Hron, K., Klammsteiner, T., Kolev, M., Lahti, L., Lopes, M. B., Moreno, V., Naskinova, I., Org, E., Paciência, I., Papoutsoglou, G., & Truu, J. (2021). Applications of Machine Learning in Human Microbiome Studies: A Review on Feature Selection, Biomarker Identification, Disease Prediction and Treatment. *Frontiers in Microbiology*, *12*, 634511. DOI: 10.3389/fmicb.2021.634511 PMID: 33737920

Mashele, S. S. (2025). Stimuli-responsive, cell-mediated drug delivery systems: Engineering smart cellular vehicles for precision therapeutics. *Pharmaceutics*, *17*(8), 1082. DOI: 10.3390/pharmaceutics17081082 PMID: 40871101

Mat Rani, N. N. I., Hussein, Z. M., Mustapa, F., Azhari, H., Sekar, M., Yi, C. X., & Mohd Amin, M. C. I. (2021). Exploring the possible targeting strategies of liposomes against methicillin-resistant Staphylococcus aureus (MRSA). *European Journal of Pharmaceutics and Biopharmaceutics, 165*, 84–105. DOI: 10.1016/j.ejpb.2021.04.021

Ma, Z., Zuo, T., Frey, N., & Rangrez, A. Y. (2024). A systematic framework for understanding the microbiome in human health and disease: From basic principles to clinical translation. *Signal Transduction and Targeted Therapy, 9*(1), 237. Advance online publication. DOI: 10.1038/s41392-024-01946-6 PMID: 39307902

McCubbin, K. D., Barkema, H. W., Babujee, A., Forseille, J., Naum, K., Buote, P., Dalton, D., Checkley, S. L., Lehman, K., Morris, T., Smilski, K., Wilkins, W. L., Anholt, R. M., Larose, S., Saxinger, L. M., Blue, D., & Otto, S. J. G. (2022). One health and antimicrobial stewardship: Where to go from here? *The Canadian Veterinary Journal = La Revue Vétérinaire Canadienne, 63*(2), 198–200.

McNair, L. (2022). Ethical and regulatory oversight of clinical research: The role of the Institutional Review Board. *Experimental Biology and Medicine (Maywood, N.J.), 247*(7), 561–566. DOI: 10.1177/15353702221078216

Meagher, K. M., Watson, S., Suh, G. A., & Virk, A. (2022). The New Precision Stewards? *Journal of Personalized Medicine, 12*(8), 1308. DOI: 10.3390/jpm12081308 PMID: 36013256

Mekki-Berrada, F., Ren, Z., Huang, T., Wong, W. K., Zhang, F., Xie, J., Tian, I. P. S., Jayavelu, S., Mahfoud, Z., Bash, D., Hippalgaonker, K., Khan, S., Buonassisi, T., Li, Q., & Wang, X. (2021). Two-step machine learning enable optimized nanoparticle synthesis. *npj Computational Materials, 7*, 55. https://doi.org/.DOI: 10.1038/s4152-021-00520-w

Memon, S., Bibi, S., & He, G. (2025). Integration of AI and ML in Tuberculosis (TB) Management: From Diagnosis to Drug Discovery. *Diseases (Basel, Switzerland), 13*(6), 184. DOI: 10.3390/diseases13060184 PMID: 40558595

Meng, F., Hennink, W. E., & Zhong, Z. (2009). Reduction-sensitive polymers and bioconjugates for biomedical applications. *Biomaterials, 30*(12), 2180–2198. DOI: 10.1016/j.biomaterials.2009.01.026 PMID: 19200596

Mennella, C., Maniscalco, U., De Pietro, G., & Esposito, M. (2024). Ethical and regulatory challenges of AI technologies in healthcare: A narrative review. *Heliyon, 10*(4), e26297. DOI: 10.1016/j.heliyon.2024.e26297

Menon, S., & Koura, K. G. (2025). Artificial intelligence for tuberculosis control: A scoping review of applications in public health. *Journal of Global Health, 15*, 04192. DOI: 10.7189/jogh.15.04192 PMID: 40709592

Metselaar, J. M., & Lammers, T. (2020). Challenges in nanomedicine clinical translation. *Drug Delivery and Translational Research, 10*(3), 721–725. DOI: 10.1007/s13346-020-00740-5

Meyer, R. A., Sunshine, J. C., & Green, J. J. (2015). Biomimetic particles as therapeutics. *Trends in Biotechnology, 33*(9), 514–524. DOI: 10.1016/j.tibtech.2015.07.001 PMID: 26277289

Miller, C., Portlock, T., Nyaga, D. M., & O'Sullivan, J. M. (2024). A review of model evaluation metrics for machine learning in genetics and genomics. *Frontiers in Bioinformatics, 4*, 1457619. Advance online publication. DOI: 10.3389/fbinf.2024.1457619

Mills, S., Stanton, C., Lane, J. A., Smith, G. J., & Ross, R. P. (2019). Precision nutrition and the microbiome, part I: Current state of the science. *Nutrients, 11*(4), 923. DOI: 10.3390/nu11040923

Mintzer, M. A., & Grinstaff, M. W. (2011). Biomedical applications of dendrimers: A tutorial. *Chemical Society Reviews, 40*(1), 173–190. DOI: 10.1039/B901839P PMID: 20877875

Mi, P. (2020). Stimuli-responsive nanocarriers for drug delivery, tumor imaging, therapy and theranostics. *Theranostics, 10*(10), 4557–4588. DOI: 10.7150/thno.38069

Mi, P., Cabral, H., & Kataoka, K. (2020). Ligand-installed nanocarriers toward precision therapy. *Advanced Materials, 32*(13), 1902604. DOI: 10.1002/adma.201902604 PMID: 31353770

Miranda Calderon, L. G., Alejo, T., Santos, S., Mendoza, G., Irusta, S., & Arruebo, M. (2023, August 30). Antibody-Functionalized Polymer Nanoparticles for Targeted Antibiotic Delivery in Models of Pathogenic Bacteria Infecting Human Macrophages. *ACS Applied Materials & Interfaces, 15*(34), 40213–40227. DOI: 10.1021/acsami.3c07367 PMID: 37596966

Mitchell, M. J., Billingsley, M. M., Haley, R. M., Wechsler, M. E., Peppas, N. A., & Langer, R. (2021). Engineering precision nanoparticles for drug delivery. *Nature Reviews. Drug Discovery, 20*(2), 101–124. DOI: 10.1038/s41573-020-0090-8

Mohsen, S., Dickinson, J. A., & Somayaji, R. (2020). Update on the adverse effects of antimicrobial therapies in community practice. *Canadian Family Physician Médecin de Famille Canadien, 66*(9), 651–659.

Molla, G., & Bitew, M. (2024). Revolutionizing Personalized Medicine: Synergy with Multi-Omics Data Generation, Main Hurdles, and Future Perspectives. *Biomedicines*, *12*(12), 2750. DOI: 10.3390/biomedicines12122750 PMID: 39767657

Mondal, S. K., Chakraborty, S., Manna, S., & Mandal, S. M. (2024). Antimicrobial nanoparticles: Current landscape and future challenges. *RSC Pharmaceutics*, *1*(3), 388–402. DOI: 10.1039/D4PM00032C

Monteiro, D. R., Gorup, L. F., Takamiya, A. S., de Camargo, E. R., Filho, A. C., & Barbosa, D. B. (2012). Silver distribution and release from an antimicrobial denture base resin containing silver colloidal nanoparticles. *Journal of Prosthodontics*, *21*(1), 7–15. DOI: 10.1111/j.1532-849X.2011.00772.x

Morar, M., & Wright, G. D. (2010). The genomic enzymology of antibiotic resistance. *Annual Review of Genetics*, *44*, 25–51. DOI: 10.1146/annurev-genet-102209-163517

Mounier, J., Ryter, A., Coquis-Rondon, M., & Sansonetti, P. J. (1990). Intracellular and cell-to-cell spread of Listeria monocytogenes involves interaction with F-actin in the enterocytelike cell line Caco-2. Infection and immunity, 58(4), 1048-1058. DOI: (Pizarro-Cerdá et al., 1990).(Mounier et al., 1990).DOI: 10.1128/iai.58.4.1048-1058.1990

Mouzakis, A., Panagopoulos, P., Papazoglou, D., & Petrakis, V. (2025). A Comprehensive Review of Nanoparticles in the Fight Against Antimicrobial Resistance. *Pathogens (Basel, Switzerland)*, *14*(11), 1090. DOI: 10.3390/pathogens14111090 PMID: 41305328

Moy, S., Irannejad, M., Manning, S. J., Farahani, M., Ahmed, Y., Gao, E., Prabhune, R., Lorenz, S., Mirza, R., & Klinger, C. (2024, April 2). Patient Perspectives on the Use of Artificial Intelligence in Health Care: A Scoping Review. *Journal of Patient-Centered Research and Reviews*, *11*(1), 51–62. DOI: 10.17294/2330-0698.2029 PMID: 38596349

Mubeen, B., Ansar, A. N., Rasool, R., Ullah, I., Imam, S. S., Alshehri, S., Ghoneim, M. M., Alzarea, S. I., Nadeem, M. S., & Kazmi, I. (2021). Nanotechnology as a novel approach in combating microbes providing an alternative to antibiotics. *Antibiotics (Basel, Switzerland)*, *10*(12), 1473. DOI: 10.3390/antibiotics10121473

Muller, E., Shiryan, I., & Borenstein, E. (2024). Multi-omic integration of microbiome data for identifying disease-associated modules. *Nature Communications*, *15*(1), 2621. Advance online publication. DOI: 10.1038/s41467-024-46888-3 PMID: 38521774

Mulvaney, P., Parak, W. J., Caruso, F., & Weiss, P. S. (2016). Standardizing nanomaterials. *ACS Nano*, *10*(11), 9763–9764. DOI: 10.1021/acsnano.6b07629 PMID: 27934085

Muñoyerro-Muñiz, D., Villegas, R., de la Oliva, V., Esteban-Medina, A., Fernández Del Valle, P., Sánchez, A., Susin, M. B., Gutierrez-Alvarez, I., Reboredo, M., Alejos, L., Loucera, C., & Dopazo, J. (2025). Ethical and secure evidence generation from regionwide clinical data through a collaborative environment for advancing predictive care. *Frontiers in Public Health*, *13*, 1630351. DOI: 10.3389/fpubh.2025.1630351

Mura, S., Nicolas, J., & Couvreur, P. (2013). Stimuli-responsive nanocarriers for drug delivery. *Nature Materials*, *12*(11), 991–1003. DOI: 10.1038/nmat3776 PMID: 24150417

Murray, C. J., Ikuta, K. S., Sharara, F., Swetschinski, L., Aguilar, G. R., Gray, A., & Tasak, N. (2022). Global burden of bacterial antimicrobial resistance in 2019: A systematic analysis. *Lancet*, *399*(10325), 629–655. DOI: 10.1016/S0140-6736(21)02724-0 PMID: 35065702

Musa, S. S., Ibrahim, A. M., Alhassan, M. Y., Musa, A. H., Jibo, A. G., Auwal, A. R., Okesanya, O. J., Othman, Z. K., Abubakar, M. S., Ahmed, M. M., Barroso, C. J. V., Sium, A. F., Garcia, M. B., Flores, J. B., Maikifi, A. S., Kouwenhoven, M. B. N., & Lucero-Prisno, D. E. (2025). Nanotechnology and machine learning: A promising confluence for the advancement of precision medicine. *Intelligence-Based Medicine*, *12*, 100267. DOI: 10.1016/j.ibmed.2025.100267

Musicha, P., Morse, T., Cocker, D., Mugisha, L., Jewell, C. P., & Feasey, N. A. (2024). Time to define One Health approaches to tackling antimicrobial resistance. *Nature Communications*, *15*(1), 8782. DOI: 10.1038/s41467-024-53057-z

Mustafa, S. S., Batool, R., Kamran, M., Javed, H., & Jamil, N. (2022). Evaluating the role of wastewaters as reservoirs of antibiotic-resistant ESKAPEE bacteria using phenotypic and molecular methods. *Infection and Drug Resistance*, *15*, 5715–5728. DOI: 10.2147/IDR.S368886

Muteeb, G., Rehman, M. T., Shahwan, M., & Aatif, M. (2023). Origin of antibiotics and antibiotic resistance, and their impacts on drug development: A narrative review. *Pharmaceuticals (Basel, Switzerland)*, *16*(11), 1615. DOI: 10.3390/ph16111615 PMID: 38004480

Nadimpalli, M., Delarocque-Astagneau, E., Love, D. C., Price, L. B., Huynh, B. T., Collard, J. M., Lay, K. S., Borand, L., Ndir, A., Walsh, T. R., Guillemot, D., De Lauzanne, A., Kerleguer, A., Tarantola, A., Piola, P., Chon, T., Lach, S., Ngo, V., Touch, S., & Abdou, A. Y. (2018). Combating Global Antibiotic Resistance: Emerging One Health Concerns in Lower- and Middle-Income Countries. *Clinical Infectious Diseases*, *66*(6), 963–969. DOI: 10.1093/cid/cix879 PMID: 29346620

Navarro-Muñoz, J. C., Selem-Mojica, N., Mullowney, M. W., Kautsar, S. A., Tryon, J. H., Parkinson, E. I., De Los Santos, E. L. C., Yeong, M., Cruz-Morales, P., Abubucker, S., Roeters, A., Lokhorst, W., Fernandez-Guerra, A., Cappelini, L. T. D., Goering, A. W., Thomson, R. J., Metcalf, W. W., Kelleher, N. L., Barona-Gomez, F., & Medema, M. H. (2020). A computational framework to explore large-scale biosynthetic diversity. *Nature Chemical Biology*, *16*(1), 60–68. DOI: 10.1038/s41589-019-0400-9

Nayeem, A., Suresh, A. S., Vellapandian, C., Singh, S., Elossaily, G. M., & Prajapati, B. G. (2024). Comprehensive insights into cephalosporins: Spectrum, generations, and clinical applications. *Current Drug Therapy*. Advance online publication. DOI: 10.2174/0115748855339119241204071146

Nazir, F., Tabish, T. A., Tariq, F., Iftikhar, S., Wasim, R., & Shahnaz, G. (2022). Stimuli-sensitive drug delivery systems for site-specific antibiotic release. *Drug Discovery Today*, *27*(6), 1698–1705. DOI: 10.1016/j.drudis.2022.02.014 PMID: 35219858

Nechaeva, A. M., Artyukhov, A. A., Svistunova, A. Y., Il'in, M., Loginova, T. P., Baranov, O. V., Shulgin, A. M., Komarova, L. G., Estifeeva, T. M., Kuzmin, P. S., Barmin, R. A., Gerasimov, V., Tsatsakis, A. M., Rezaee, R., Rudakovskaya, P. G., & Mezhuev, Y. O. (2026). Nanoparticles for doxorubicin delivery: Advances in carrier design and synergistic cancer therapy. *Topics in Current Chemistry (Cham)*, *384*(2), 9. Advance online publication. DOI: 10.1007/s41061-026-00547-x

Negut, I., & Bita, B. (2023, March 17). Polymeric Micellar Systems-A Special Emphasis on "Smart" Drug Delivery. *Pharmaceutics*, *15*(3), 976. DOI: 10.3390/pharmaceutics15030976 PMID: 36986837

Newman, M., Barabási, A.-L., & Watts, D. J. (2011). *The structure and dynamics of networks*. Princeton University Press., DOI: 10.1515/9781400841356

Nicolete, R., dos Santos, D. F., & Faccioli, L. H. (2011). The Uptake of PLGA Micro or Nanoparticles by Macrophages Provokes a Distinct in Vitro Inflammatory Response. *International Immunopharmacology*, *11*(10), 1557–1563. DOI: 10.1016/j.intimp.2011.05.014 PMID: 21621649

Nithyanand, P., Boya, B. R., Lee, J. H., & Lee, J. (2025). Polymicrobial biofilms: Interkingdom interactions, resistance and therapeutic strategies. *Microbial Biotechnology, 18*(8), e70218. Advance online publication. DOI: 10.1111/1751-7915.70218

Nouruzi, E., Hosseini, S. M., Asghari, B., Mahjoub, R., Zare, E. N., Shahbazi, M. A., Kalhori, F., & Arabestani, M. R. (2023). Effect of poly (lactic-co-glycolic acid) polymer nanoparticles loaded with vancomycin against Staphylococcus aureus biofilm. *BMC Biotechnology, 23*(1), 39. DOI: 10.1186/s12896-023-00811-8 PMID: 37723466

Noury, H., Rahdar, A., Ferreira, L. F. R., & Jamalpoor, Z. (2025). AI-driven innovations in smart multifunctional nanocarriers for drug and gene delivery: A mini-review. *Critical Reviews in Oncology/Hematology, 210*, 104701. DOI: 10.1016/j.critrevonc.2025.104701 PMID: 40086770

Nsairat, H., Khater, D., Sayed, U., Odeh, F., Al Bawab, A., & Alshaer, W. (2022). Liposomes: Structure, composition, types, and clinical applications. *Heliyon, 8*(5), e09394. DOI: 10.1016/j.heliyon.2022.e09394 PMID: 35600452

Nsubuga, M., Galiwango, R., Jjingo, D., & Mboowa, G. (2024). Generalizability of machine learning in predicting antimicrobial resistance in *E. coli*: A multi-country case study in Africa. *BMC Genomics, 25*(1), 287. Advance online publication. DOI: 10.1186/s12864-024-10214-4

Nuhn, L. (2023). Artificial Intelligence Assists Nanoparticles to Enter Solid Tumors. *Nature Nanotechnology, 18*(6), 550–551. DOI: 10.1038/s41565-023-01382-7 PMID: 37081083

Nunziata, G., Borroni, A., & Rossi, F. (2025). Advanced microfluidic strategies for core-shell nanoparticles: The next-generation of polymeric and lipid-based drug nanocarriers. *Chemical Engineering Journal Advances, 22*, 100759. DOI: 10.1016/j.ceja.2025.100759

Nusrat, S., Aliyu, M., & Zohora, F. T. (2025). Mechanisms of antimicrobial resistance: From genetic evolution to clinical manifestations. *AIMS Microbiology, 11*(4), 1007–1034. DOI: 10.3934/microbiol.2025045

Nwobodo, D. C., Ugwu, M. C., Anie, C. O., Al-Ouqaili, M. T. S., Ikem, C. J., Uchenna, V. C., & Saki, M. (2022). Antibiotic resistance: The challenges and some emerging strategies for tackling a global menace. *Journal of Clinical Laboratory Analysis, 36*(9), e24655. Advance online publication. DOI: 10.1002/jcla.24655

Nyandoro, V. O., Ismail, E. A., Tageldin, A., Gafar, M. A., Peters, X. Q., Mautsoe, R., Omolo, C. A., & Govender, T. (2025). Potential of nanocarrier-mediated delivery of vancomycin for MRSA infections. *Expert Opinion on Drug Delivery*, *22*(3), 347–365. DOI: 10.1080/17425247.2025.2459756 PMID: 39949087

Okeke, I. N., de Kraker, M. E. A., Van Boeckel, T. P., Kumar, C. K., Schmitt, H., Gales, A. C., Bertagnolio, S., Sharland, M., & Laxminarayan, R. (2024). The scope of the antimicrobial resistance challenge. *Lancet*, *403*(10442), 2426–2438. DOI: 10.1016/S0140-6736(24)00876-6 PMID: 38797176

Okkeh, M., Bloise, N., Restivo, E., De Vita, L., Pallavicini, P., & Visai, L. (2021). Gold nanoparticles: Can they be the next magic bullet for multidrug-resistant bacteria? *Nanomaterials (Basel, Switzerland)*, *11*(2), 312. DOI: 10.3390/nano11020312

Olatunji, I., Bardaji, D. K. R., Miranda, R. R., Savka, M. A., & Hudson, A. O. (2024). Artificial intelligence tools for the identification of antibiotic resistance genes. *Frontiers in Microbiology*, *15*, 1437602. DOI: 10.3389/fmicb.2024.1437602 PMID: 39070267

Operti, M. C., Bernhardt, A., Grimm, S., Engel, A., Figdor, C. G., & Tagit, O. (2021). PLGA-based nanomedicines: Challenges and opportunities in industrial scale-up. *International Journal of Pharmaceutics*, *605*, 120807. DOI: 10.1016/j.ijpharm.2021.120807 PMID: 34144133

Ortiz-Perez, A., Tilborg, D. V., Meel, R. V. D., Grisoni, F., & Albertazzi, L. (2024). Machine learning-guided high throughput nanoparticle design. *Digital Discovery*, *3*(7), 1280–1291. DOI: 10.1039/D4DD00104D

Osthoff, M., Gürtler, N., Bassetti, S., Balestra, G., Marsch, S., Pargger, H., Weisser, M., & Egli, A. (2017). Impact of MALDI-TOF-MS-based identification directly from positive blood cultures on patient management: A controlled clinical trial. *Clinical Microbiology and Infection*, *23*(2), 78–85. DOI: 10.1016/j.cmi.2016.08.009 PMID: 27569710

Otaigbe, I. I. (2025). Mitigating inequitable access to appropriate antibiotics in low-and middle-income countries. *JAC-Antimicrobial Resistance*, *7*(2), dlaf061. Advance online publication. DOI: 10.1093/jacamr/dlaf061 PMID: 40276025

Oyejide, A. J., Adekunle, Y. A., Abodunrin, O. D., & Atoyebi, E. O. (2025). Artificial intelligence, computational tools and robotics for drug discovery, development, and delivery. *Intelligent Pharmacy*, *3*(3), 207–224. DOI: 10.1016/j.ipha.2025.01.001

Ozbek, O., Genc, D. E., & Ulgen, O., K. (. (2024). Advances in Physiologically Based Pharmacokinetic (PBPK) Modeling of Nanomaterials. *ACS Pharmacology &amp. Translational Science*, *7*(8), 2251–2279. DOI: 10.1021/acsptsci.4c00250 PMID: 39144562

Paczesny, J., & Bielec, K. (2020). Application of Bacteriophages in Nanotechnology. *Nanomaterials (Basel, Switzerland)*, *10*(10), 1944. DOI: 10.3390/nano10101944 PMID: 33003494

Palasantzas, V. E. J. M., Tamargo-Rubio, I., Le, K., Slager, J., Wijmenga, C., Jonkers, I. H., Kumar, V., Fu, J., & Withoff, S. (2023). iPSC-derived organ-on-a-chip models for personalized human genetics and pharmacogenomics studies. *Trends in Genetics*, *39*(4), 268–284. DOI: 10.1016/j.tig.2023.01.002 PMID: 36746737

Paliwal, R., Babu, R. J., & Palakurthi, S. (2014). Nanomedicine scale-up technologies: Feasibilities and challenges. *AAPS PharmSciTech*, *15*(6), 1527–1534. DOI: 10.1208/s12249-014-0177-9 PMID: 25047256

Palme, J. B., Kristiansson, E., & Larsson, D. G. J. (2008). Environment factors influencing the development and spread of antibiotic resistance. *FEMS Microbiology Reviews*, *42*(1), 68–80. DOI: 10.1093/femsre/fux053

Panchpuri, M., Painuli, R., & Kumar, C. (2025). Artificial intelligence in smart drug delivery systems: A step toward personalized medicine. *RSC Pharmaceutics*, *2*(5), 882–914. DOI: 10.1039/D5PM00089K

Papo, D., & Buldú, J. M. (2026). A complex network perspective on brain disease. *Biological Reviews of the Cambridge Philosophical Society*, *101*(1), 364–399. DOI: 10.1111/brv.70086

Parvin, N., Joo, S. W., & Mandal, T. K. (2025). Nanomaterial-Based Strategies to Combat Antibiotic Resistance: Mechanisms and Applications. *Antibiotics (Basel, Switzerland)*, *14*(2), 207. DOI: 10.3390/antibiotics14020207 PMID: 40001450

Patel, T. K. (2025). Nanotechnology in Herbal-Based Wound Care. Applications of Herbal Excipients and Polymers in Wound Healing, 247–280. DOI: 10.4018/979-8-3373-3977-1.ch010

Pathak, K., Sarma, M., Sahariah, M., Shankarishan, P., Sahariah, J. J., Deka, S., Das, A., Islam, M. A., Pramanik, P., Borthakur, P. P., Bharadwaj, A. R., & Bhaumik, R. (2026). Nanoparticles in the fight against antimicrobial challenges: A comprehensive review. *Next Nanotechnology*, *9*, 100420. Advance online publication. DOI: 10.1016/j.nxnano.2026.100420

Pathinarupothi, R. K., Durga, P., & Rangan, E. S. (2019). IoT-based smart edge for global health: Remote monitoring with severity detection and alerts transmission. *IEEE Internet of Things Journal*, *6*(2), 2449–2462. DOI: 10.1109/JIOT.2018.2870068

Patil, S., Singh, I., Verma, I. K., Kumar, A., Sharma, J., Ratn, A., Dhakad, M. S., & Sharma, D. (2025). Vaccines as Potential Frontliners Against Antimicrobial Resistance (AMR): A Focused Review. *Infection and Drug Resistance*, *18*, 5023–5041. DOI: 10.2147/IDR.S544665 PMID: 41020145

Pati, R., Sahu, R., Panda, J., & Sonawane, A. (2016). Encapsulation of zinc-rifampicin complex into transferrin-conjugated silver quantum-dots improves its antimycobacterial activity and stability and facilitates drug delivery into macrophages. *Scientific Reports*, *6*, 24184. Advance online publication. DOI: 10.1038/srep24184

Patra, J. K., Das, G., Fraceto, L. F., Campos, E. V. R., Rodriguez-Torres, M. D. P., Acosta-Torres, L. S., Diaz-Torres, L. A., Grillo, R., Swamy, M. K., Sharma, S., Habtemariam, S., & Shin, H. S. (2018). Nano based drug delivery systems: Recent developments and future prospects. *Journal of Nanobiotechnology*, *16*(1), 71. Advance online publication. DOI: 10.1186/s12951-018-0392-8

Patri, A. K., Kukowska-Latallo, J. F., & Baker, J. R. (2002). Targeted drug delivery with dendrimers: Comparison of the release kinetics of covalently conjugated drug and non-covalent drug inclusion complex. *Advanced Drug Delivery Reviews*, *54*(2), 299–314. DOI: 10.1016/j.addr.2005.09.014 PMID: 16290254

Pattapulavar, V., S, S. K., Ramanujam, S., Subburaj, S., & John, G. C. (2026). Probiotic-based therapeutics for a One Health future: Redefining antibiotic dependency to combat antimicrobial resistance. *Frontiers in Microbiology, 17*, 1736436. https://doi.org/DOI: 10.3389/fmicb.2026.1736436

Pattni, B. S., Chupin, V. V., & Torchilin, V. P. (2015). New developments in liposomal drug delivery. *Chemical Society Reviews*, *115*(19), 10938–10966. DOI: 10.1021/acs.chemrev.5b00046 PMID: 26010257

Paul, D., & Das, B. (2022). Gut microbiome in the emergence of antibiotic-resistant bacterial pathogens. *Human Microbiome in Health and Disease - Part B*, 1–31. DOI: 10.1016/bs.pmbts.2022.07.009

Pawar, S. S., & Rohane, S. H. (2021). Review on Discovery Studio: An important tool for molecular docking. *Asian Journal of Research in Chemistry*, *14*(1), 86–88. DOI: 10.52711/0974-4150.2021.00015

Peer, D., Karp, J. M., Hong, S., Farokhzad, O. C., Margalit, R., & Langer, R. (2020). Nanocarriers as an emerging platform for cancer therapy. Nano-enabled medical applications, 61-91.

Pelgrift, R. Y., & Friedman, A. J. (2013). Nanotechnology as a therapeutic tool to combat microbial resistance. *Advanced Drug Delivery Reviews*, *65*(13–14), 1803–1815. DOI: 10.1016/j.addr.2013.07.011 PMID: 23892192

Pennisi, F., Pinto, A., Ricciardi, G. E., Signorelli, C., & Gianfredi, V. (2025). The role of artificial intelligence and machine learning models in antimicrobial stewardship in public health: A narrative review. *Antibiotics (Basel, Switzerland)*, *14*(2), 134. DOI: 10.3390/antibiotics14020134

Perini, H. F., Matos, B. S., de Oliveira, C. J. F., & da Silva, M. V. (2024). Biomimetic nanocarriers: Integrating natural functions for advanced therapeutic applications. *Beilstein Journal of Nanotechnology*, *15*, 1619–1626. DOI: 10.3762/bjnano.15.127

Pham, N.-P., Gingras, H., Godin, C., Feng, J., Groppi, A., Nikolski, M., Leprohon, P., & Ouellette, M. (2024). Holistic understanding of trimethoprim resistance in *Streptococcus pneumoniae* using an integrative approach of genome-wide association study, resistance reconstruction, and machine learning. *mBio*, *15*(9), e0136024. DOI: 10.1128/mbio.01360-24

Pisani, S., Tufail, S., Rosalia, M., Dorati, R., Genta, I., Chiesa, E., & Conti, B. (2024). Antibiotic-Loaded Nano-Sized Delivery Systems: An Insight into Gentamicin and Vancomycin. *Journal of Functional Biomaterials*, *15*(7), 194. DOI: 10.3390/jfb15070194 PMID: 39057315

Pochet, A., Bourguignon, T., Grandé, A., Godinez-Leon, J. A., Li, X., Fine, J., Sencio, V., Chatagnon, J., Dremierre, B., Vandeputte, A., Mascarau, R., Piveteau, C., Brodin, P., Machelart, A., & Gref, R. (2026). Selective accumulation of nanoparticles in infected cells for targeted drug delivery. *Journal of Controlled Release : Official Journal of the Controlled Release Society*, *389*, 114415. DOI: 10.1016/j.jconrel.2025.114415

Polianciuc, S. I., Gurzău, A. E., Kiss, B., Ştefan, M. G., & Loghin, F. (2020). Antibiotics in the environment: Causes and consequences. *Medicine and Pharmacy Reports*, *93*(3), 231–240. DOI: 10.15386/mpr-1742

Ponkarthika, M., Prabhu, A., Ruba, M., Kishore, Y. R., & GV, M. (2026). Bridging Health Equity in Times of Crisis: Ensuring Inclusive Care During Emergencies and Disasters. In Advancing Equity in Global Healthcare Systems (pp. 27-54). IGI Global Scientific Publishing.

Poovithaa, M., Ruba, M., Ponkarthika, M., Ragul, N., Ragul, V., & Sangeetha, K. (2025, November). IoT Smart Traffic Control System for Hill Roads with Emergency Vehicle Priority. In 2025 IEEE 7th International Conference on Computing, Communication and Automation (ICCCA) (pp. 1-6). IEEE.

Popa, S. L., Pop, C., Dita, M. O., Brata, V. D., Bolchis, R., Czako, Z., Saadani, M. M., Ismaiel, A., Dumitrascu, D. I., Grad, S., David, L., Cismaru, G., & Padureanu, A. M. (2022). Deep learning and antibiotic resistance. *Antibiotics (Basel, Switzerland)*, *11*(11), 1674. Advance online publication. DOI: 10.3390/antibiotics11111674

Popova, M., Isayev, O., & Tropsha, A. (2018). Deep reinforcement learning for de novo drug design. *Science Advances*, *4*(7), eaap7885. Advance online publication. DOI: 10.1126/sciadv.aap7885

Powers, J. H.III. (2024). Antimicrobial stewardship. *BMJ (Clinical Research Ed.)*, *385*, q1170. DOI: 10.1136/bmj.q1170 PMID: 38839082

Prabaharan, M. (2011). Chitosan-based nanoparticles for tumor-targeted drug delivery. *International Journal of Biological Macromolecules*, *48*(2), 177–183. DOI: 10.1016/j.ijbiomac.2014.10.052 PMID: 25450550

Prabhu, A. E., & Selvakumar, V. S. (2025). A Review of AI-Driven Methods for Cytotoxicity Analysis and Predictive Modeling in the Treatment of Cancer Using Nanomaterials. In *2025 International Conference on Electronics, Computing, Communication and Control Technology (ICECCC)* (pp. 1-6). IEEE. https://doi.org/DOI: 10.1109/ICECCC65144.2025.11064249

Price, W. N.II, & Cohen, I. G. (2019). Privacy in the age of medical big data. *Nature Medicine*, *25*(1), 37–43. DOI: 10.1038/s41591-018-0272-7 PMID: 30617331

Qadri, Y. A., Shaikh, S., Ahmad, K., Choi, I., Kim, S. W., & Vasilakos, A. V. (2025). Explainable artificial intelligence: A perspective on drug discovery. *Pharmaceutics*, *17*(9), 1119. DOI: 10.3390/pharmaceutics17091119 PMID: 41012456

Qin, H., Li, Z., Li, S., Huang, J., Ren, J., & Wu, X. (2025). Advances in hydrogels for capturing and neutralizing inflammatory cytokines. *Journal of Tissue Engineering*, *16*, 20417314251342175. DOI: 10.1177/20417314251342175 PMID: 40575355

Qin, S., Xiao, W., Zhou, C., Pu, Q., Deng, X., Lan, L., Liang, H., Song, X., & Wu, M. (2022). Pseudomonas aeruginosa: Pathogenesis, virulence factors, antibiotic resistance, interaction with host, technology advances and emerging therapeutics. *Signal Transduction and Targeted Therapy*, *7*(1), 199. DOI: 10.1038/s41392-022-01056-1 PMID: 35752612

Quan, W., Zhou, J., Wang, J., Huang, J., & Du, L. (2025). Machine learning-driven precision nutrition: A paradigm evolution in dietary assessment and intervention. *Nutrients*, *18*(1), 45. DOI: 10.3390/nu18010045

Rabaan, A. A., Alhumaid, S., Al Mutair, A., Garout, M., Abulhamayel, Y., Halwani, M. A., Alestad, J. H., Al Bshabshe, A., Sulaiman, T., Alfonaisan, M. K., Almusawi, T., Albayat, H., Alsaeed, M., Alfaresi, M., Alotaibi, S., Alhashem, Y. N., Temsah, M. H., Ali, U., & Ahmed, N. (2022). Application of Artificial Intelligence in Combating High Antimicrobial Resistance Rates. *Antibiotics (Basel, Switzerland)*, *11*(6), 784. DOI: 10.3390/antibiotics11060784 PMID: 35740190

Rahman, M. U., Shah, J. A., Khan, M. N., Bilal, H., Zhu, D., Du, Z., & Mu, D. S. (2025). Innovative approaches to combat antimicrobial resistance: A review of emerging therapies and technologies. *Probiotics and Antimicrobial Proteins*, •••, 1–18. DOI: 10.1007/s12602-025-10676-2 PMID: 41026406

Ramos, T. I., Villacis-Aguirre, C. A., López-Aguilar, K. V., Santiago Padilla, L., Altamirano, C., Toledo, J. R., & Santiago Vispo, N. (2022). The Hitchhiker's guide to human therapeutic nanoparticle development. *Pharmaceutics*, *14*(2), 247. DOI: 10.3390/pharmaceutics14020247 PMID: 35213980

Rani, N., Singh, P., Kumar, S., Kumar, P., Bhankar, V., & Kumar, K. (2023). Plant-mediated synthesis of nanoparticles and their applications: A review. *Materials Research Bulletin*, *163*, 112233. DOI: 10.1016/j.materresbull.2023.112233

Rao, L., Yuan, Y., Shen, X., Yu, G., & Chen, X. (2024). Designing nanotheranostics with machine learning. *Nature Nanotechnology*, *19*(12), 1769–1781. DOI: 10.1038/s41565-024-01753-8 PMID: 39362960

Rashid, M., Rabbi, M. A., Ara, T., Hossain, M. M., Islam, M. S., Elaissari, A., Ahmad, H., & Rahman, M. M. (2021). Vancomycin conjugated iron oxide nanoparticles for magnetic targeting and efficient capture of Gram-positive and Gram-negative bacteria. *RSC Advances*, *11*(57), 36319–36329. DOI: 10.1039/D1RA04390K PMID: 35492746

Rath, B. P., Kar, S., Pattnaik, S., & Mohanty, C. (2025). Computational methods to explore the therapeutic potential of Triphala, a polyherbal formulation, against LuxS protein of *Helicobacter pylori*. *Microbe*, *9*, 100559. DOI: 10.1016/j.microb.2025.100559

Redgrave, L. S., Sutton, S. B., Webber, M. A., & Piddock, L. J. (2014). Fluoroquinolone resistance: Mechanisms, impact on bacteria, and role in evolutionary success. *Trends in Microbiology*, *22*(8), 438–445. DOI: 10.1016/j.tim.2014.04.007

Ren, X., Chakraborty, T., Doijad, S., Falgenhacer, L., Falgenhauer, J., Woesmann, A., Houschild, A. C., Schwengers, O., & Heider, D. (2022). Prediction of antimicrobial resistance based on whole-genome sequencing and machine learning. *Bioinformatics (Oxford, England)*, *38*(2), 325–334. DOI: 10.1093/bioinformatics/btab681 PMID: 34613360

Reyed, M. R., & Prabhakar, P. K. M. (2025). Polybiome derived therapies and fermentobiometrics: Novel strategies for blocking cancer pathways and progression. In *Cancer Etiology, Diagnosis and Treatments* (pp. 145–212). Nova Science Publishers. https://doi.org/DOI: 10.52305/JKMU3682

Reyed, M. R. (2019). Computational biotechnology: An approach in silico based modeling bioprocess. *International Journal of Research Studies in Microbiology and Biotechnology*, *5*(4), 115. DOI: 10.22259/2454-9428.0504001

Reyed, M. R. (2020). Investigation of antimicrobial resistance in wastewater based on public health: An integrated methodology for coronavirus disease 2019 (COVID-19) outbreak tracking schemes and real-time detection of reoccurrence is needed. *Journal of Infectious Diseases & Case Reports*, *1*(2), 115. DOI: 10.47363/JIDSCR/2020(1)115

Reyed, M. R. (2023). Simulate the dynamic interactions across nutritional supplements, the polybiome, and the biochemical approach dependent on patient-derived gastrointestinal knowledge. *Journal of Clinical Epidemiology & Toxicology*, *4*(4), 133. DOI: 10.47363/JCET/2023(4)133

Reyed, M. R. (2026a). Advancing diabetes reclassification using the NexDi AI system with multiomic insights: Egypt's living laboratory. In Reyed, R., Prabhakar, P., & Haghi, A. (Eds.), *Advancing Personalized Medicine With AI-Driven Microbiome and Nutrigenomics: Innovations and Future Perspectives* (pp. 1–92). IGI Global., DOI: 10.4018/979-8-3373-3121-8.ch001

Reyed, M. R. (2026c). Digital nanoplastic science (DNPS) paradigm: Computational intelligence and proteostasis disruptions. *Zenodo*. https://doi.org/DOI: 10.5281/zenodo.18435353

Reyed, M. R. (2026e). Polybiome systems medicine (PSM): A systems-level biodiversity conversion framework for precision nutrition and predictive health 2040. *Zenodo*. https://doi.org/DOI: 10.5281/zenodo.18727664

Reyed, M. R. (2026f). Polybiome systems medicine (PSM) – Volume III: AI nutrigenomics, precision fermentation, and genotype-aligned biomanufacturing. *Zenodo*. https://doi.org/DOI: 10.5281/zenodo.18871656

Reyed, M. R. (2026g). RPCF-AMR: From resistance surveillance to predictive control—A precision convergence framework for integrated antimicrobial resistance governance. *Zenodo*. https://doi.org/DOI: 10.5281/zenodo.19260312

Reyed, M. R., Patel, R., Patel, S., & Prajapati, B. G. (2026b). Application of synthetic biology in genetic disorder. In *Synthetic Biology and Its Consequences in Present Healthcare* (pp. 201–213). Elsevier., DOI: 10.1016/B978-0-443-33649-2.00001-0

Reyed, M. R., & Prabhakar, P. K. (2026d). Modulating neuroinflammation and autophagy in psychiatric disorders through mechanisms and therapies: Therapeutic strategies and pathways. In Mokhtari, T., & Uludag, K. (Eds.), *Autophagy and Inflammation in Neuropsychological Disorders* (pp. 61–124). IGI Global., DOI: 10.4018/979-8-3693-5908-2.ch003

Rezasoltani, S., Yadegar, A., Hatami, B., Asadzadeh Aghdaei, H., & Zali, M. R. (2020). Antimicrobial Resistance as a Hidden Menace Lurking Behind the COVID-19 Outbreak: The Global Impacts of Too Much Hygiene on AMR. *Frontiers in Microbiology*, *11*, 590683. DOI: 10.3389/fmicb.2020.590683 PMID: 33384670

Rezvantalab, S., Mihandoost, S., & Rezaiee, M. (2024). Machine learning assisted exploration of the influential parameters on the PLGA nanoparticles. *Scientific Reports*, *14*(1), 1114. DOI: 10.1038/s41598-023-50876-w PMID: 38212322

Ringaci, A., Shevchenko, K. G., Zelepukin, I. V., Popova, A. V., & Nikitin, M. P. (2022). Phage-mimicking nanoagents for rapid depolymerase specificity screening against multidrug resistant bacteria. *Biosensors & Bioelectronics*, *213*, 114444. DOI: 10.1016/j.bios.2022.114444 PMID: 35691082

Rizwan, M., Yahya, R., Hassan, A., Yar, M., Azzahari, A. D., Selvanathan, V., Sonsudin, F., & Abouloula, C. N. (2017). pH-sensitive hydrogels in drug delivery: Brief history, properties, swelling, and release mechanism, material selection, and applications. *Polymers*, *9*(4), 137. DOI: 10.3390/polym9040137 PMID: 30970818

Rodríguez-Almonacid, C. , Salazar-Bravo, J. ;, Karamysheva, Z. N., Soares, H., Nolasco, S., Tarannum, A., Camilo Rodríguez-Almonacid, C., Salazar-Bravo, J., & Karamysheva, Z. N. (2023). Molecular Mechanisms of Persistence in Protozoan Parasites. *Microorganisms 2023, Vol. 11, Page 2248, 11*(9), 2248. DOI: 10.3390/microorganisms11092248

Rodríguez-Pérez, R., & Bajorath, J. (2022). Evolution of support vector machine and regression modeling in chemoinformatics and drug discovery. *Journal of Computer-Aided Molecular Design*, *36*(5), 355–362. DOI: 10.1007/s10822-022-00442-9 PMID: 35304657

Rolff, J., Bonhoeffer, S., Kloft, C., Leistner, R., Regoes, R., & Hochberg, M. E. (2024). Forecasting antimicrobial resistance evolution. *Trends in Microbiology, 32*(8), 736–745. DOI: 10.1016/j.tim.2023.12.009

Roy R, Tiwari M, Donelli G, Tiwari V. (2022) Strategies for combating bacterial biofilms: A focus on anti-biofilm agents and their mechanisms of action. Virulence. 2018 Jan 1;9(1):522-554. DOI: . PMID: 28362216; PMCID: PMC5955472.DOI: 10.1080/21505594.2017.1313372

Ruba, M., Ashwin Prabhu, G., Robinson Joel, M., Vishnu Priya, K., Dharshani, L., & Dharani, R. (2026). Integrating emerging technologies for organizational agility: An interdisciplinary approach to digital transformation. In Roache, D. (Ed.), *High performance leadership for organizational excellence: Innovation, technology, and resilience* (pp. 63–92). IGI Global Scientific Publishing.

Rudd, K. E., Johnson, S. C., Agesa, K. M., Shackelford, K. A., Tsoi, D., Kievlan, D. R., Colombara, D. V., Ikuta, K. S., Kissoon, N., Finfer, S., Fleischmann-Struzek, C., Machado, F. R., Reinhart, K. K., Rowan, K., Seymour, C. W., Watson, R. S., West, T. E., Marinho, F., Hay, S. I., & Naghavi, M. (2020). Global, regional, and national sepsis incidence and mortality, 1990–2017: Analysis for the Global Burden of Disease Study. *Lancet, 395*(10219), 200–211. DOI: 10.1016/S0140-6736(19)32989-7 PMID: 31954465

Rusu, A. V., Trif, M., & Rocha, J. M. (2023). Microbial secondary metabolites via fermentation approaches for dietary supplementation formulations. *Molecules (Basel, Switzerland), 28*(16), 6020. DOI: 10.3390/molecules28166020

Rybak, M. J., Le, J., Lodise, T. P., Levine, D. P., Bradley, J. S., Liu, C., & Tamma, P. D. (2020). Therapeutic monitoring of vancomycin for serious methicillin-resistant Staphylococcus aureus infections: A revised consensus guideline and review by the American Society of Health-System Pharmacists, the Infectious Diseases Society of America, the Pediatric Infectious Diseases Society, and the Society of Infectious Diseases Pharmacists. *American Journal of Health-System Pharmacy, 77*(11), 835–864. DOI: 10.1093/ajhp/zxaa036 PMID: 32191793

Sabih Ur Rehman, S., Nasar, M. I., Mesquita, C. S., Al Khodor, S., Notebaart, R. A., Ott, S., Mundra, S., Arasardanam, R. P., Muhammad, K., & Alam, M. T. (2025). Integrative systems biology approaches for analyzing microbiome dysbiosis and species interactions. *Briefings in Bioinformatics, 26*(4), bbaf323. Advance online publication. DOI: 10.1093/bib/bbaf323

Sabnis, A., & Edwards, A. M. (2023). Lipopolysaccharide as an antibiotic target. Biochimica et Biophysica Acta (BBA)-. *Biochimica et Biophysica Acta. Molecular Cell Research, 1870*(7), 119507. DOI: 10.1016/j.bbamcr.2023.119507 PMID: 37268022

Safari, M. (2023). *Recent Advances in Quantum Dots-Based Biosensors*. Quantum Dots - Recent Advances, New Perspectives and Contemporary Applications., DOI: 10.5772/intechopen.108205

Sagar, A., Kolluru, V., Jaiswal, U., Kumavat, G., Hole, S. R., & Kumar, A. (2025). Machine Learning and Artificial Intelligence for Predictive Modeling in Antimicrobial Resistance Data Sets, Challenges, and Future Directions. 2025 3rd International Conference on Smart Systems for Applications in Electrical Sciences (ICSSES), 1–6. DOI: 10.1109/ICSSES64899.2025.11009316

Saha, S. (2025). How nanoinformatics could pave the way to safer design of engineered nanomaterials? *Frontiers in Nanotechnology, 7*, 1559053. DOI: 10.3389/fnano.2025.1559053

Sakagianni, A., Koufopoulou, C., Feretzakis, G., Kalles, D., Verykios, V. S., Myrianthefs, P., & Fildisis, G. (2023). Using machine learning to predict antimicrobial resistance—A literature review. *Antibiotics (Basel, Switzerland), 12*(3), 452. DOI: 10.3390/antibiotics12030452

Sakagianni, A., Koufopoulou, C., Koufopoulos, P., Feretzakis, G., Anastasiou, A., Theodorakis, N., & Myrianthefs, P. (2025). Influence of Microbiome Interactions on Antibiotic Resistance Development in the ICU Environment: Insights and Opportunities with Machine Learning. *Acta Microbiologica Hellenica, 70*(2), 14. DOI: 10.3390/amh70020014

Saleem, N., Kumar, N., & El-Omar, E.. (2026). Nanoantimicrobial peptides (nano-AMPs) to combat resistant gram-negative bacteria. *Drug Delivery and Translational Research*. Advance online publication. DOI: 10.1007/s13346-026-02085-x

Salehi, M., Khayami, R., Akbari, R., & Mirmozaffari, M. (2025). An integrated predictive impact-enhanced process mining framework for strategic oncology workflow optimization: Case study in Iran. *Bioengineering (Basel, Switzerland), 12*(12), 1288. DOI: 10.3390/bioengineering12121288

Saleh, Y., Abu Talib, M., Nasir, Q., & Dakalbab, F. (2025). Evaluating large language models: A systematic review of efficiency, applications, and future directions. *Frontiers of Computer Science, 7*, 1523699. Advance online publication. DOI: 10.3389/fcomp.2025.1523699

Salmani-Zarchi, H., Mousavi-Sagharchi, S. M. A., Sepahdoost, N., Ranjbar-Jamalabadi, M., Gross, J. D., Jooya, H., & Samadi, A. (2024). Antimicrobial feature of nanoparticles in the antibiotic resistance era: From mechanism to application. *Advanced Biomedical Research, 13*, 113. DOI: 10.4103/abr.abr_92_24

Salo-Ahen, O. M. H., Alanko, I., Bhadane, R., Alexandre, A. M., Honorato, R. V., Hossain, S., Juffer, A. H., Kabedev, A., Lahtela-Kakkonen, M., Larsen, A. S., Lescrinier, E., Marimuthu, P., Mirza, M. U., Mustafa, G., Nunes-Alves, A., Pantsar, T., Saadabadi, A., Singaravelu, K., & Vanmeert, M. (2021). Molecular dynamics simulations in drug discovery and pharmaceutical development. *Processes (Basel, Switzerland), 9*(1), 71. DOI: 10.3390/pr9010071

Samathoti, P., Kumarachari, R. K., Bukke, S. P. N., Rajasekhar, E. S. K., Jaiswal, A. A., & Eftekhari, Z. (2025). The role of nanomedicine and artificial intelligence in cancer health care: Individual applications and emerging integrations-a narrative review. *Discover. Oncology, 16*(1), 697. DOI: 10.1007/s12672-025-02469-4 PMID: 40338421

Sambaza, S. S., & Naicker, N. (2023). Contribution of wastewater to antimicrobial resistance: A review article. *Journal of Global Antimicrobial Resistance, 34*, 23–29. DOI: 10.1016/j.jgar.2023.05.010

Sanapalli, B. K. R., Palit, S., Deshpande, A., Tokala, R., Sigalapalli, D. K., & Sanapalli, V. (2026). Artificial Intelligence and the Discovery of Antibiotics: Reinventing with Opportunities, Challenges, and Clinical Translation. *Antibiotics (Basel, Switzerland), 15*(2), 233. DOI: 10.3390/antibiotics15020233 PMID: 41750530

Santhini, C., Ruba, M., Sakthi Priya, V., & Hemalatha, R. (2025). Intercultural communication and assessment practices in the integration of AI with education. In *Generative AI Approaches to Sustainable Development in Higher Education*. IGI Global. DOI: 10.4018/979-8-3693-5623-4.ch010

Sarkar, D., Majumder, S., Giri, K., & Sabnam, N. (2023). In silico characterization, molecular docking, and dynamic simulation of a novel fungal cell-death suppressing effector, MoRlpA as potential cathepsin B-like cysteine protease inhibitor during rice blast infection. *Journal of Biomolecular Structure & Dynamics, 41*(18), 9039–9056. DOI: 10.1080/07391102.2022.2139763

Sarker, I. H. (2021). Machine learning: Algorithms, real-world applications and research directions. *SN Computer Science, 2*(3), 160. DOI: 10.1007/s42979-021-00592-x

Sarode, R. J., & Mahajan, H. S. (2024). Dendrimers for drug delivery: An overview of its classes, synthesis, and applications. *Journal of Drug Delivery Science and Technology*, *98*, 105896. DOI: 10.1016/j.jddst.2024.105896

Satheeskumar, R. (2025). Enhancing drug discovery with AI: Predictive modeling of pharmacokinetics using graph neural networks and ensemble learning. *Intelligent Pharmacy*, *3*(2), 127–140. DOI: 10.1016/j.ipha.2024.11.002

Sawant, R. R., & Torchilin, V. P. (2012). Challenges in development of targeted liposomal therapeutics. *The AAPS Journal*, *14*(2), 303–315. DOI: 10.1208/s12248-012-9330-0 PMID: 22415612

Scaglione, G., Mastroianni, N., Rizzo, A., Palomba, E., Carcione, D., Brigante, G., Principe, L., Colaneri, M., Gori, A., & Borgonovo, F. (2026). Integrating artificial intelligence with genome sequencing against antimicrobial resistance: A narrative review. *Frontiers in Public Health*, *14*, 1757161. DOI: 10.3389/fpubh.2026.1757161

Schinkel, M., Paranjape, K., Nannan Panday, R. S., Skyttberg, N., & Nanayakkara, P. W. B. (2019). Clinical applications of artificial intelligence in sepsis: A narrative review. *Computers in Biology and Medicine*, *115*, 103488. Advance online publication. DOI: 10.1016/j.compbiomed.2019.103488 PMID: 31634699

Seal, S., Mahale, M., García-Ortegón, M., Joshi, C. K., Hosseini-Gerami, L., Beatson, A., Greenig, M., Shekhar, M., Patra, A., Weis, C., Mehrjou, A., Badré, A., Paisley, B., Lowe, R., Singh, S., Shah, F., Johannesson, B., Williams, D., Rouquie, D., & Bender, A. (2025). Machine learning for toxicity prediction using chemical structures: Pillars for success in the real world. *Chemical Research in Toxicology*, *38*(5), 759–807. DOI: 10.1021/acs.chemrestox.5c00033 PMID: 40314361

Sedighi, O., Bednarke, B., Sherriff, H., & Doiron, A. L. (2024). Nanoparticle-based strategies for managing biofilm infections in wounds: A comprehensive review. *ACS Omega*, *9*(26), 27853–27871. DOI: 10.1021/acsomega.4c02343 PMID: 38973924

Segurado, O. G., Jiang, R., & Pipe, S. W. (2022, September). Challenges and opportunities when transitioning from in vivo gene replacement to in vivo CRISPR/Cas9 therapies - a spotlight on hemophilia. *Expert Opinion on Biological Therapy*, *22*(9), 1091–1098. DOI: 10.1080/14712598.2022.2090241 PMID: 35708146

Senjab, R. M., AlSawaftah, N., Abuwatfa, W. H., & Husseini, G. A. (2024). Advances in liposomal nanotechnology: From concept to clinics. *RSC Pharmaceutics*, *1*(5), 928–948. DOI: 10.1039/D4PM00176A

Sercombe, L., Veerati, T., Moheimani, F., Wu, S. Y., Sood, A. K., & Hua, S. (2015). Advances and challenges of liposome-assisted drug delivery. *Frontiers in Pharmacology*, *6*, 286. DOI: 10.3389/fphar.2015.00286 PMID: 26648870

Serrano, D. R., Luciano, F. C., Anaya, B. J., Ongoren, B., Kara, A., Molina, G., Ramirez, B. I., Sánchez-Guirales, S. A., Simon, J. A., Tomietto, G., Rapti, C., Ruiz, H. K., Rawat, S., Kumar, D., & Lalatsa, A. (2024). Artificial intelligence (AI) applications in drug discovery and drug delivery: Revolutionizing personalized medicine. *Pharmaceutics*, *16*(10), 1328. DOI: 10.3390/pharmaceutics16101328 PMID: 39458657

Shah, A., Aftab, S., Nisar, J., Ashiq, M. N., & Iftikhar, F. J. (2021). Nanocarriers for targeted drug delivery. *Journal of Drug Delivery Science and Technology*, *62*, 102426. DOI: 10.1016/j.jddst.2021.102426

Shalaby, K. S., Soliman, M. E., Casettari, L., Bonacucina, M., Cespi, G., Palmieri, G. F., Sammour, O. A., & El Shamy, A. A. (2014). Determination of factors controlling the particle size and entrapment efficiency of noscapine in PEG/PLA nanoparticles using artificial neural networks. *International Journal of Nanomedicine*, *9*, 4953–4964. DOI: 10.2147/IJN.S68737 PMID: 25364252

Shandilya, R., Pathak, N., Lohiya, N. K., Sharma, R. S., & Mishra, P. K. (2020). Nanotechnology in reproductive medicine: Opportunities for clinical translation. *Clinical and Experimental Reproductive Medicine*, *47*(4), 245–262. DOI: 10.5653/cerm.2020.03650 PMID: 33227186

Shanmugaraja, T., Ruba, M., Priya, V. S., Venkatesh, T., Supriya, M., & Nallamuthu, M. (2023, September). ECG noise elimination using an FPGA version of a modified error normalized LMS adaptive filter. In AIP Conference Proceedings (Vol. 2764, No. 1, p. 060033). AIP Publishing LLC. DOI: 10.1063/5.0144367

Sharma, D., Gautam, S., Singh, S., Srivastava, N., Khan, A. M., & Bisht, D. (2025). Unveiling the nanoworld of antimicrobial resistance: Integrating nature and nanotechnology. *Frontiers in Microbiology*, *15*, 1391345. DOI: 10.3389/fmicb.2024.1391345

Sheikhpour, M., Barani, L., & Kasaeian, A. (2017). Biomimetics in drug delivery systems: A critical review. *Journal of Controlled Release*, *253*, 97–109. DOI: 10.1016/j.jconrel.2017.03.026 PMID: 28322976

Shen, C., Zhang, M., Lu, M., Chang, E., Gao, Z., Ban, W., Liu, Q., Zuo, Z., & Jiang, C. (2026). Machine learning empowered formulation design, optimization and characterization of nanoparticulate drug delivery systems: Current applications, challenges, and future perspectives. *Acta Pharmaceutica Sinica. B*, *16*(2), 665–685. DOI: 10.1016/j.apsb.2025.12.011 PMID: 41685160

Shen, F., Lu, Z., Yan, K., Luo, K., Pei, S., & Xiang, P. (2025). Synthesis and properties of carbon quantum dots as an antimicrobial agent and detection of ciprofloxacin. *Scientific Reports*, *15*(1), 28535. Advance online publication. DOI: 10.1038/s41598-025-14383-4 PMID: 40764737

Shen, Y., Fan, N., Ma, S. X., Cheng, X., Yang, X., & Wang, G. (2025). Gut Microbiota Dysbiosis: Pathogenesis, Diseases, Prevention, and Therapy. *MedComm*, *6*(5), e70168. DOI: 10.1002/mco2.70168 PMID: 40255918

Shen, Y., Zhang, P., Luo, J., Chen, S., Gu, S., Lin, Z., & Tang, Z. (2026). Artificial intelligence drives advances in multiomics analysis and precision medicine for sepsis. *Biomedicines*, *14*(2), 261. DOI: 10.3390/biomedicines14020261

Shepherd, M. J., Fu, T., Harrington, N. E., Kottara, A., Cagney, K., Chalmers, J. D., Paterson, S., Fothergill, J. L., & Brockhurst, M. A. (2024). Ecological and evolutionary mechanisms driving within-patient emergence of antimicrobial resistance. *Nature Reviews. Microbiology*, *22*(10), 650–665. DOI: 10.1038/s41579-024-01041-1

Shi, J., Kantoff, P. W., Wooster, R., & Farokhzad, O. C. (2017). Cancer nanomedicine:progress, challenges and opportunities. *Nature Reviews. Cancer*, *17*(1), 20–37. DOI: 10.1038/nrc.2016.108 PMID: 27834398

Shin, M. K., Lee, B., Kim, S. T., Yoo, J. S., & Sung, J. S. (2022). Designing a novel functional peptide with dual antimicrobial and anti-inflammatory activities *via* in silico methods. *Frontiers in Immunology*, *13*, 821070. DOI: 10.3389/fimmu.2022.821070

Shinu, P., Mouslem, A. K. A., Nair, A. B., Venugopala, K. N., Attimarad, M., Singh, V. A., Nagaraja, S., Alotaibi, G., & Deb, P. K. (2022). Progress report: Antimicrobial drug discovery in the resistance era. *Pharmaceuticals*, *15*(4), 413. DOI: 10.3390/ph15040413

Shi, Y., van der Meel, R., Chen, X., & Lammers, T. (2020). The EPR effect and beyond: Strategies to improve tumor targeting and cancer nanomedicine treatment efficacy. *Theranostics*, *10*(17), 7921–7924. DOI: 10.7150/thno.49577 PMID: 32685029

Shrestha, A., Sadeyen, J. R., & Iqbal, M. (2018). Enhancing protective efficacy of poultry vaccines through targeted delivery of antigens to antigen-presenting cells. *Vaccines*, *6*(4), 75. DOI: 10.3390/vaccines6040075 PMID: 30445683

Sibiya, S., Ruba, M., Ponkarthika, M., Vijayasabariswari, M., Sheba, S., Vignesh, S., . . . Sankar, S. S. (2025, October). Integrated Wearable Health Monitoring and Emergency Response System With IOT Sensors, AED and Mobile App for Enhanced Patient Outcomes. In 2025 IEEE 2nd International Conference on Green Industrial Electronics and Sustainable Technologies (GIEST) (pp. 1-6). IEEE.

Sill, T. J., & Recum, H. A. (2008). Electrospinning: Applications in drug delivery and tissue engineering. *Biomaterials*, *29*(13), 1989–2006. DOI: 10.1016/j.biomaterials.2008.01.011 PMID: 18281090

Silva, P. M. A., & Queirós, O. (2025). Advances in nanodrug delivery for tumor microenvironment and drug resistance-insights from the special issue " Nanodrug delivery systems for targeting the tumor microenvironment and simultaneously overcoming drug resistance properties. *Pharmaceutics*, *17*(7), 942. DOI: 10.3390/pharmaceutics17070942

Silva, Y. P., Bernardi, A., & Frozza, R. L. (2020). The role of short-chain fatty acids from gut microbiota in gut-brain communication. *Frontiers in Endocrinology*, *11*, 25. DOI: 10.3389/fendo.2020.00025

Singhal, K., Azizi, S., Tu, T., Mahdavi, S. S., Wei, J., Chung, H. W., Scales, N., Tanwani, A., Cole-Lewis, H., Pfohl, S., Payne, P., Seneviratne, M., Gamble, P., Kelly, C., Babiker, A., Schärli, N., Chowdhery, A., Mansfield, P., Demner-Fushman, D., & Natarajan, V. (2023). Large language models encode clinical knowledge. *Nature*, *620*(7972), 172–180. DOI: 10.1038/s41586-023-06291-2

Singh, H., Balusamy, S. R., Sukweenadhi, J., Saravanan, M., Aruchamy, M., Mijakovic, I., & Singh, P. (2025). Smart hybrid nanomaterials for chronic infections: Microbiome-responsive and sustainable therapeutic platforms. *Journal of Nanobiotechnology*, *23*(1), 698. DOI: 10.1186/s12951-025-03712-4 PMID: 41184906

Smits, W. K., Garey, K. W., Riley, T. V., & Johnson, S. (2025). Clostridioides difficile is a bacterial priority pathogen. *Anaerobe*, *93*, 102965. DOI: 10.1016/j.anaerobe.2025.102965 PMID: 40288746

Smyth, C., Leigh, R. J., Do, T. T., & Walsh, F. (2025). Communities of plasmids as strategies for antimicrobial resistance gene survival in wastewater treatment plant effluent. *npj Antimicrobials and Resistance, 3*(1), 78. https://doi.org/DOI: 10.1038/s44259-025-00151-x

Sodagari, H. R., Ghasemi, M., Varga, C., & Habib, I. (2025). Machine learning prediction of multidrug resistance in swine-derived *Campylobacter* spp. using United States antimicrobial resistance surveillance data (2013-2023). *Veterinary Sciences*, *12*(10), 937. DOI: 10.3390/vetsci12100937

Solanki, R., Makwana, N., Kumar, R., Joshi, M., Patel, A., Bhatia, D., & Sahoo, D. K. (2024). Nanomedicines as a cutting-edge solution to combat antimicrobial resistance. *RSC Advances, 14*(45), 33568–33586. DOI: 10.1039/D4RA06117A PMID: 39439838

Spasic, A. M. (2018). Classification and methodology—finely dispersed systems. In Spasic, A. M. (Ed.), *Rheology of Emulsions* (Vol. 22, pp. 41–49). Elsevier., DOI: 10.1016/B978-0-12-813836-6.00003-9

Sraphet, S., & Javadi, B. (2022). Application of hierarchical clustering to analyze solvent-accessible surface area patterns in Amycolatopsis lipases. *Biology (Basel), 11*(5), 652. DOI: 10.3390/biology11050652

Sriram, T., Chakraborty, T., & Prasanna, P. M. (2024). Artificial intelligence powered insights into nanotoxicology. *Int. J. Adv. Life Sci. Res, 7*(2), 68–80. DOI: 10.31632/ijalsr.2024.v07i02.005

Srivastav, A. K., Mishra, M. K., Lillard, J. W.Jr, & Singh, R. (2025). Transforming Pharmacogenomics and CRISPR Gene Editing with the Power of Artificial Intelligence for Precision Medicine. *Pharmaceutics, 17*(5), 555. DOI: 10.3390/pharmaceutics17050555 PMID: 40430848

Srivastava, P., & Kowshik, M. (2013). Mechanisms of metal resistance and homeostasis in haloarchaea. *Archaea : an International Microbiological Journal, 732864.* Advance online publication. DOI: 10.1155/2013/732864

Stokes, J. M., Yang, K., Swanson, K., Jin, W., Cubillos-Ruiz, A., Donghia, N. M., MacNair, C. R., French, S., Carfrae, L. A., Bloom-Ackermann, Z., Tran, V. M., Chiappino-Pepe, A., Badran, A. H., Andrews, I. W., Chory, E. J., Church, G. M., Brown, E. D., Jaakkola, T. S., Barzilay, R., & Collins, J. J. (2020). A Deep Learning Approach to Antibiotic Discovery. *Cell, 180*(4), 688–702.e13. DOI: 10.1016/j.cell.2020.01.021 PMID: 32084340

Sulimov, V. B., Kutov, D. C., & Sulimov, A. V. (2019). Advances in docking. *Current Medicinal Chemistry, 26*(42), 7555–7580. DOI: 10.2174/0929867325666180904115000

Sultana, A., Zare, M., Thomas, V., Kumar, T. S. S., & Ramakrishna, S. (2022). Nanobased drug delivery systems: Conventional drug delivery routes, recent developments and future prospects. *Medicine in Drug Discovery, 15*, 100134. DOI: 10.1016/j.medidd.2022.100134

Sum, D. K. C., Chong, Y. Y., & Tan, J. L. (2025). Comparative analyses suggest genome stability and plasticity in *Stenotrophomonas maltophilia. International Journal of Molecular Sciences*, *26*(21), 10477. DOI: 10.3390/ijms262110477

Sun, M., Zhao, S., Gilvary, C., Elemento, O., Zhou, J., & Wang, F. (2020). Graph convolutional networks for computational drug development and discovery. *Briefings in Bioinformatics*, *21*(3), 919–935. DOI: 10.1093/bib/bbz042

Suriyaamporn, P., Pamornpathomkul, B., Wongprayoon, P., Rojanarata, T., Ngawhirunpat, T., & Opanasopit, P. (2024). The artificial intelligence and design of experiment assisted in the development of progesterone-loaded solid-lipid nanoparticles for transdermal drug delivery. *Farmatsiia (Sofia)*, *71*, 1–12. DOI: 10.3897/pharmacia.71.e123549

Su, S., & Kang, M, P. (2020). Recent advances in nanocarrier-assisted therapeutics delivery systems. *Pharmaceutics*, *12*(9), 837.

Sushnitha, M., Evangelopoulos, M., Tasciotti, E., & Taraballi, F. (2020). Cell membrane–based biomimetic nanoparticles and the immune system: Immunomodulatory interactions to therapeutic applications. *Frontiers in Bioengineering and Biotechnology*, *8*, 627. DOI: 10.3389/fbioe.2020.00627 PMID: 32626700

Svenson, S., & Tomalia, D. A. (2012). Dendrimers in biomedical applications—Reflections on the field. *Advanced Drug Delivery Reviews*, *64*, 102–115. DOI: 10.1016/j.addr.2012.09.030 PMID: 16305813

Taha, B. A., Ahmed, N. M., Talreja, R. K., Haider, A. J., Al Mashhadany, Y., Al-Jubouri, Q., Huddin, A. B., Mokhtar, M. H. H., Rustagi, S., Kaushik, A., Chaudhary, V., & Arsad, N. (2024). Synergizing Nanomaterials and Artificial Intelligence in Advanced Optical Biosensors for Precision Antimicrobial Resistance Diagnosis. *ACS Synthetic Biology*, *13*(6), 1600–1620. DOI: 10.1021/acssynbio.4c00070 PMID: 38842483

Talkington, A. M., Cao, Y., Kearsley, A. J., & Lai, S. K. (2025). Opportunities for machine learning and artificial intelligence in physiologically-based pharmacokinetic (PBPK) modeling. *Advanced Drug Delivery Reviews*, *227*, 115716. Advance online publication. DOI: 10.1016/j.addr.2025.115716 PMID: 41093240

Tang, Y., & Wang, G. (2021). NIR light-responsive nanocarriers for controlled release. *Journal of Photochemistry and Photobiology C, Photochemistry Reviews*, *47*, 100420. DOI: 10.1016/j.jphotochemrev.2021.100420

Tan, L., Zhou, Z., Liu, X., Li, J., Zheng, Y., Cui, Z., Yang, X., Liang, Y., Li, Z., Feng, X., Zhu, S., Yeung, K. W. K., Yang, C., Wang, X., & Wu, S. (2020). Overcoming multidrug-resistant MRSA using conventional aminoglycoside antibiotics. *Advanced Science (Weinheim, Baden-Wurttemberg, Germany)*, *7*(14), 1902070. Advance online publication. DOI: 10.1002/advs.201902070

Tan, R., Jin, M., Li, J., & Yang, D. (2025). The dissemination, health risks, and mitigation approaches of antibiotic resistance genes in the gut microbiome. *Journal ofHazardous Materials Advances*, *17*, 100634. DOI: 10.1016/j.hazadv.2025.100634

Tatonetti, N. P., Ye, P. P., Daneshjou, R., & Altman, R. B. (2012). Data-driven prediction of drug effects and interactions. *Science Translational Medicine*, *4*(125), 125ra31. Advance online publication. DOI: 10.1126/scitranslmed.3003377

Tavares Luiz, M., Santos Rosa Viegas, J., Palma Abriata, J., Viegas, F., Testa Moura de Carvalho Vicentini, F., Lopes Badra Bentley, M. V., Chorilli, M., Maldonado Marchetti, J., & Tapia-Blácido, D. R. (2021). Design of experiments (DoE) to develop and to optimize nanoparticles as drug delivery systems. *European Journal of Pharmaceutics and Biopharmaceutics : Official Journal of Arbeitsgemeinschaft für Pharmazeutische Verfahrenstechnik e.V*, *165*, 127–148. DOI: 10.1016/j.ejpb.2021.05.011

Teixeira, M. C., Carbone, C., Sousa, M. C., Espina, M., Garcia, M. L., Sanchez-Lopez, E., & Souto, E. B. (2020). Nanomedicines for the Delivery of Antimicrobial Peptides (AMPs). *Nanomaterials (Basel, Switzerland)*, *10*(3), 560. DOI: 10.3390/nano10030560 PMID: 32244858

Thakur, B., Kaur, S., Rani, N., Kaur, R., Upadhyay, S. K., & Tripathi, M. (2025). Exploring microbial contributions to nutraceutical production: From natural to designed foods. *Molecular Biotechnology*, *67*(8), 2946–2961. DOI: 10.1007/s12033-023-00937-2

Thambi, T., Park, J. H., & Lee, D. S. (2016). Stimuli-responsive polymersomes for cancer therapy. *Biomaterials Science*, *4*(1), 55–69. DOI: 10.1039/C5BM00268K PMID: 26456625

Thawong, N., Srilohasin, P., Phelan, J. E., Phornsiricharoenphant, W., Tongsima, S., Suriyaphol, P., Prammananan, T., Faksri, K., Sawaengdee, W., Wang, L., Hinthong, W., Hibberd, M. L., Campino, S., Wattanapokayakit, S., Mahasirimongkol, S., Chaiprasert, A., & Clark, T. G. (2025). Genomic decoding of drug-resistant tuberculosis transmission in Thailand over three decades. *Scientific Reports 2025 15:1*, *15*(1), 1–11. DOI: 10.1038/s41598-025-15093-7

Theophilus, R. J., & Taft, D. H. (2023). Antimicrobial Resistance Genes (ARGs), the Gut Microbiome, and Infant Nutrition. *Nutrients, 15*(14), 3177. DOI: 10.3390/nu15143177 PMID: 37513595

Timbrook, T. T., Morton, J. B., McConeghy, K. W., Caffrey, A. R., Mylonakis, E., & LaPlante, K. L. (2017). The effect of molecular rapid diagnostic testing on clinical outcomes in bloodstream infections: A systematic review and meta-analysis. *Clinical Infectious Diseases*, *64*(1), 15–23. DOI: 10.1093/cid/ciw649

Tiwari, P. C., Pal, R., Chaudhary, M. J., & Nath, R. (2023). Artificial intelligence revolutionizing drug development: Exploring opportunities and challenges. *Drug Development Research*, *84*(8), 1652–1663. DOI: 10.1002/ddr.22115 PMID: 37712494

Tokuda, M., & Shintani, M. (2024). Microbial evolution through horizontal gene transfer by mobile genetic elements. *Microbial Biotechnology, 17*(1), e14408. https://doi.org/DOI: 10.1111/1751-7915.14408

Topol, E. J. (2019). High-performance medicine: The convergence of human and artificial intelligence. *Nature Medicine*, *25*(1), 44–56. DOI: 10.1038/s41591-018-0300-7

Torchilin, V. P. (2005). Recent advances with liposomes as pharmaceutical carriers. *Nature Reviews. Drug Discovery*, *4*(2), 145–160. DOI: 10.1038/nrd1632 PMID: 15688077

Torchilin, V. P. (2006). Multifunctional nanocarriers. *Advanced Drug Delivery Reviews*, *58*(14), 1532–1555. DOI: 10.1016/j.addr.2006.09.009

Torchilin, V. P. (2014). Multifunctional, stimuli-sensitive nanoparticulate systems for drug delivery. *Nature Reviews. Drug Discovery*, *13*(11), 813–827. DOI: 10.1038/nrd4333 PMID: 25287120

Törős, G., Gulyás, G., El-Ramady, H., Alibrahem, W., Muthu, A., Gangakhedkar, P., Atieh, R., & Prokisch, J. (2025). Sustainable nanotechnology strategies for modulating the human gut microbiota. *International Journal of Molecular Sciences*, *26*(12), 5433. DOI: 10.3390/ijms26125433

Tousif, S., & Ahmad, S. (2015). Challenges of Tuberculosis Treatment with DOTS: An Immune Impairment Perspective. *Journal of Cell Science & Therapy*, *06*(05). Advance online publication. DOI: 10.4172/2157-7013.1000223

Trafton, A. (2020). Artificial intelligence yields new antibiotic. *MIT News*. https://news.mit.edu/2020/artificial-intelligence-identifies-new-antibiotic-0220

Trasca, D.-M., Dorin, P. I., Carmen, S., Varut, R.-M., Singer, C. E., Radivojevic, K., & Stoica, G. A. (2025). Artificial intelligence in biomedicine: A systematic review from nanomedicine to neurology and hepatology. *Pharmaceutics*, *17*(12), 1564. DOI: 10.3390/pharmaceutics17121564

Trucillo, P. (2021). Drug carriers: Classification, administration, release profiles, and industrial approach. *Processes (Basel, Switzerland)*, *9*(3), 470. DOI: 10.3390/pr9030470

Tsirogianni, A. G., Chountoulesi, M., & Pippa, N. (2025). Recent Advances in pH-Responsive Liposomes: Lessons Learnt and New Directions in Nanomedicine Development. *Materials (Basel)*, *18*(18), 4295. DOI: 10.3390/ma18184295 PMID: 41010139

Tyssen, D., Henderson, S. A., Johnson, A., Sterjovski, J., Moore, K., La, J., Zanin, M., Sonza, S., Karellas, P., Giannis, M. P., Krippner, G., Wesselingh, S., McCarthy, T., Gorry, P. R., Ramsland, P. A., Cone, R., Paull, J. R. A., Lewis, G. R., & Tachedjian, G. (2010). Structure Activity Relationship of Dendrimer Microbicides with Dual Action Antiviral Activity. *PLoS One*, *5*(9), e12309. DOI: 10.1371/journal.pone.0012309 PMID: 20808791

Tzavellas, N. P., Atzemoglou, N., Bozidis, P., & Gartzonika, K. (2026). The Role of Environmental and Climatic Factors in Accelerating Antibiotic Resistance in the Mediterranean Region. *Acta Microbiologica Hellenica*, *71*(1), 1. DOI: 10.3390/amh71010001

Vagena, I.-A., Malapani, C., Gatou, M.-A., Lagopati, N., & Pavlatou, E. A. (2025). Enhancement of EPR Effect for Passive Tumor Targeting: Current Status and Future Perspectives. *Applied Sciences (Basel, Switzerland)*, *15*(6), 3189. DOI: 10.3390/app15063189

Vahabi, N., & Michailidis, G. (2022). Unsupervised Multi-Omics Data Integration Methods: A Comprehensive Review. *Frontiers in Genetics*, *13*, 854752. Advance online publication. DOI: 10.3389/fgene.2022.854752 PMID: 35391796

Van Hoek, A. H., Mevius, D., Guerra, B., Mullany, P., Roberts, A. P., & Aarts, H. J. (2011). Acquired antibiotic resistance genes: An overview. *Frontiers in Microbiology*, *2*, 203. DOI: 10.3389/fmicb.2011.00203 PMID: 22046172

van Pinxteren, D. J. M., & Jespers, W. (2025). Integrating machine learning into free energy perturbation workflows. *Journal of Chemical Information and Modeling*, *65*(19), 9856–9864. DOI: 10.1021/acs.jcim.5c01449

Ventola, C. L. (2015). The antibiotic resistance crisis: part 1: causes and threats. *P&T : a Peer-Reviewed Journal for Formulary Management, 40*(4), 277. PMID: 25859123

Villanueva-Miranda, I., Xiao, G., & Xie, Y. (2025). Artificial intelligence in early warning systems for infectious disease surveillance: A systematic review. *Frontiers in Public Health, 13*, 1609615. DOI: 10.3389/fpubh.2025.1609615

Vivarelli, S., De Francesco, C., Paba, E., Giambò, F., & Fenga, C. (2026). The resistome bridge between livestock and workers: Novel frameworks for early detection and monitoring of antimicrobial resistance. *Frontiers in Public Health, 14*, 1746385. DOI: 10.3389/fpubh.2026.1746385

Vo, D.-K., & Trinh, K. T. L. (2025). Advances in Wearable Biosensors for Wound Healing and Infection Monitoring. *Biosensors (Basel), 15*(3), 139. DOI: 10.3390/bios15030139 PMID: 40136936

Voter, A. F. (1997). Hyperdynamics: Accelerated molecular dynamics of infrequent events. *Physical Review Letters, 78*(20), 3908–3911. DOI: 10.1103/PhysRevLett.78.3908

Vyas, H., Bhardwaj, M., Das, A., Agarwal, J., & Tiwari, A. K. (2026). Safety and regulatory considerations in nano-bio-medicine. In *Nano-Bio-Medicine for Gastrointestinal Diseases* (pp. 311–328). Academic Press., DOI: 10.1016/B978-0-443-34039-0.00002-0

Wadan, A. S., Ali, M. H., & Ellakwa, D. E. (2026). Sustainable nanomaterials for precision dental medicine: Green synthesis, therapeutic applications, and future directions. *Journal of Nanobiotechnology, 24*(1), 190. DOI: 10.1186/s12951-025-04008-3

Waghule, T., Dabholkar, N., Gorantla, S., Rapalli, V. K., Saha, R. N., & Singhvi, G. (2021). Quality by design (QbD) in the formulation and optimization of liquid crystalline nanoparticles (LCNPs): A risk based industrial approach. *Biomedicine and Pharmacotherapy, 141*, 111940. DOI: 10.1016/j.biopha.2021.111940 PMID: 34328089

Wang, X. W., Wang, T., & Liu, Y. Y. (2026). Artificial intelligence for microbiology and microbiome research. *Cell Systems, 17*(2). https://doi.org/DOI: 10.1016/j.cels.2026.101531

Wang, X., Shan, M., Zhang, S., Chen, X., Liu, W., Chen, J., & Liu, X. (2022). Stimuli-responsive antibacterial materials: molecular structures, design principles, and biomedical applications. *Advanced Science, 9*(13), 2104843. https://doi.org/ DOI: 10.1002/advs.202104843

Wang, C., Kumar, G. A., & Rajapakse, J. C. (2025). Drug discovery and mechanism prediction with explainable graph neural networks. *Scientific Reports, 15*(1), 179. Advance online publication. DOI: 10.1038/s41598-024-83090-3

Wang, C., Yang, Y., Cao, Y., Liu, K., Shi, H., Guo, X., Liu, W., Hao, R., Song, H., & Zhao, R. (2023). Nanocarriers for the delivery of antibiotics into cells against intracellular bacterial infection. *Biomaterials Science, 11*(2), 432–444. DOI: 10.1039/D2BM01489K PMID: 36503914

Wang, F., Harker, A., Edirisinghe, M., & Parhizkar, M. (2025). Advanced experimental design strategies for drug development. *Advanced Intelligent Discovery, 2025*, e202500087. DOI: 10.1002/aidi.202500087

Wang, H., Huang, Q., Chang, H., Xiao, J., & Cheng, Y. (2016). Stimuli-responsive dendrimers in drug delivery. *Biomaterials Science, 4*(3), 375–390. DOI: 10.1039/C5BM00532A PMID: 26806314

Wang, H., Liu, Y., Li, Y., Zhao, Y., & Li, X. (2025, August 28). Ultrasound-Responsive Drug Delivery System for Cancer Therapy. *Macromolecular Bioscience, e00082*(11), e00082. Advance online publication. DOI: 10.1002/mabi.202500082 PMID: 40873253

Wang, J., Li, B., Qiu, L., Qiao, X., & Yang, H. (2022). Dendrimer-based drug delivery systems: History, challenges, and latest developments. *Journal of Biological Engineering, 16*(1), 18. Advance online publication. DOI: 10.1186/s13036-022-00298-5 PMID: 35879774

Wang, K., Huang, Y., Wang, Y., You, Q., & Wang, L. (2024). Recent advances from computer-aided drug design to artificial intelligence drug design. *RSC Medicinal Chemistry, 15*(12), 3978–4000. DOI: 10.1039/d4md00522h

Wang, L., Hu, C., & Shao, L. (2017). The antimicrobial activity of nanoparticles: Present situation and prospects for the future. *International Journal of Nanomedicine, 12*, 1227–1249. DOI: 10.2147/IJN.S121956 PMID: 28243086

Wang, Q., Sun, B., Yi, Y., Velkov, T., Shen, J., Dai, C., & Jiang, H. (2025). Progress of AI-Driven Drug–Target Interaction Prediction and Lead Optimization. *International Journal of Molecular Sciences, 26*(20), 10037. DOI: 10.3390/ijms262010037 PMID: 41155330

Wang, S., Yu, S., Lin, Y., Zou, P., Chai, G., Yu, H. H., Wickremasinghe, H., Shetty, N., Ling, J., Li, J., & Velkov, T. (2018). Co-delivery of ciprofloxacin and colistin in liposomal formulations with enhanced in vitro antimicrobial activities against multidrug-resistant *Pseudomonas aeruginosa. Pharmaceutical Research, 35*(10), 187. DOI: 10.1007/s11095-018-2464-8 PMID: 30094660

Wang, T., Wu, C., Hu, Y., Zhang, Y., & Ma, J. (2023). Stimuli-responsive nanocarrier delivery systems for Pt-based antitumor complexes: A review. *RSC Advances, 13*(24), 16488–16511. DOI: 10.1039/D3RA00866E

Wang, W., Feng, S., Ye, Z., Gao, H., Lin, J., & Ouyang, D. (2022a). Prediction of lipid nanoparticles for mRNA vaccines by the machine learning algorithm. *Acta Pharmaceutica Sinica. B, 12*(6), 2950–2962. DOI: 10.1016/j.apsb.2021.11.021 PMID: 35755271

Wang, W., Ye, Z., Gao, H., & Ouyang, D. (2021). Computational pharmaceutics—A new paradigm of drug delivery. *Journal of Controlled Release, 338*, 119–136. DOI: 10.1016/j.jconrel.2021.08.030

Wang, X., Li, C., Wang, Y., Chen, H., Zhang, X., Luo, C., Zhou, W., Li, L., Teng, L., Yu, H., & Wang, J. (2022b). Smart drug delivery systems for precise cancer therapy. *Acta Pharmaceutica Sinica. B, 12*(11), 4098–4121. DOI: 10.1016/j.apsb.2022.08.013 PMID: 36386470

Wang, Y., Huang, R., Feng, S., & Mo, R. (2025). Advances in nanocarriers for targeted drug delivery and controlled drug release. *Chinese Journal of Natural Medicines, 23*(5), 513–528. DOI: 10.1016/S1875-5364(25)60861-2 PMID: 40383609

Wang, Y., Li, T., Zhou, H., & Liu, C. (2025, May 29). Advancements in Nanocarrier Delivery Systems for Photodynamic Therapy in Lung Cancer. *International Journal of Nanomedicine, 20*, 6853–6874. DOI: 10.2147/IJN.S521444 PMID: 40458748

Wang, Y., Zhao, Y., Bollas, A., Wang, Y., & Au, K. F. (2021). Nanopore sequencing technology, bioinformatics and applications. *Nature Biotechnology, 39*(11), 1348–1365. DOI: 10.1038/s41587-021-01108-x PMID: 34750572

Wang, Z., Wang, Y., Zhang, S., Wang, S., Xu, Z., & Feng, Z. J. (2024). Trend analysis and prediction of gonorrhea in mainland China based on a hybrid time series model. *BMC Infectious Diseases, 24*(1), 113. Advance online publication. DOI: 10.1186/s12879-023-08969-4

Watkins, R. R. (2022). Antibiotic stewardship in the era of precision medicine. *JAC-Antimicrobial Resistance, 4*(3), dlac066. Advance online publication. DOI: 10.1093/jacamr/dlac066

Watson, J. L., Juergens, D., Bennett, N. R., Trippe, B. L., Yim, J., Eisenach, H. E., & Baker, D. (2023). De novo design of protein structure and function with RFdiffusion. *Nature, 620*(7976), 1089–1100. DOI: 10.1038/s41586-023-06415-8 PMID: 37433327

Waykar, R., & Kulkarni, Y. A. (2025). Informed consent: Navigating the bioethical complexities of clinical research. *Bioinformation, 21*(9), 2950–2955. DOI: 10.6026/973206300212950

Wegener Parfrey, L. (2015). Mock communities highlight the diversity of host-associated eukaryotes. *Molecular Ecology, 24*(17), 4337–4339. DOI: 10.1111/mec.13311

Wheeler, N. E., Price, V., Cunningham-Oakes, E., Tsang, K. K., Nunn, J. G., Midega, J. T., Anjum, M. F., Wade, M. J., Feasey, N. A., Peacock, S. J., Jauneikaite, E., & Baker, K. S. (2023). Innovations in genomic antimicrobial resistance surveillance. *The Lancet Microbe, 4*(12), e1063–e1070. DOI: 10.1016/S2666-5247(23)00285-9 PMID: 37977163

Wickramage, I., Spigaglia, P., & Sun, X. (2021). Mechanisms of antibiotic resistance of Clostridioides difficile. *The Journal of Antimicrobial Chemotherapy, 76*(12), 3077–3090. DOI: 10.1093/jac/dkab231 PMID: 34297842

Wimalawansa, S. J. (2025). Reforming food, drug, and nutraceutical regulations to improve public health and reduce healthcare costs. *Foods, 14*(13), 2328. DOI: 10.3390/foods14132328

Wörheide, M. A., Krumsiek, J., Kastenmüller, G., & Arnold, M. (2021). Multiomics integration in biomedical research - A metabolomics-centric review. *Analytica Chimica Acta, 1141*, 144–162. DOI: 10.1016/j.aca.2020.10.038

World Health Organization. (2021). *Ethics and governance of artificial intelligence for health.* https://www.who.int/publications/i/item/9789240029200

World Health Organization. (2022). Global Antimicrobial Resistance and Use Surveillance System (GLASS) report: Antibiotic use data for 2022. https://www.who.int/publications/i/item/9789240108127

World Health Organization. (2023). Antimicrobial resistance. https://www.who.int/news-room/fact-sheets/detail/antimicrobial-resistance

World Health Organization. (2023). *GLASS manual for antimicrobial resistance surveillance in common bacteria causing human infection*. World Health Organization.

World Health Organization. (2024a, February 7). Antimicrobial resistance. WHO Fact Sheets. URL:https://www.who.int/news-room/fact-sheets/detail/antimicrobial-resistance

World Health Organization. (2024b, May 17). WHO bacterial priority pathogens list, 2024. URL:https://www.who.int/publications/i/item/9789240093461

World Health Organization. (2024c, May 14). 2023 Antibacterial agents in clinical and preclinical development: an overview and analysis. URL:https://www.who.int/publications/i/item/9789240094000

Wu, J., & Koelzer, V. H. (2024). Toward generative digital twins in biomedical research. *Computational and Structural Biotechnology Journal*, *23*, 3481–3488. DOI: 10.1016/j.csbj.2024.09.030

Wu, S., Zhou, H., Chen, D., Lu, Y., Li, Y., & Qiao, J. (2024). Multiomic analysis tools for microbial metabolites prediction. *Briefings in Bioinformatics*, *25*(4), bbae264. Advance online publication. DOI: 10.1093/bib/bbae264

Wu, Y., Wang, N., Xiong, P., Wang, R., Deng, J., & Ouyang, D. (2025). Artificial intelligence for drug delivery: Yesterday, today and tomorrow. *Acta Pharmaceutica Sinica. B*. Advance online publication. DOI: 10.1016/j.apsb.2025.09.022

Wu, Z., Yao, T., & Shen, N. (2025). To face Disease X: Building resilient futures in the age of emergent threats. *BMJ Global Health*, *10*(7), e020479. DOI: 10.1136/bmjgh-2025-020479

Xiao, C., Choi, E., & Sun, J. (2018). Opportunities and challenges in developing deep learning models using electronic health records data: A systematic review. *Journal of the American Medical Informatics Association : JAMIA*, *25*(10), 1419–1428. DOI: 10.1093/jamia/ocy068 PMID: 29893864

Xiaolin, X., Xiaozhi, L., Guoping, H., Hongwei, L., Jinkuo, G., Xiyun, B., Zhen, T., Xiaofang, M., Yanxia, L., Na, X., Chunyan, Z., Rui, G., Kuan, W., Cheng, Z., Cuancuan, W., Mingyong, L., & Xinping, D. (2023). Overfit deep neural network for predicting drug-target interactions. *iScience*, *26*(9), 107646. DOI: 10.1016/j.isci.2023.107646

Xiao, X., Trinh, T. X., Gerelkhuu, Z., Ha, E., & Yoon, T. H. (2024). Automated machine learning in nanotoxicity assessment: A comparative study of predictive model performance. *Computational and Structural Biotechnology Journal*, *25*, 9–19. DOI: 10.1016/j.csbj.2024.02.003 PMID: 38414794

Xia, S., Chen, E., & Zhang, Y. (2023). Integrated molecular modeling and machine learning for drug design. *Journal of Chemical Theory and Computation, 19*(21), 7478–7495. DOI: 10.1021/acs.jctc.3c00814

Xie, J., Sun, L., & Zhao, Y. F. (2025). On the data quality and imbalance in machine learning-based design and manufacturing—A systematic review. *Engineering (Beijing), 45*(18), 105–131. DOI: 10.1016/j.eng.2024.04.024

Xie, Y., Lu, H., Liu, Y., Hu, G., Lian, S., Liu, J., Pang, S., Zhu, G., & Ding, X. (2025). Unveiling the Mechanisms of Bacterial Resistance and Countermeasures. *Pathogens (Basel, Switzerland), 14*(11), 1085. DOI: 10.3390/pathogens14111085 PMID: 41305322

Xuan, L., Ju, Z., Skonieczna, M., Zhou, P. K., & Huang, R. (2023). Nanoparticles-induced potential toxicity on human health: Applications, toxicity mechanisms, and evaluation models. *MedComm, 4*(4), e327. DOI: 10.1002/mco2.327 PMID: 37457660

Xu, Z., Xie, Y., Chen, W., & Deng, W. (2025). Nanocarrier-based systems for targeted delivery: Current challenges and future directions. *MedComm, 6*(9), e70337. DOI: 10.1002/mco2.70337 PMID: 40859956

Yakobi, S. H., & Nwodo, U. U. (2025). *AI-driven modelling, antimicrobial discovery, and precision therapeutics for targeting bacterial persisters. In Silico Research in Biomedicine, 100062.*, DOI: 10.1016/j.insi.2025.100062

Yang, G., Chen, S., & Zhang, J. (2019). Bioinspired and biomimetic nanotherapies for the treatment of infectious diseases. *Frontiers in Pharmacology, 10*, 751. DOI: 10.3389/fphar.2019.00751 PMID: 31333467

Yang, H., Wang, W., Wang, Y., Ke, L., Wang, L., & Li, R. (2017). Enhanced intracellular delivery and antibacterial efficacy of enrofloxacin-loaded docosanoic acid solid lipid nanoparticles against intracellular Salmonella. *Scientific Reports, 7*(1), 41104. DOI: 10.1038/srep41104 PMID: 28112240

Yang, J., des Rieux, A., & Malfanti, A. (2025). Stimuli-Responsive Nanomedicines for the Treatment of Non-cancer Related Inflammatory Diseases. *ACS Nano, 19*(16), 15189–15219. DOI: 10.1021/acsnano.5c00700 PMID: 40249331

Yang, L., Pijuan-Galito, S., Rho, H. S., Vasilevich, A. S., Eren, A. D., Ge, L., Habibović, P., Alexander, M. R., de Boer, J., Carlier, A., van Rijn, P., & Zhou, Q. (2021). High-throughput methods in the discovery and study of biomaterials and materiobiology. *Chemical Reviews, 121*(8), 4561–4677. DOI: 10.1021/acs.chemrev.0c00752 PMID: 33705116

Yang, S. Y., Han, S. M., Lee, J. Y., Kim, K. S., Lee, J. E., & Lee, D. W. (2025). Advancing gut microbiome research: The shift from metagenomics to multiomics and future perspectives. *Journal of Microbiology and Biotechnology*, *35*, e2412001. DOI: 10.4014/jmb.2412.12001

Yang, S., Hu, X., Zhu, J., Zheng, B., Bi, W., Wang, X., Wu, J., Mi, Z., & Wu, Y. (2025). Aspects and Implementation of Pharmaceutical Quality by Design from Conceptual Frameworks to Industrial Applications. *Pharmaceutics*, *17*(5), 623. DOI: 10.3390/pharmaceutics17050623 PMID: 40430914

Yang, Y., Jiang, X., Lai, H., & Zhang, X. (2022). Smart bacteria-responsive drug delivery systems in medical implants. *Journal of Functional Biomaterials*, *13*(4), 173. Advance online publication. DOI: 10.3390/jfb13040173

Yan, H., Wen, P., Tian, S., Zhang, H., Han, B., Khan, J., Xue, Y., Chen, X., Li, X., & Li, Y. (2024). Enhancing biofilm penetration and antibiofilm efficacy with protein nanocarriers against pathogenic biofilms. *International Journal of Biological Macromolecules*, *256*(Pt 1), 128300. Advance online publication. DOI: 10.1016/j.ijbiomac.2023.128300

Yarahmadi, A., Najafiyan, H., Yousefi, M. H., Khosravi, E., Shabani, E., Afkhami, H., & Aghaei, S. S. (2025). Beyond antibiotics: Exploring multifaceted approaches to combat bacterial resistance in the modern era: A comprehensive review. *Frontiers in Cellular and Infection Microbiology*, *15*, 1493915. Advance online publication. DOI: 10.3389/fcimb.2025.1493915

Yeh, Y. C., Huang, T. H., Yang, S. C., Chen, C. C., & Fang, J. Y. (2020, April 24). Nano-Based Drug Delivery or Targeting to Eradicate Bacteria for Infection Mitigation: A Review of Recent Advances. *Frontiers in Chemistry*, *8*, 286. DOI: 10.3389/fchem.2020.00286 PMID: 32391321

Yetgin, A. (2025). Revolutionizing multi-omics analysis with artificial intelligence and data processing. *Quantitative Biology*, 13(3). Portico. DOI: 10.1002/qub2.70002

Yetisen, A. K., Martinez-Hurtado, J. L., Unal, B., Khademhosseini, A., & Butt, H. (2018). Wearables in medicine. *Advanced Materials*, *30*(8), 1706910. Advance online publication. DOI: 10.1002/adma.201706910

Yin, S. (2025). Artificial Intelligence-Assisted Nanosensors for Clinical Diagnostics: Current Advances and Future Prospects. *Biosensors (Basel)*, *15*(10), 656. DOI: 10.3390/bios15100656 PMID: 41149307

Yoon, M. Y., & Yoon, S. S. (2018). Disruption of the gut ecosystem by antibiotics. *Yonsei Medical Journal*, *59*(1), 4–12. DOI: 10.3349/ymj.2018.59.1.4

Younis, M. A., Tawfeek, H. M., Abdellatif, A. A. H., Abdel-Aleem, J. A., & Harashima, H. (2022). Clinical translation of nanomedicines: Challenges, opportunities, and keys. *Advanced Drug Delivery Reviews*, *181*, 114083. DOI: 10.1016/j.addr.2021.114083 PMID: 34929251

Youshia, J., Ali, M. E., & Lamprecht, A. (2017). Artificial neural network-based particle size prediction of polymeric nanoparticles. *European Journal of Pharmaceutics and Biopharmaceutics*, *119*, 333–342. DOI: 10.1016/j.ejpb.2017.06.030 PMID: 28694160

Yu, X., Gao, D., Gao, L., Lai, J., Zhang, C., Zhao, Y., Zhong, L., Jia, B., Wang, F., Chen, X., & Liu, Z. (2017). Inhibiting metastasis and preventing tumor relapse by triggering host immunity with tumor-targeted photodynamic therapy using photosensitizer-loaded functional nanographenes. *ACS Nano*, *11*(10), 10147–10158. DOI: 10.1021/acsnano.7b04736

Yu, Y., Mei, L., Shi, Y., Zhang, X., Cheng, K., Cao, F., Zhang, L., Xu, J., Li, X., & Xu, Z. (2020). Ag-conjugated graphene quantum dots with blue light-enhanced singlet oxygen generation for ternary-mode highly efficient antimicrobial therapy. *Journal of Materials Chemistry. B*, *8*(7), 1371–1382. DOI: 10.1039/c9tb02300c

Yu, Y., Si, X., Hu, C., & Zhang, J. (2019). A Review of Recurrent Neural Networks: LSTM Cells and Network Architectures. *Neural Computation*, *31*(7), 1235–1270. DOI: 10.1162/neco_a_01199 PMID: 31113301

Zagajewski, A., Turner, P., Feehily, C., El Sayyed, H., Andersson, M., Barrett, L., Oakley, S., Stracy, M., Crook, D., Nellåker, C., Stoesser, N., & Kapanidis, A. N. (2023). Deep learning and single-cell phenotyping for rapid antimicrobial susceptibility detection in Escherichia coli. *Communications Biology*, *6*(1), 1164. Advance online publication. DOI: 10.1038/s42003-023-05524-4 PMID: 37964031

Zdrazil, B. (2025). Fifteen years of ChEMBL and its role in cheminformatics and drug discovery. *Journal of Cheminformatics*, *17*(1), 32. DOI: 10.1186/s13321-025-00963-z

Zeng, L., An, L., & Wu, X. (2011). Modeling drug-carrier interaction in the drug release from nanocarriers. *Journal of Drug Delivery*, *370308*. Advance online publication. DOI: 10.1155/2011/370308

Zeng, Z., Jiang, J., Sun, Y., Li, W., Zheng, D., Lin, H., & Zhao, C. (2025). Infected microenvironment responsive nanoprodrug for the specific therapeutics of Helicobacter pylori clearance. *Biomaterials*, *123415*, 123415. Advance online publication. DOI: 10.1016/j.biomaterials.2025.123415 PMID: 40411983

Zhang N. (2023). Promoting the bench-to-bedside translation of nanomedicines. *Medical review (2021), 3*(1), 1–3. https://doi.org/DOI: 10.1515/mr-2023-0007

Zhang, C. Y., Lin, W., Gao, J., Shi, X., Davaritouchaee, M., Nielsen, A. E., Mancini, R. J., & Wang, Z. (2019). pH-responsive nanoparticles targeted to lungs for improved therapy of acute lung inflammation/injury. *ACS Applied Materials & Interfaces, 11*(18), 16380–16390. DOI: 10.1021/acsami.9b04051

Zhang, C., Li, J., Tang, Q., Li, L., & Cao, D. (2025). Targeting proteostasis for cancer therapy: Current advances, challenges, and future perspectives. *Molecular Cancer, 24*(1), 265. DOI: 10.1186/s12943-025-02472-x

Zhang, D., Yin, C., Hunold, K. M., Jiang, X., Caterino, J. M., & Zhang, P. (2021). An interpretable deep-learning model for early prediction of sepsis in the emergency department. *Patterns (New York, N.Y.), 2*(2), 100196. Advance online publication. DOI: 10.1016/j.patter.2020.100196 PMID: 33659912

Zhang, J., Li, H., Zhang, Y., Huang, J., Ren, L., Zhang, C., Zou, Q., & Zhang, Y. (2025). Computational toxicology in drug discovery: Applications of artificial intelligence in ADMET and toxicity prediction. *Briefings in Bioinformatics, 26*(5), bbaf533. Advance online publication. DOI: 10.1093/bib/bbaf533 PMID: 41052279

Zhang, J., Lin, G., Xie, N., Ye, Y., Zhao, H., & Lu, J. (2026). Alternations along gut-brain axis in response to hypo- and hyperosmotic stress in yellowfin seabream (*Acanthopagrus latus*). *Aquaculture Reports, 48*, 103533. DOI: 10.1016/j.aqrep.2026.103533

Zhang, J., Tang, W., Zhang, X., Song, Z., & Tong, T. (2023). An Overview of Stimuli-Responsive Intelligent Antibacterial Nanomaterials. *Pharmaceutics, 15*(8), 2113. DOI: 10.3390/pharmaceutics15082113 PMID: 37631327

Zhang, L., Chen, J., Zhang, F., Xu, J., Bi, N., Gou, J., & Jia, L. (2022). Silicon quantum dots and MOFs hybrid multicolor fluorescent nanosensor for ultrasensitive and visual intelligent sensing of tetracycline. *Colloids and Surfaces. A, Physicochemical and Engineering Aspects, 652*, 129853. DOI: 10.1016/j.colsurfa.2022.129853

Zhang, T., Wang, J., Feng, Q., Xu, X., Zhu, W., Mao, S., & Liu, J. (2026). Age- and diet-driven assembly of the gut antibiotic resistome in humans and food-producing animals. *Gut Microbes, 18*(1), 2610052. DOI: 10.1080/19490976.2025.2610052

Zhang, Y., Hu, Y., Jiang, N., & Yetisen, A. K. (2023). Wearable artificial intelligence biosensor networks. *Biosensors & Bioelectronics, 219*, 114825. Advance online publication. DOI: 10.1016/j.bios.2022.114825

Zhang, Y., Lin, S., Fu, J., Zhang, W., Shu, G., Lin, J., Li, H., Xu, F., Tang, H., Peng, G., Zhao, L., Chen, S., & Fu, H. (2022). Nanocarriers for combating biofilms: Advantages and challenges. *Journal of Applied Microbiology*, *133*(3), 1273–1287. DOI: 10.1111/jam.15640

Zhang, Z., Chen, K., Zhang, S., He, Y., Lei, G., Zhao, Y., & Liang, J. (2025). Identification of fecal microbiota and related metabolites associated with feed efficiency in DLY pigs. *Animals (Basel)*, *15*(20), 3026. DOI: 10.3390/ani15203026

Zhang, Z., Chen, L., Zhong, F., Wang, D., Jiang, J., Zhang, S., Jiang, H., Zheng, M., & Li, X. (2022). Graph neural network approaches for drug target interactions. *Current Opinion in Structural Biology*, *73*, 102327. Advance online publication. DOI: 10.1016/j.sbi.2021.102327

Zhao, Y., Yin, J., Zhang, L., Zhang, Y., & Chen, X. (2024). Drug–drug interaction prediction: Databases, web servers and computational models. *Briefings in Bioinformatics*, *25*(1), bbad445. Advance online publication. DOI: 10.1093/bib/bbad445

Zheng, Y., Koh, H. Y., Ju, J., Yang, M., May, L. T., Webb, G. I., Li, L., Pan, S., & Church, G. (2025). Large language models for drug discovery and development. *Patterns (New York, N.Y.)*, *6*(10), 101346. Advance online publication. DOI: 10.1016/j.patter.2025.101346

Zhitlov, M. Y., Korshun, V. A., & Alferova, V. A. (2025). Microfluidics to meet antibiotic resistance: A growing research frontier. *Antibiotics (Basel, Switzerland)*, *14*(12), 1232. DOI: 10.3390/antibiotics14121232

Zhou, R., Ng, S. K., Sung, J. J. Y., Goh, W. W. B., & Wong, S. H. (2023). Data pre-processing for analyzing microbiome data – A mini review. *Computational and Structural Biotechnology Journal*, *21*, 4804–4815. DOI: 10.1016/j.csbj.2023.10.001 PMID: 37841330

Zhou, Y., Wang, X., Tian, X., Zhang, D., Cui, H., Du, W., Yang, Z., Li, J., Li, W., Xu, J., Duanmu, Y., Yu, T., Cai, F., Li, W., Jin, Z., Wu, W., & Huang, H. (2025). Stealth missiles with precision guidance: A novel multifunctional nano-drug delivery system based on biomimetic cell membrane coating technology. *Materials Today. Bio*, *33*, 101922. DOI: 10.1016/j.mtbio.2025.101922 PMID: 40528836

Zhu, H. (2020). Big data and artificial intelligence modeling for drug discovery. *Annual Review of Pharmacology and Toxicology*, *60*, 573–589. DOI: 10.1146/annurev-pharmtox-010919-023324

Zhu, Y., Tang, Y., Qi, X., & Zhu, X. (2026). Transformer models, graph networks, and generative AI in gut microbiome research: A narrative review. *Bioengineering (Basel, Switzerland), 13*(2), 144. DOI: 10.3390/bioengineering13020144

Zielinska, A., Carreiró, F., Oliveira, A. M., Neves, A., Pires, B., Venkatesh, D. N., & Souto, E. B. (2020). Polymeric nanoparticles: Production, characterization, toxicology and ecotoxicology. *Molecules (Basel, Switzerland), 25*(16), 3731. DOI: 10.3390/molecules25163731 PMID: 32824172

Zielińska, K., Udekwu, K. I., Rudnicki, W., Frolova, A., & Łabaj, P. P. (2025). Healthy microbiome-moving toward functional interpretation. *GigaScience, 14*, giaf015. Advance online publication. DOI: 10.1093/gigascience/giaf015

Zolnik, B. S., González-Fernández, Á., Sadrieh, N., & Dobrovolskaia, M. A. (2010). Minireview: Nanoparticles and the immune system. *Endocrinology, 151*(2), 458–465. DOI: 10.1210/en.2009-1082 PMID: 20016026

Zou, D., Du, X., Liu, Y., Fang, M., & Chu, M. (2026). Integrated DIA proteomics and transcriptomics unravel molecular mechanisms of uterine prolificacy in Yunshang Black goats during the estrous cycle. *BMC Genomics, 27*(1), 314. DOI: 10.1186/s12864-026-12634-w

About the Contributors

Reyed M. Reyed is an Associate Professor of Microbial Biotechnology at the Genetic Engineering and Biotechnology Research Institute (GEBRI), City of Scientific Research and Technological Applications (SRTA-City), Egypt, and serves as the Editor of *AI, Nanobiotechnology, and the Future of Precision Antibiotics*. His research operates at the convergence of microbial biotechnology, artificial intelligence, and nanobiological systems, with a focus on advancing next-generation antimicrobial strategies. His work integrates microbiome science, computational modeling, and nanotechnology to address antimicrobial resistance through precision-driven approaches. Dr. Reyed has an established international publication record, contributing to edited volumes and scholarly works with leading academic publishers, including Springer Nature, IGI Global, Bentham Science, and Nova Science Publishers. He is the developer of Polybiome Systems Medicine and the originator of the Reyed Precision Convergence Framework for Antimicrobial Resistance (RPCF-AMR). Through these contributions, his work is positioned within the emerging paradigm of precision antibiotics, where data-driven intelligence and multiscale biological integration shape the future of antimicrobial innovation and targeted therapeutic development..

***

Mariam Sayed A. Ahmed is a medical graduate from the Faculty of Medicine, Cairo University, Egypt, with a focus on clinical research and translational medicine. Her academic interests center on understanding disease mechanisms and improving diagnostic and therapeutic strategies through integrative approaches. She has contributed to research examining clinical and molecular aspects of disease, particularly in infectious diseases and antimicrobial resistance. She is adept at applying emerging biomedical technologies and is dedicated to translating scientific advances into improvements in patient care. Her work reflects a balanced perspective that combines clinical insight with evidence-based research.

Hagar Sayed Abdelgwad is affiliated with the Faculty of Medicine, Banha University, Egypt. Her academic interests lie in clinical medicine and biomedical research, with a focus on improving therapeutic strategies and patient outcomes. She has been involved in clinical and research activities focused on disease diagnosis and management, particularly in infectious diseases. Her work reflects an interest in integrating clinical practice with emerging therapeutic approaches, including those informed by advances in biotechnology and nanomedicine. is committed to evidence-based medicine and collaborative research, contributing to the development of effective and sustainable healthcare solutions. Her work emphasizes the importance of aligning clinical insight with scientific innovation.

Arun Agrawal is an Assistant Professor at the Institute of Technology and Management, Gwalior, India. He holds an M.Tech. in Information Technology and has over 19 years of teaching experience in academic institutions, along with 5 years of professional experience in the software industry. His research focuses on Artificial Intelligence, Vehicular Ad Hoc Networks (VANET), and Digital Image Processing. He has authored more than 48 research articles published in Scopus-indexed journals, reflecting his active contribution to the academic and research community. His work primarily emphasizes the development of intelligent systems, advanced communication networks, and image-based analytical solutions. His current research interests include the application of AI-driven techniques to real-world problems in smart transportation systems and data-driven image analysis.

Dini Agustina, M.Biomed. born in Jember, August 1, 1983, completed his medical profession in 2007 at the Faculty of Medicine, Universitas Brawijaya, Malang. Furthermore, in 2011-2014 he took a Master's Program in Biomedical Sciences at Brawijaya University, Malang with a GPA of 3.77 and a laude predicate. The author has been active as a lecturer at the Faculty of Medicine, University of Jember since 2008 in the Department of Microbiology. A lot of research in the health field with a focus on tropical infections and on vaccine development has been conducted by the authors with funding from internal and external grants from the Ministry of Education, Culture, Research and Technology. Many publications and books in the field of tropical infections and molecular medicine have been produced by authors, both nationally and internationally. Throughout his career, the author has actively participated in various international seminar activities both at home and abroad, including Vietnam and Malaysia.

Mohamed Ihab Al Refaei, Department of Zoology, Faculty of Education, Ain Shams University, Egypt, conducts research in microbiology, environmental biology, and host–pathogen interactions within the biological sciences. His research focuses on elucidating how specific environmental and biological factors shape disease dynamics and microbial adaptation. He employs experimental and conceptual methodologies to investigate microbial systems, emphasizing precise mechanisms underlying host–pathogen interactions and environmental microbiology. His interdisciplinary work integrates traditional biological sciences with technological tools, such as nanobiotechnology and systems modeling, to advance understanding of microbial ecology and its direct implications for health and disease.

Nirbanya Devi is a research scholar at Assam University, Silchar-788011, Assam, India. She holds a M.Sc. degree in Chemistry from Rajiv Gandhi University, Doimukh-79112, Arunachal Pradesh, India. Her research focuses on Electrochemical detection of vitamins and hormones. She has published four research papers and presented her work at three national conferences and one international conference. Her current work explores the Development of conducting polymer nanocomposite-modified electrodes for sensitive biomolecule detection.

Dhanushiya R. is a PhD scholar in the Department of Biotechnology, affiliated to Stella Maris College (Autonomous), Chennai. She holds a bachelor's degree in bioinformatics, which provides a strong computational foundation for her interdisciplinary research. Her ongoing doctoral work focuses on the degradation of environmental pollutants using microbial and nanotechnology-based approaches. Her research integrates microbiology, environmental biotechnology, and nanotechnology to develop sustainable and efficient solutions for pollution mitigation. She is also interested in exploring innovative bioremediation strategies that can be applied across diverse ecological environments.

Veena Gayathri, Dr., is an associate professor at Stella Maris College (Autonomous), University of Madras specializing in Environmental Biotechnology. Her research and area of interest is in integrated method for recalcitrant removal and wastewater treatment. They have published research in various peer – reviewed journal both in national and international journals. They also have a patent for biodegradation of chlorophenols by halotolerant bacterial consortium.

Sarbani Giri is involved in teaching and research for last 26 years in Molecular and Cell Biology, Genetics, Toxicology and Environmental Health at Department of Life Science and Bioinformatics, Assam University, Silchar . She is recipient of the DBT Overseas Associateship and Indo-US Science and Technology Forum Fellowship by Govt. of India.She received research training in Metabolomics and

Cancer Biology from Laboratory of Metabolism, National Cancer Institute, National Institute of Health, Maryland USA (2005- 2006) and Department of Molecular and Cell Biology, University of California, Berkeley, USA,(2010) respectively. Her research focus is mainly cancer biology with special emphasis on role of tight junction integrity, downstream signalling pathways involved and environmental health assessment with major focus on genetic toxicology.Besides she served as the departmental Head for four times and as Dean, School of Life Sciences besides being member of several academic bodies.She has published nearly 65 papers in various reputed scientific journals and guided nearly 26 Ph.D. scholars under her supervision.

Deepak Gupta is a distinguished academician with over 20 years of teaching experience and expertise in Computer Science and Engineering. He holds a Ph.D. in Computer Engineering and has an extensive research portfolio, including many international journal papers, multiple conference proceedings, and several books. His research interests span machine learning, artificial intelligence, and data mining. Currently, Dr. Gupta serves as an Associate Professor and Head of Department of CSE (Emerging Branches) at ITM Gwalior, MP, India. He has also contributed significantly to the field through patents and copyrights and actively engages as a reviewer, conference chair and editor.

Halah Abdulrahman Hafiz is a professor of clinical nutrition at the Faculty of Applied Medical Sciences, Umm Al-Qura University, Saudi Arabia. She brings extensive academic and clinical experience in nutritional science, with a research focus on the role of diet and bioactive compounds in modulating disease risk and progression. Her work explores the interaction between nutrition, metabolic regulation, and inflammatory pathways, with particular attention to functional foods and therapeutic dietary strategies. She has contributed to advancing the understanding of how nutritional interventions can be integrated into broader biomedical and preventive healthcare frameworks. She is also interested in emerging intersections between nutrition and technology, including nanonutrition and personalized health approaches. Through her academic and research contributions, she supports the development of evidence-based strategies to improve clinical outcomes and long-term health. Her work reflects a commitment to interdisciplinary research that connects nutrition with evolving concepts in precision medicine.

Bagus Hermansyah, M.Biomed, was born in Pasuruan on April 5, 1983. He graduated as a doctor in 2007 from the Faculty of Medicine, Universitas Brawijaya. Then in 2011 he continued his master's education at the Biomedical Science Study Program, Faculty of Medicine, University of Brawijaya with Research in the field of

Medical Parasitology (graduated in 2014). The author has been active as a lecturer at the Department of Medical Parasitology, Faculty of Medicine, University of Jember since 2008 until now. Several scientific research and publications in the field of Medical Pathology and Biomedical Science have been conducted by the author. Many of the resulting scientific publications have been published in national and international accredited Scientific Journals. In addition to teaching, researching and publishing scientific publications, the author also actively participates in various training activities, workshops and seminars at the national and international levels. Various community service has also been carried out by the author.

Adil Husain is an Assistant Professor of Biochemistry at Babu Banarasi Das University, Lucknow, India. He holds a Ph.D. in Biochemistry with research focused on liquid biopsy biomarkers, particularly circulating cell-free DNA and IDH1 mutations in glioma. He qualified UGC-NET (JRF) and has extensive research experience in molecular oncology and translational cancer research. Dr. Husain has authored several peer-reviewed publications and book chapters in reputed international journals and publishers. He actively serves as a reviewer for leading journals and is dedicated to integrating research with teaching in biomedical sciences.

K. Hussain is currently working as Associate Professor and Head, Department of Electrical Engineering, at Sharad Institute of Technology College of Engineering, Yadrav, Ichalkaranji, Kolhapur, Maharashtra. Dr. K. Hussain is an esteemed electrical engineer with 22+ years of teaching, administrative, and research experience. He successfully led the department to achieve NBA Accreditation twice. He is an active reviewer for prestigious journals and has held multiple administrative roles, including Chairman of the Board of Studies and member of the Academic Council. Dr. K Hussain is life member of ISTE, Member of IE(I), senior member of IAEEEE, Member of IAENG and has one grant patent and published three patents on innovative technologies and published more than 28 papers in reputed journals and conferences. He co-authored a book on "Protection and Switchgear" and has secured significant research and consultancy funds, including a DBATU Vice Chancellor Research Promotion Grant and an AICTE MODROB project. His dedication to excellence in education and innovation has earned him numerous accolades.

Mona M. Khalifa is a biochemistry graduate from the Faculty of Science, Cairo University, Egypt, with a growing research trajectory at the interface of nanobiotechnology, antimicrobial resistance (AMR), and AI-enabled precision therapeutics. Her work centers on the rational design of stimuli-responsive nanocarriers that can adapt to pathological microenvironments and enhance targeted antimicrobial delivery, particularly against multidrug-resistant pathogens. Her research approach combines experimental insight with computational modeling, including molecular docking, data-driven formulation strategies, and predictive analytics for nanomedicine optimization. She is particularly interested in how artificial intelligence can reshape formulation science by improving design efficiency and reducing reliance on empirical trial-and-error. She contributed to interdisciplinary scientific writing focused on emerging paradigms in precision antimicrobial therapy, emphasizing the integration of conceptual innovation and practical application. Her work aligns with global One Health priorities, aiming to develop scalable, translational strategies to address AMR across clinical and environmental contexts. She continues to explore integrated frameworks that connect nanoscale engineering with systems-level therapeutic design.

Ahmed Ibrahim Mohammed is a researcher in pharmaceutical biotechnology at the Faculty of Biotechnology, Badr University in Cairo, Egypt. His academic work focuses on advanced drug delivery systems, with particular emphasis on the development of nanostructured carriers for controlled and site-specific therapeutic release. His research involves the formulation and characterization of nanoparticle-based systems, including polymeric and lipid-based platforms, designed to improve drug stability, bioavailability, and targeting precision. He has contributed to studies investigating the physicochemical behavior of nanocarriers and their performance in biologically relevant environments. Work reflects a strong interest in translating nanotechnology from laboratory-scale innovation into clinically relevant applications. He actively engages in multidisciplinary collaborations to enhance therapeutic outcomes, particularly in infectious diseases and antimicrobial resistance. His research contributes to the broader effort of integrating biotechnology and nanomedicine into next-generation pharmaceutical solutions.

Eman Ibrahim Mohammed is a researcher in biotechnology at the Faculty of Biotechnology, Badr University in Cairo, Egypt. Her research focuses on nanobiotechnology and its applications in medical and pharmaceutical sciences. She has been involved in the development and evaluation of nanomaterials for drug delivery and therapeutic applications, with particular attention to improving targeting efficiency and treatment outcomes. Her work integrates molecular biology techniques with nanotechnological approaches to address contemporary biomedical challenges. is especially interested in the potential of nanotechnology to enhance antimicrobial therapies and overcome resistance mechanisms. She continues to contribute to research efforts aimed at advancing innovative and effective treatment strategies.

Srinivasan Nagaraj is currently working in Chitanya Bharathi Institute of Technology, Proddatur as Professor in Computer Science and Engineering Department. He completed his M.Tech(CSE) from Acharya Nagarjuna University and was awarded with Ph.D. from Annamalai University in 2023. He published more than 20 research papers in National and International journals, he published 25 patents to his credit and 2 Scopus book chapters and also participated in 15 National and International Conferences. He taught all Computer science and Machine Learning subjects for Undergraduate and Post graduate students. He has a stint of 24 years of experience in various reputed colleges and universities at various academic and administrative positions.

Tarun Kumar Patel is a distinguished academician and researcher in the field of biotechnology, currently serving as an Assistant Professor and Head of the Department of Biotechnology at Sant Guru Ghasidas Government P.G. College, Kurud, Dhamtari, Chhattisgarh, India. With a robust academic background and a decade-long teaching career, Dr. Patel has made significant contributions to fungal biology, mass spectrometry, and microbial biotechnology. Educational Background and Qualifications Dr. Patel earned his Postgraduation and subsequently his Ph.D. in Biotechnology with research focused on detecting and analyzing aflatoxin production in Aspergillus flavus strains using advanced mass spectrometry techniques. His academic excellence is further demonstrated by qualifying the CSIR-NET (Life Sciences) and the GATE examination. Research Contributions Dr. Patel's research encompasses a wide array of topics within biotechnology. His notable works include studies on the biosynthesis of aflatoxin B1 in Aspergillus flavus, identification of genetic relationships in Chlorophytum species, and the exploration of entomopathogenic fungi as biocontrol agents in agriculture. He has contributed to several peer-reviewed journals and book chapters, including publications in Biologia, Biotechnology and Bioprocess Engineering, and contributions to books published by Elsevier and Springer. Professional Experience and Teaching With over ten years of teaching experience at the undergraduate level and seven years at the postgraduate level, Dr. Patel has been instrumental in shaping the academic journey of numerous students. His commitment to education is evident in his active participation in various workshops and training programs, enhancing both his knowledge and teaching methodologies. Professional Affiliations and Recognitions Dr. Patel is an active member of esteemed scientific societies, including the Asian Federation of Biotechnology (AFOB) and the Association of Microbiologists of India (AMI). His research has garnered citations and recognition within the scientific community, reflecting the impact of his work on the field of biotechnology. His dedication to research and education in biotechnology has significantly contributed to advancements in fungal biology and microbial biotechnology. His work continues to inspire students and researchers alike, fostering innovation and excellence in the scientific community.

Khemnath Patir is an Assistant Professor at Department of Applied Science and Humanities, Assam University, Silchar-788011, Assam, India. He holds a Ph.D. in Chemistry from Gauhati University, Guwahati-781014, Assam, India. His research focuses on Luminescent nanoparticles, Photocatalysis, Adsorption, Food Chemistry and Wastewater Remediation. He has contributed to 21 publications, attended several Conferences and Workshops. His current work explores Green synthesis of nanoparticles, Luminescent and Colorimetric sensors, Wastewater treatment by Catalytic process.

Y Reddy is currently working in Chitanya Bharathi Institute of Technology, Proddatur as Professor in Computer Science and Engineering Department. He completed his M.Tech(CSE) from Acharya Nagarjuna University and was awarded with Ph.D. from Bharath University in 2016. He published more than 20 research papers in National and International journals, he published 5 patents to his credit and also participated in 15 National and International Conferences. He taught all Computer science and Machine Learning subjects for Undergraduate and Post graduate students. He has a stint of 22 years of experience in various reputed colleges and universities at various academic and administrative positions.

Debanjali Roy is a distinguished physicist who has made significant contributions to the field of nanoparticles synthesis, characterization, and application. She earned her Ph.D. in Physics from ICFAI University, Dehradun, in 2022, specializing in the fabrication and characterization of iron oxide nanomaterials and nanocomposites for advanced device applications. With over nine years of teaching experience, Dr. Roy has established herself as a dedicated educator and researcher. Her research portfolio includes numerous publications in reputable international journals, books, and conference proceedings. Her expertise extends beyond nanoparticles to AI-based research, showcasing her versatility and commitment to interdisciplinary exploration. Dr. Roy has edited books with prominent publishers, including IGI Global and Springer Nature, demonstrating her ability to collaborate and lead in academic publishing. As a mentor, she has guided hundreds of students, providing invaluable guidance and support in their academic pursuits. Through her work, Dr. Debanjali Barman Roy continues to inspire and educate future generations of scientists and researchers, leaving a lasting impact on the field of physics and beyond.

Dhanya K. S. is a research scholar at Stella Maris College (Autonomous), University of Madras specializing in Biotechnology. Their research focuses on Environmental biotechnology and different aspects of bioremediation for sustainability of the environment. Their research integrates microbiology and biotechnology along with bioinformatics. They have published research in peer-reviewed journal and are currently working on nanoparticles and its benefits.

Yashodhan Sonwane is currently working as an Assistant Professor, Mechanical Engineering Department, in Ajeenkya DY Patil University, Pune. He has done his masters in mechanical engineering from SRM University, Chennai and contributed his knowledge and skills in his area of expertise i.e computer aided design. He has patented two designs based on his innovation and keeps an interest in doing research in multidisciplinary aspects in order to exchange knowledge and skills.

Jharna Tamuli is a research scholar at Tezpur University, Napaam-784028, Assam, India. She holds a M.Sc. degree in Chemistry from Dibrugarh University, Dibrugarh-786004, Assam, India. Her research focuses on Functional molecular materials. She has contributed to Metal-organic frameworks during her master's degree. Her current work explores switchable magnetic materials.

Jyotishman Tamuli is a Ph.D. research scholar at Department of Life Science and Bioinformatics, Assam University, Silchar-788011, Assam, India. He holds a M.Sc. degree in Life Science and Bioinformatics from Assam University, Silchar. His research focuses on Cancer Biology, Genotoxicity and Nanotechnology. He has contributed to few publications and presented his works at four international conferences. His current work explores Green synthesized nanoparticles with anticancer properties and exploring different Ayurvedic formulations for their antioxidant, anti-inflammatory and antidiabetic properties.

Index

A

B

N

O

P

R

S

T

www.ingramcontent.com/pod-product-compliance
Lightning Source LLC
LaVergne TN
LVHW062332300926
849161LV00007B/1661

* 9 7 9 8 3 3 7 3 5 2 6 8 8 *